Energy, Environment, and Sustainability

Series Editor

Avinash Kumar Agarwal, Department of Mechanical Engineering, Indian Institute of Technology Kanpur, Kanpur, Uttar Pradesh, India

AIMS AND SCOPE
This books series publishes cutting edge monographs and professional books focused on all aspects of energy and environmental sustainability, especially as it relates to energy concerns. The Series is published in partnership with the International Society for Energy, Environment, and Sustainability. The books in these series are edited or authored by top researchers and professional across the globe. The series aims at publishing state-of-the-art research and development in areas including, but not limited to:

- Renewable Energy
- Alternative Fuels
- Engines and Locomotives
- Combustion and Propulsion
- Fossil Fuels
- Carbon Capture
- Control and Automation for Energy
- Environmental Pollution
- Waste Management
- Transportation Sustainability

Review Process

The proposal for each volume is reviewed by the main editor and/or the advisory board. The chapters in each volume are individually reviewed single blind by expert reviewers (at least four reviews per chapter) and the main editor.

Ethics Statement for this series can be found in the Springer standard guidelines here https://www.springer.com/us/authors-editors/journal-author/journal-author-helpdesk/before-you-start/before-you-start/1330#c14214

More information about this series at https://link.springer.com/bookseries/15901

P. A. Lakshminarayanan · Avinash Kumar Agarwal
Editors

Handbook of Thermal Management of Engines

Springer

Editors
P. A. Lakshminarayanan
Indian Institute of Technology Kanpur
Kanpur, India

Avinash Kumar Agarwal
Department of Mechanical Engineering
Indian Institute of Technology Kanpur
Kanpur, India

ISSN 2522-8366 ISSN 2522-8374 (electronic)
Energy, Environment, and Sustainability
ISBN 978-981-16-8569-9 ISBN 978-981-16-8570-5 (eBook)
https://doi.org/10.1007/978-981-16-8570-5

This Springer imprint is published by the registered company Springer Nature Singapore Pte Ltd.
The registered company address is: 152 Beach Road, #21-01/04 Gateway East, Singapore 189721, Singapore

Top and underhood freewheeling, Thermal and Flow Simulation *Courtesy* Dassault

Preface

The International Society for Energy, Environment, and Sustainability (ISEES) was founded at the Indian Institute of Technology Kanpur (IIT Kanpur), India, in January 2014 to spread knowledge/awareness and catalyse research activities in the fields of Energy, Environment, Sustainability, and Combustion. Society's goal is to contribute to the development of clean, affordable, and secure energy resources and a sustainable environment for society and spread knowledge in the areas mentioned above, and create awareness about the environmental challenges the world is facing today. The unique way adopted by ISEES was to break the conventional silos of specializations (engineering, science, environment, agriculture, biotechnology, materials, fuels, etc.) to tackle the problems related to energy, environment, and sustainability in a holistic manner. This is quite evident by the participation of experts from all fields to resolve these issues. The ISEES is involved in various activities such as conducting workshops, seminars, conferences, etc., in the domains of its interests. The society also recognizes the outstanding works of young scientists, professionals, and engineers for their contributions in these fields by conferring awards under various categories.

The Fifth International Conference on Sustainable Energy and Environmental Challenges (V-SEEC) was organized under the auspices of ISEES during 19–21 December 2020, in virtual mode due to restrictions on travel because of the ongoing COVID-19 pandemic situation. This conference provided a platform for discussions between eminent scientists and engineers from various countries, including India, Spain, Austria, Bangladesh, Mexico, USA, Malaysia, China, UK, Netherlands, Germany, Israel, and Saudi Arabia. At this conference, eminent international speakers presented their views on energy, combustion, emissions, and alternative energy resources for sustainable development and a cleaner environment. The conference presented two high-voltage plenary talks by Dr. V. K. Saraswat, Honourable Member, NITI Aayog, on *Technologies for Energy Security and Sustainability* and Prof. Sandeep Verma, Secretary, SERB, on *New and Equitable R&D Funding Opportunities at SERB.*

The conference included nine technical sessions on topics related to energy and environmental sustainability. Each session had 6–7 eminent scientists from all over the world, and they shared their opinion and discussed the trends for the future.

The technical sessions in the conference included Emerging Contaminants: Monitoring and Degradation Challenges; Advanced Engine Technologies and Alternative Transportation Fuels; Future Fuels for Sustainable Transport; Sustainable Bioprocessing for Biofuel/Non-biofuel Production by Carbon Emission Reduction; Future of Solar Energy; Desalination and Wastewater Treatment by Membrane Technology; Biotechnology in Sustainable Development; Emerging Solutions for Environmental Applications and Challenges and Opportunities for Electric Vehicle Adoption. More than 500 participants and speakers attended the three-day conference.

The conference concluded with a high-voltage panel discussion on *Challenges and Opportunities for Electric Vehicle Adoption*, where the panellists were Prof. Gautam Kalghatgi of the University of Oxford, Prof. Ashok Jhunjhunwala of IIT Madras, Dr. Kelly Senecal of Convergent Science, Dr. Amir Abdul Manan of Saudi Aramco, and Dr. Sayan Biswas of University of Minnesota, USA. Prof. Avinash Kumar Agarwal, ISEES, moderated the panel discussion. This conference laid out the roadmap for technology development, opportunities, and challenges in Energy, Environment, and Sustainability domain. All these topics are very relevant for the country and the world in the present context. We acknowledge the support received from various agencies and organizations for the successful conduct of the Fifth ISEES conference V-SEEC, where these books germinated. The conference was conducted with the support of SERB and our publishing partner SpringerNature. Special thanks are due to Dr. Sandeep Verma, Secretary, SERB, and Ms. Swati Maharishi of SpringerNature for their consistent efforts.

This book deals with the vast subject of thermal management of an engine and the vehicle by collecting the current research on the subject from the universities and industries applicable mainly to diesel and natural gas engines and present them in six logical parts:

1. The Gestalt of Thermal Management (Chaps. 1 and 2)
2. Thermal Management employing Turbocharging and Insulation Between Aftertreatment Systems and the Engine (Chaps. 3, 4, and 5)
3. Techniques for Early Light-Off of Aftertreatment Systems (Chaps. 6, 7, 8, and 9)
4. The Methane Conundrum (Chap. 10)
5. Simulation of Heat Underbody and Underhood (Chaps. 11, 12, and 13)
6. Fuels, Oils, and Equipment for Managing at Ultra-Low Temperatures (Chaps. 14 and 15)

All effort is concentrated in minimizing the harmful effects on fuel economy and life of aftertreatment systems strongly affected by thermal management techniques. Introducing the subject in Chap. 1, ARAI experts address the reduction in parasitic losses in the engine cooling system during part-load operation, engine lubrication system, and reduction in wear and friction when a simultaneous decrease in exhaust emissions and improvement in fuel consumption is achieved for optimum thermal management aimed at quicker light off of the aftertreatment devices, that include post-injection of fuel, gas exchange control, cylinder deactivation as well as waste heat recovery. The important role of simulation in studying the complex thermal activity within the engine and outside is mentioned with a case study presented at the end.

In Chap. 2, the experts from Cummins foresee the importance of thermal management of diesel exhaust aftertreatment system with evolving standards for ultra-low NO_x and particulate number. They summarize the developments in thermal management of diesel exhaust aftertreatment with special emphasis on the effects of the technologies on exhaust heat, emissions, and fuel efficiency. The subjects treated are engine-based thermal management, introducing fuel before aftertreatment, as well as heat generation and retention, and control. With the advent of hybrids and alternative propulsion systems, thermal management now must be integrated into system-level strategies, increasing complexity and requiring advanced integration and controls, bringing in rich possibilities for further research and development.

In Chap. 3, the dependence of the performance of the aftertreatment devices on their working temperature and in turn on the turbine-out temperature is discussed by the professors from the University of Valencia. The optimum conversion efficiency and regeneration can be achieved by choosing a strategy to increase the temperature at the inlet of the devices, at the same time addressing concerns on the engine fuel economy. Diameters of exhaust and intake valves, valve timings as well as the use of multi-step openings were theoretically studied to predict the temperature at the turbine outlet, coupled to external models for heat transfer and friction losses in steady and transient conditions. The potential of many proposals is deliberated as a function of the engine operating range. The layout of insulation is guided by the trade-off between the turbine outlet temperature and fuel consumption.

Chapter 4 by Turbo Energy Private Limited describes the important role played by turbochargers in increasing the low-end torque and rated power by up to 20% when engines are downsized. Though the turbocharger with fixed turbine geometry is widely used in medium engine-power intensity applications and the turbocharger with variable turbine geometry (VTG) copes with the high engine-power intensity requirements. Also, a VTG is useful for optimizing the exhaust gas heat energy for the catalyst operation during cold start and emission management. Insulated turbine housing and exhaust manifold go a long way in taming the hydrocarbon and carbon monoxide emissions as well as early warming up the aftertreatment systems.

Chapter 5 by Mahindra and Mahindra Limited mentions thermal safety and performance as the key attributes of a car, which decide the successful launch and customer acceptance of a product platform. Thermal management becomes challenging with the stringent second-stage BSVI standard, CAFE norms that require the vehicle systems to be highly efficient, and other safety requirements. The vehicle packaging and insulation strategies should lead to the safe functioning of the vehicle system throughout the real-world as well as the extreme conditions. This chapter includes concepts and validation aspects of thermal insulation design encompassing the basics of insulation, thermal design considerations, material selection, vehicle-level thermal assessment, and manufacturing.

Chapter 6 by TTM, ETH, BUWAL, STARFILTER, and AFHB of Switzerland is divided into four parts for easy reading of the probably most versatile method for increasing the exhaust temperature during cold start or in city drive, namely throttling the inlet or the exhaust. In the first part, throttling compressor and turbine upstream or downstream are thoroughly discussed. The use of coated and fuel-borne catalysts is deliberated. The combustion of soot in particulate traps and CRT is not triggered easily in a city bus in dense traffic or in other applications where only 150–200 °C can be reached. Then, active support is needed for augmenting passive regeneration. The convenient method applicable to any diesel engine is throttling the airflow where λ can be reduced to increase the exhaust temperature to a high 300 °C. Using a computational simulation as well as an experimental verification, intake throttling is shown as an active tool to increase the exhaust temperature needed for regeneration of a particulate filter. It is concluded that throttling downstream of the turbocharger compressor is advisable. The influence on fuel economy is negligible because of the short operation time.

Also, simpler solutions are possible to retrofit engines using mechanical fuel injection systems. It is recommended to use air intake throttling in combination with a catalytic coating of particle filters or fuel-borne catalysts to cover a wide range of engine applications and operating conditions.

In the second part of Chap. 6, Transport Refrigeration Units (TRU) powered by small diesel engines that emit high PM, are treated. A fuel-borne catalyst (FBC) facilitates the regeneration of honeycomb ceramic traps, by lowering the exhaust temperature requirements under almost all operating conditions. Therefore, the Swiss development team together with industrial partners developed a fully automatic active regeneration system for CARB that augments the FBC strategy by incorporating a fast-acting intake air throttle valve, which raises the exhaust gas temperature by more than 250 °C when closed, and when opened provides high oxygen immediately to the heated filter thus decoupling the availability of temperature and oxygen. This investigation describes the development and prototype testing of a typical 26 kW TRU unit with a reduction in PM less than 300 nm by more than 99%, EC-mass by 97%, PM by 86%, HC, and NO_2 on average by about 60%. This system is cost-effective even for retrofitting such small engines and applicable for other off-road retrofits.

The increase in particle mass in winter and ground-level ozone in summer is mainly due to heavy-duty diesel vehicles such as garbage trucks, city buses, and construction vehicles. The third part of Chap. 6 describes how equipping Euro 3 and older vehicles with a universal retrofit kit can alleviate the problem to a large extent. In the fourth part of Chap. 6, an optimal configuration for EGR plus DPF is evaluated in a passenger car engine. The soot in the DPF is burned with the oxygen contained in the exhaust gas rather than with NO_2. The investigations showed that low-pressure EGR for reduction of NO_x and intake throttling for regeneration of DPF are the best solutions for a retrofit. To increase the EGR rate at low loads, a second throttle is inserted on the low-pressure side in front of the compressor. The NO_x is reduced by more than 50%, and the filtration efficiency is better than 99%.

The important parameters for the regeneration of soot-laden DPFs are oxygen and temperature as described by the Arrhenius equation as well as the composition of the soot. Usually in engine exhaust, temperature and oxygen content develop in opposite directions. The technique for decoupling temperature and oxygen for

DPF regeneration mentioned briefly in the previous chapter is dealt with at length in Chap. 7 by the professors from TTM and CFS of Switzerland as the known techniques, like injection after the engine, static throttling, and dynamic throttling have some limitations. The transient temperature behaviour is simulated for studying the effect of different conflicting parameters and optimizing the parameters for fuel economy and emissions.

The aftertreatment systems function well when there is a combination of city, rural, and highway operations with the latter two helping in maintaining high operating temperatures above light off. However, when the vehicles operate solely under urban conditions, the engine-out temperatures in the exhaust rarely reach 150 °C. Injecting fuel late in the cycle to produce heat is uneconomical since the problem extends throughout the city drive. Chapter 8 by experts from TTM, Switzerland, and STARFILTER, Germany explain how the difficulty is resolved without heating, by various schemes to recover the exhaust heat after the exothermic reactions in the DOC and the DPF. Ingenious schemes that integrate the catalysts in a compact enclosure, achieve both energy recuperation and reduction in heat losses to the surroundings. However, the endothermic reactions in an SCR require a small intermediary heat source within the recuperation loop to keep the temperature level sufficiently high.

Though the current NO_x emission limits for heavy-duty engines are already very strict, it is expected that these limits would be further squeezed in near future. Therefore, the present SCR technology should be further improved with a strong focus on reducing NO_x in low-temperature conditions in real-road operations. Chapter 9 by Doosan aims to provide an insight into the best next-generation SCR concept and a proper thermal management strategy for it. The article also shows with an example that the simulation technique is a useful tool in the concept design and thermal management of SCR systems.

The emissions of the natural gas engines have a major component of methane in the total hydrocarbons. Since methane is a key contributor to climate change as a greenhouse gas, it needs urgent attention. As it is the most stable of all the hydrocarbons due to its strong C–H bond, the light-off temperature for methane at a catalyst is higher than for other hydrocarbons. In the cold phase, if not treated or controlled properly, methane in the tailpipe can defeat the new emission standards. In Chap. 10 by Mahindra and Mahindra Limited, the reasons for high methane emission particularly during the cold phase, and the strategies to reduce it are discussed.

The next three chapters describe the simulation of the underbody and underhood heat transfer that can substantially reduce expensive and repetitive experiments on a vehicle during development. Both internal and external heat transfer are key to provide a better understanding of the underbody heat transfer, cold start warm-up, and thermal ageing of the catalytic converter for gasoline engines and adequate thermal protection for the underbody components.

Chapter 11 by Ashok Leyland and SRM Institute first explains briefly the functioning of the cooling system and the components. Then the factors to be considered in the design of a cooling system are listed, and the use of computational fluid dynamics (CFD) is described, to simulate the heat transfer and hence predict the temperatures in the components related to the engine. Simulation of convective heat transfer requires accurate prediction of the flow of the fluid considering the important turbulence model. The boiling phenomenon in engine coolant flow requires special models, as well. The conjugate heat transfer method is used for solving conduction in solid regions in conjunction with convection. Radiation models effectively predict radiative heat transfer which is important in components that reach very high temperatures. Next, the underhood thermal analysis, which involves airflow through the heat exchangers, is explained. Virtual simulation methods using the CFD evaluate the performance of the cooling system even before prototyping. The use of CFD in industrial problems for design and optimization is possible only due to parallel processing and higher computing power.

Theoretical and experimental studies of flow and heat transfer in exhaust systems, internal as well as external, are important to understand the underbody heat transfer, cold start warm-up, and ageing of the catalytic converters and to provide sufficient thermal protection for the underbody components. Chapter 12 by MEDA Technical Engineering Services of Canada describes how the internal flow in a typical automobile exhaust system can be simplified using a 1-D model considering friction and heat transfer to the inner surfaces; the heat transfer from the outside surfaces due to the external flow and underbody is highly complex as treated in the next chapter by the use of a full-scale 3-D model of a vehicle. The predicted external heat transfer coefficients from this model are then used as boundary conditions for the 1-D model iteratively until convergence ensures an accurate prediction of the skin temperatures.

Despite all the efforts that vehicle manufacturers take, physical tests have their limitations as they can only test a handful of model variations and driving conditions. Some of the driving conditions that are critical to ensure vehicle safety may not be tested reliably. Traditionally, the use of CFD in the vehicle development process is limited due to the difficulty of solvers in handling production-level models and the lack of accuracy to make bold design decisions. However, there has been significant progress in these areas in the past decade to replace testing during vehicle development. In addition, the leaps in computational speed and a reduction in computational costs are two of the enablers. In Chap. 13 by Dassault, the simulation workflows used for thermal design are explained in detail. The article begins with a discussion of design challenges and a review of the simulation methods for thermal design. Next, cooling airflow simulations are discussed, where simulations of the heat exchangers and fans are detailed using the CFD solver PowerFLOW™. Following that, the chapter discusses the surface temperature map of the entire vehicle relevant to the design. This workflow takes into account convection, conduction, and radiation heat transfer. Subsequently, the simulation process is described for a more generalized testing environment like thermal duty cycle and key-off and soak. The discussion moves onto automation of the thermal workflows, optimization of design, and post-processing. Finally, the 1-D and system tools used for modelling vehicle components used in thermal workflows are explained.

Engine performance deteriorates as the ambient temperature drops below 10 °C. Fuel, engine oil, and battery are carefully selected for the efficient functioning of the engine in cold climates. Chapter 14 penned by the authors from the Indian Oil Company and the University of Birmingham gives details of the winter-grade fuels with better flowability and the ability to finely atomize when injected as well as the flow-improved lubricating oils with low frictional and pumping losses for easier starting of the engine.

EU Emission regulation at a cold ambient temperature of –7 °C only speaks about hydrocarbons and CO emissions from all engines and particle numbers only from GDI and diesel engines. However, Chap. 15 by the IIT Kanpur indicates that all emissions including particle numbers increase disproportionately at low temperatures from Euro 5 onwards, referring to the tests on SI and CI engines conducted elsewhere as per WHTC or WLTD.

We hope that the book would greatly interest the professionals, post-graduate students involved in the design and development of advanced diesel and CNG engines.

The editors would like to express their sincere gratitude to the authors from all over the world for submitting their high-quality work in time and revising it appropriately at short notice. We would like to express our special gratitude to our prolific set of reviewers of different chapters and for their valuable suggestions to improve them:

- Ashok Leyland, Chennai: Mr. Ganesh Prasad, and Mr. Sahaya Surendira Babu
- Mahindra Research Valley, Chennai: Dr. Srinivas Gunti, Dr. Shankar Venugopal, Mr. Ravi Ranjan, Mr. Parvej Khan, Mr. Mutta Surendranath, Mr. Ashwani Verma, and Mr. Paradarami Udaya Kumar
- Mahindra University, Hyderabad: Dr. Tamma Bhaskar

Kanpur, India
September 2021

P. A. Lakshminarayanan
Avinash Kumar Agarwal

Contents

Editors and Contributors

About the Editors

P. A. Lakshminarayanan studied at the Indian Institute of Technology (IIT), Madras for his B.Tech., M.S., and Ph.D. degrees. He worked at the Loughborough University of Technology and Kirloskar Oil Engines Ltd. for 25 years, before moving to Ashok Leyland in 2002 to head the Engine R&D. From 2011 till 2019, he was the CTO, and the Technical Adviser at Simpson and Co. Ltd. He is now an adjunct professor at IIT Kanpur. He has developed more than eight diesel and CNG engine platforms and 150 engine types commercially successful for efficiency and cost-effectiveness. Two engine designs received prizes from the Institute of Directors (India). Twelve ideas were patented during the development of engines over 40 years. He has authored 54 research papers in journals and conferences of international repute. Four of them received prizes for integrity and quality of contents from the SAE (Arch Colwell Award), Combustion Society (India), AVL (Graz), and AVL (India) in 1983, 1993, 2005, and 2011. He is elected to the fellowships of SAE (2009), INAE (2013), and ISEES (2018).

Prof. Avinash Kumar Agarwal joined IIT Kanpur in 2001. He worked at the Engine Research Center, UW@Madison, the USA as a Post-Doctoral Fellow (1999–2001). His interests are IC engines, combustion, alternate and conventional fuels, lubricating oil tribology, optical diagnostics, laser ignition, HCCI, emissions, and particulate control, 1D and 3D Simulations of engine processes, and large-bore engines. Professor Agarwal has published 435+ peer-reviewed international journal and conference papers, 70 edited books, 92 books chapters, and 12,200+ Scopus and 19,000+ Google Scholar citations. He is the associate principal editor of FUEL. He has edited *Handbook of Combustion* (5 volumes; 3168 pages), published by Wiley VCH, Germany. Professor Agarwal is a Fellow of SAE (2012), Fellow of ASME (2013), Fellow of ISEES (2015), Fellow of INAE (2015), Fellow of NASI (2018), Fellow of Royal Society of Chemistry (2018), and a Fellow of American Association of Advancement in Science (2020). He is the recipient of several prestigious awards such as Clarivate Analytics India Citation Award-2017 in Engineering and Technology, NASI-Reliance Industries Platinum Jubilee Award-2012; INAE Silver Jubilee Young Engineer Award-2012; Dr. C. V. Raman Young Teachers Award: 2011; SAE Ralph R. Teetor Educational Award-2008; INSA Young Scientist Award-2007; UICT Young Scientist Award-2007; INAE Young Engineer Award-2005. Professor Agarwal received Prestigious CSIR Shanti Swarup Bhatnagar Award-2016 in Engineering Sciences. Professor Agarwal is conferred upon Sir J C Bose National Fellowship (2019) by SERB for his outstanding contributions. Professor Agarwal was a highly cited researcher (2018) and was in the top ten HCR from India among 4000 HCR researchers globally in 22 fields of inquiry.

Contributors

A. Amstutz ETH, Zürich, Switzerland

Sudharsan Annur Balasubramanian CFD Engineering Analyst, MEDA Engineering and Technical Services, Windsor, Canada

Chin-Wei Chang Dassault Systemes, Simulia, Waltham, USA

Hitesh B. Chaudhari The Automotive Research Association of India, Pune, India

Young-Chang Cho Dassault Systemes, Simulia, Waltham, USA

J. Czerwinski AFHB, Abgasprüfstelle der Fachhochschule Biel, Port, Switzerland

Subhankar Dev The Automotive Research Association of India, Pune, India

V. Dolz Universitat Politécnica de Valencia. CMT-Motores Térmicos, Valencia, Spain

J. Galindo Universitat Politécnica de Valencia. CMT-Motores Térmicos, Valencia, Spain

Z. Gerald Liu Cummins Inc., Madison, WI, USA

Srinivas Gunti Mahindra & Mahindra Ltd., Mahindra Research Valley, Mahindra World City, Chengalpattu, Tamil Nadu, India

L. Guzzella ETH, Zürich, Switzerland

J. Gómez Universitat Politécnica de Valencia. CMT-Motores Térmicos, Valencia, Spain

P. T. Haridas Ashok Leyland Technical Center, Ashok Leyland Ltd, Chennai, Tamil Nadu, India

Y. Hohl Liebherr Machines Bulle SA, Bulle, Switzerland

F. Jaussi Liebherr Machines Bulle SA, Bulle, Switzerland

Simhachalam Juttu The Automotive Research Association of India, Pune, India

Satheesh Kandasamy Dassault Systemes, Simulia, Waltham, USA

S. Kany TSH, Hohentengen, Germany

Parvej Khan Mahindra & Mahindra Ltd., Mahindra Research Valley, Mahindra World City, Chengalpattu, Tamil Nadu, India

P. A. Lakshminarayanan Indian Institute of Technology Kanpur, Kalyanpur, Kanpur, India

F. Legerer VERT, Verification of Emission Reduction Technologies, Brugg, Switzerland

Han Li Dassault Systemes, Simulia, Waltham, USA

J. M. Luján Universitat Politécnica de Valencia. CMT-Motores Térmicos, Valencia, Spain

Th. Lutz ETH, Zürich, Switzerland

Chr. Lämmle Computer Flow Solutions, Zürich, Switzerland; Combustion and Flow Solutions GmbH, Zürich, Switzerland

Neelkanth V. Marathe The Automotive Research Association of India, Pune, India

A. Mayer TTM, Niederrohrdorf, Switzerland

Devadatta Mukutmoni Dassault Systemes, Simulia, Waltham, USA

Achuth Munnannur Cummins Inc., Madison, WI, USA

P. Nöthiger PNE, Olten, Switzerland

Nathan Ottinger Cummins Inc., Madison, WI, USA

P. Piqueras Universitat Politécnica de Valencia. CMT-Motores Térmicos, Valencia, Spain

Parthiban Rajamani Mahindra & Mahindra Ltd., Mahindra Research Valley, Mahindra World City, Chengalpattu, Tamil Nadu, India

A. S. Ramadhas R&D Centre, Indian Oil Corporation, Faridabad, India

K. Ramesh Turbo Energy Private Limited (TEL), Chennai, India

Ravi Ranjan Mahindra & Mahindra Ltd., Mahindra Research Valley, Mahindra World City, Chengalpattu, Tamil Nadu, India

U. G. Remesh Ashok Leyland Technical Center, Ashok Leyland Ltd, Chennai, Tamil Nadu, India

Mohak P. Samant The Automotive Research Association of India, Pune, India

Patro Sambit Kumar SRM Institute of Science and Technology, Chennai, Tamil Nadu, India

M. Sathya Prasad Ashok Leyland Technical Center, Ashok Leyland Ltd, Chennai, Tamil Nadu, India

J. R. Serrano Universitat Politécnica de Valencia. CMT-Motores Térmicos, Valencia, Spain

R. Sneha Priya Ashok Leyland Technical Center, Ashok Leyland Ltd, Chennai, Tamil Nadu, India

H. Stieglbauer STARFILTER, Penzberg, Germany

D. A. Subramani Turbo Energy Private Limited (TEL), Chennai, India

C. Vijay Ram Ashok Leyland Technical Center, Ashok Leyland Ltd, Chennai, Tamil Nadu, India

Nagesh H. Walke The Automotive Research Association of India, Pune, India

Huhu Wang Dassault Systemes, Simulia, Waltham, USA

Tae Joong Wang Aftertreatment System Development Part, Engine Business Group, HYUNDAI DOOSAN INFRACORE (Co., Ltd.), Incheon, Republic of Korea

M. Wyser BUWAL, Bundesamt für Umwelt, Bern, Switzerland

Hongming Xu University of Birmingham, Birmingham, England

Part I
The Gestalt of Thermal Management

Chapter 1
Introduction to Thermal Management Techniques

Neelkanth V. Marathe, Nagesh H. Walke, Simhachalam Juttu, Hitesh B. Chaudhari, Subhankar Dev, and Mohak P. Samant

1.1 Background

The term 'thermal management' refers to optimizing the thermal balance in an engine and vehicle subsystems, such that their operating temperatures are maintained at an optimum level. Thermal management has become a very crucial part of vehicle development, because it directly or indirectly affects engine performance, fuel economy, safety, reliability, engine component life, driver & passenger comfort, materials selection, emissions, maintenance, aerodynamics, etc. Thermal management of the engine has become critical in optimization for energy efficiency and emissions reduction. Air handling system, combustion, exhaust system, cooling system, lubrication system, and their interactions are very important in the efficient thermal management of the engine. Significant improvements in engine efficiency can be achieved by optimizing the warm-up phase with a proper coolant flow rate, resulting in reduced frictional losses. Electric water pump, split cooling, map-controlled thermostat, variable flow oil pump, and other technology are applied to improve fuel consumption, reliability, and passenger comfort in cold conditions. Passenger comfort through thermal management includes technologies to regulate the temperature within the passenger cabin such as heating, ventilation, and air-conditioning (HVAC) systems, glazing, etc.

Engine thermal loads are created through the conversion of chemical energy to thermal energy and mechanical energy and the transfer of that energy through the powertrain. It becomes essential to maintain proper engine component temperatures and prevent failures that result from excessive thermal stresses, fatigue cracking, combustion knocking, and lubricant degradation. Additionally, for Internal Combustion Engines (ICEs), increasingly stringent exhaust regulations are making their

N. V. Marathe · N. H. Walke (✉) · S. Juttu · H. B. Chaudhari · S. Dev · M. P. Samant
The Automotive Research Association of India, Pune 411038, India
e-mail: walke.nvh@araiindia.com; walke.edl@araiindia.com

P. A. Lakshminarayanan and A. K. Agarwal (eds.), *Handbook of Thermal Management of Engines*, Energy, Environment, and Sustainability,
https://doi.org/10.1007/978-981-16-8570-5_1

compliance ever more difficult. Tailpipe emissions are dependent on the catalyst temperature and its corresponding conversion efficiency. When the catalyst temperature deviates from its optimum temperature range, its conversion efficiency drops worsening tailpipe emissions. An additional challenge is posed due to the introduction of new testing cycles designed to better match real driving conditions and behaviors. In this case, also thermal management becomes very important for a drastic reduction in CO_2 and pollutant emissions. The engine mix between diesel, gasoline, and electrified vehicles in future fleets will depend on the cost of new emission optimization devices and thermal management strategies and the fuel consumption they achieve. The modern diesel engine has a crucial role to play in this scenario because it is needed to keep overall fleet consumption and CO_2 emissions low. For hybrid/electric vehicles, whose technology is based on insulated-gate bipolar transistors (IGBTs), the high heat fluxes cause higher operating temperatures severely affecting device reliability. For Fuel Cells powered vehicles, the operating performance of the stack depends on thermal management keeping the system within optimum temperature range.

The requirements of meeting very low tailpipe NO_x limits like 0.027 g/kWh [1] and CO_2 emission reduction targets to 15% and 30% reduction in 2025 and 2030 respectively with respect to the 2019 CO_2 emission levels [2] have put tremendous challenges on the researchers. Thermal management is considered as one of the strategies to help to achieve these requirements. Figure 1.1 describes the thermal management strategies commonly used.

The following are the summarized key benefits of thermal management [3].

- Reduce parasitic power losses
- Decrease exhaust emissions
- Improve flexibility in component packaging
- Enhance fuel economy
- Improve cooling system control
- Quicker engine warm-up during cold start
- Reduce engine wear and friction
- Increase lubricant life

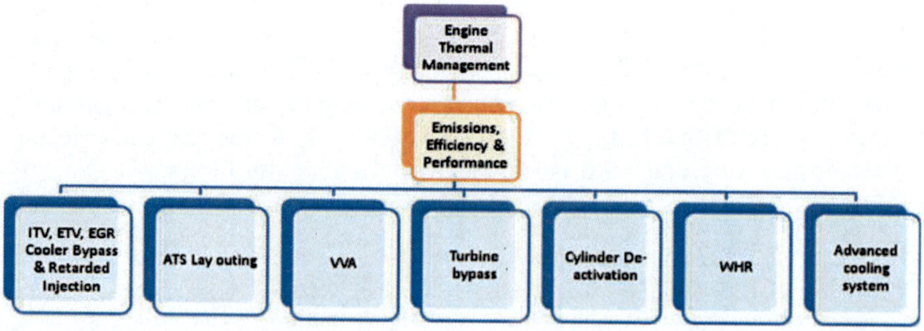

Fig. 1.1 Engine thermal management strategies

- Eliminating overcooling during part-load operation
- Reduce system pressure drop and hose losses
- Eliminating hot soak after the engine stop
- Increase the average combustion temperature.

1.1.1 Engine Cooling System

Typically, about 40% of combustion energy in an IC Engine is utilized to produce the effective power output, 30% leaves through the exhaust and 30% leaves through the cooling system. The engine cooling system accomplishes the necessary heat rejection by circulating air or liquid coolant around the combustion chamber, absorbing heat from the engine, and dissipating to the ambient. The performance of the conventional engine cooling system has always been constrained by its passive nature and design for the required heat rejection capability at the maximum power. This leads to considerable losses in the cooling system at part load conditions where vehicles operate most of the time. The automotive cooling system has the potential to improve ICE performance through enhanced coolant temperature control and reduced parasitic losses. Advanced automotive thermal management systems use controllable actuators such as smart thermostat-valve, variable speed water-pump, and electric radiator-fan in place of conventional mechanical cooling system components to improve engine temperature tracking over entire operating ranges covering vehicle duty cycle. In the smart cooling system, the electro-mechanical actuators are required to be precisely calibrated to control engine temperature close to the target operating temperature for the best efficiency. The design and placement of cooling components are required to be considered when attempting to maximize the performance. A thorough understanding of heat transfer modes and pathways is necessary for the design of an effective engine cooling system.

In the novel thermal management system, rather than being circulated, the coolant remains still during the warm-up phase so that the engine oil quickly reaches its operating temperature of between 80 and 120 °C. This significantly shortens the phase of greater frictional resistance due to viscous oil. In the case of a split cooling thermal management system, the crankcase and cylinder head have their cooling water loops connected via a valve. During the warm-up phase, the coolant in the block is not circulated. The oil cooler is also bypassed during this phase. The water in the crankcase is often not circulated, even at a low load when the engine is warm. The coolant, which circulates through the heads, also flows through the cabin heater as well as the EGR cooler.

1.2 Engine Lubrication System

Engine thermal management involving optimization of energy flow for various engine sub-systems like lubrication systems can reduce the parasitic losses and thereby improve the engine efficiency and performance. Heat rejection to the lubricating system also plays an important role in engine performance and durability. Engine oil loses its lubricating qualities when temperatures are exceeded and as a result, excessive engine wear occurs. Fixed displacement oil pumps worked the same regardless of the oil viscosity or demands of the engine and would be oversized to handle the harshest engine operating conditions. They operate at peak performance and it's up to the pressure regulator to bleed off the excess pressure. This excess pressure that is bled off is wasted energy. Map-controlled variable displacement oil pumps (VDOP) and map-controlled piston cooling jets (PCJ) are two kinds of technologies to improve efficiency. It is observed that the overall fuel efficiency improvement is in the range of 3–5% [4]. Another advantage of controlling the oil pressure and volume is heat management. By regulating the flow of the oil with engine temperature, loads, and engine speeds, heat transfer can be optimized in the head and the pistons. Viscosity and flow characteristics are very important in lubrication oil properties. Also, heat transfer is a strong function of these parameters which contribute to Nusselt, Reynolds, and Prandtl numbers. With the information programmed into the engine control module (ECM), modern engines know their oil. The ECM knows what the oil pressure should be for a given engine speed and coolant temperature. If the numbers do not match, it will set a code and put the engine into a reduced power mode.

VDOP is normally mounted directly on the crankshaft, eliminating the need for an intermediary shaft and the associated risk of its failure. VDOP are 'Gerotor' designs, having trochoid gears that allow for smooth operation, low noise, and excellent suction. Figure 1.2 shows the vane-type variable displacement oil pump. The centrically seated drive gear drives an external eccentrically seated annular gear. The result of which is cavities inside the pump compress and enlarge to create the suction and feed effect. The inner rotor sits on the crankshaft and drives the outer rotor. Since the inner and outer rotors have different rotating axes, more space is created on the suction side due to the rotating motion. The oil is drawn in and transported to the pressure side. On the pressure side, the space between the gear teeth becomes smaller again, and oil is forced into the oil circuit under pressure. A VDOP changes the rotating axis of the outer gear. To achieve this, the gears of the inner rotor are replaced with variable-length vanes. The outer gear pivots on an axis; opposite the pivot is an electronic actuator. On mechanical versions, a spring replaces the actuator and the opposite side of the housing has oil or a piston that pushes against the spring to regulate pressure. Oil pressure sensors are positioned in the oil gallery to monitor the overall pressure in the system. Oil temperature is normally calculated using inputs of various other sensors. Oil temperature plays a critical role in calculating the actuator position during cold start-up.

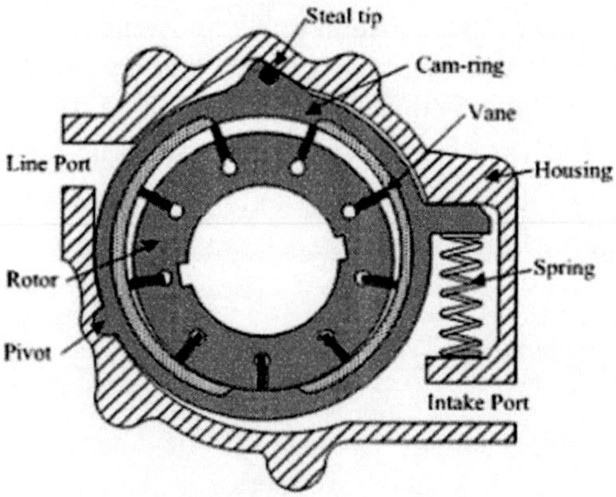

Vane Type Pump

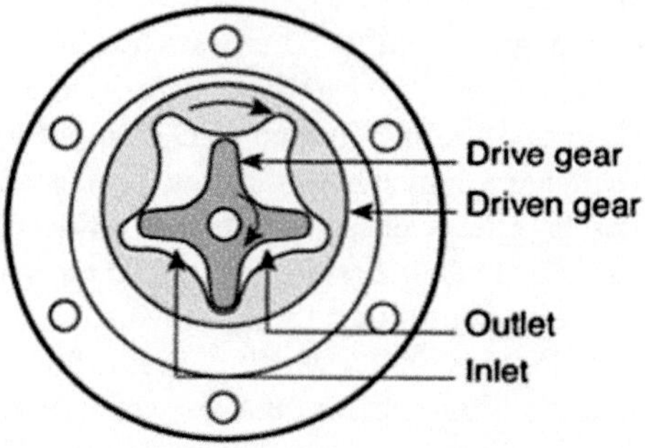

Gerotor Type Pump

Fig. 1.2 Variable displacement oil pump (VDOP)

The engine control module (ECM) looks at data including engine coolant temperature, engine load, calculated oil temperature, and other monitors to determine the position of the oil pump actuator and oil pressure. If the system detects an overheating condition or a problem with one or more of the inputs, it may put the system into a reduced power mode to prevent damage.

Modern heavy-duty engines are with high power density for enhancing performance and meeting strict emissions thereby putting more and more thermal stresses on various components including pistons. To avoid component damage, engines are equipped with oil cooling jets and these accounts for an additional increase in engine parasitic losses as piston cooling jet drives a major chunk of power demand by the oil pump. Understanding the crown temperature profiles over the engine operating map can help in developing the control strategies for oil flow meant for cooling the piston. Variable flow oil and water pumps are making a significant contribution to the Super Truck program [5]. Similarly, there is also a potential benefit to fuel economy improvement when controlling the delivery of oil through map-based control for either lubrication or heat removal via a piston cooling jet (PCJ). Also, map-controlled PCJ operations can take care of deterioration of combustion by lowering the temperatures with unwanted parasitic losses and fuel penalty at low load conditions or even at cold start. Burke et al. [6] observed a fuel consumption improvement of up to 3.4% over the New European Drive cycle and up to 5.8% over the urban phase, with the use of a variable displacement oil pump in the assessment on a 2.4 litre diesel.

1.3 Combustion System

1.3.1 Thermal Management for After-Treatment

Modern engines are equipped with catalytic exhaust after-treatment systems that require a suitable exhaust temperature window to operate at their best efficiency levels. The engine after-treatment systems comprise typically the oxidation catalyst, a selective reduction catalyst, and a particulate filter. Diesel Particulate Filter (DPF) and Selective Catalytic Reduction (SCR) are used in the exhaust line to drastically reduce Particulate Matter (PM) and Oxides of Nitrogen (NO_x) emissions to fulfill emission limits. Exhaust after-treatment systems have different processes occurring at different temperatures and also the effectiveness of one catalyst depends on the performance of the other catalyst. For example, typically a DOC system, widely known for CO and HC conversion, also contributes to the conversion of the engine out nitric oxides (NO) to nitrogen dioxides (NO_2) which is useful for both, during passive regeneration in the particulate filter and for NOx conversion in SCR. The oxidation catalyst carries the most weightage in thermal management. Temperature impacts SCR performance in addition to light off, as it is required for extracting gaseous ammonia from aqueous urea, for which there is a cut-off dosing temperature of roughly 180–200 °C depending on catalyst material. For the effective regeneration process of Diesel Particulate Filter (DPF), thermal management of the exhaust flow before and inside Diesel Oxidation Catalyst (DOC) is necessary. At medium or small loads, which are relatively common during actual diesel engine operation, the exhaust temperature can be too low (150–350 °C) for DPF active regeneration to initiate. The engine control strategy needs adjustment to raise the exhaust temperature before DPF to initiate active regeneration. Table 1.1 shows the typical temperature ranges required for various processes.

As shown in Table 1.1, the efficient reduction of engine-out emissions requires a certain after-treatment system temperature range, which is achieved by thermal management via control of engine exhaust flow and temperature. The role of exhaust gas thermal management for the after-treatment system is to enable the catalysts to rapidly attain activation temperatures so that catalysts can function at their better efficiency levels thereby keeping the tailpipe emission at the lowest possible levels. One of the main issues is the performance of the SCR system during cold start and warm-up phases of the engine where the exhaust temperatures are too low to (a) allow thermal activation of the reactor and (b) promote high conversion efficiency and significant NO_x concentration reduction. The dependence of conversion efficiency on temperature has become increasingly problematic when dealing with modern

Table 1.1 Typical exhaust temperature ranges favoring different processes

Sr. No	Process	Optimum exhaust temperature range
1	DOC CO-HC light-off	200–270 °C
2	DOC NO_2 conversion	180–450 °C
3	HC desorption	> 200 °C
4	Desulfation	400 °C to 650 °C
5	Passive regeneration	240–250 °C
6	Active regeneration	> 550 °C
7	SCR peak efficiency	280–350 °C
8	Urea deposition	125–175 °C
9	Ammonia desorption	> 450 °C
10	NO_2 dissociation	> 400 °C

diesel engines. Their higher fuel efficiency yield is highly advantageous, but this also means that their exhaust gases are relatively cold under partial load, especially in city traffic. These colder gases cool the SCR converter unit as they pass through it and reduce its conversion efficiency.

The art of managing exhaust temperatures while maintaining the benefits of effective combustion efficiencies is called thermal management. The main constraint in thermal management is to limit the potential fuel consumption increase. There are different technologies available for thermal management; however, each one has its magnitude of energy cost and efficiency losses. The selection depends on the engine type, duty cycle, and level of exhaust emission control required. For example, engines operating with their duty cycles mostly at high load for a longer time will already have high exhaust temperatures and may require relatively less challenging temperature management. Whereas, applications, where duty cycles are mostly at low loads with lower exhaust temperatures, may require complicated temperature management systems and may result in a considerable penalty on fuel efficiency. Many parameters are important during thermal management, including exhaust enthalpy, temperatures, NO_x emission reduction required, back pressure, intake restriction, fuel consumption, air–fuel ratio, etc. The optimization depends on several considerations including as mentioned below:

- Engine out emission levels over operating zone—low, moderate, or high.
- Exhaust temperatures available over the operating map, during emission cycles both steady and transient operations.
- Preferred regions of peak engine efficiency as per the duty cycle and type of application.
- A kind of regeneration method is used and their regeneration pattern is based on the duty cycle.
- The type of catalyst material used.

While selecting the type of thermal management strategy, there has to be clarity on the requirement as to whether an exhaust enthalpy rise is needed for rapid heat of the engine exhaust and after-treatment system or only exhaust temperature rise is sufficient to prevent the after-treatment system from cooling down. Airflow reductions generally result in higher exhaust gas temperatures and lower exhaust flow rates, which are beneficial for maintaining already elevated component temperatures. Airflow reductions also reduce pumping work, which improves fuel efficiency.

1.3.2 Cylinder Gas Exchange Control

Intake throttling and exhaust restriction are the most commonly used strategies for attaining required exhaust temperatures. An electromechanically controlled valve train system for internal Exhaust Gas Recirculation (internal EGR) can also be used for thermal management. In internal EGR, the exhaust valve re-opens for a second lift during the intake stroke, rebreathing part of the hot exhaust gases, thus increasing the temperature inside the combustion chamber and thereby of the exhaust gases before the next combustion cycle. The main drawback is that this may increase the engine-out soot. Phase change material (PCM) and electrically heating are also used to keep the converter hot beyond light-off temperature.

In the work by Lauren et al. [7], both intake and exhaust throttling were compared for a Stage V off-road diesel engine; temperature rise with associated fuel consumption penalty was assessed over steady-state points as well as transient emission cycles. It is observed that at part loads exhaust throttling incurred a higher percentage of fuel consumption penalty compared to intake throttle, for all speed and load range for every 100 °C rises in exhaust temperatures. However, for intake throttling, the PM emissions are higher than exhaust throttling for a similar rise in exhaust temperatures. The author further measured the impact of intake throttling along with waste-gate and retarded timings on exhaust temperature rise and its adverse effect on PM emissions and fuel consumption. They witnessed a 50–150 °C rise in exhaust gas temperature with a 140% rise in the particulate emissions and a 3% rise in fuel consumption measured over the NRTC. Figures 1.3 and 1.4 show a comparison of the rise in exhaust temperatures and relative increase of particulate emissions respectively, as measured over NRTC with thermal management using intake air restriction and turbocharger waste-gate control as against the base engine operation [7].

Different thermal management techniques like intake throttling exhaust throttling, retarded injection timing, VVA (Variable Valve Actuation) based thermal management, application of air gap manifold (AGM), and turbine bypass were evaluated by Kovacs et al. [8] for increasing exhaust enthalpy and decreasing engine out NOx over the portion of cold start in FTP cycle, with model-based analysis. A further effect of EGR cooler bypass and charge air cooler bypass were also investigated. In VVA, LIVC (Late Intake Valve Closing) and EIVC (Early Intake Valve Closing), and EEVO (Early Exhaust Valve Opening) were used to increase both exhaust gas temperature and enthalpy. Impact on Exhaust temperature, exhaust enthalpy, fuel

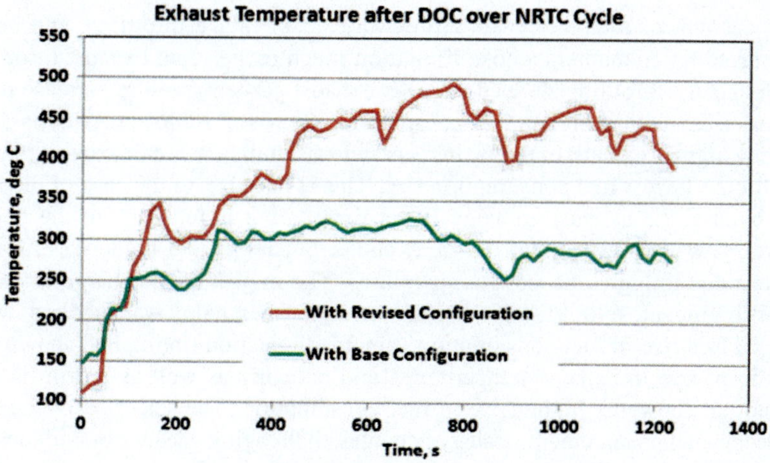

Fig. 1.3 Exhaust temperatures over NRTC [7]

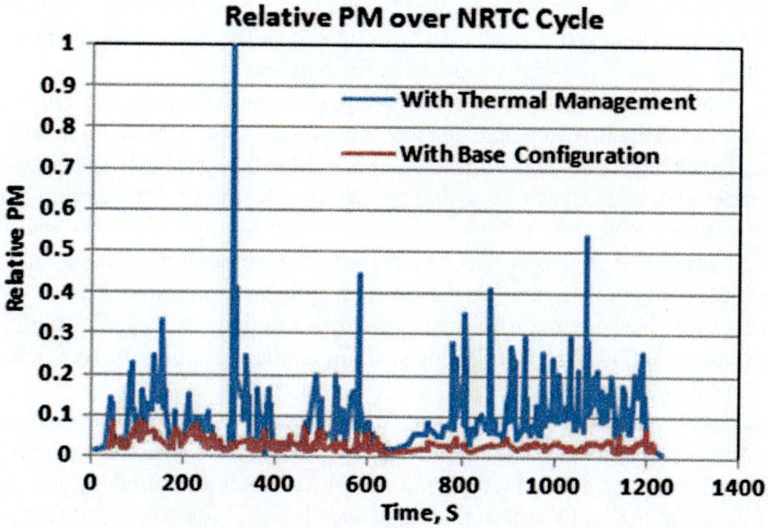

Fig. 1.4 Relative PM over NRTC [7]

consumption, and cumulative NO_x emissions was investigated over the first 300 s of the cold start HD-FTP cycle. It was observed that with both LIVC and intake throttling, exhaust temperatures were increased but due to a large decrease in the exhaust flow rate, the effect on exhaust enthalpy was negligible. However, it was observed that fuel consumption with LIVC is lower compared to intake throttling. Exhaust throttling showed the second-highest increase in exhaust temperature with a reduction in exhaust flow rate. The larger drop in exhaust enthalpy, as explained, was on account of a direct cut on airflow in case of intake throttling, and hence, intake throttling encounters smoke limitation much earlier than exhaust throttling. Though exhaust throttling showed a higher exhaust gas temperature increase it was also associated with a higher fuel consumption increase. EGR cooler bypass had shown the highest exhaust enthalpy increase and a significant rise in exhaust temperatures for the lowest fuel consumption rise. This is because, in the case of the EGR cooler bypass, the exhaust enthalpy is not drained but is fed back to the engine. However, it was observed that the EGR cooler bypass gave a higher rise in NOx emissions due to higher process temperatures. The impact of the charge air cooler was not significant. With EEVO, the effect on exhaust enthalpy was moderate while it also led to a rise in fuel consumption. Retarded injection timing had shown both a significant rise in exhaust temperatures and enthalpy as well as a drop in NO_x emission but showed a higher rise in fuel consumption. The selection of the most suitable thermal management strategy becomes challenging because of both positive and negative effects. However, retarded SOI, intake throttling, and exhaust throttling were considered more suitable, as they had offered the highest enthalpy and temperatures rise with a reduction in NO_x emissions. Further, these three thermal management strategies were investigated for the first 300 s of the HD-FTP transient cycle and it was observed that retarded timing shows the highest temperature rise followed by exhaust throttling compared to the baseline.

In the stepwise investigations by Kovacs et al. [8], retarded injection timings under load were used in the first step, exhaust throttling in the second, retarded timing even at no load in the third, and intake throttling was used in the fourth. Retarded injection timing under load and exhaust throttling could obtain the highest increase in exhaust enthalpy with associated fuel consumption. Retarded timing even at no load in the third step further had an insignificant increase in the temperatures. Intake throttling increased the temperatures and prevented temperatures from going below 185 °C, but there was no increase in the exhaust enthalpy. Cumulatively there was a 96 K rise in temperatures, a 65% rise in enthalpy with an almost 70% drop in NO_x emission during the cold part of the cycle.

In internal combustion engines, variable valve actuation (VVA) is the process of altering the valve actuation timing and valve lift to increase exhaust temperatures, improve fuel economy, and reduce emissions. VVA enables the cylinder gas exchange process to be optimized for different engine speeds and loads. There are many ways in which variable valve timing (VVT) and variable valve lift (VVL) can be achieved, ranging from mechanical devices to electro-hydraulic and cam-less systems. The simplest form of VVT is cam phasing, whereby the phase angle of the camshaft is adjusted relative to the crankshaft, accordingly valve opening and closing can be adjusted with respect to engine speed. However, valve lift and valve opening duration cannot be altered solely with a cam-phasing system. Achieving variable duration requires a more complex system, such as multiple cam profiles or oscillating cams.

Early intake valve closing (EIVC) is a way to decrease the pumping losses associated with low engine speed and low load operations by closing the intake valve earlier than its normal timing. Early intake valve opening (EIVO) has significant potential to reduce emissions. EIVO allows the backflow of some of the combusted gas into the intake manifold, which is an internal EGR for controlling the peak combustion temperature of the cylinder and hence a reduction in NO_x emissions. Early closure of the exhaust valve can give control on NO_x emissions through residual exhaust gas left in the cylinder. Early opening of the exhaust valve gives effective control over exhaust gas temperatures.

1.3.3 Post Fuel Injection

An increase in the late post-injection quantity can produce a significant rise in the temperature inside DOC, though its impact on the exhaust temperature before DOC is relatively limited. With an optimized injection quantity, a considerable temperature increase inside the DOC can be obtained with relatively low HC emission. For example, if the intake throttle valve is partially closed to achieve DOC light-off temperature and the late post-injection quantity is increased, the exhaust temperature after DOC can be made greater than 550 °C which is adequate for DPF active regeneration.

Hiranuma et al. [9], investigated exhaust thermal management techniques with both passive and active regeneration for a commercial vehicle. It was found that in the city driving adequate PM oxidation was not possible and consequently continuous passive regeneration did not occur and it was necessary to utilize active regeneration. Four systems namely only particulate filter, a pre-catalyst with a particulate filter, coated particulate filter, and pre-catalyst with coated particulate filter were assessed for their effect on PM oxidation rate for both low as well as the high-temperature region. It was observed that the use of a pre-catalyst significantly improved the PM oxidation rate at temperatures around 300 °C compared to only a PM filter. Pre-catalyst with coated PM filter improved further the PM oxidation rate at lower temperatures, however at a higher temperature region (about 600 °C), it was observed that both only PM filter and PM filter with pre-catalyst showed similar behavior as O_2 related oxidation becomes dominant at high temperatures. It was also observed that pre-catalyst with a coated PM filter unit has a wider operating region available for continuous regeneration compared to pre-catalyst with PM filter. For urban operations, active regeneration was necessary since PM accumulated levels were quite high (refer to Fig. 1.5). The optimized system consisted of a pre-catalyst for enabling NO_2 generation and the temperature rising for active regeneration, an uncoated SiC-based PM filter, and a post-catalyst for the oxidation of CO during active regeneration, as shown in Fig. 1.6.

The active regeneration methods are useful under low load or urban driving conditions but reflect in fuel consumption deterioration. It is the optimization between fuel consumption and reliability as for lower fuel consumption impact the active regeneration is kept at longer intervals. However, longer active regeneration intervals can cause implications on the filter life due to exposure to very high temperatures during spontaneous oxidations.

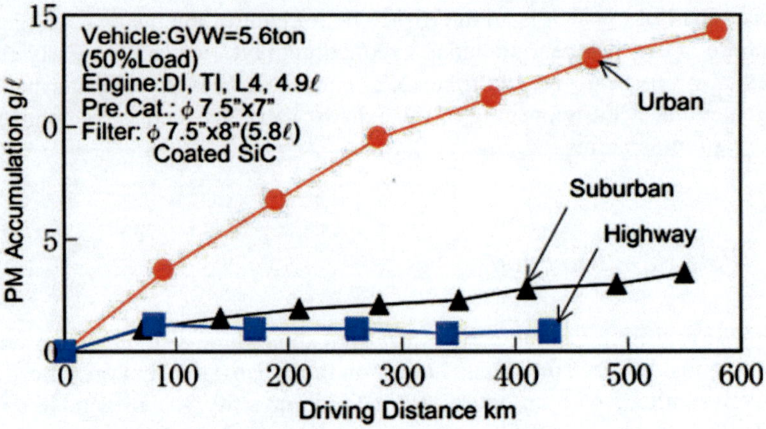

Fig. 1.5 PM accumulation for various driving conditions [9]

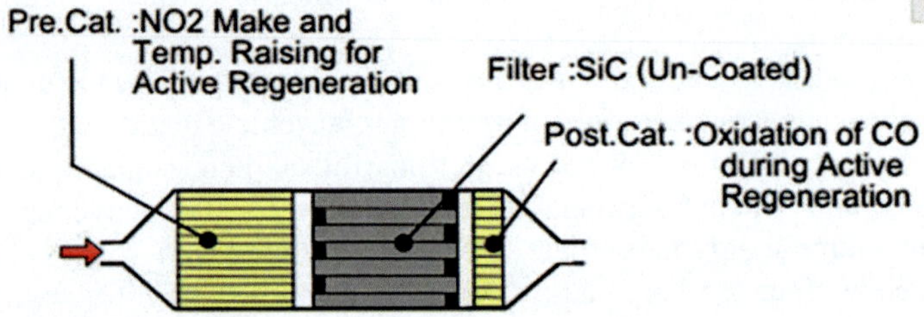

Fig. 1.6 Optimized DPF system [9]

1.4 Cylinder Deactivation (CDA)

Cylinder deactivation (CDA) can be used for thermal management as well as to improve fuel efficiency. Typically, a low proportion (around 30%) of engine full load power is used during light-load driving conditions resulting in poor fuel efficiency. Due to the deactivation of some of the cylinders, the remaining active cylinders work at increased cylinder pressures resulting in improved fuel consumption and increases exhaust temperatures. The methods of fuel efficiency improvement include reduced heat transfer and pumping losses, increased combustion efficiency, and reduced mechanical losses. It has been observed that cylinder deactivation can give fuel consumption improvement in the range of 10–30% in highway operating conditions [10].

Cylinder deactivation is achieved by keeping the intake and exhaust valve closed for the particular cylinders, where there is no gas exchange, creating an air spring in the combustion chamber. The trapped exhaust gases, from the previous charge burn, are compressed during the piston upward stroke and expanded on the piston downward stroke. The compression and expansion of the trapped exhaust gases have an equalizing effect with no net extra work by the engine. Also fuel delivery to the disabled cylinders is cut-off by the engine management system. The transition between normal engine operation and cylinder deactivation is required to be smooth, using changes in ignition timing, cam timing, and throttle position.

Cylinder deactivation (CDA) technology has been identified as a key enabler to reduce diesel engine fuel consumption as well as increase the exhaust temperature and has been considered to be an important part of the strategy for reaching future low NO_X targets. The increased exhaust temperature improves conversion efficiency for NO_X reduction and has the potential to lower the fuel consumption from post-injection catalyst heating and minimize associated oil dilution issues.

1.5 Waste Heat Recovery (WHR)

Modern engines convert approximately 40% of the fuel energy into mechanical power. Residual part of heat energy emitted as a waste to the environment through engine exhaust, engine cooling, EGR cooler, charge air cooler, and oil cooler. A part of this heat energy can be converted into mechanical power by using a Waste Heat Recovery (WHR) system. Among the various WHR systems, a well-known way of recuperating this waste energy is by employing an Organic Rankine Cycle (ORC). Table 1.2 shows the typical proportion of fuel thermal energy that goes waste and potential energy that can be recuperated using the Organic Rankine Cycle.

The typical layout of the ORC system is shown in Fig. 1.7. A key component of the ORC system is the expansion mechanism. There are mainly two ways as using a piston machine or using a turbine, selections of which are mainly based on efficiency, weight, packaging, and influence on entire process efficiency, durability, and vibrations. The mechanical energy generated by the Rankine process can be delivered to the engine either directly or via a belt transmission. Rankine Cycle systems are already in use in off-highway applications, such as stationary engines or marine power packs. In the steam power process, a working fluid is delivered by a fluid pump from a lower pressure to a higher pressure level and is conveyed through a distributor valve into the evaporator, where it evaporates with the heat from the exhaust gas. The expansion machine generates power. The steam from the outlet is re-liquefied in the condenser and the residual heat is released to the cooling system.

Table 1.2 The typical proportion of fuel thermal energy available and recoverable by ORC [11]

S. No.	Engine sub-system	Total fuel thermal energy proportion available	Potential recovery by ORC
1	Turbine outlet	20–24%	Around 2%
2	Engine cooling	14–28%	0.5–1%
3	EGR cooler	3–18%	Around 1.5%
4	Charge air cooler	2–11%	0.5–1.0%

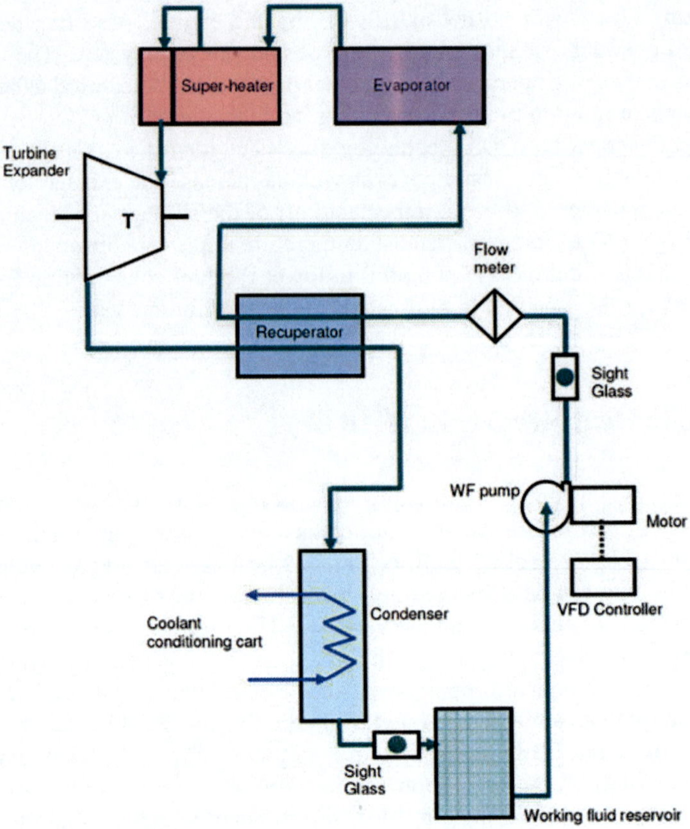

Fig. 1.7 Layout for the ORC WHR system [12]

For conditions such as overrun mode, an operation without energy transformation is necessary, since an extremely low rate of fuel injection is employed. In this case, the steam is released via a bypass valve directly to the condenser. A condensate pump delivers the working fluid to the vessel. After the vessel, the liquid working fluid is again conveyed to the fluid pump. The system is controlled by a Control Unit, which can also be integrated into the engine control unit.

The amount of heat recovery from any available heat source depends to a great extent on the quality of the heat source. The higher the heat source temperatures, the larger is the portion of heat recovery. Various factors like the choice of a working fluid, design of evaporator, condenser, and expander, and the operating point at which the WHR is being utilized are important considerations for making an ORC WHR efficient.

In the bottoming ORC simulation-based assessment done by Shu et al. [13] for a heavy-duty diesel engine for potential heat recovery from exhaust and cooling system, two different working fluids R245fa and R601a were investigated at intermediate and

Table 1.3 Thermal efficiency at two operating points utilizing ORC

	Thermal efficiency at intermediate speed (%)	Thermal efficiency at rated speed (%)
Original diesel engine	43.3	40.3
With ORC integrated	47.2	43.5

rated speeds engine operations. Parameters investigated include waste heat recovered, expansion power, system efficiency for variation in mass flow of working fluid, and evaporating pressure. The maximum efficiency of the ORC system obtained was about 9.6% with R245fa working fluid at intermediate speed. The thermal efficiency improvement achieved in the assessment utilizing the bottoming ORC is as shown in Table 1.3.

1.6 Thermal Management Assessment by Simulation

The scope of thermal management can never be summed up simply as increasing exhaust gas temperatures. Increasing temperatures are not always beneficial for optimum after-treatment performance. Catalyst chemical kinetics is a complex phenomenon, needs extra attention and a more critical understanding of catalyst chemistry. Thermal management simulation is very effective to understand the energy flow, fuel consumption, and catalyst performance with vehicle behavior under different operating conditions, as the physical testing costs are very high. Assessment and calibration of after-treatment systems are required to be done on steady-state and transient legislative cycles, with hot as well as cold start conditions. Also, they are required to be tuned for off-cycle emissions, not-to-exceed (NTE) limits, and real-world operations. Also, there are onboard diagnostics (OBD) and In-service compliance (ISC) requirements. Due to stringent emission norms, conversion efficiency requirements from the after-treatment systems are very high and hence their control requirements are quite complex. Also, there could be other constraints like commonality for different power ratings, packaging space available, the total cost of ownership (TCO), ease of service, etc. While evaluating and calibrating for all these aspects, there are always limitations from available time, budget, and testing resources for the development. As such, the simulation will be a very effective tool to balance the available resources and development tasks. Performance assessment does not depend on weather conditions, location, and prototype availability, hence extensive assessment is possible. Effective use of simulation significantly minimizes the requirement of expensive test facilities and development time as well as improves the quality and robustness of the product.

Catalyst material plays a major role in NOx reduction. Different catalyst material exhibits a different rate of conversion at given temperatures. Figure 1.8 shows

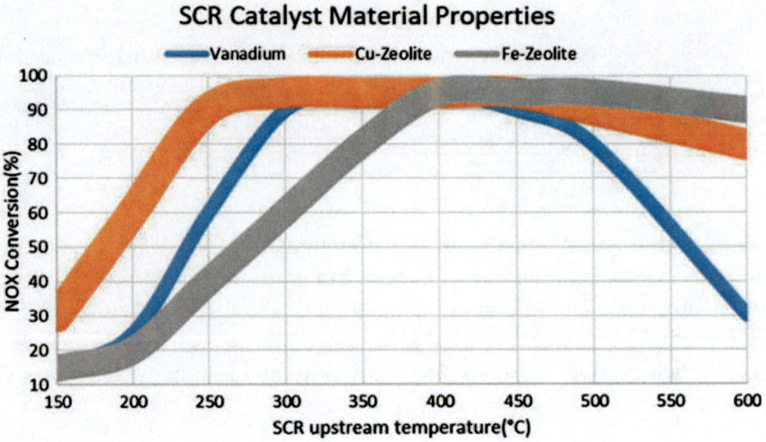

Fig. 1.8 Comparison of operational characteristics selective catalysts

representative generic trends of different catalyst material NO_x conversion efficiencies versus operating temperatures. Any selective catalyst has an operation window of temperatures for targeted conversion efficiencies and low, as well as too high operating temperatures, reduce catalyst performance.

The fundamental difference can be observed for the different catalyst materials. Copper exchanged zeolites are best suited for low catalyst temperatures and comparatively more stable at higher operating temperatures, whereas Iron exchanged zeolites are the best choice for high-temperature operations. Vanadium-based catalysts are the most cost-effective solutions and also offer the best resistance against Sulphur contamination but on the other hand, are limited with a smaller temperature window for high efficiency. The role of the thermal management system is to maintain the operating temperatures within the most efficient operation zone as per the particular selective catalyst material.

Engine out NO_x consists of majorly NO and NO_2. Another factor impacting significantly the SCR performance is the NO_2 proportion in the exhaust, as the presence of NO_2 can boost the NO_x conversion at a much higher rate. However, NO_2 proportion in the NO_x if exceeds 50%, impacts NO_x conversion efficiency adversely. Figure 1.9 shows a general trend of the impact of NO_2 on SCR efficiency. To increase NO_2 proportion in the exhaust, some portion NO is oxidized to NO_2 in the oxidation catalyst upstream of SCR. This in turn is dependent on DOC upstream temperatures. DOC has its characteristics for oxidation as shown in Fig. 1.10. Too low or too high DOC upstream temperatures adversely affect the NO_2 proportion upstream of SCR and in turn can deteriorate SCR conversion efficiency. Hence, close control over DOC upstream temperature and NO_2 yield from DOC catalyst material is a crucial calibration parameter in the case of SCR after-treatment. Hence, DOC and in turn SCR performance depends upon the temperature upstream of DOC. In the case of cold operations, earlier the catalyst light-off better the after-treatment performance. On the other hand, during hot operations such as high speed and load operations, where

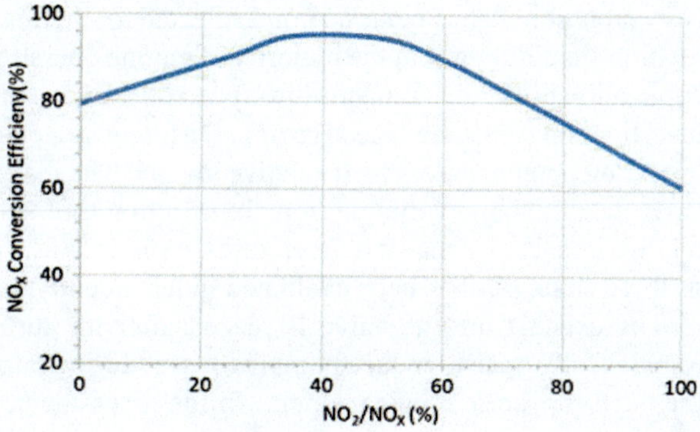

Fig. 1.9 Impact of NO_2 yield on SCR conversion efficiency

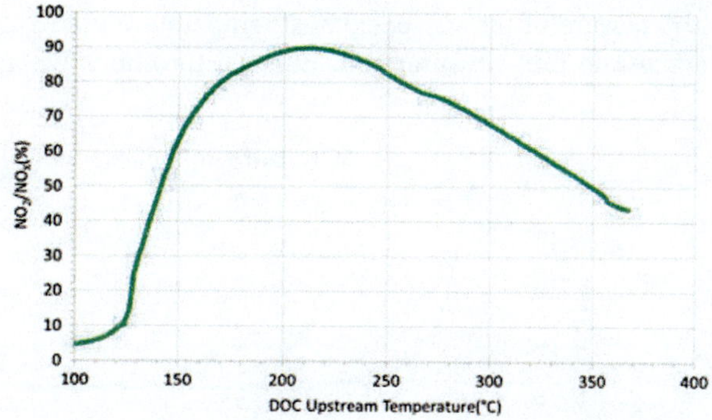

Fig. 1.10 Thermal characteristics of the oxidation catalyst

exhaust temperatures are excessively high, after-treatment devices may completely inverse their characteristics. Hence, close control over the thermal characteristics of exhaust after-treatment is of paramount importance for precise tailpipe emissions.

1.6.1 *Case Study I—On-Road Heavy Duty Application*

To explain the different goals of thermal management, the first case demonstrates the method for the selection of relevant thermal management strategies based on application requirements for a heavy-duty long-haul truck. For this application, the world harmonized steady-state cycle (WHSC) and world harmonized transient cycle (WHTC) are the legislative cycles for emissions measurement. See Appendix for details of the modes of WHSC and other references. In the cold portion of WHTC, after-treatment devices take time to reach to catalyst light off and this period impacts overall cycle average emissions. Hence, cold start light-off is the most challenging portion of emission control for this application. The main objective of assessment in this case study is to assess the possibility to attain SCR light-off, to achieve dosing temperature in the legislative cycle as early as possible, to curb cycle averaged tailpipe

NO_X emissions. The operating point selected for thermal management strategy investigation should represent the majority of engine operation. As the Heavy-duty engine application primarily demands low-speed operations, the operating point selected for investigation is in a low-speed region. Different strategies have been evaluated at this operating point to assess their relative impact. The main strategies as exhaust throttle valve actuation and variation in main injection timing have been evaluated for their relative impact on exhaust temperatures, emissions as well as fuel consumption. Each of these strategies has been evaluated independently one at a time.

The exhaust throttle valve is placed after the turbine in the exhaust pipe and remains fully open to reduce pumping losses during normal engine operation without the need for thermal management. In the assessment, the exhaust throttle opening is reduced progressively. As shown in Fig. 1.11, initially temperature remains fairly constant with the closure of the exhaust throttle valve and beyond approximately 80% closure of the valve, exhaust temperature increases sharply with an associated increase in fuel consumption. Exhaust throttle valve actuation has little impact on

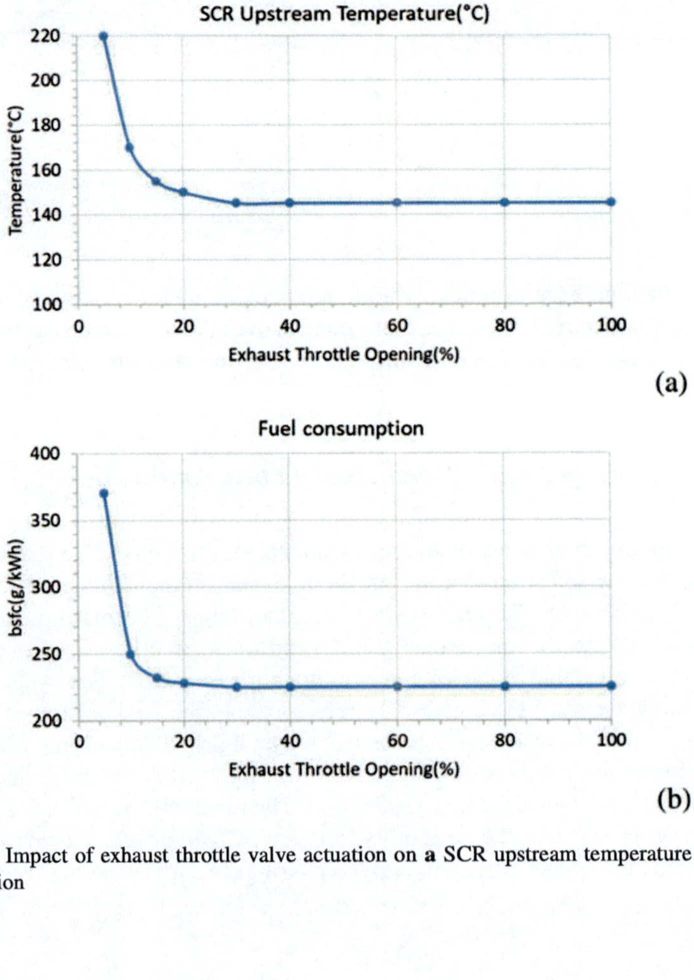

Fig. 1.11 Impact of exhaust throttle valve actuation on **a** SCR upstream temperature and **b** fuel consumption

the combustion as compared to the intake throttle valve actuation, as it has a lower impact on the fresh air mass flow.

Exhaust throttle valve actuation has a significant impact on the exhaust temperature, which can be seen from the SCR upstream temperature graph. There is a certain fuel consumption penalty associated with the throttle closure. This strategy is most suitable for cold operations to light-off DOC and can be combined with other injection strategies to achieve a balance of fuel consumption penalty for temperature rise over most of the operating zone. The other way to increase the exhaust temperature is by retarding the combustion with a delay in the main injection timing. This has a significant impact on the combustion process, as the proportion of premixed combustion and diffusion combustion changes. This results in higher fuel consumption and increased THC and CO emissions. Fuel injection retard also reduces in-cylinder temperatures, which results in reduced thermal NO_x. Figure 1.12 shows the sweep of main injection timing retard from 6° before TDC to 6° after TDC and its impact on exhaust temperatures and brake-specific fuel consumption (BSFC). The trends show that with the main injection retard from 6° before TDC to 6° after TDC, there is a temperature rise of 30 °C upstream of SCR with an associated increase in BSFC. The CO and HC trends show reduction with retard near TDC and increase exponentially with the main injection retard beyond TDC. Temperature increase by main injection retard is limited by combustion instability, which is depicted by a high increase in HC emissions at retarded timings.

The main injection retard along with a small pilot quantity for torque correction is also one of the thermal management strategies, however, it has a relatively lower impact on an increase in exhaust temperatures significantly, and also injection retard is limited by combustion instability resulting in higher THC and CO emissions. Similarly, a remarkable increase in the temperature across exhaust after-treatment can be achieved with post-injection. However, this strategy is only functional if the DOC light-off temperature is attained by some other means. Hence, post-injection is a popular thermal management strategy for DPF regeneration used along with throttle actuation for DOC light-off. Several other strategies are available for exhaust thermal management, which involve advanced hardware such as variable valve actuation and Variable geometry turbine actuation. Variable valve actuation is generally preferred in the case of passenger car applications, where the engine operates over a wide range of engine speeds, and presently it is not cost-effective for Heavy-duty applications. Variable geometry turbine actuation is yet another effective method of raising exhaust temperatures but this is also more popular in the case of passenger car applications, however, due to limited cost-effectiveness it is not that popular in the case of heavy-duty applications.

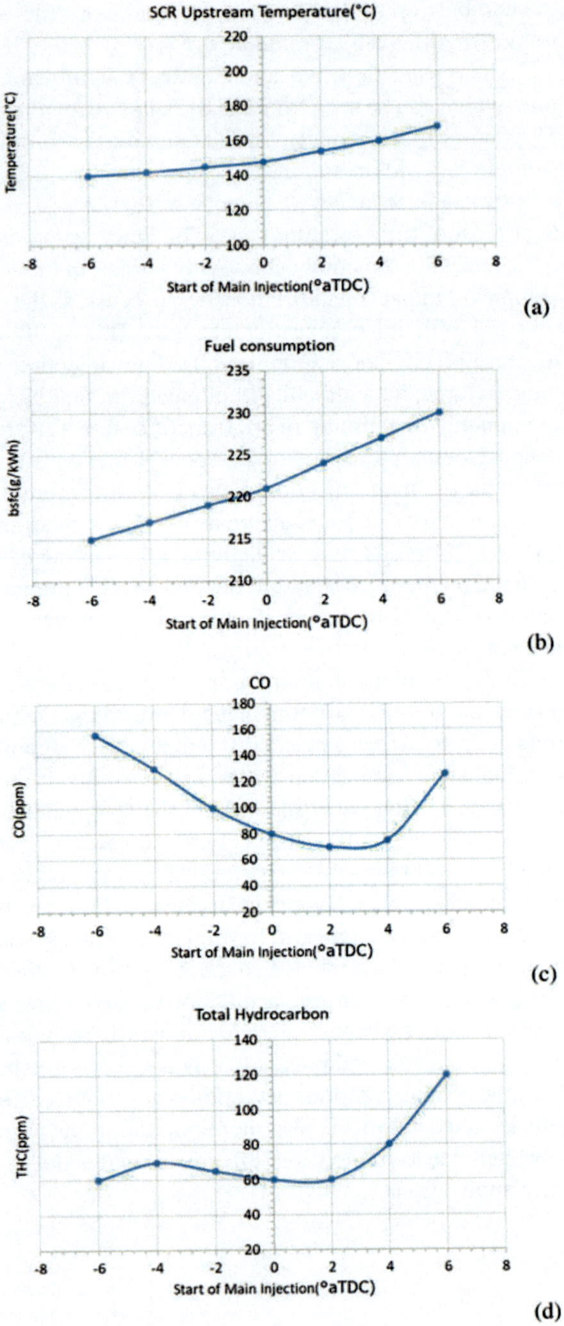

Fig. 1.12 Impact of the main injection retard on **a** SCR upstream temperature, **b** BSFC, **c** HC, and **d** CO emission

1.6.2 Case Study II—Off-Road Application

The second case study is from the off-road application, where the legislative cycles are different from those for on-road application. In this case, the steady-state eight-mode cycle is critical, whereas the transient cycle is relatively less demanding from the thermal management requirement perspective. Figure 1.13 shows exhaust temperatures and NO_2 to NO_x ratio upstream SCR system with and without exhaust gas recirculation (EGR). Initially, engine combustion tuned for the target engine-out NO_x without EGR. As shown in Fig. 1.13, exhaust temperature at mode 1 was found much higher than the catalyst's best zone of operation and NO_2 yield in DOC also drops drastically at temperatures beyond 450 °C in addition to the sharp drop in catalyst conversion efficiency. It was also observed that temperatures at mode 4 and mode 8 are below dosing trigger temperature, which means that there is no NO_x

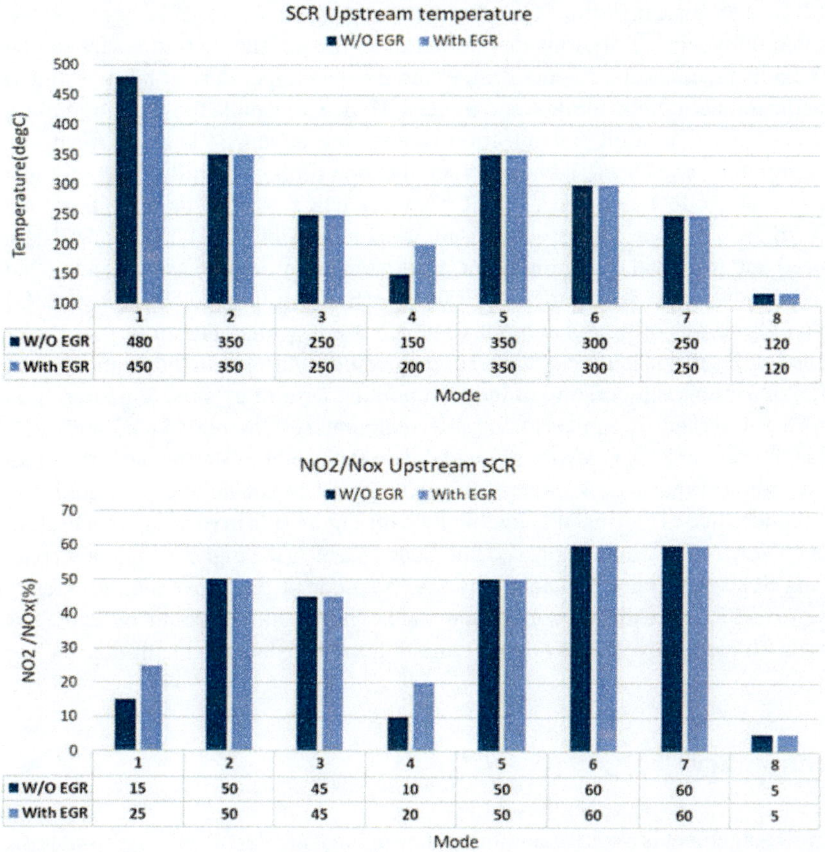

(a)

(b)

Fig. 1.13 **a** SCR upstream temperatures, **b** NO_2/NO_x upstream SCR with and without EGR

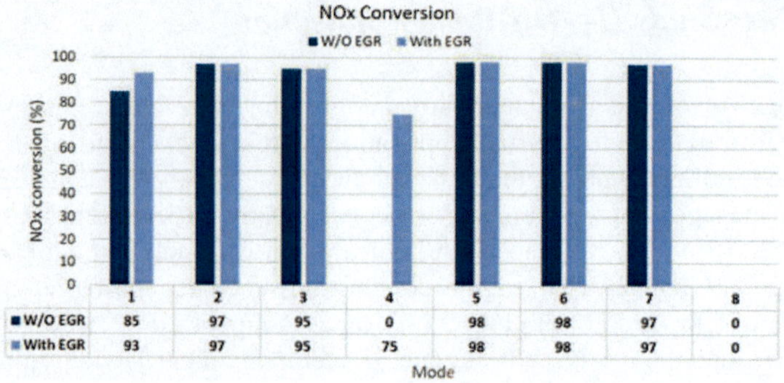

Fig. 1.14 SCR conversion efficiency with and without EGR

conversion possible in the SCR system at these two modes. This resulted in low overall NO_x conversion in the SCR over the steady-state cycle and consequently, very high tailpipe NO_x emissions that do not meet the legislative limits. Hence, the thermal management requirements were to reduce the temperature at mode 1 and to increase the temperature at mode 4 and mode 8. Hence, to handle these contradicting requirements, EGR is used and injection timings are advanced keeping fairly the same engine-out NO_x. Due to advanced fuel injection timing and mild EGR, exhaust temperature at mode 1 reduced to SCR high-efficiency zone. This also improved the NO_2 to NO_x ratio at this operating point, as shown in Fig. 1.13. Mode 4 is a high-speed and low-load operation point, where temperature increase was achieved with a relatively higher percentage of EGR, which also improved the NO_2 to NO_x ratio. No change was observed at mode 8, which is a low idle operation.

The use of this thermal management strategy could achieve both the requirements, reduction of exhaust temperature at mode 1, and increase of exhaust temperature at mode 4. So it served to achieve favorable temperatures for both DOC and SCR operation, moreover, BSFC also improved due to the use of EGR and advanced fuel injection timing. Figure 1.14 shows the impact on SCR conversion efficiencies at different modes due to thermal management. With a reduction in exhaust temperature at mode 1, NO_2/NO_x ratio is improved and SCR operates at a higher efficiency zone, improving SCR conversion efficiency by 8%. At mode 4, the conversion efficiency increases to 75% from the no conversion state. This results in better overall SCR conversion efficiency over the cycle meeting tailpipe legislative NO_x limits.

1.7 Summary

Thermal management is about maintaining the engine and related vehicle subsystems at an optimum exhaust temperature level and has become very crucial in vehicle development as it directly or indirectly affects engine performance, fuel economy, safety, emissions, etc. Significant improvements can be achieved by efficient engine thermal management, optimizing air handling, combustion, Exhaust, cooling, and lubrication systems. Thermal management is an essential part of the strategies for meeting future pollutant emission norms and CO_2 emission reduction requirements.

Conventional engine cooling system performance is always constrained by its passive nature and leads to considerable losses. The advanced cooling system has quite a high potential to improve ICE performance through enhanced coolant temperature control and reduced parasitic losses. Also, an advanced lubrication system has the potential to reduce parasitic losses and thereby improve engine efficiency and performance. Modern engines are equipped with catalytic exhaust after-treatment systems that require a suitable narrow exhaust temperature window to operate at their best efficiency levels. The role of exhaust gas thermal management for the after-treatment system is to enable the catalysts to rapidly attain the best-operating temperatures and maintain them so that catalysts function at their better efficiency levels thereby keeping the tailpipe emission at the lowest possible levels. The main constraint in thermal management is to limit the potential fuel consumption increase. There are different technologies available for thermal management; however, each one has its magnitude of energy cost and efficiency losses. The selection depends on the engine application, duty cycle, and level of exhaust emission conversion required.

The situation in heavy-duty vehicles is different from passenger car applications as the engines are tested to a different duty cycle than the passenger car test cycle. Maintaining the engine at the correct exhaust gas temperatures is now even more critical for modern diesel engines than before. A cold engine operation harms both fuel consumption and exhaust emissions. Hence, thermal management is equally applicable to the modern high compression ratio, turbocharged direct injection gasoline engines that face challenges with exhaust emissions of similar nature.

Thermal management is going to be of equal significance to the Battery-operated Electric Vehicle (BEV) segment. The performance of BEV can vary significantly depending on its operating temperature. A case of serious overheating can lead to severe safety concerns. Keeping the EV battery at an optimum temperature can help to preserve its capacity, life and ensure greater range on a single charge. In addition to the battery, thermal management for BEVs will also include other components like the electric motor, inverters, etc.

Simulation of thermal management is very effective to understand the energy flow, fuel consumption, and catalyst performance with vehicle behavior under different operating conditions. Conversion efficiency requirements from the after-treatment systems are very high under different operating requirements. Thermal management simulation has become a very effective tool for selecting the best suitable thermal management strategy according to the engine application requirements. Effective use of simulation significantly minimizes the requirement of expensive test facilities and development time as well as improves the quality and robustness of the product.

Table 1.4 Description of the WHSC

Mode	% Rated torque	% Rated speed	% Weight	Duration, seconds
1	0	Idle	8.5	210
2	100	55	2.0	50
3	25	55	10.0	250
4	75	55	3.0	75
5	100	35	2.0	50
6	25	25	8.0	200
7	75	45	3.0	75
8	25	45	6.0	150
9	50	55	5.0	125
10	100	75	2.0	50
11	50	35	8.0	200
12	25	35	10.0	250
13	0	Idle	8.5	210

Appendix: WHSC and WHTC

A steady-state engine dynamometer test schedule defined by the global technical regulation (GTR) No. 4 (https://unece.org/transport/standards/transport/vehicle-regulations-wp29/global-technical-regulations-gtrs) developed by the UN ECE GRPE group (https://unece.org/transportvehicle-regulations/working-party-pollution-and-energy-introduction) is known as WHSC. See Table 1.4 given below. The GTR discusses a worldwide harmonized heavy-duty certification (WHDC) procedure for engine exhaust emissions. A hot start steady-state test cycle (WHSC) and a transient test cycle (WHTC) with both cold and hot start requirements have been created imitating typical driving conditions in Australia, the USA, Japan, and the EU.

References

1. https://ww2.arb.ca.gov/our-work/programs/heavy-duty-low-nox/about
2. https://www.europarl.europa.eu/news/en/press-room/20190412IPR39009/meps-approve-new-co2-emissions-limits-for-trucks
3. Nessim W, Zhang F (2012) Powertrain warm-up improvement using thermal management systems. Int J Sci Technol Res 1(4), May 2012. ISSN 2277-8616
4. Arici O, Johnson JH, Kulkarni AJ (1999) The vehicle engine cooling system simulation. SAE paper 1999-01-0240
5. https://mb.cision.com/Main/1643/9402426/113949.pdf
6. Burke RD et al (2014) Investigation into the benefits of reduced oil flows in internal combustion engines. Int J Engine Res
7. Lauren M et al, Different methods to improve the exhaust gas temperature in modern stage V off-road diesel engine over transient emission cycles. SAE technical paper 2020-01-0903

8. Kovacs D et al, Modelling heavy-duty engine thermal management technologies to meet future cold start requirements. SAE technical paper 2019-01-0731
9. Hiranuma S et al, Development of DPF system for commercial vehicle—basic characteristics and active regenerating performance. SAE technical paper 2003-01-3182
10. Pillai S, LoRusso J, Van Benschoten M, Cylinder deactivation reborn—part 1. Analytical and experimental evaluation of cylinder deactivation on a diesel engine. SAE technical paper 2015-01-2809
11. Folkson R, Alternative fuels and advanced vehicle technologies for improved environmental performance. Woodhead Publishing
12. Teng H, Waste heat recovery concept to reduce fuel consumption and heat rejection from a diesel engine. SAE technical paper 2010-01-1928
13. Shu G-Q et al, Simulations of a bottoming organic rankine cycle (ORC) driven by waste heat in a diesel engine (DE). SAE technical paper 2013-01-085

Chapter 2
Thermal Management of Exhaust Aftertreatment for Diesel Engines

Achuth Munnannur, Nathan Ottinger, and Z. Gerald Liu

2.1 Introduction

The modern diesel engine is a complex system with multiple integrated components that work together to make it most acceptable to the customer in terms of power, emissions, fuel efficiency, performance, reliability, and durability. Although performance, reliability, and durability have seen big advances, the largest change in the modern diesel engine is the introduction of exhaust aftertreatment coupled with ultralow sulfur fuels. Considerations of fuel efficiency have been driven by customer needs, declining fossil fuel resources, and the impact of greenhouse gas emissions on climate change. Another important technology-forcing factor is emissions. Since the 1970s, awareness of the harmful effects of pollutants produced from industrial sources had started to rise and regulations were deliberated on and implemented in several parts of the world. Almost a decade later, concern over global air pollution emerged, after research showed that pollutants can cause widespread damage beyond their place of origin and are not mere "local problems." This gave emissions issues a global dimension, and as expected there have been regulatory initiatives from around the world to reduce the environmental and health impact of emissions from diesel vehicles as well. With such varied and often contradictory demands to be met, engine manufacturers must seek an integrated perspective on diesel engine development. In this respect, exhaust thermal management is an important strategy that merits discussion and exploration.

This chapter on thermal management of diesel exhaust aftertreatment starts by providing an appreciation for the challenges posed to modern diesel engines that create a need for thermal management. Then the suite of existing and promising strategies and technologies available to implement thermal management are described.

A. Munnannur · N. Ottinger · Z. Gerald Liu (✉)
Cummins Inc., Madison, WI, USA
e-mail: jerry.z.liu@cummins.com

P. A. Lakshminarayanan and A. K. Agarwal (eds.), *Handbook of Thermal Management of Engines*, Energy, Environment, and Sustainability,
https://doi.org/10.1007/978-981-16-8570-5_2

Special attention is paid to the implications of the technologies on exhaust heat, emissions, and fuel efficiency.

2.2 Need for Aftertreatment Thermal Management

In this section, the various requirements that drive the thermal management of diesel aftertreatment are described. First, the global emissions regulations which have driven the introduction of diesel aftertreatment and necessitate thermal management are outlined. The performance of state-of-the-art catalyst and filter technologies and how thermal management is utilized to optimize their real-world operation are then discussed. Next, thermal management's role in diesel exhaust fluid (DEF) introduction and deposit mitigation and removal is examined. Finally, the case for thermal management from an aftertreatment packaging perspective is provided.

2.2.1 *Regulatory Requirements*

Governments worldwide regulate emissions from diesel engines. Whereas the US and Europe initially had stricter emissions standards than the rest of the world, other countries such as China, India, and Brazil have rapidly increased the stringency of their emissions standards in the last 3–5 years. Examples of on-highway heavy-duty diesel NO_x emissions limits and emissions test cycles for the US, Europe, China, and India are shown in Table 2.1.

To meet these regulations, a variety of engine and exhaust aftertreatment technologies are required. Exhaust aftertreatment for diesel engines has been under development for many decades, with the first widespread implementation on heavy-duty diesel vehicles occurring in 2005 to comply with Euro IV regulations. At that time, a selective catalytic reduction (SCR) catalyst was used to meet emissions standards. Figure 2.1 shows a schematic of the catalyst technologies implemented in one example of a diesel aftertreatment configuration in use around the world recently. Catalysts are organized in the direction of exhaust flow, from left to right. The catalyst components include a diesel oxidation catalyst (DOC) for the oxidation of CO

Table 2.1 Global on-highway and US non-road diesel exhaust NO_x limits, g/kW-h

Region (application)	NO_x limit	Test cycles
US (on-highway)	0.27	FTP, SET
US (non-road)	0.41	NRTC, ISO 8178
Europe (on-highway)	0.46	WHTC
China (on-highway)	0.46	WHTC
India (on-highway)	0.46	WHTC

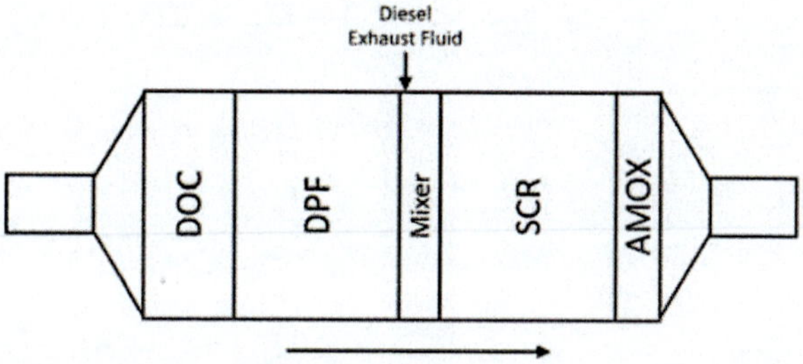

Fig. 2.1 Catalyst and filter configuration of a representative diesel exhaust aftertreatment system. The arrow indicates the direction of exhaust flow

(Eq. 2.1), unburned hydrocarbons (Eq. 2.2), NO, and the soluble organic fraction (SOF) of soot; a diesel particulate filter (DPF) for the reduction of particulate matter (PM); an SCR catalyst for NO_x control; an ammonia oxidation (AMOX) catalyst to limit emissions of NH_3 (Eq. 2.3). The general chemical composition, or formulation, of each of these catalysts, is:

- DOC: platinum group metals (PGM) impregnated onto high surface area γ-Al_2O_3 and wash coated onto a flow-through cordierite substrate.
- DPF: Cordierite or silicon carbide (SiC) wall-flow substrate. May contain a wash coat consisting of PGM impregnated onto high surface area γ-Al_2O_3.
- SCR: There are two primary types of SCR catalysts.

  1. Vanadium-based, which consist of vanadium and tungsten oxides impregnated on a high surface area TiO_2 support and either extruded or wash coated onto a flow-through cordierite substrate, broadly referred to as V_2O_5-WO_3/TiO_2 in the literature, even though the oxidation states of V and W are constantly changing due to the redox nature of this catalyst.
  2. Metal exchanged zeolites, which consist primarily of either Cu or Fe exchanged zeolites, e.g., SSZ-13, wash coated onto a flow-through cordierite substrate.

- AMOX: This type of catalyst almost always has a dual-layer wash coat applied to a flow-through cordierite substrate, even though single-layer catalysts have also been used. The bottom layer generally consists of PGM impregnated onto high surface area γ-Al_2O_3, while the top layer consists of either Cu/zeolite or Fe/zeolite.

Other configurations of these technologies can be applied to meet emissions requirements, such as placing the SCR catalyst upstream of the DOC and DPF, and catalyst technologies can be combined, such as SCRF (selective catalytic reduction on the filter), but these general catalyst types are used regardless of the configuration. Of these catalysts, thermal management is most critical for NO_x conversion over the SCR. NO_x is converted through a combination of the standard, fast, and slow SCR reactions that are shown in Eqs. 2.4–2.6. These reactions rely on the introduction of NH_3 into the diesel exhaust stream, which is accomplished via injection of an aqueous solution of urea (DEF, AUS 32, AdBlue, etc.). This process is discussed in detail in *Diesel Exhaust Fluid (DEF) Dosing and SCR Solid Deposits.*

$$CO + O_2 \rightarrow CO_2 + 0.5O_2 \quad (2.1)$$

$$2HC + O_2 \rightarrow H_2O + 0.5CO_2 \quad (2.2)$$

$$4NH_3 + 3O_2 \rightarrow 2N_2 + 6H_2O \quad (2.3)$$

$$4NO + 4NH_3 + O_2 \rightarrow 4N_2 + 6H_2O \quad (2.4)$$

$$NO + NO_2 + 2NH_3 \rightarrow 2N_2 + 3H_2O \quad (2.5)$$

$$6NO_2 + 8NH_3 \rightarrow 7N_2 + 12H_2O \quad (2.6)$$

In addition to regulations for NO_x and other criteria pollutants (i.e., PM, CO, and total hydrocarbon (THC)), the other critical regulations which drive requirements for thermal management are greenhouse gas or CO_2 emissions requirements. Greenhouse gas regulations have been promulgated to reduce the impact of transportation on climate change. This began with Phase 1 Greenhouse Gas Emission Standards in the US, which took effect between 2014 and 2018. The Phase 2 standards apply to engines and vehicles manufactured after 2021. Table 2.2 shows the CO_2 emission (g/bhp-h) and fuel consumption (gallon/100 bhp-h) requirements for medium heavy-duty and heavy heavy-duty tractor engines operating over the SET regulatory test cycle [1]. Most of the exhaust thermal management strategies that will be discussed in this chapter increase fuel consumption and, as a result, CO_2 emissions. Thus, any CO_2 penalty resulting from thermal management may need to be offset by engine or vehicle efficiency improvements.

The US Department of Energy's (DOE) SuperTruck programs continue to promote advancements in engines and associated technologies. SuperTruck I set a goal of 50% brake thermal efficiency, which was demonstrated to be achievable with improvements to the base engine including increased compression ratio and power cylinder friction reductions, as well as improvements to ancillary systems including the

Table 2.2 US GHG phase-2 heavy-duty tractor engine standards for engines over the SET cycle [1]

Model year	Standard	Units	Very heavy-duty	Medium heavy-duty
2021–2023	CO_2	g/bhp-h	447	473
	Fuel consumption	Gallon/100 bhp-h	4.3910	4.6464
2024–2026	CO_2	g/bhp-h	436	461
	Fuel consumption	Gallon/100 bhp-h	4.2829	4.5285
2027 and later	CO_2	g/bhp-h	432	457
	Fuel consumption	Gallon/100 bhp-h	4.2436	4.4892

turbocharger and powertrain and incorporation of waste heat recovery. The ongoing SuperTruck II program aims to develop and demonstrate cost-effective technologies that more than double the freight efficiency of Class 8 trucks, and effective thermal management will continue to play an important role in this initiative as well.

2.2.2 *Catalyst Performance*

As discussed above, modern diesel aftertreatment systems rely on a system of catalysts and filters to comply with global emissions regulations. Also, sophisticated mixers, injectors, sensors (NO_x, NH_3, PM, temperature, and delta pressure), and adaptive controls are all utilized to meet emissions requirements over the mandated useful life of the diesel engine and emissions control system. US heavy heavy-duty diesel engines and emissions components certified for sale in 2021 have a regulated useful life of 435,000 miles, and Europe, China, and India similarly mandate a useful life of 700,000 km for similarly sized engines. The California Air Resources Board (CARB) is extending useful life requirements for heavy heavy-duty diesel engines to 1 million miles, and other regulatory bodies will likely increase useful life requirements in the future. In addition to the catalysts, delta pressure, temperature, NO_x, and particulate matter (PM) sensors are all required to control and diagnose any issues in real-time. As with the Euro IV aftertreatment, a pump and injector are required to introduce an aqueous urea solution as a source of NH_3 for the SCR reactions, and a mixer ensures that the NH_3 is uniformly distributed throughout the exhaust stream.

The aftertreatment catalyst technologies require elevated exhaust temperatures to operate efficiently. The aqueous urea solution is approximately 67.5% deionized water, which requires significant energy input to vaporize. Similarly, each of the catalytic reactions in Eqs. 2.1–2.6 is temperature-dependent. The Arrhenius equation (Eq. 2.7) describes the dependence of reaction rates on temperature, where k is the rate constant, A is the pre-exponential factor, E_a is the activation energy, R is the universal gas constant, and T is the temperature. The rate of these reactions increases exponentially as a function of temperature.

$$k = A\,e^{(-E_a/RT)} \tag{2.7}$$

This exponential relationship between reaction rate and temperature can necessitate thermal management. Figure 2.2 shows steady-state NO_x conversion via Eq. 2.4 over a state-of-the-art Cu/zeolite SCR catalyst. Between 200 and 400 °C, greater than 95% of the inlet NO_x is converted. However, as the catalyst temperature decreases below 200 °C, the rate of NO_x conversion rapidly decreases. At 150 °C, only approximately 13% of the inlet NO_x is converted to N_2. Figure 2.3 shows the DOC inlet temperature for a 2017 Cummins X15 diesel engine during a cold-start FTP certification cycle. The DOC inlet temperature is below 200 °C for the majority of the initial 400 s of the cycle. The SCR temperature would be even lower than the DOC inlet temperature due to heat transfer to the DOC and DPF catalysts, both upstream of the

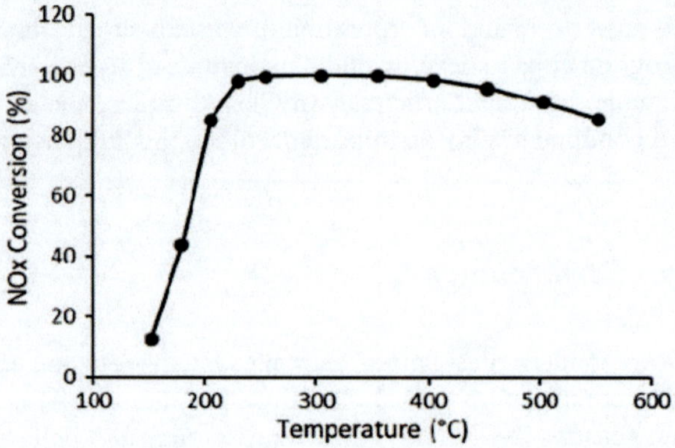

Fig. 2.2 NO_x conversion as a function of temperature for a state-of-the-art Cu/zeolite SCR catalyst

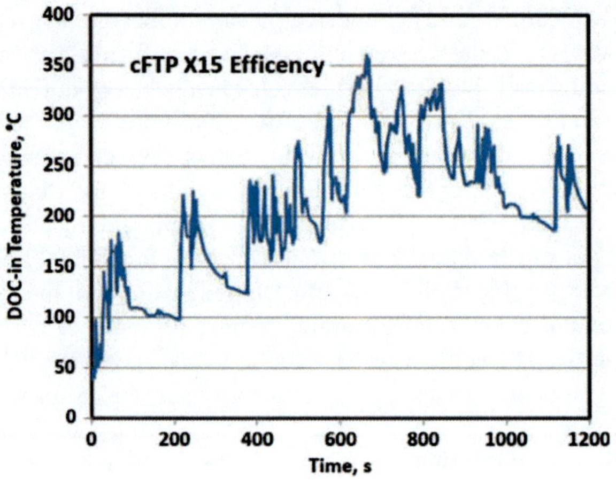

Fig. 2.3 DOC inlet temperature of a Cummins X15 diesel engine during a cold-start FTP cycle

SCR catalyst. During cold-start and other low exhaust temperature conditions such as idle and low load, higher NO_x conversion can be achieved if thermal management strategies are implemented to increase the temperature of the SCR catalyst. While the data in Fig. 2.2 is for a Cu/zeolite SCR catalyst, other SCR catalyst types as well as DOC and AMOX catalysts have similarly low conversions at low exhaust temperatures.

2.3 Aftertreatment Recovery

2.3.1 Diesel Particulate Filter (DPF) Regeneration

Thermal management is required for diesel particulate filter (DPF) regeneration. With increasing runtime, particulate matter (PM) accumulates in the channels of the wall-flow DPF, elevating the backpressure on the engine. Real-time PM loading on a vehicle can be estimated based on engine and aftertreatment operating conditions, or it can be inferred based upon a pressure drop measurement taken across the DPF. Once PM loading in the DPF has reached a pre-defined level, PM must be removed from the DPF via either active or passive regeneration. Active regeneration refers to the oxidation of PM with O_2, whereas passive regeneration refers to the use of NO_2 to oxidize PM. To accomplish an active DPF regeneration, thermal management techniques are used to increase the exhaust temperature to greater than 500 °C for O_2 to fully oxidize PM trapped in the DPF. Figure 2.4 shows CO and CO_2 evolution during temperature-programmed oxidation (TPO) of particulate matter collected from a diesel engine operating on ultra-low sulfur diesel (ULSD) fuel. Oxidation of PM with O_2 begins at approximately 300 °C. Between 300 and 500 °C, the organic carbon (OC) fraction of PM is oxidized to CO and CO_2. OC contains semi-volatile hydrocarbon species. Above 500 °C, the elemental carbon (EC) fraction of PM is oxidized. It is clear that the exhaust temperature must be raised above 500 °C to remove the EC, and that the rate of EC oxidation increases significantly as the temperature increases above 500 °C. The amount of particulate matter loading and DPF temperature are the key factors that determine the length of an active DPF regeneration event. Once the active DPF regeneration has been removed the PM, soot begins to re-accumulate, and the process must be periodically repeated.

Passive DPF regeneration is an alternative to active regeneration which utilizes NO_2, present in nascent diesel exhaust and created over the DOC, to oxidize PM. Because the kinetics of NO_2 soot oxidation is faster than soot oxidation by O_2 below 500 °C, this form of DPF regeneration can be accomplished without thermal management in some duty cycles. Figure 2.5 shows results from a TPO experiment

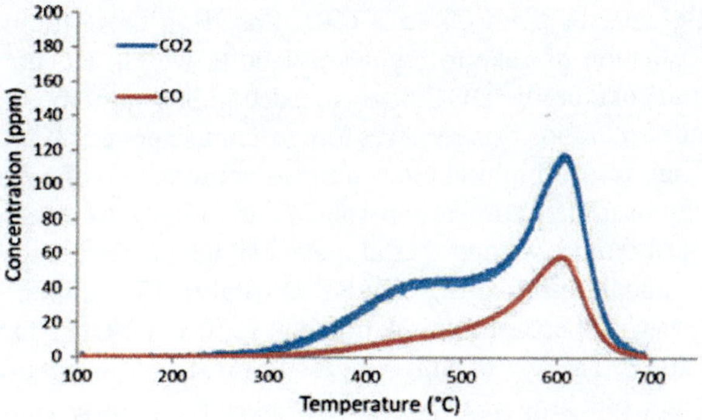

Fig. 2.4 CO and CO_2 evolution during temperature-programmed oxidation (TPO) of particulate matter collected from a diesel engine operating on ULSD. O_2 is used as the oxidizer in this case

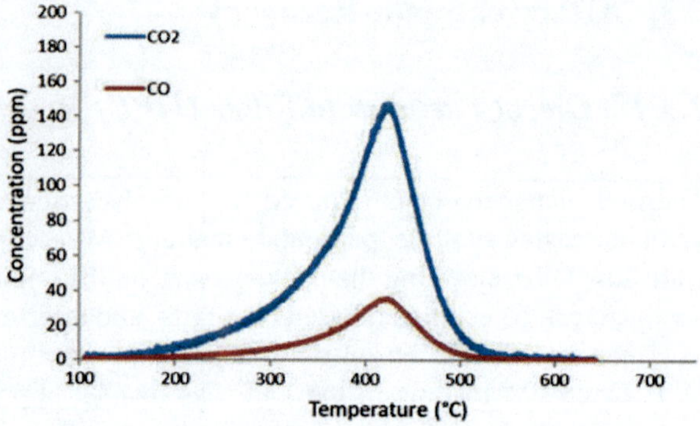

Fig. 2.5 CO and CO_2 evolution during temperature-programmed oxidation (TPO) of particulate matter collected from a diesel engine operating on ULSD. NO_2 is used as the oxidizer in this case

where NO_2 is used as the oxidizer. The PM is from the same engine and collected over the same duty cycle as the PM in Fig. 2.4. In comparison to oxidation with O_2, NO_2 fully oxidizes PM at a significantly lower temperature. With NO_2 as the oxidizer, all PM is oxidized below 500 °C. However, passive regeneration requires a significant amount of NO_2, and it is still temperature-dependent. In most duty cycles, thermal management will still be required to elevate the DPF temperature and generate sufficient NO_2 for passive DPF regeneration.

2.3.2 *Catalyst Recovery*

Beyond the removal of particulate matter from the DPF, DOC and SCR catalysts may also require periodic regeneration to remove contaminants. Both DOC and SCR catalysts may, under low-load conditions, accumulate hydrocarbon and the soluble organic fraction (SOF) of particulate matter. This is colloquially sometimes referred to as "coke." Coke can reduce catalyst performance by blocking active sites, may lead to face-plugging of the catalyst (Fig. 2.6), and can increase the risk of catalyst thermal damage due to rapid oxidation and the accompanying exotherm. Nakano et al. have studied the removal of carbonaceous deposits from a DOC [2]. The authors analyzed exhaust gas before the DOC catalyst with a GC-MS, and they found a diversity of HC species from C7 to > C19. Based on these findings, the authors attribute the formation of coke to higher molecular weight HC compounds. The authors report that coke on the DOC begins to decompose at temperatures below 200 °C, but that full decomposition requires temperatures above 500 °C in an oxidizing environment. Their results highlight the complex composition of coke, which contains both readily combustible hydrocarbon species in addition to more refractory particulate matter. Furthermore, Ottinger et al. have evaluated the impact of carbonaceous deposits on a vanadium-based SCR (V-SCR) catalyst [3]. In their study, the authors found that real-world coke accumulation can reduce V-SCR NO_x conversion by up to 40% at 250 °C. The loss of NO_x conversion due to carbonaceous deposits has been attributed to the blocking of active catalyst sites. NO_x conversion was fully recovered, and the

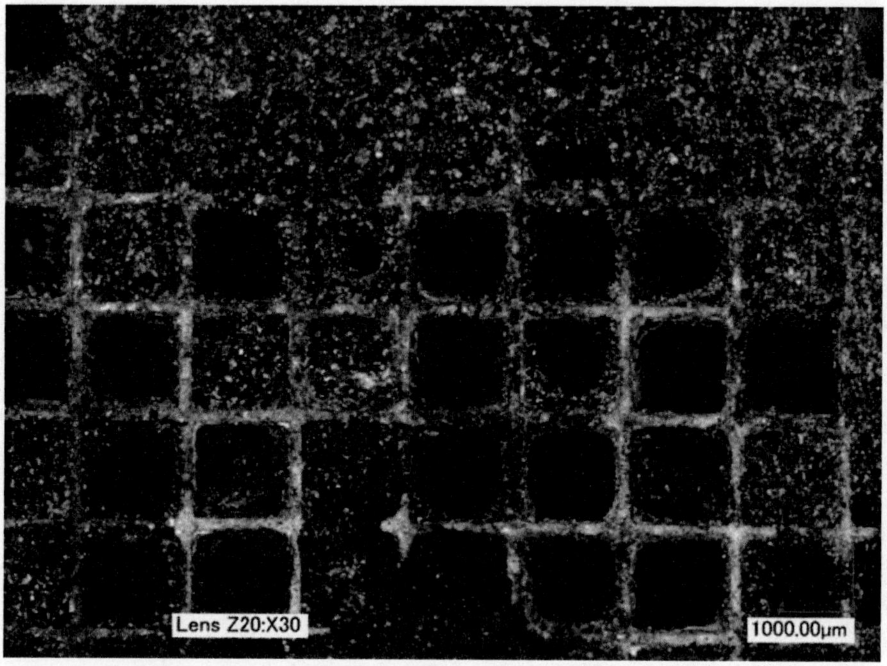

Fig. 2.6 Partial face-plugging of a diesel aftertreatment catalyst

carbonaceous deposits were fully oxidized, following a 500 °C-1 h thermal treatment, illustrating the benefit of thermal management.

Periodic sulfur removal, i.e., desulfation, is also required to maintain high levels of catalytic activity over DOC, SCR, and AMOX catalysts. Sulfur contained in diesel fuel is oxidized during combustion and emitted primarily as SO_2, with approximately 5% emitted as SO_3 [4]. Some SO_2 is then oxidized to SO_3 by the DOC catalyst, depending on the operating conditions. Hamzehlouyan et al. have shown that up to 100% of engine-out SO_2 can be oxidized to SO_3 at 372 °C by a model Pt/γ-Al_2O_3 DOC catalyst [5]. Thermodynamic equilibrium favors SO_3 at low temperatures, with the kinetics of SO_2 oxidation limiting SO_3 formation. Above approximately 400 °C, the formation of SO_3 is limited by the SO_2:SO_3 thermodynamic equilibrium. Sulfur oxides, but especially SO_3, poison active catalyst sites on the DOC, SCR, and AMOX catalysts and lead to a significant deterioration in emissions reduction performance. To minimize this deterioration, sulfur concentrations in diesel fuel have systematically been reduced through regulatory action. Countries and regions with advanced emissions norms all have regulations in place that limit the maximum diesel fuel sulfur level to less than 10 or 15 parts per million by weight (ppmw). Before these regulations, diesel fuel for mobile applications could contain up to 5000 ppmw sulfur. Marine diesel fuel still contains high levels of sulfur compounds, but the International Maritime Organization (IMO) has recently reduced the permissible diesel fuel sulfur concentration limit from 35,000 to 5000 ppmw effective January 1, 2020.

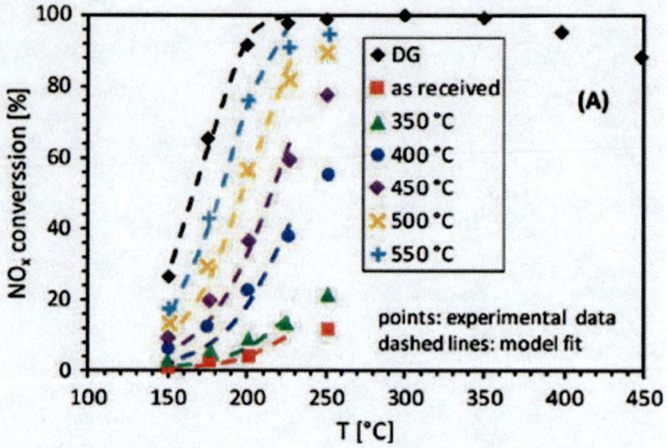

Fig. 2.7 NO_x conversion of an HSD exposed Cu-SSZ-13 catalyst upon reactivation at increasing temperatures [6]

To remove sulfur compounds and recover the performance of diesel aftertreatment catalysts, their temperature must be periodically increased. Xi et al. have investigated the desulfation of a state-of-the-art Cu/zeolite catalyst that was exposed to high sulfur diesel (HSD) fuel in a real-world application [6]. Figure 2.7 highlights the results. The as-received HSD exposed sample has approximately 90% lower NO_x conversion at 250 °C than the degreened (DG) catalyst which has not been exposed to sulfur. Following desulfation at progressively higher temperatures from 350 to 550 °C, the catalyst gradually recovers NO_x conversion performance as sulfur is removed from active sites. However, even after exposure to temperatures as high as 550 °C, the full performance of the degreened catalyst is not recovered in this case due to the extremely high levels of sulfur to which the catalyst was exposed. Thermal management is required to desulfate SCR catalysts if the duty cycle of the vehicle does not include periodic high exhaust gas conditions.

2.3.3 *Diesel Exhaust Fluid (DEF) Dosing and SCR Solid Deposits*

Typically, in urea SCR technology, aqueous urea solution of eutectic composition (32.5 wt% solution of urea in water)—known as Diesel Exhaust Fluid (DEF)—is injected into hot exhaust gases leading to a series of thermal, fluid dynamic, and reactive processes that eventually produce the ammonia necessary for NO_x reduction reactions within monolithic catalytic substrates. Marine systems may use a 40 wt% solution of urea in water. It has been widely accepted that the following processes take place as the 32.5% urea DEF is injected into hot exhaust gases:

- Drops produced by atomization of the injected liquid exchange mass, momentum, and energy with surroundings, leading to near-complete vaporization of water:

$$NH_2 - CO - NH_{2(aq.)} \rightarrow NH_2 - CO - NH_{2(s/l)} + 6.9H_2O \quad (2.8)$$

- Solid urea left behind melts and undergoes thermolysis to form ammonia and isocyanic acid:

$$NH_2 - CO - NH_{2(l)} \rightarrow NH_{3(g)} + HNCO_{(g)} \tag{2.9}$$

- Isocyanic acid undergoes hydrolysis to form ammonia and carbon dioxide:

$$HNCO_{(g)} + H_2O_{(g)} \rightarrow NH_{3(g)} + CO_{2(g)} \tag{2.10}$$

Water vaporization and thermolysis processes are endothermic, while the hydrolysis reaction is exothermic. Homogeneous hydrolysis occurring in the gas phase is reported to be a possibility, but in the absence of any catalytic acceleration, the effect of gas-phase hydrolysis under temperatures and residence times as experienced typically in mobile SCR systems appears to be negligible. Under optimal conditions, assuming completeness of thermolysis, an equimolar mixture of ammonia and isocyanic acid may be available at the inlet of the SCR catalyst. Isocyanic acid can then undergo hydrolysis after entering the SCR catalyst. Thermal implications of these processes will be discussed in detail in the section *Impact of DEF dosing*.

The DEF injector introduces a precisely controlled amount of DEF spray with the injection rate being controlled by engine operating conditions. A downstream mixer is usually employed to facilitate droplet-breakup and mixing processes. Through their synergetic effect, these components are expected to help deliver a well-mixed stream of reductant and NOx to the downstream SCR catalyst, with the least amount of solid deposits, and by maintaining the engine backpressure within acceptable limits. Design considerations of the aftertreatment system are heavily influenced by customer requirements that drive system-level choices such as space available to install the system and the total cost, and sub-system level choices such as availability of compressed air for air-assisted DEF atomization.

The DEF injector is a critical component of the aftertreatment system and characteristics of the injector must be matched well with those of other components to achieve optimal performance under the wide range of operating conditions encountered in real applications. A good DEF injector provides stable atomization characteristics and spray structure over the range of DEF flow rates of interest. Since DEF will be injected directly into the high-velocity exhaust gases, the atomizer should also be designed to be capable of producing a fine spray with adequate penetration and dispersion so that it mixes well with the exhaust gases while minimizing wall-wetting. The most common DEF injectors available in the markets rely on liquid-only pressure atomization (airless injection) or twin-fluid atomization using air (air-assisted injection). In airless pressure swirl and pressure jet atomizers, DEF is pressurized typically to 5–10 bars so that the high relative velocity created between the injected DEF jet or sheet and the surroundings breaks up the liquid once it exits the discharge orifice. Air-assisted twin-fluid atomizers are promising for DEF injection applications since they can be designed rather economically to provide finer drops than those produced by a pressure atomizer. The challenges will be, however, to provide

Table 2.3 Typical droplet size derived parameters and their definitions [7]

Derived parameter	Definition
Sauter mean diameter (SMD)	Diameter of a droplet with the same volume to surface area ratio as the entire spray
Dv10	Droplet diameter below which 10% of the spray volume is contained
Dv50	Droplet diameter below which 50% of the spray volume is contained
Dv90	Droplet diameter below which 90% of the spray volume is contained
Relative span factor (RSF)	= (Dv90 – Dv10)/Dv50

a source of air on-board and to maintain the quality of atomization acceptable over the wide range of liquid flow rates with the air that is available.

The heterogeneous nature of the atomization process produces a spectrum of droplet sizes distributed about a defined mean value, at any given operating condition. This spectrum is typically represented as either number-based (representing the number of droplets) or volume-based (representing volume of droplets) frequency distribution histograms for equivalent sphere diameters. Also, useful derived parameters to assess spray uniformity can be obtained, some of which are listed in Table 2.3 [7].

Spray characteristics can change with operating conditions, such as temperature, injection pressure, and air to liquid ratio (for air-assisted injection). Figure 2.8a shows the variation of SMD with air to the liquid ratio for a typical air-assisted injector. An increase in the air to liquid ratio results in improved atomization and reduced SMD. Figure 2.8b shows the variation of SMD with injection pressure for a typical airless assisted injector. An increase in injection pressure reduces SMD.

A major performance constraint faced by SCR systems is the incomplete decomposition of urea ahead of the catalyst. This can occur due to incomplete water vaporization (Eq. 2.8) or thermolysis (Eq. 2.9). The reasons for incomplete thermolysis include short residence times due to space constraints of the aftertreatment system, low exhaust gas temperatures, and impingement of DEF spray of particles on solid surfaces. Impingement of DEF particles on low-temperature walls and resulting incomplete decomposition can lead to the formation of solid substances usually referred to as SCR solid deposits. These deposits consist of urea or polymeric complexes as shown in Fig. 2.9. Once initiated, deposits can lower the temperature of the wall on which they reside and act as favoured sites for the accumulation of additional deposit mass. Beyond a certain level of accumulation, deposits can adversely affect system performance by increasing the engine backpressure, reducing the deNO_x efficiency, lowering the overall fuel economy, and potentially triggering a service event. Non-EGR applications, where the dosing rates could be several times higher than those for corresponding EGR applications, pose additional challenges for deposit robustness due to higher dosing rates.

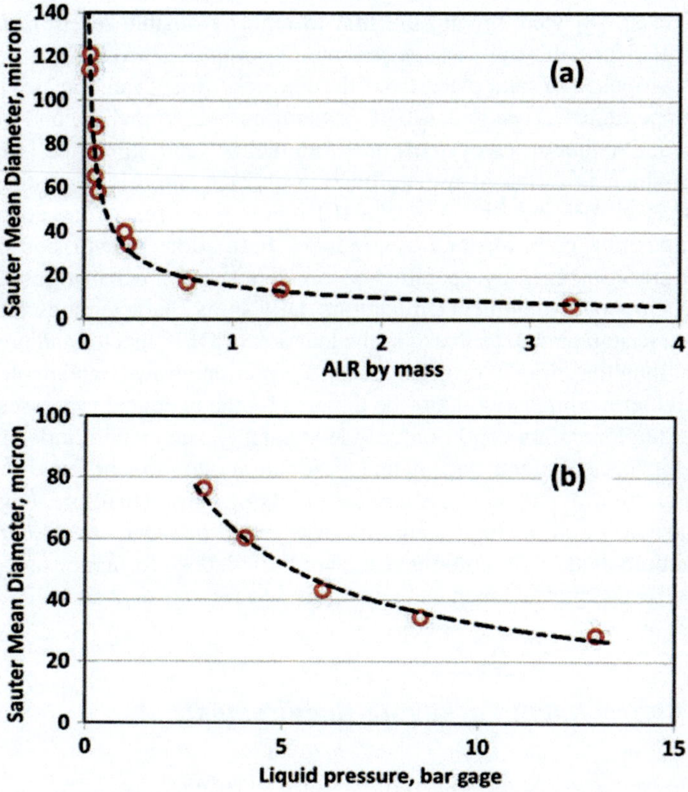

Fig. 2.8 Effect of operating parameters on mean drop-size **a** SMD versus ALR (air-to-liquid flow ratio) and **b** SMD versus injection pressure

Fig. 2.9 Results of thermal gravimetric analysis (TGA) of deposits found in real AT systems (thin lines) and reference pure materials (thick lines)

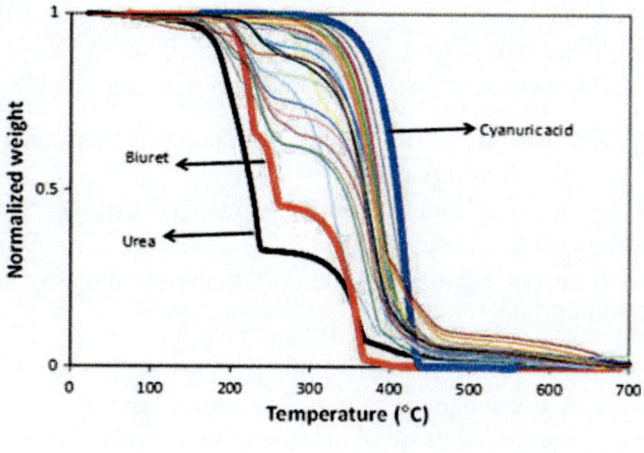

Schaber et al. [8] were among the first to report plausible reaction paths that can lead to the formation of various polymeric complexes from urea. Formation of biuret is proposed to take place from the reaction of isocyanic acid with intact urea. At temperatures exceeding 190 °C, cyanuric acid, ammelide, and ammeline are produced. Production of cyanuric acid, ammeline, and ammelide is complete at 250 °C, which is followed by sublimation and decomposition of these products. Brack et al. [9] and Zhang et al. [10] report more recent investigations of urea decomposition pathways and by-products. Formation, growth, and depletion of deposits are influenced by operating parameters such as exhaust gas flow rate, exhaust gas temperature, ambient temperature, DEF spray characteristics, DEF injection rate, and design parameters such as the location of DEF injector and presence of solid walls along the spray or drop trajectory. Formation, growth, and depletion of a specific polymeric complex will also be dictated by the chemical processes favored under prevailing operating conditions. Understandably, such a multitude of influential parameters necessitates a systematic investigation into the root cause of deposits from thermal, fluid dynamic, and chemical points of view. Therefore, engineering focus is placed on optimizing system design to avoid or reduce the occurrence of thermal and fluid dynamic conditions that favor DEF deposit formation and retention and to use effective thermal management options to remove deposits once formed.

2.3.4 Aftertreatment Packaging Requirements

Modern aftertreatment systems need to conform to stringent packaging constraints and emission performance targets. These systems include several pipes, tubular geometries, and expansions and contractions. The use of double-walled pipes with insulation is essential to keep the heat loss to lower levels. DOC, DPF, SCR, or ASC catalysts also include a mounting mat that provides the force to retain the catalytic component in place but also acts as an additional insulator.

The considerations in thermal performance include

- The durability of exhaust components including doors, sensor tables, and electronic components
- Light-off of catalysts and keeping the catalyst temperature within the desired temperature window
- Transient response of the system under changing exhaust flow and temperature conditions.

Therefore, efforts aim at reducing the heat loss across the system but retaining skin temperature and temperature of electronic components within acceptable limits. With the increased use of simulations to reduce product development time and testing costs, 3D computational fluid dynamics (CFD) simulations and finite element analysis (FEA) simulations are often used. Meda et al. [11] outline the typical steps involved in this process. CFD analysis can be used to improve the flow characteristics without significantly affecting the backpressure. The CFD analysis is further used to calculate the temperature distribution on all surfaces of components such as DOC, DPF, and SCR and the outside surface of the system. The temperatures are obtained during regeneration at several critical locations on the system. Structural design evaluation of such a diesel system is done to make sure that under cyclical hot and cold regimes, the design features are provided which will allow the components and system to be thermally flexible and at the same time keeping in view the ability of the design to withstand the mechanical loads alternating hot regeneration and cooling loads. The analysis involves mapping the temperature distribution obtained from CFD onto an FEA mesh.

2.4 Engine-Based Thermal Management

In this section on engine-based thermal management, we discuss various thermal management strategies that primarily utilize the combustion of fuel in the engine to generate additional heat for the aftertreatment system. Five options are discussed in detail, and their pros and cons are clarified. The strategies in this section are some of the most widely used in commercial applications today. The four strategies include:

- Active Duty Cycle Adjustment
- In-cylinder Post-injection of Fuel
- Cylinder Deactivation
- Exhaust Restriction or Intake Throttling.

2.4.1 Active Duty Cycle Adjustment

From a hardware perspective, the simplest form of engine-based thermal management is the active-duty cycle adjustment. With today's electronically controlled diesel engines, it is possible to choose from an array of engine operating conditions to meet a power output requested by a vehicle operator. Furthermore, it is possible to control engine operating conditions independent of the operator. The engine can be run at a less efficient engine operating mode to generate the necessary heat for thermal management. Examples of this are idle strategies intended to desorb water vapor or hydrocarbons that can accumulate on the diesel aftertreatment catalysts at low temperatures when an engine is idling for an extended duration of time. In this case, the engine speed may be increased periodically to increase the exhaust temperature to a point where adsorbed hydrocarbons can desorb. The exhaust temperature increase will be limited without a way to significantly increase the torque requirement and fueling. Two methods to increase engine torque for thermal management are discussed later in this section.

2.4.2 *In-Cylinder Post-Injection of Fuel*

In-cylinder diesel injection following the primary fuel combustion event is used commercially for exhaust aftertreatment thermal management. Depending on the proximity to top dead center (TDC), post-injection and late post-injection of diesel fuel can support thermal management in two ways:

1. Post-injection—Increases exhaust temperatures.
2. Late post-injection—Generates hydrocarbon species that are transferred to the DOC, where heat is generated through HC oxidation.

Both strategies are shown as a function of crank angle in Fig. 2.10 from Tan et al. [12]. It should be noted that this is a simplified conceptual representation, and real-world diesel engines may operate with a different fuel injection strategy. Following the pilot- and pre-injections and the main injection event centered on TDC, the post-injection occurs close to the main combustion event. The injected fuel is oxidized in-cylinder by the oxygen not consumed in the main combustion event. Because diesel engines operate lean, there is sufficient O_2 available for complete oxidation of post-injected diesel. Directly following the diesel combustion event at TDC, the products of combustion are at a sufficiently high temperature and pressure to promote rapid oxidation of the post-injected fuel. However, Tan et al. have shown that complete oxidation of post-injected diesel fuel is not achieved under real-world conditions [12]. The authors showed that post-injection directly increases the heat of the exhaust gases by an amount equivalent to the amount of fuel post-injected.

Late post-injection is further separated from the main injection event. Whereas post-injection occurs between 10 and 30° after TDC, a late post-injection occurs at least 40 °C after TDC [12]. At this point, the temperature and pressure in the cylinder are no longer sufficient to oxidize the same fraction of injected diesel fuel as in the post-injection strategy. Oxidation still occurs, but the level of oxidation is reduced with increasing injection delay. CO and total hydrocarbon (THC) emissions resulting from 10 mm^3 of late post-injected diesel increase from a negligible amount at 40° after TDC to more than 1000 ppm by 80° after TDC [12]. Both CO and THC are products of incomplete diesel oxidation. If late post-injection is conducted 120° after TDC, diesel fuel can be discharged from the cylinder during the exhaust stroke as a liquid. The late post-injected diesel fuel is vaporized, partially oxidized, and

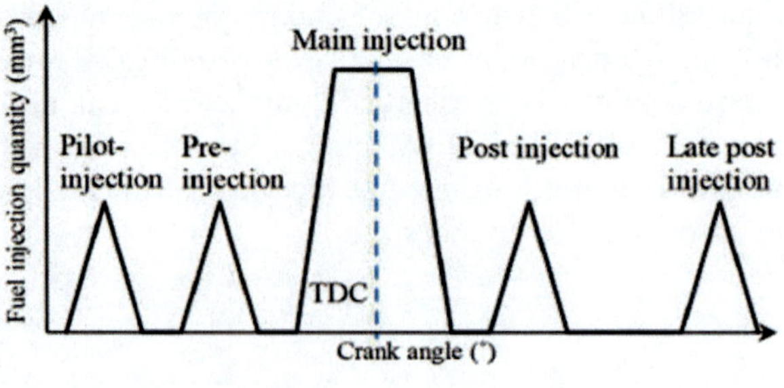

Fig. 2.10 Schematic of possible fuel injection strategies [12]

transported to the DOC with the exhaust gases. At the DOC, the products of diesel fuel partial oxidization, along with unburned diesel fuel, are oxidized by the DOC, creating additional heat for thermal management.

Parks et al. have suggested another approach called individual-cylinder delayed and extended main (IC DEM) injection [13]. In this case, an individual cylinder is enriched during the main injection event. As opposed to the post-injection strategy discussed above, the IC DEM strategy achieves net rich combustion conditions in the single enriched cylinder. This net rich combustion produces exhaust with light reductant species such as H_2 and CO. These species are not produced during lean combustion or as a result of post-injection. Figure 2.5 in their article shows that more than 0.3% H_2 can be produced with a fuel penalty of 100%. The other strategies evaluated by the authors (including late post-injection) produced less than ~ 0.05% H_2 at a similar fuel penalty. The H_2, light hydrocarbon species, and CO produced during rich combustion facilitated by the IC DEM strategy are oxidized by an oxidation catalyst at a significantly lower temperature than the nascent diesel fuel supplied by the late post-injection (Fig. 2.11).

A combination of post-injection and late post-injection can also be used. At temperatures below the DOC light-off temperature, discussed in the upcoming section titled *DOC Light off and Quenching*, post-injection can be used to raise the temperature of the DOC to the point where a late post-injection can be effective. Primary advantages of this method of thermal management include its low cost of implementation and high reliability. The implementation cost is low because no additional hardware is required to add diesel injection events following the main injection. And, the lack of additional hardware increases the reliability of this thermal management solution, in comparison to other options discussed herein. One challenge with in-cylinder post-injection is oil dilution by the injected and unburned fuel. When diesel fuel is injected after the main combustion event, a fraction of the injected fuel

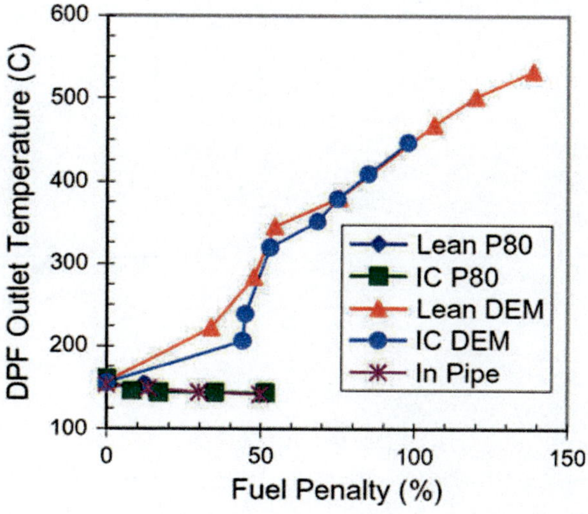

Fig. 2.11 Exhaust temperature downstream of a DPF as a function of fuel penalty and post-injection strategy [13]

comes into contact with the cylinder wall and can lead to oil dilution. Parks et al. proposed that an advantage of their individual cylinder strategies is that the cylinder receiving the post-injection can be alternated, thereby limiting fuel accumulation on the cylinder walls and oil dilution.

Another challenge for in-cylinder post-injection is that this method of thermal management has been shown to increase PM and particle number (PN) emissions, in comparison to PM and PN emissions under non-regeneration conditions [14, 15]. Figure 2.12 shows PN emissions before, during, and after a DPF regeneration accomplished with in-cylinder post-injection of diesel fuel [15]. The data is from a light-duty diesel vehicle operating over the New European Driving Cycle (NEDC). On the left side of Fig. 2.12, the non-volatile fraction of PN, similar to elemental carbon (EC), increases approximately 50% during and after the DPF regeneration. PN emissions are higher directly following the regeneration due to the removal of the soot cake layer, which increases PM and PN filtration efficiency. Total PN emissions, shown on the right side of Fig. 2.12, are two orders of magnitude higher than PN emissions before the regeneration event. This implies that the increase in PN emissions during regeneration is primarily due to volatile and semi-volatile particles. Giechaskiel et al. reported that these particles can be attributed to the release of volatile material stored on the DPF combining with unburnt lubricant and the post-injected diesel fuel [15]. The authors found peak particle number emissions at particle mobility diameters of either 12 or 35 nm, depending on the conditions leading up to the regeneration and during the regeneration.

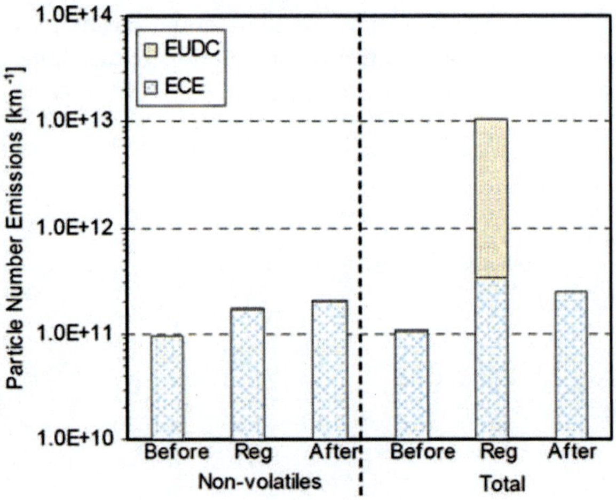

Fig. 2.12 Particle number (PN) emissions before, during regeneration, and after DPF regeneration during the urban (ECE) and extra-urban (EUDC) portions of the New European Driving Cycle (NEDC) [16]

2.4.3 Cylinder Deactivation

Cylinder deactivation (CDA), also known as active cylinder control (ACC), active fuel management, and variable displacement, is a technology that was first commercialized in mobile applications in the 1980s to improve fuel economy and exhaust emissions of light-duty, gasoline vehicles. At idle and low-load conditions, the power requirement is less than 50% of an engine's maximum power, and pumping losses are significant when the throttle is in a closed position (e.g., idle). Under these conditions, cylinders can be deactivated to significantly reduce fuel consumption. This is accomplished by mechanically disabling the intake and exhaust valves. According to the US EPA, 13% of all MY 2019 passenger cars and light-duty trucks were equipped with CDA technology (https://www.epa.gov/automotive-trends/highlights-automotive-trends-report). These include GM's V-8 engine in which more than 4 cylinders can be deactivated down to the Ford 3-cylinder engine equipped with single cylinder CDA.

CDA has more recently begun to receive attention from heavy-duty diesel engine manufacturers as a technology that can improve fuel economy while simultaneously improving emissions performance at otherwise challenging operating conditions (i.e., cold start, extended idle, urban driving, and cold ambient conditions) through thermal management [16, 17]. Similar to gasoline CDA, diesel CDA reduces airflow, which generally increases exhaust temperatures and helps maintain exhaust temperatures once they are elevated, and pumping work, increasing fuel efficiency of combustion events. Figure 2.13 shows the impact of diesel CDA on exhaust temperature, fuel flow, engine-out (EO) NO_x emissions, and air-to-fuel (AFR) ratio for a 500 hp in-line six-cylinder engine operating at 1000 rpm and up to 3 bar BMEP (brake mean effective pressure) [17]. During CDA conditions, all valves on deactivated cylinders were closed, and fueling to these cylinders was disabled. The number of cylinders shut-off was varied as a function of the operating condition. Implementation of CDA successfully increased exhaust temperatures up to 200 °C at the maximum engine load of 3 bar BMEP by decreasing the AFR. At lower engine loads, the exhaust temperature increased more moderately. Importantly, temperatures were increased into the range where DEF introduction can be enabled and SCR $DeNO_x$ efficiency is high, except at the 0 bar BMEP test condition. The fuel flow decreased more than 30% at 0 bar BMEP, and EO NO_x and AFR were also both lower as a result of CDA. Also, Ramesh et al. evaluated one-third (2-cylinder), half (3-cylinder), and two-third (4-cylinder) CDA strategies of a Cummins in-line, six-cylinder, medium-duty, diesel engine [16]. Their results show that CDA can be used to increase DPF inlet temperatures from 500 to 600 °C, sufficient for DPF active regeneration. The authors reported these findings at cruise conditions (1200 rpm/7.6 bar BMEP). Challenges still exist for the widespread implementation of diesel CDA. These include the impact of CDA on transient operating conditions as well as mitigating the noise, vibration, and harshness (NVH) effects of various CDA strategies.

In addition to full cylinder deactivation (CDA), other variable valve actuation strategies have been proposed in the literature. One example is an early exhaust

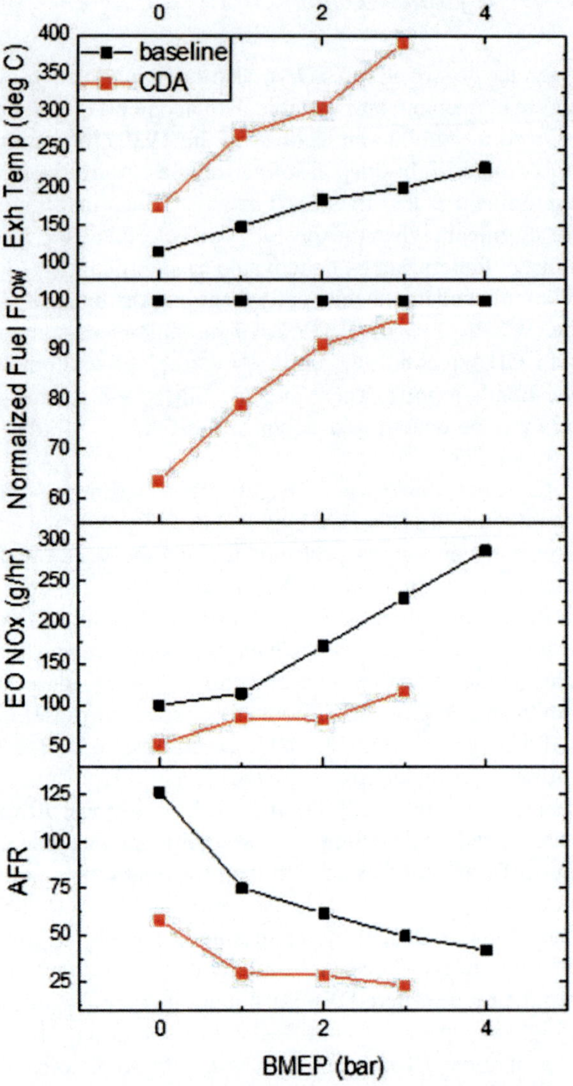

Fig. 2.13 Exhaust temperature, fuel flow, engine-out NO_x, and air/fuel ratio under CDA and baseline engine operation at 1000 rpm [17]

valve opening, as described by Roberts et al. [18]. With this strategy, the exhaust valve(s) of one or more cylinders may be opened earlier in the compression stroke than nominal to transfer more heat of combustion into the exhaust system. However, there is a fuel penalty trade-off that must be considered as part of this strategy.

2.4.4 *Exhaust Restriction or Intake Throttling*

Exhaust restriction, i.e., exhaust braking, can be used for diesel aftertreatment thermal management. While this technology is most commonly used to improve the braking performance of a diesel vehicle, it can, in certain situations, be useful for thermal management. When an exhaust restriction is added, backpressure on the engine is increased, forcing the engine to work harder to remove exhaust gases during the exhaust stroke. Especially in extended idle conditions, a restriction can be applied, increasing the load on the engine and putting it into a higher power operating mode. In this way, exhaust temperatures can be increased significantly above those experienced at idle. This implementation of exhaust braking can be used to periodically remove unburned hydrocarbons which can accumulate on diesel aftertreatment catalysts at low exhaust temperatures, as discussed in the section on *Catalyst Recovery*. An example of an exhaust brake manufactured by PacBrake™ is shown in Fig. 2.14. In this configuration, a butterfly valve is positioned in the exhaust flow. In the open position, the valve is parallel to the flow and does not significantly restrict the flow of exhaust gases. However, when the exhaust brake is applied, the valve is pneumatically or hydraulically turned perpendicular to the flow, closing off the majority of the exhaust line and dramatically increasing back pressure on the engine. Holes in the butterfly valve allow exhaust to escape, preventing the engine from abruptly shutting off. Exhaust brakes can also be integrated into the variable geometry turbocharger (VGT). The biggest concern for exhaust brake technology is durability. Exhaust brake components are exposed to diesel exhaust throughout their lifetime, and a failed exhaust brake can prevent a diesel engine from operating at load.

Ozel et al. have shown that a diesel engine intake throttle can also be used for thermal management [19]. While prior generations of diesel engines were not equipped with intake throttles, many modern diesel engines use intake throttling to help control exhaust gas recirculation rates and improve engine shutdown. Reducing fresh airflow to the intake manifold lowers the manifold pressure, enabling higher EGR rates. EGR will be discussed in more detail in the following section. The authors evaluated the impact of intake throttling on a 2.0-L light-duty diesel engine

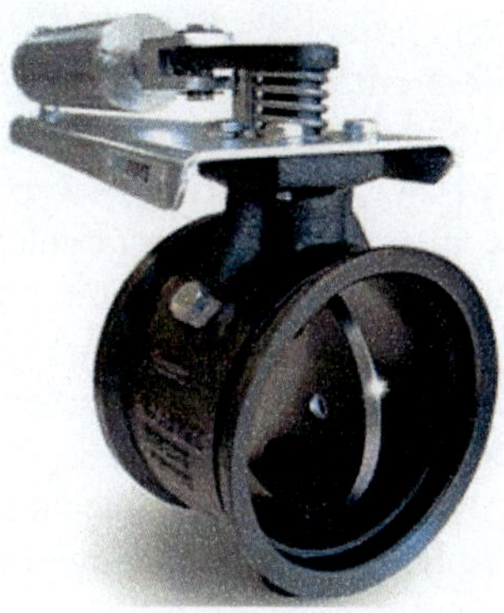

Fig. 2.14 A diesel exhaust brake manufactured by Pacbrake

at idle conditions. They showed that heavy throttling was able to increase the exhaust temperature by 60 °C above the baseline exhaust temperature of 100–105 °C. The addition of a post-injection 30–45° after TDC resulted in a 20–30 °C temperature increase on top of that obtained with intake throttling. Please refer to the section on *In-cylinder Post-injection of Fuel* for additional discussion on this topic.

2.4.5 Combustion and Fuel Strategies

Exhaust gas recirculation (EGR) is one of the key emission control technologies for reducing NO_x emissions and it achieves this through directly controlling combustion temperature. Hence it is an important technology to survey in the context of engine thermal management. Developments in the area of advanced combustion technologies and fuels are also of interest and it should be noted that many of these technologies use EGR as a crucial enabler.

EGR technology relies on returning a part of the exhaust gas to the combustion chamber and using this cooled exhaust gas as a diluent. This lowers the oxygen concentration in the combustion chamber and increases the mixture specific heat thereby reducing the combustion temperature and in-cylinder NO_x. Along with electronic high-pressure fuel systems and aftertreatment technology, cooled Exhaust Gas Recirculation (where the diluent is returned through the intake system after cooling) is considered a major step in the evolution of heavy-duty diesel technology [20]. An EGR system can be operated as high pressure or low pressure. In the high-pressure EGR system, a portion of the exhaust used for recirculation is extracted from a location upstream of the turbine, while in a low-pressure EGR system, it is extracted from a location downstream of a diesel particulate filter. To capitalize on the advantages offered by each, and to avoid their disadvantages, hybrid systems have been investigated where low-pressure EGR is used at high engine loads, and high-pressure EGR at lower engine loads, and some combination of the two in the transition region [21].

In conventional diesel combustion, in-cylinder conditions that tend to reduce NO_x (such as the presence of EGR or delaying the injection) can increase chances of PM formation due to local rich equivalence ratios, by the behavior known as the "PM-NOx trade-off". Therefore, it is difficult to achieve the simultaneous reduction of these criteria pollutants to the desired low levels, and an aftertreatment system is often necessary. The addition of SCR technology effectively allows one to avoid the PM-NO_x trade-off providing more flexibility. Based on the choice of combustion recipes and emission targets, applications could use EGR without NO_x aftertreatment, use EGR in conjunction with SCR aftertreatment or use SCR aftertreatment only solutions. For example, the mix of technologies used in Heavy duty diesel and Locomotive segments to meet progressively stricter NO_x emission targets is shown in Fig. 2.15 [22].

Investigations on cleaner combustion strategies that can significantly reduce the levels of emissions within the cylinder itself are an area of ongoing research.

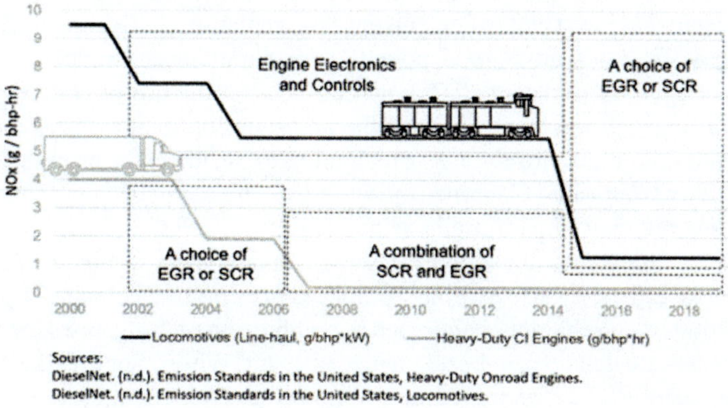

Fig. 2.15 A mix of technologies used in heavy-duty diesel and locomotive segments to meet NO_x emission targets [22]

Figure 2.16 shows several non-conventional diesel combustion methods and conventional diesel combustion on the Equivalence ratio—Temperature map [23].

Most promising combustion strategies rely on achieving the simultaneous reduction of multiple regulated pollutants either by operating the engine in non-traditional diesel engine operating regimes or by controlling fuel composition such that significant levels of NOx and soot are avoided. In some of these Low-Temperature combustions (LTC) strategies, fuel injection is significantly advanced to improve the level

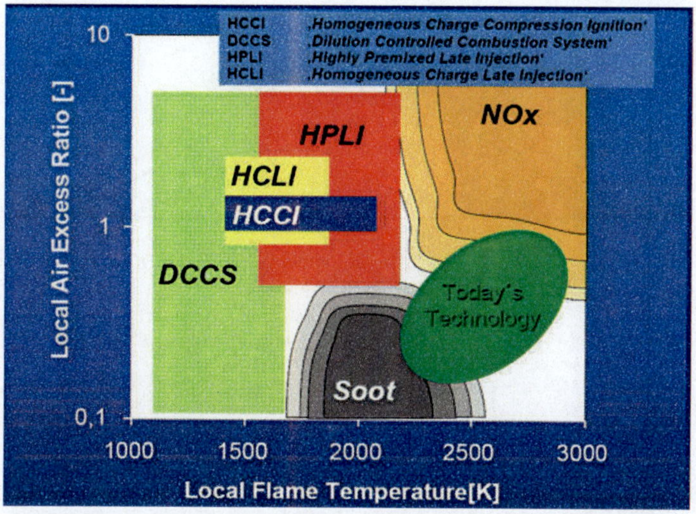

Fig. 2.16 Comparison of several non-conventional diesel combustion methods with conventional diesel combustion. "Today's Technology" refers to conventional diesel combustion [23]

of pre-mixing and higher levels of exhaust gas recirculation (EGR) are used. Consequently, these strategies have good potential for simultaneous reduction of NO_x and PM. A related, more recent LTC approach exploits the use of combinations of fuels with different properties to control the auto-ignition temperatures and hence combustion kinetics more closely. These technologies fall under the broad category of "Fuel Reactivity Controlled Compression Ignition" and a typical approach involves injecting a low reactivity fuel initially as port injection, and then using direct injection of diesel to initiate combustion [24]. However, as shown in Fig. 2.17 (adopted from Asad et al. [25]) they also exhibit differences in terms of challenges, operability limits, and combustion characteristics such as combustion phasing, pressure rise rate, and heat release rate that need to be taken in to account. While adopting the advanced combustion technologies, the integration with the aftertreatment system and other components need to be considered, and following a holistic systems approach is crucial [26]. More details on the potential of such combustion technologies to meet advanced emission regulations can be found elsewhere (e.g., Stanton [26], Dev et al. [27]).

Also, alternatives that can at least partially replace diesel fuel in the combustion process including liquid or gaseous fossil fuels (e.g., natural gas), and renewable fuels (e.g., biofuels from indigenous sources) are being investigated. Natural gas has established itself as a key source of energy in various markets especially with the advent of the hydraulic fracturing process and it has been used in both spark

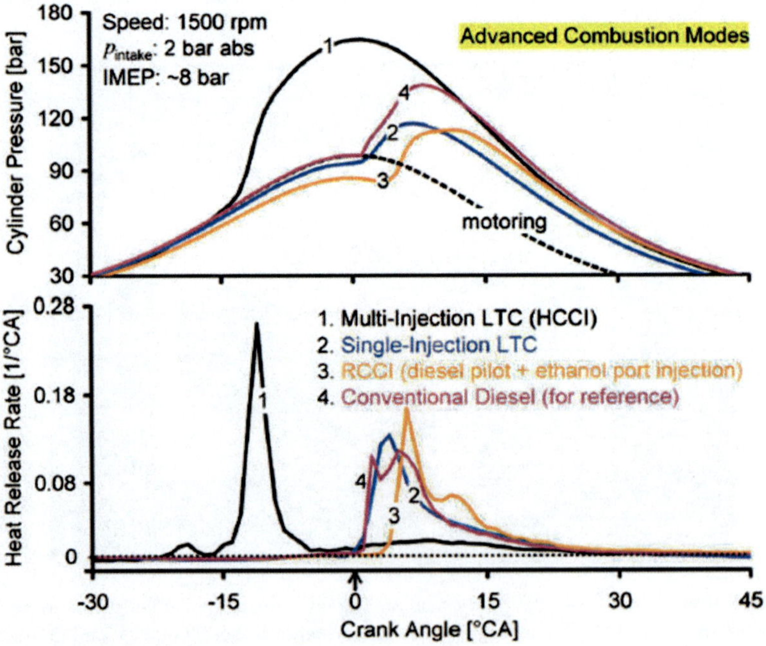

Fig. 2.17 Advanced combustion modes for diesel engines [25]

ignited and compression ignited engines (see Korakianitis et al. [28] for a detailed survey). In compression ignited engines, natural gas is introduced into the intake manifold, and small quantities of diesel fuel are used as the ignition source. A significant advantage is offered by the possibility of using a simpler Three-Way Catalyst for emissions reduction rather than more complex aftertreatment systems. However, increased engine-out methane emissions are a concern. Königsson et al. [29] investigated a port injected Diesel Dual Fuel concept with methane injected in the intake manifold with closer control over air–fuel ratio. They found that advanced combustion phasing can be used to improve combustion efficiency and extend the lean limit of the dual-fuel diesel engine, though, in practice, this results in a penalty with regards to NO_x emissions and thermal efficiency. The use of high levels of EGR helped reduce the complexity of aftertreatment and a three-way catalyst may suffice. May et al. [30] investigated Premixed Dual-Fuel Combustion (PDFC) with natural gas on an HD single-cylinder research engine. Here, part of the diesel introduction is done earlier to allow vaporization and partial pre-mixing with the homogenously mixed NG in the cylinder, while the second diesel introduction happens around TDC. In comparison with conventional natural gas-diesel combustion, PDFC showed significant reductions in methane slip as well as CO emissions. Another approach is the high-pressure direct injection (HPDI) of natural gas, where pilot diesel fuel ignites directly injected natural gas, with both fuels introduced through a single injector [31], though cost and complexity pose challenges.

2.5 Aftertreatment Fuel Introduction

Another method of aftertreatment thermal management is to inject diesel fuel into the exhaust gases upstream of an oxidation catalyst, typically a DOC. The diesel fuel is then oxidized over the DOC, creating heat that is then convected to the downstream catalyst elements (i.e., DPF, SCR, etc.). Figure 2.18 shows a schematic representation of a retrofit diesel aftertreatment system utilizing hydrocarbon injection upstream of a DOC for active DPF regeneration [32]. As shown in the figure, there are many components of a hydrocarbon injection system. They include an electronic controller, a manifold or diesel control valve, fluid transfer lines, an injection nozzle, and control and diagnostic systems. Also, mixing elements may be an additional required component. When thermal management is requested by the electronic control unit, the diesel control valve is opened, permitting diesel to flow through the fluid transfer lines to the dosing nozzle. Liquid diesel is then introduced into the exhaust stream through the dosing nozzle, where it mixes with the exhaust gases. In some situations, where aftertreatment geometry requires it, mixing devices may be included after the nozzle to enhance mixing and improve the uniformity of diesel at the face of the DOC.

A limitation of thermal management with diesel introduction is that this method cannot be used at low exhaust temperatures where diesel fuel will remain in a liquid state and the diesel oxidation catalyst is inactive for hydrocarbon oxidation reactions. Diesel fuel consists of a mixture of hydrocarbons from approximately C9 to

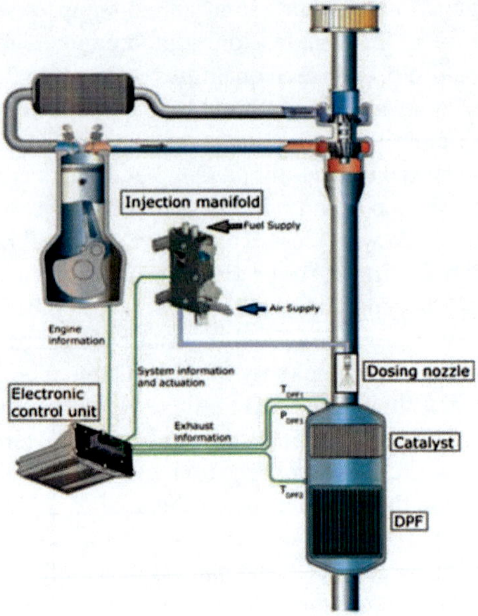

Fig. 2.18 Schematic of a retrofit diesel aftertreatment system utilizing hydrocarbon injection for thermal management [32]

C23, as seen in Fig. 2.19 [33]. These hydrocarbons have a wide range of boiling points. For example, while decane ($C_{10}H_{22}$) boils at 173 °C, the boiling point of 1-decylnapthalene ($C_{20}H_{28}$) is significantly higher, at 379 °C. The initial boiling point of diesel is approximately 130 °C, and 50% of diesel boils by approximately 275 °C, as shown in the distillation profile of Fig. 2.19. Aftertreatment diesel introduction can also lead to an increase in PM and PN emissions if not well-optimized, similar

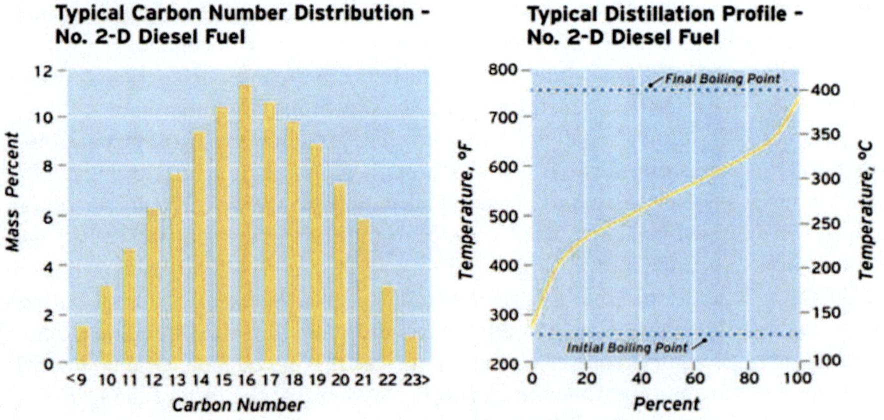

Fig. 2.19 No. 2 diesel carbon number distribution and distillation profile [33]

to thermal management with in-cylinder post-injection. For a detailed discussion of this phenomenon, see the section on *In-cylinder Post-injection of Fuel.*

2.5.1 DOC Light off and Quenching

In addition to considerations regarding the volatility of diesel fuel discussed above, the DOC must be at a sufficiently high temperature to oxidize the diesel fuel introduced into the exhaust. Kumar et al. have studied the oxidation of diesel over a state-of-the-art DOC utilizing a catalyst core reactor [34]. Diesel conversion as a function of temperature at concentrations from 0.2 to 1.8% C1 is shown in Fig. 2.20. Below 200 °C, the DOC is practically inactive for diesel oxidation, preventing thermal management. At lower inlet concentrations (i.e., 0.2% C1), diesel conversion starts at lower temperatures, achieving approximately 60% conversion at 230 °C. However, at higher diesel concentrations (i.e., 1.8% C1), such as those which would be required to generate significant heat for thermal management, oxidation of diesel by the DOC is still below 50% at 250 °C. Thus, thermal management via the introduction of diesel would not be advisable below 250 °C. The DOC must be at a sufficiently high temperature before diesel injection is started, or it is possible that hydrocarbon oxidation could be quenched, resulting in aftertreatment fouling and excessive emissions of hydrocarbons. Kozlov et al. studied DOC hydrocarbon quenching over a variety of state-of-the-art DOC catalysts [35]. The authors reported a minimum DOC quenching temperature of approximately 285 °C but found that the quenching temperature can increase to 335 °C, depending on the DOC aging condition and inherent hydrocarbon oxidation activity.

The temperature dependence of diesel oxidation over a DOC catalyst is a significant constraint when utilizing diesel injection for thermal management. However, the DOC light-off temperature can be reduced by increasing the PGM (platinum group

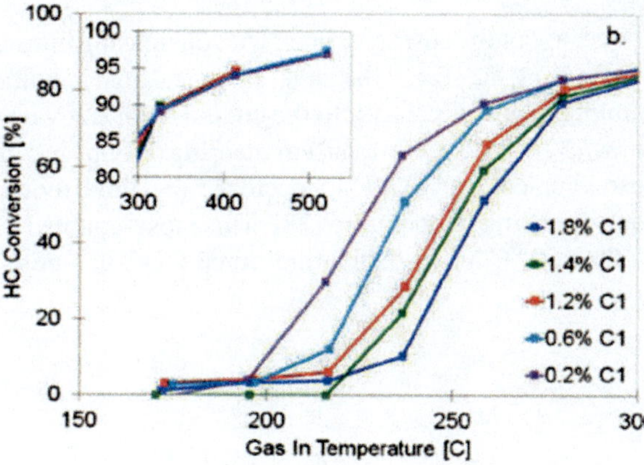

Fig. 2.20 Steady-state diesel conversion over a DOC catalyst as a function of diesel concentration from 0.2 to 1.8% C1 [34]

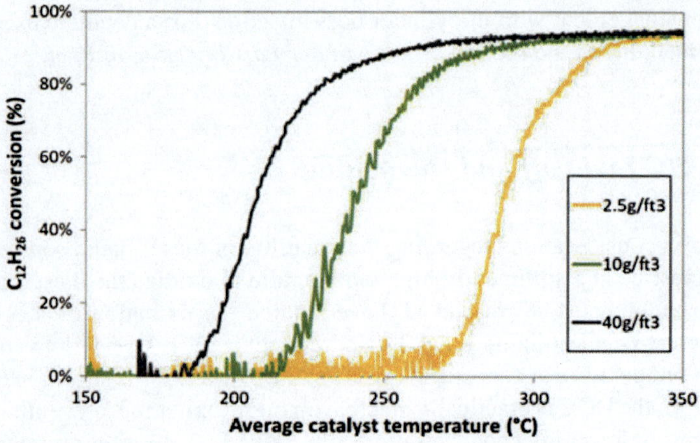

Fig. 2.21 $C_{12}H_{26}$ light off as a function of temperature for DOC catalysts with 2.5, 10, and 40 g/ft^3 PGM loading

metal) loading of the DOC catalyst. Figure 2.21 shows the impact of PGM concentrations of 2.5, 10, and 40 g/ft^3 on the light off of n-$C_{12}H_{26}$ (n-dodecane). $C_{12}H_{26}$ is used as a surrogate for diesel fuel in this case. Increasing the PGM loading from 2.5 to 40 g/ft^3 shifts the T_{50} temperature (the temperature at which 50% conversion is achieved) about 75 °C lower, but such an increase in PGM would also significantly increase the cost of the aftertreatment system. Another method used in the industry to decrease the light-off temperature of a DOC is to change the relative ratio of platinum to palladium (Pt:Pd) on the catalyst. Shakya et al. studied the effect of DOC Pt:Pd loading on $C_{10}H_{22}$ (decane) oxidation, another surrogate for diesel fuel (Fig. 2.22) [36]. The authors showed that a DOC catalyst with a 10:1 Pt:Pd ratio achieved 50% $C_{10}H_{22}$ conversion 50 °C sooner than the same DOC with a Pt:Pd ratio of 1:3. Finally, the DOC light-off temperature can be reduced by increasing the physical size of the DOC catalyst, but this again increases the aftertreatment cost as well as its physical size.

Because of the cost associated with decreasing intrinsic DOC light-off temperature or increasing the size of the DOC catalyst, some manufacturers have suggested close-coupling the DOC catalyst to the engine to take advantage of higher exhaust temperatures [37]. IAV, a German engineering group, has developed a pre-turbocharger close-coupled DOC-DPF to decrease the time required for these components to reach operating temperature [38]. This close-coupled architecture keeps the DOC up to 100–150 °C hotter than a traditional DOC in a post-turbocharger location.

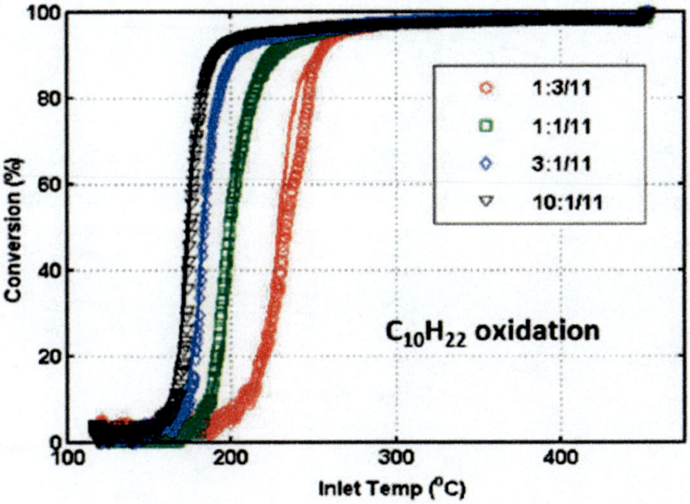

Fig. 2.22 n-$C_{10}H_{22}$ oxidation by a DOC catalyst with 11 g/ft^3 PGM and the Pt:Pd ratios shown in the legend [36]

2.5.2 DOC Degradation

Another challenge with aftertreatment diesel introduction for thermal management is the potential for hydrothermal degradation of the DOC. For diesel particulate filter regeneration and SCR desulfation, DPF and SCR temperatures of greater than 500 °C are required. An even higher DOC outlet temperature is required to achieve these temperatures. And, although unintended, temperatures exceeding this may occasionally be reached due to engine transients (e.g., drop-to-idle) during regeneration or the rapid light-off of pre-stored hydrocarbons. Figure 2.23 shows the impact of accelerated hydrothermal aging on $C_{12}H_{26}$ oxidation over a commercial DOC catalyst. HC oxidation is only minimally degraded by hydrothermal aging of the DOC, even at temperatures far exceeding those anticipated in real-world operation (> 700 °C). However, NO oxidation to NO_2 is much more sensitive to DOC hydrothermal aging (not shown). NO oxidation at 250 °C decreases from 72% for the same commercial catalyst treated at 650 °C-2 h to 53% after treating at 800 °C-2 h. Aging the DOC up to 1100 °C-2 h further reduces the NO oxidation to only 19%. As mentioned in the introduction to this chapter, the DOC oxidizes NO to NO_2 for passive DPF regeneration and increased NO_x conversion at low temperatures via the fast SCR reaction. If the DOC produces less NO_2, less soot will be oxidized via passive soot oxidation and additional active DPF regenerations will be required, exacerbating the issue of DOC thermal aging.

Contamination also referred to as catalyst poisoning, is the other primary degradation mode of DOC catalysts, affecting light off of diesel fuel dosed onto the catalyst. DOC catalysts are typically the first catalyst element in the aftertreatment system, and

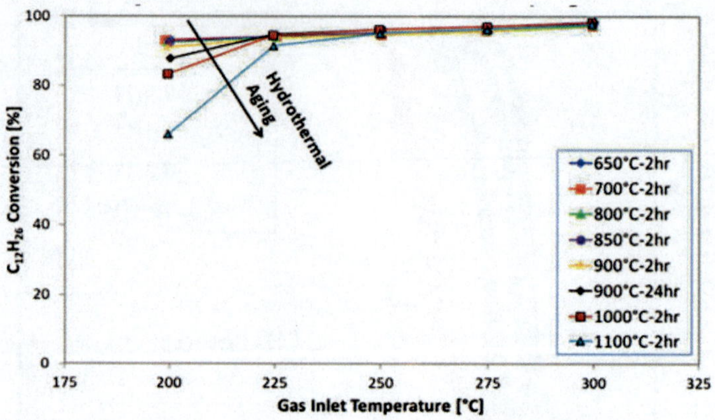

Fig. 2.23 Impact of hydrothermal aging on HC conversion of a commercial DOC catalyst

are, as a result, exposed to nascent diesel exhaust which may contain catalyst poisons such as sulfur oxides, phosphorous, zinc, sodium, potassium, etc. These are predominantly derived from diesel fuel (sulfur), lube oil (phosphorous and zinc), or biodiesel (sodium, potassium, and other contaminants). Figure 2.24 displays the impact of sodium on the temperature at which 80% hydrocarbon conversion is achieved (T-80) for a Pd-only, a Pt:Pd non-zeolite, and a Pt:Pd zeolite DOC catalyst [39]. The most broadly used type of DOC catalyst is the Pt:Pd non-zeolite type. For this catalyst, 1 wt% sodium increased the T-80 from 206 to 273 °C, and 3 wt% sodium contamination further increased the T-80 to 554 °C. For this reason, diesel fuel, biodiesel, and lube oil must conform to global specifications limiting the concentrations of these catalyst poisons.

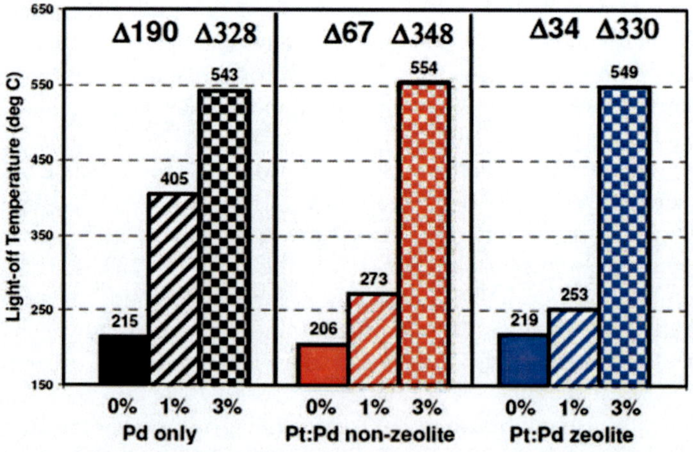

Fig. 2.24 T-80 HC light-off performance of various DOC catalyst types following 0, 1, and 3 wt% Na contamination [39]

2.5.3 *DPF Temperature Gradients During Active Regeneration*

DPF temperature gradients must be minimized during active regeneration. A consistent DPF temperature ensures that soot is oxidized uniformly and minimizes the thermal stress on the DPF during regeneration events. A large temperature gradient can lead to incomplete soot oxidation, which leads to lower soot capacity, and an increase in regeneration frequency. As discussed above, increasing regeneration frequency can hurt the DOC catalyst which must generate the heat required for regeneration. Large temperature gradients can also lead to DPF cracking due to localized thermal stress. A DPF crack will be detected by the engine and aftertreatment diagnostics and will require replacement before the vehicle can operate as intended. This will result in significant vehicle downtime.

Tong et al. have studied the effect that liquid diesel introduction onto a DOC can have on the temperature gradient of a DPF during active regeneration [40]. The authors first measured the temperature gradient without liquid diesel introduction. In this case, the maximum temperature gradient at the DPF inlet was 3 °C. Figure 2.25 shows the DPF temperature gradient measured in the same aftertreatment system with 3.1 kg/h diesel introduction onto the upstream DOC. The maximum radial gradient at the inlet of the DPF is now 40 °C. At the outlet of the DPF, the maximum gradient is even higher at 53 °C. In the presence of 3 g/L soot loading, with the same diesel injection rate, the maximum gradient at the outlet is 58 °C. The authors attributed these gradients to the non-uniform distribution of diesel fuel at the DOC inlet. In a case such as this, a mixing device can be added to increase the uniformity of diesel at the DOC inlet and minimize the temperature gradient in the DOC and DPF.

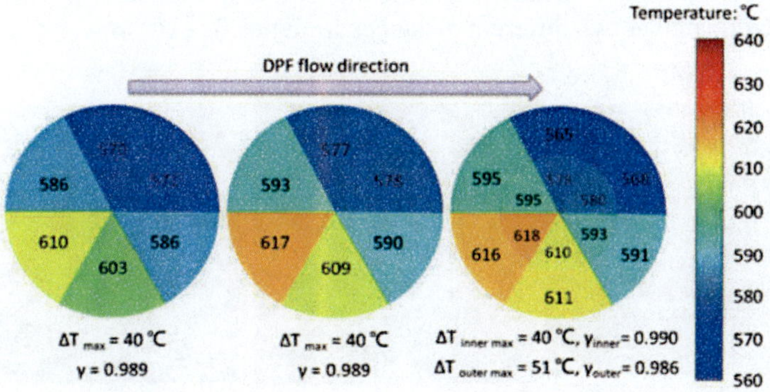

Fig. 2.25 DPF axial and radial temperature distribution with liquid diesel injection onto the upstream DOC catalyst [40]

2.5.4 Hydrocarbon Injector Selection

The spray characteristics produced by the aftertreatment HC injector significantly influence the performance of the system. It is desirable to have good atomization quality so that the droplets undergo fast vaporization and the vapor can mix well with the exhaust to avoid potential non-uniformities at the DOC inlet. Injection pressures are usually maintained between 7 and 12 bar. Typically, the performance of the system can be assessed by:

- Hydrocarbon oxidation efficiency and hydrocarbon slip across DOC: These are calculated from measurements with a hydrocarbon measurement system for different temperatures and space velocities.
- A uniformity index for temperature: This is calculated from a DOC instrumented with thermocouples at specified axial and radial locations.

One of the challenges encountered by HC doser systems is tip carboning. This is due to the combined effect of fuel leakage at the injector tip and temperature, and the exposure to engine-out soot. An additional issue is the face plugging of the DOC mentioned before. Relatively coarser sprays will produce large droplets that do not vaporize sufficiently by the time they reach the DOC and eventually contribute to the plugging of DOC channels.

A suitably designed HC injector that gives acceptable atomization quality and avoids factors leading to tip carboning is essential to ensure reliability. Figure 2.26 compares the spray structures produced by three different injector designs, acquired by a high-speed shadowgraph. The Sauter Mean Diameters for these designs at 70 mm from nozzle exit are given in Table 2.4.

The operating pressure also has a significant impact on the quality of atomization. An example is shown in Fig. 2.27 where the Sauter Mean Diameters for two different doses are compared at two different distances from nozzle exit, for various injection

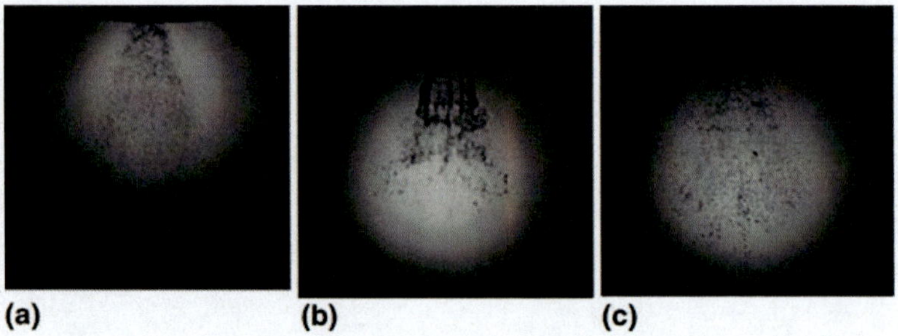

Fig. 2.26 Spray structures produced by three different HC doser designs

Table 2.4 Sauter mean diameters at 70 mm distance, for three HC injectors

Injector	A	B	C
SMD (μ)	45.4	129.8	62.7

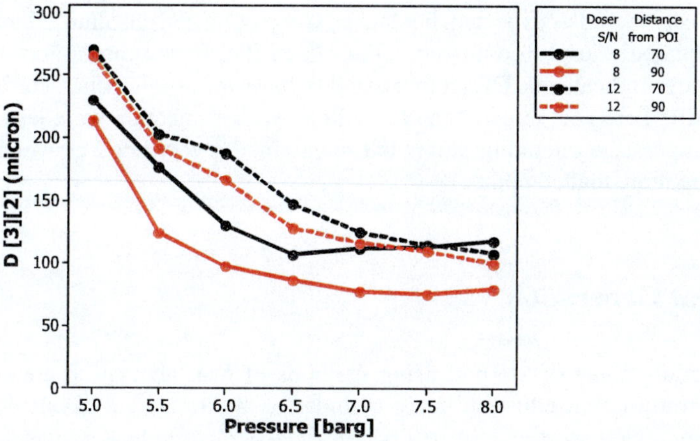

Fig. 2.27 Effect of injection pressure on SMD

pressures. An increase in injection pressure decreases the SMD but the quantitative impact differs based on the doser design. For a given doser, the drop size is lower at a distance farther away from the nozzle exit, because of the additional breakup.

Frobert et al. [41] report a study on the effect of injection of different reducing agents into a DOC. They compared Diesel fuel in cylinder late post-injection against injection in the exhaust line through an exhaust port injector, of diesel fuel, B10 (diesel fuel containing 10% of esters), and gasoline. The hydrocarbon conversion efficiency results are reported in Fig. 2.28. With the increase in space velocity, the

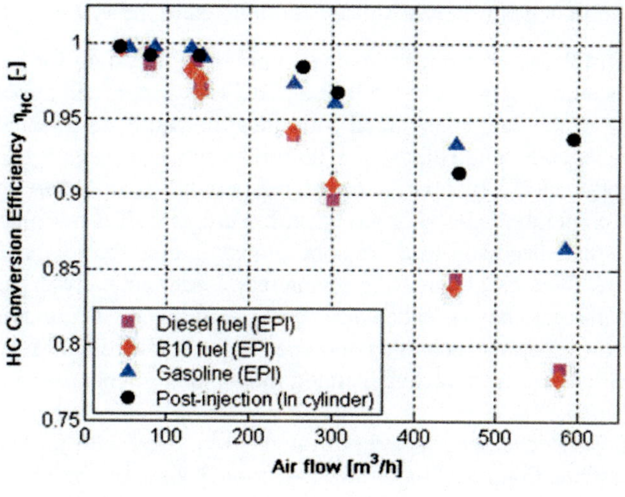

Fig. 2.28 HC conversion efficiency as a function of the flow rate [41]

conversion decreased as expected, but late post-injection and gasoline exhaust injection showed better results compared to diesel and B10 fuels exhaust port injection. Overall results indicate that Diesel fuel and B10 have very similar characteristics and impact, while Late post-injection and gasoline exhaust injection were less efficient for low temperature operating points but maintained a very good conversion level even for the more loaded points.

2.6 Heat Generation

In this section, some of the promising methods of heat generation are surveyed, different from fuel introduction in the cylinder or aftertreatment discussed before. These include the use of a dedicated burner device, electric heater, plasma burner, microwave heater, and some combinations of these. As expected, these technologies have different levels of maturity and offer specific advantages and disadvantages.

2.6.1 Burner Device

A burner can be used to raise the exhaust gas temperature on demand and control the gas temperature more closely to:

- regenerate diesel particulate filter under any engine operating point, reducing the fuel penalty associated with active regeneration
- raise the exhaust gas temperature and maintain it at values that allow efficient operation of catalysts
- remove SCR solid deposits and/or sulfur accumulated in the system.

Zelenka et al. [42] describe a full flow fuel burner DPF system for heavy-duty applications. This burner operates in the lean air/fuel ratio range, and the diesel fuel from the fuel tank of the vehicle is mixed with the atomization air injected into the combustion chamber and ignited there using ignition electrodes. Thus, by harnessing the chemical energy of the fuel, exhaust gas temperature is raised to a range of approx. 600 °C. Temperatures, exhaust gas backpressure, as well as information from the vehicle on engine speed and load are continuously monitored by the electronic control unit (ECU). The ECU starts the regeneration automatically depending on freely programmable parameters (exhaust gas backpressure, DPF loading time, etc.). Despite higher complexity, the technology is appealing to obtain reliable regeneration under all circumstances (e.g. vehicles operated in stop and go manner in the inner-cities).

Kimura et al. [43] developed a burner system to be used with an exhaust gas aftertreatment system. Figure 2.29a shows a sectional view of the aftertreatment system, while Fig. 2.29b shows the internal details of the combustion chamber. The air needed for combustion is supplied by the turbocharger and controlled by the

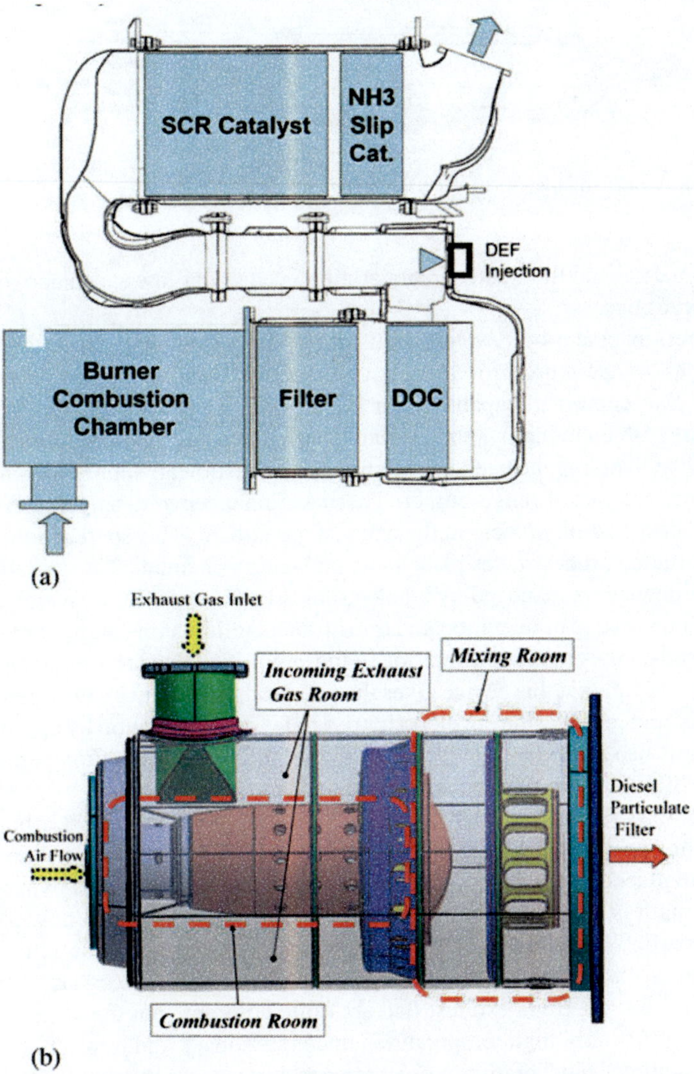

Fig. 2.29 Burner system for a medium-duty aftertreatment [43] **a** sectional view **b** combustion chamber

combustion air valve. The atomization module mixes diesel fuel delivered by the fuel pump and air from the air tanks, supplying this mixture to the nozzle at the base of the burner combustion chamber. The ignition coil elevates battery voltage to a high enough level to generate a spark across the electrodes located within the burner combustion chamber to produce a flame. This burner system was found to be effective in improving NO_x reduction in the cold FTP cycle, and in completing

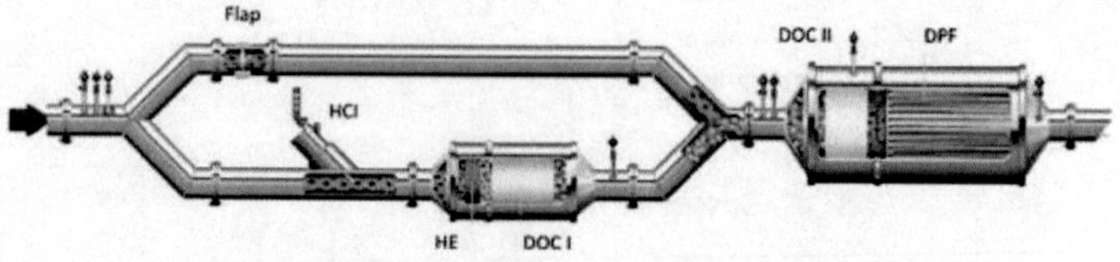

Fig. 2.30 Two-stage electrothermal supported HC conversion device [45]

the diesel particulate filter active regeneration even under low speed and low torque engine conditions.

Ramsbottom et al. [44] developed a Thermal Enhancer device for actively raising exhaust gas temperatures for the above-mentioned purposes, and this Thermal Enhancer also combined capability for HC dosing. This device is fitted axially in the exhaust path and employs the available exhaust oxygen for partial range burner functionality, avoiding the need for additional combustion air supply. When required to function as a partial range burner, Thermal Enhancer initiates an automatically controlled combustion process in the exhaust gas stream. This spark is initiated, and the diesel-fueled process takes place in a combustion chamber. The operation of the Thermal Enhancer is managed by an electronic control unit (ECU) which functions to determine the state of operation and transition handling of the burner or HC dosing modes based on the operating condition of the engine and the Aftertreatment System. The authors report that this device gives the potential to introduce a DPF regeneration 47% earlier and increase the envelope for possible soot oxidization by approximately 138%. They also state that the device increases the catalyst inlet temperatures to a minimum of 300 °C for over an additional 50% of the FTP cycle.

Baier et al. [45] report a two-stage electro-thermal supported hydrocarbon conversion system. In this system, a heating element, the hydrocarbon injector, and the first catalyst are placed in a bypass, and the mass flow ratio between this bypass and the main flow path is controlled by an exhaust flap, placed in the main flow path. Only one hydrocarbon injector is used for both catalysts. Figure 2.30 shows the details of this system. Admittedly, the processes of fuel evaporation and catalytic two-stage heat release controlled via exhaust flap are quite complex, but the system was able to achieve sufficiently high temperatures under stationary and transient conditions. The system allows regeneration even at temperatures lower than 200 °C.

2.6.2 *Electric Heater*

Heat for thermal management may also be generated with an electric resistance heater. The basic operating concept is the same as in other electric resistance heating applications. Electrical energy is passed through a resistive material which converts that electrical energy into heat energy. The amount of heat generated over a given time can be calculated with Joule's law: $Q = I^2Rt$, where Q is the heat generated, I

is the current passing through the circuit, R is the electrical resistance of the material, and t is the time. Common metallic materials for electric resistance heaters include Nichrome, an alloy of 80% nickel and 20% chromium, Kanthal, a ferritic iron-chromium-aluminum alloy, and Cupronickel, an alloy of copper and nickel. Exhaust electric heaters have been commercialized for stationary and light-duty gasoline and diesel applications, and are currently being evaluated for widespread use in heavy-duty diesel exhaust aftertreatment. For each unique application, the following three primary factors must be considered:

1. Configuration (in-line or electrically heated catalyst)
2. Voltage and power
3. Location.

Figure 2.31 shows an in-line, 24 V DC electric heater developed by Watlow for diesel aftertreatment applications [46]. The key components of an exhaust electric heater are the power supply, heater controller/switch, temperature sensors, and resistance heating element. The alternator and batteries function as the sources of power for electric heaters. The controller monitors the temperature, current, and voltage of the heater to control the heater on/off switch. The power consumption, and thus, the heat generation, is managed by modulating the pulse width of current to the heater. In this way, the heater's duty cycle, power consumption, and heat output can be easily adjusted depending on the heating requirement and available power (i.e., alternator output and available stored energy in the battery). In their study, the authors found that the use of the electric heater resulted in an average temperature increase of 75 °C at the heater outlet location in comparison to the case without electric heating [46]. The average power applied was 2.3 kW, corresponding to 85 A at 27 V DC. This is the approximate output voltage of an alternator used on a diesel engine with a 24 V DC electrical system.

Another electric heater configuration that has been proposed is an electrically heated catalyst (EHC). In this arrangement, the heater is an integral part of the

Fig. 2.31 Electric resistance exhaust heater [46]

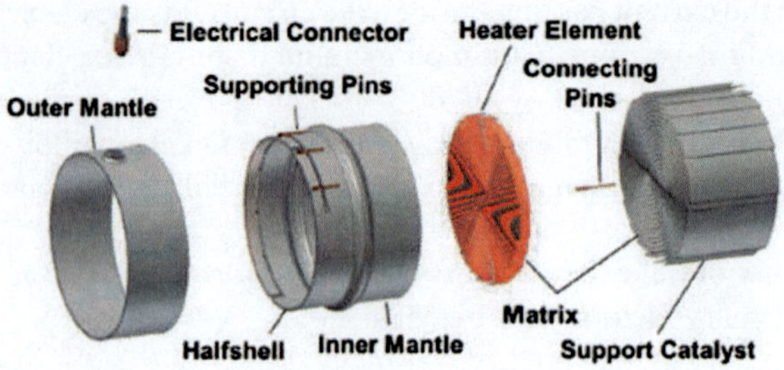

Fig. 2.32 Design of an electrically heated catalyst (EHC) [47]

catalyst instead of a standalone device. Pfahl et al. have illustrated one permutation of the EHC concept in Fig. 2.32 [47]. In their case, the electric heater has an input power of 1.8 kW, and the catalyst substrate is a conductive metallic foil that can be coated with a DOC or SCR wash coat material. The authors showed that only 1/3 of the primary fuel energy is required to generate the same energy at the catalyst position as with engine-based thermal management. The primary advantage of an EHC, in comparison to engine-based thermal management and the in-line heater discussed above, is that heat generated is transferred directly to the active catalyst sites. With the other options, heat must be generated, convected to the aftertreatment system, and then transferred to the active catalyst sites, a less efficient process.

The voltage and power output of the electrical heating system must also be considered. In the past, exhaust heating systems were limited to 12 V DC, since this was the prevailing electrical system voltage of light-duty automobiles. This voltage severely limits the practical power output of the exhaust heater. For example, a 4 kW exhaust heater would require a current of 333 A at 12 V. However, the same 4 kW heater would only require a current of 167 A at 24 V. Furthermore, Culbertson et al. state that 48 V would be preferred over 24 V from an electrical standpoint because this would further reduce the current demand and the size of the requisite power cables [46]. Their heater is easily adaptable to either 24 or 48 V DC. Today's heavy-duty diesel engine electrical systems are either 12 or 24 V DC, but 48 V DC systems are being considered, and hybrid applications will require even higher electrical system voltages, exceeding 300 V DC.

The nominal heater power will depend on the specific application requirements. Thus far we have discussed heaters with 1.8 and 4 kW outputs. But, many other power output options have been proposed in the literature. For example, Sharp et al. have proposed a 5 kW in-line heater in their study of diesel aftertreatment solutions to meet proposed ultra-low NO_x regulations [48]. Also, the authors have suggested that a 2 kW electrically heated catalyst could be combined with the diesel fuel introduction system discussed in the section titled Aftertreatment Fuel Introduction. Figure 2.33 shows a schematic of this proposed thermal management device. In this case, electric power is used to directly heat the catalyst to the diesel light-off temperature before fuel is introduced to further heat the aftertreatment system.

Finally, the location of the electric heater in the exhaust aftertreatment system is important. Its location will depend on the primary intended function. If DPF

Fig. 2.33 Electrically heated catalyst (EHC) assisted catalytic fuel combustion [48]

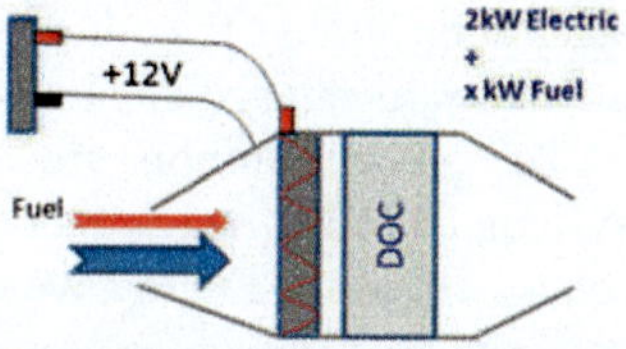

regeneration is the primary objective, the exhaust heater will be positioned before the DOC catalyst. However, if improving low-temperature NO_x conversion is the primary objective, it can be advantageous to place the exhaust heater directly before the SCR catalyst. Without heat loss to the DOC and DPF, the SCR catalyst can be heated more rapidly. As discussed above, the benefit of the electrically heated catalyst is that the active sites can be directly heated. It may be mentioned here that there is a need to increase the ampere-hour rating of the battery to accommodate this.

2.6.3 Microwave Heater

Another method to generate heat in a diesel aftertreatment system is with microwave radiation. Figure 2.34 shows a schematic of such a system proposed by Ning and He [49]. It is known that diesel PM readily absorbs microwave radiation, increasing the kinetic energy within the molecule, and ultimately manifesting as heat. Microwave radiation is generated by a magnetron and is directed to the DPF by the walls of the aftertreatment system, which are not microwave-absorbing. In the system proposed by Ning and He, glass is used to isolate the microwave source from the exhaust stream [49]. Once heated to the desired temperature by the microwave source, the PM is then oxidized to CO_2 by either O_2 and/or NO_2, as in other DPF regeneration strategies. Unlike PM, cordierite, the material of which DPFs are predominantly constructed, does not absorb microwave radiation. This allows the radiation to target the PM on the filter, without also heating the filter substrate, significantly reducing the energy requirement for PM oxidation. On the other hand, another popular DPF material, silicon carbide, does absorb microwave radiation. Thus, oxidation of PM on silicon carbide DPFs requires additional microwave radiation energy in comparison to PM on a cordierite filter. An advantage of the silicon carbide DPF's microwave

Fig. 2.34 Schematic diagram of a diesel aftertreatment system utilizing microwave radiation for DPF regeneration [49]

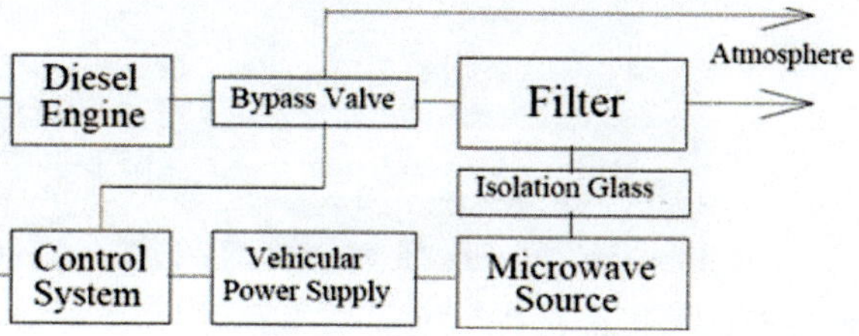

absorption is that the filter itself can be heated and used as a source of heat for downstream catalyst elements such as the SCR.

The present authors are not aware of a commercialized microwave radiation system, although research continues today into this technology. Challenges include robustness of the microwave source, microwave cavity design, power consumption, and regulatory concerns around mobile microwave systems.

2.6.4 Plasma Burner

A final option for generating heat in an exhaust aftertreatment system is the use of a thermal plasma [50–52]. An example of this is the rotating arc thermal plasma burner developed by Pyun et al. [51]. A schematic of their plasma burner is shown in Fig. 2.35. The primary components are the power source, high voltage (HV) and low voltage (LV) electrodes, the plasma reactor, the diesel fuel introduction system including the fuel atomizing nozzle, the fresh air inlet and pump, and the flame holder. Air and diesel fuel are introduced into the plasma reactor in a swirling flow pattern to promote mixing of the mixture. Fuel can be introduced at the high voltage electrode in small-scale burner applications, or it can be introduced through the ground electrode through narrow holes for large-scale applications. In this instance, the narrow holes function as nozzles. Continuously discharging plasma created between the high and low voltage electrodes and the ground electrode ignites the air/fuel mixture. The heat created by the combustion of the diesel fuel is then convected to the downstream filter or catalyst elements. The plasma burner can be contained in the center of the exhaust stream, or it may be perpendicular to the exhaust stream, as suggested by Lee et al. [52]. Challenges associated with using thermal plasma for heat generation include the

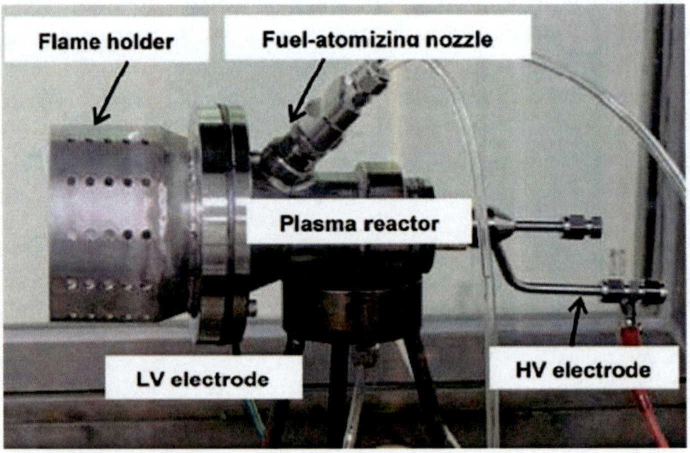

Fig. 2.35 Schematic of a plasma burner [51]

need for fresh air introduction, the cost associated with power electronics for control, the power consumption, and maintaining stable plasma. As of writing, a plasma burner has not been commercialized for diesel exhaust aftertreatment applications.

In addition to the thermal plasma described above, non-thermal plasma has also been suggested for application in diesel aftertreatment systems [50]. As its name implies, non-thermal plasma does not generate significant heat. Instead, non-thermal plasma can oxidize NO to NO_2 and simultaneously reduce emissions of CO and HC species, via the creation of O and OH radicals which interact with exhaust gas species. In contrast to the heat generation of thermal plasma, non-thermal plasma reduces the need for thermal management. The formation of NO_2 increases passive soot oxidation, reducing the frequency of active DPF regeneration, which requires thermal management. And, because NO_2 creation by non-thermal plasma is not dependent on exhaust temperature, this NO_2 can be used to improve SCR NO_x conversion without the need to rapidly increase the DOC temperature through thermal management. Feasibility for cost-effective application to vehicle engines can be worked out based on the volumes.

2.6.5 *Aftertreatment Heat Retention*

Modern diesel aftertreatment is a complex system with several catalytic and non-catalytic components. Retaining the heat of exhaust gases is an important strategy to achieve the desired temperatures at the catalysts. This, in turn, has a significant influence on steady-state and transient emissions performance due to the impact on catalyst activity. This section will describe the most common approaches for retaining exhaust heat and associated considerations. First, the effect of insulation and air gaps will be described, and this will be followed by an investigation into close-coupled and pre-turbo aftertreatment systems. The use of phase change materials will be described in the final subsection.

2.6.6 *Insulation and Aftertreatment Design*

Most common engineering methods for heat retention include the use of:

- External insulation: insulation applied outside the exhaust flow path, such as removable insulation blankets
- Internal Insulation: Insulation that has some exhaust gas exposure
- Airgaps: using air as the insulating medium.

These passive measures of heat retention have been previously demonstrated in catalytic converters for light-duty gasoline engines, and these studies are forerunners to the approaches followed in longer modern diesel aftertreatment systems. Robertson [53] reported that the warm-up time of a catalytic converter can be reduced by effectively insulating the exhaust pipe to produce an increase in the temperature of the exhaust gas entering the converter, and consequently the emissions of HC and CO were reduced. The author used a single bend and cross-over pipe configurations and compared results from ceramic insulation against fibrous moist-pack insulation. His studies indicated the fibrous insulation to be effective in conserving thermal energy while the ceramic insulation was not effective due to its high thermal conductivity. Dual wall pipes that use air as an insulation material have been used in aftertreatment applications for efficient heat retention, particularly during transient cycles. De Sousa [54] reports a dual wall pipe with a thin inner pipe (that lowers thermal storage capacity) and a structurally stronger thicker outer pipe. They propose the use of circumferential convolutions in the inner pipe to increase its durability. Lee et al. [55] conducted a study using a design that had both dual-walled air-gap exhaust manifold and dual-walled air gap pipe. Their design was shown to shorten the light-off time in the converter and reduce emissions. Also, a reduction in surface temperatures was achieved due to the insulation effect of air.

Modern diesel aftertreatment systems use a combination of dual wall airgap and insulating materials to achieve the desired level of heat retention. It is desirable to use predictive models to achieve the right balance between efficiency and cost. These models may include the effect of internal and external flow, forced and natural convection, conduction through walls, insulation and mats, and radiation between surfaces. One-dimensional models with different levels of detail have been effectively used for assessing various scenarios [53, 55]. Subsequently, integrated system-level models for overall aftertreatment modeling framework comprising dedicated models for single or dual wall pipes, oxidation catalysts, wall-flow particulate filters, and selective catalytic converters have also been developed [56]. More complex CFD models that include chemical reactions are also reported [57] for assessing transient emission performance.

Figure 2.36 shows the results from thermal imaging of a typical diesel aftertreatment system with DOC, DPF, decomposition tube, SCR, and ASC. The highest heat loss is from uninsulated areas, including clamps and sections of the can.

Figure 2.37 shows a comparison of different insulation levels on the temperature drop in the aftertreatment system shown in Fig. 2.36 for an FTP cycle. Insulating the system from the DOC inlet to the SCR outlet improves heat retention, but the best results are obtained when the full system is insulated. In comparison with baseline, the different levels of insulation increased NO_x conversion efficiency in the cycle by up to 30% without any other means of thermal management.

Though the effectiveness of insulating the system with insulation of appropriate thermal properties and thickness has been well known, the choice of insulation material and its integration should consider multiple factors including:

- Long- and short-term temperature exposure limits should be higher than what is expected in the application
- The maximum thermal conductivity within the operating range should be sufficiently low (typically two orders of magnitude lower than metals)

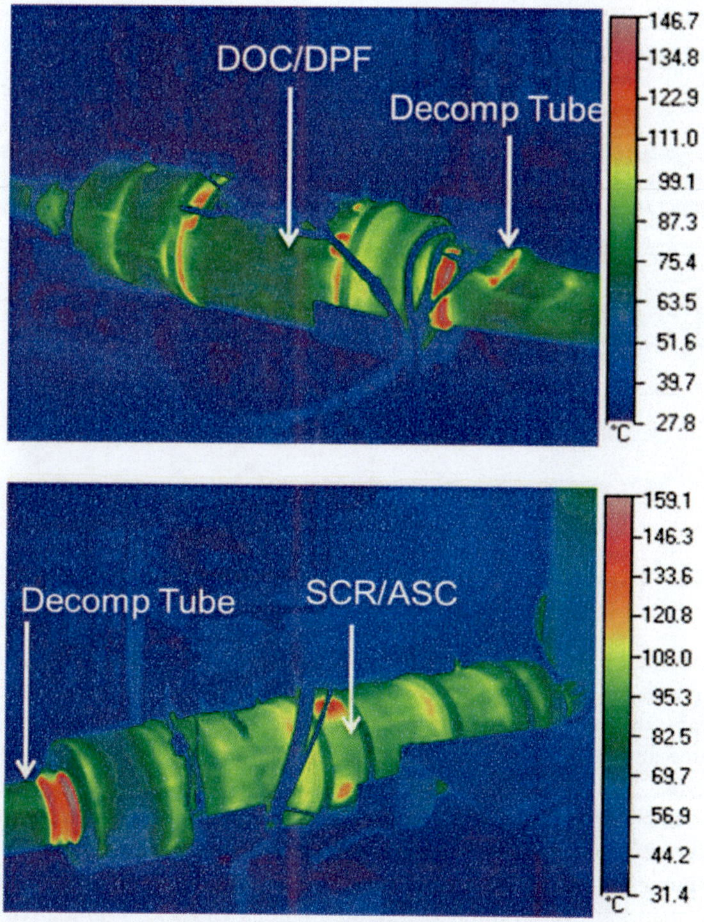

Fig. 2.36 Thermal imaging results of a typical aftertreatment system

- The vibration levels experienced within the application should not cause damage or material loss
- Internal insulations should fit tightly within the installation space to provide efficient heat retention
- External insulations should be able to withstand harsh weather conditions, water, and debris.

2.6.7 Close-Coupled and Pre-turbo Aftertreatment Designs

Modern and next generation aftertreatment systems need to not only meet stringent emission regulations but also address critical non-emission needs, such as weight

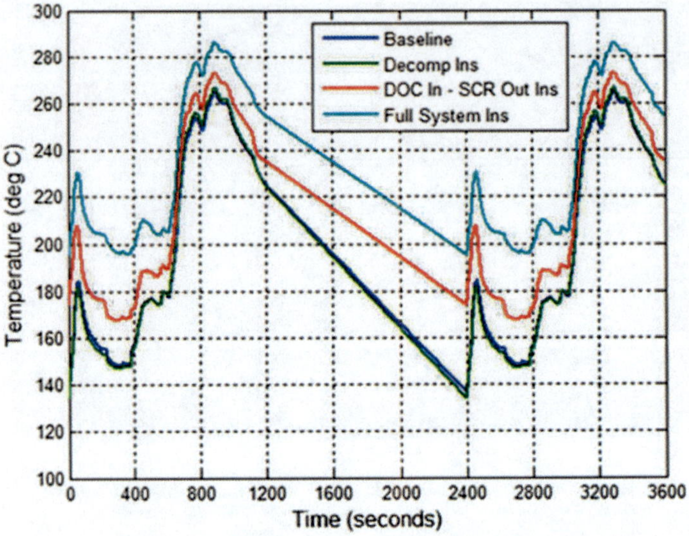

Fig. 2.37 Temperature drop in a typical aftertreatment system with different insulation levels

reduction, space constraints, and ease of maintenance. Consequently, space claims are getting reduced. Su et al. [58] report the details of the development work done for an advanced compact aftertreatment system that meets advanced regulations with 60% volume and 40% weight reductions compared to standard modular aftertreatment systems. The reduction in space claim certainly helps with heat retention. Locating the catalysts closer to the turbocharger outlet in a close-coupled position is an option to effectively use the exhaust heat, avoiding additional temperature losses within a downpipe. Host et al. [59] investigated the thermal performance of exhaust system hardware options using simulations and used cumulative sensible enthalpy as the measure of the catalyst light-off improvements. A large relative benefit of close-coupling the catalyst was observed, but additional benefits were realized with improved wastegate opening/flow area, directed wastegate flow, and scroll configurations with the reduced wetted area. Another related option to achieve rapid light off and efficient conversion of pollutant emissions is the use of a small, low thermal inertia catalyst upstream of the turbocharger. A larger catalyst placed downstream can complete the conversion of any remaining pollutants. Joergl et al. [60] list the following requirements for the pre-turbo catalysts: packageability into exhaust manifold area, withstand thermal stress without material loss, minimal damping of pressure fluctuations, low thermal inertia, minimal pressure drop, and elevated temperature limits. They investigated pre-turbo catalyst DOCs of different sizes to understand the effects on engine steady-state and transient emissions and performance with the focus on the use of larger substrates and the implication for the engine's turbocharger. The study showed the advantages of the pre-turbo catalyst for reducing cold start HC and CO emissions in all cases. The authors also discussed the implications of choosing small, medium, or large size pre-turbo catalysts on emissions performance with and without a post-turbo DOC, and the impact of the catalyst size on system response times. Subramaniam et al. [61] reported a feasibility study on a pre-turbo catalyst-based aftertreatment system for non-road, large-bore applications that operate in a largely steady-state manner. They concluded that the pre-turbo aftertreatment system offers benefits in terms of compact, cost-effective emissions reduction, with the additional advantage of reduced fuel consumption. However, as stated before, the durability of close-coupled or pre-turbo catalyst aftertreatment systems demands special attention from a durability standpoint. In addition to catalyst durability, improved durability of the mat used in the packaging of catalytic substrates needs to be ensured. Watson and Gonzalez [62] report an engineered mat mount consisting of a multicomponent construction of intumescent and non-intumescent mats. The pressures for their engineered mat solution were between the intumescent and non-intumescent mat mounts, and the engineered solution was expected to be better at holding the substrate at lower temperatures than the traditional intumescent or non-intumescent mats. This technology also showed improved erosion resistance and hot vibration performance.

The cost of pre-turbo catalyst is relatively less and its impact on back pressure though marginal must be taken into account.

2.6.8 Use of Phase Change Materials

In automotive applications, it is desirable to have the means to retain heat in catalysts between engine-off and engine-on events and to reject heat during high-temperature operations to avoid catalyst durability issues. The duration between engine off and on events could often be of the order of several hours. The use of vacuum insulation for heat retention is well demonstrated in several general applications and can be extended to aftertreatment systems. Benson et al. [63] proposed a novel means to achieve variable thermal conductance in vacuum insulation by using metal hydrides. The pressure of the hydrogen is a predictable function of the hydride temperature and the gas is readily reabsorbed whenever the hydride can cool. In their design, the temperature of the metal hydride is closely controlled, and with changes in the temperature, the hydride reversibly absorbs or desorbs hydrogen to decrease or increase the thermal conductivity. A logical extension relies on thermal energy storage through latent heat. Burch et al. [64] used a high-temperature phase-change material within the vacuum insulation and realized the ability to maintain catalyst temperature above light off for 10 h. Burch and Biel [65] provide more details of a design that features thermal energy storage and heat retention options as shown in Fig. 2.38. In this vacuum-insulated catalytic converter (VICC) system, vacuum insulation is used around the monoliths with metal bellows and thin sections of the uncoated monolith at the ends to block heat loss by conduction and radiation. A metal or salt phase-change material (PCM) is packaged between the monoliths and vacuum insulation. To prevent overheating of the converter during periods of long,

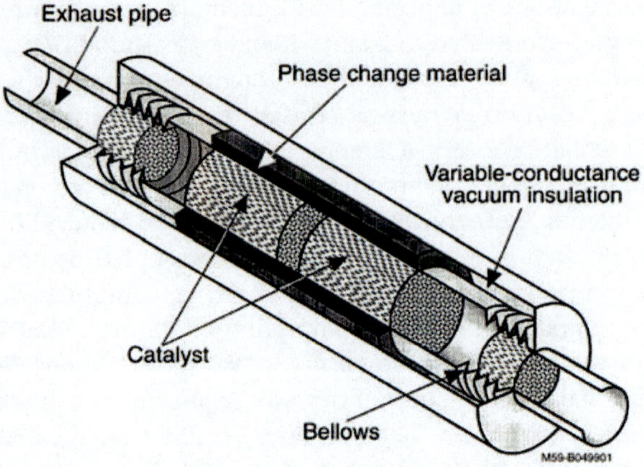

Fig. 2.38 Major features of vacuum insulated catalytic converter [65]

heavy engine use, a few grams of metal hydride charged with hydrogen are attached to the hot side of the vacuum insulation. When a critical temperature is reached, the hydride releases about 1/40th of an atmosphere of hydrogen into the vacuum space. This hydrogen increases the effective thermal conductivity of the insulation by more than 100 times, allowing heat to flow out of the converter. Early converter prototypes using this concept demonstrated the excellent heat retention capability of this approach, requiring 18–24 h to cool from 600–250 °C versus 20–30 min for conventional converters, but questions remained on durability and real-world emission benefits. Burch and Biel [65] demonstrated that durability improvements can be made to the design with excellent thermal and emission reduction performance. Further, a significant increase in average converter temperatures resulting in a significant decrease in emissions was shown to be possible in a cycle representative of real-world driving.

Gaiser and Seethaler [66] investigated two arrangements of a Latent Heat Storage (LHS) device. In the sequential arrangement, the LHS device was placed upstream of the catalyst. During normal engine operation, the LHS is heated up by the exhaust gas and stores both sensible and latent heat. During engine-off time the latent heat capacity maintains the temperature of the LHS at the melting temperature of the phase change material. At cold-start, the exhaust gas passes the LHS and thereby is heated up to above the light-off temperature of the catalytic converter. In the integrated arrangement, the LHS and catalyst were integrated into one common substrate. One half of the channels in the combined substrate are filled with phase change material, while the other half of the channels are passed by the exhaust gas. The sequential arrangement shows significant benefits to catalyst light-off resulting from fast heat up. The integrated setup showed the best benefits as it offers an immediate catalyst light-off right from the start of the engine. Recently, Hamedi et al. [67] computationally investigated different materials for PCM additives and catalyst substrates to improve the thermal behavior of a light-duty diesel aftertreatment system. In their design, an annular layer of PCM was integrated around the oxidation catalyst. They found that using a PCM-graphite mixture and substituting a ceramic catalyst substrate with a comparable metallic substrate offer improvements to the aftertreatment system thermal behavior. Jankowski and McKluskey [68] provide a review of phase change materials for thermal buffering of vehicle systems. They include a comprehensive overview of phase change materials covering the relevant operating range for different vehicle applications and conclude that there is potential for making significant improvements in both emissions reduction and overall energy efficiency by further exploration of PCM thermal buffering on vehicles.

2.7 Exhaust Heat Management

This section will highlight three important exhaust heat management considerations, namely waste heat recovery, accounting for the impact of DEF dosing, and maintaining the maximum exhaust temperature threshold. Though internal combustion engine technology has advanced significantly over the past several decades, a substantial portion of the fuel energy is still lost as heat to the surroundings through the exhaust. Several waste heat recovery technologies of varying levels of maturity and complexity are available, and the first subsection will touch on the promising candidates and their implications for thermal management. In aftertreatment systems equipped with urea SCR technology, temperature reduction can happen due to DEF dosing leading to undesired performance. Heat management considerations relevant to this issue will be discussed. In modern aftertreatment systems equipped with catalysts, a maximum exhaust gas temperature threshold may exist, or a lower exhaust temperature may be preferable under some conditions. This aspect will be covered in the subsection on the maximum exhaust temperature threshold. Finally, a brief description is provided of the thermal management considerations in advanced propulsion systems using electric or fuel cell technology.

2.7.1 Waste Heat Recovery

A waste heat recovery system that converts engine exhaust heat to mechanical or electrical energy is quite attractive from an efficiency standpoint. The research into waste heat recovery in the 1970s was prompted by increasing fuel costs and diminishing fuel supplies. Leising et al. [69] investigated various waste recovery options for truck engines. The options included were preheating, regeneration/recuperation, turbocharging, turbo compounding, and Rankine engine compounding, and the authors drew conclusions on the relative effectiveness of these options.

In turbo compounding, the energy recovered from the exhaust is used to increase engine output, and it is a key enabler for power density growth and fuel economy. However, this technology also increases pumping losses and lowers exhaust gas temperatures. Mechanical turbo compounding involves placing a turbine in the engine exhaust and mechanically coupling it to the engine crankshaft using the gear

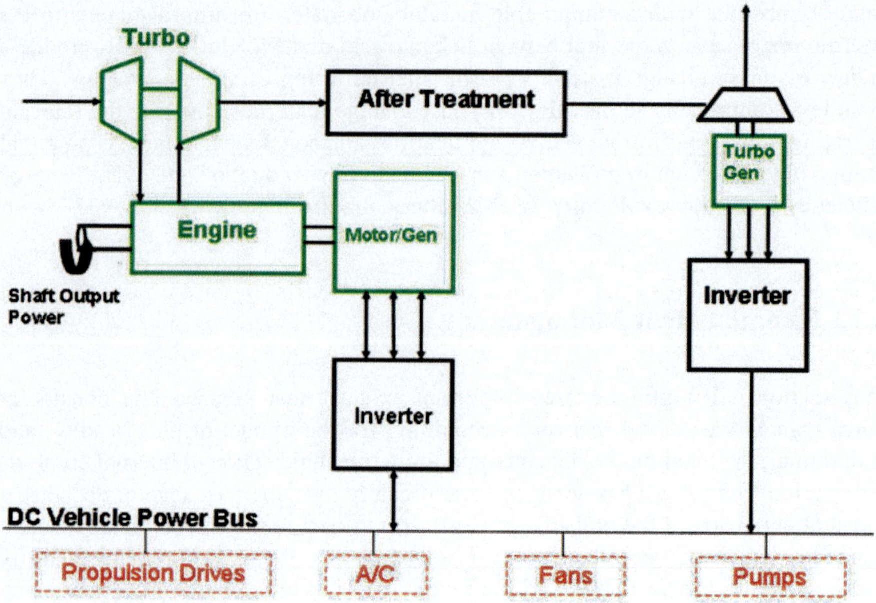

Fig. 2.39 A proposed electrical turbo compounding architecture [72]

train. The technology is commercialized (e.g., see Tennant and Walsham [70] for a review of early work), but the coupling of the turbine to the crankshaft introduces additional complexity. Electric turbo compounding avoids the need for direct mechanical coupling and uses an electric generator/motor system [71, 72], but this concept needs further technological advances to be competitive. Figure 2.39 shows a typical architecture. Hountalas et al. [73] conducted an analytical study using an engine simulation model of a heavy-duty diesel engine with mechanical and electrical turbo compounding and its impact on engine performance and emissions. Though mechanical and electrical turbocharging reduced the overall engine BSFC, relatively small improvements were observed at low and part engine loads. A reduction in NO_x emissions and an increase in soot were also observed.

The use of thermodynamic cycles to extract heat energy from the exhaust gases is another option and the use of the Organic Rankine Cycle is highly promising [74, 75] since it can give a larger reduction in BSFC with lower sensitivity to engine load. Best efficiency and maximum power output may be obtained by using an organic fluid that has good heat transfer capacity, thermal stability, and low viscosity, and examples of working fluids include R134a and R245fa. Figure 2.40 shows a layout proposed for a heavy-duty truck, that recovers heat from EGR and exhaust gases [76]. Indeed, the system is complex and requires many subcomponents. Efforts are needed to reduce the cost and complexity of the system, select the appropriate working fluids considering factors including safety and cost, reduce the volume and weight of the system, and develop transient controls for drivability.

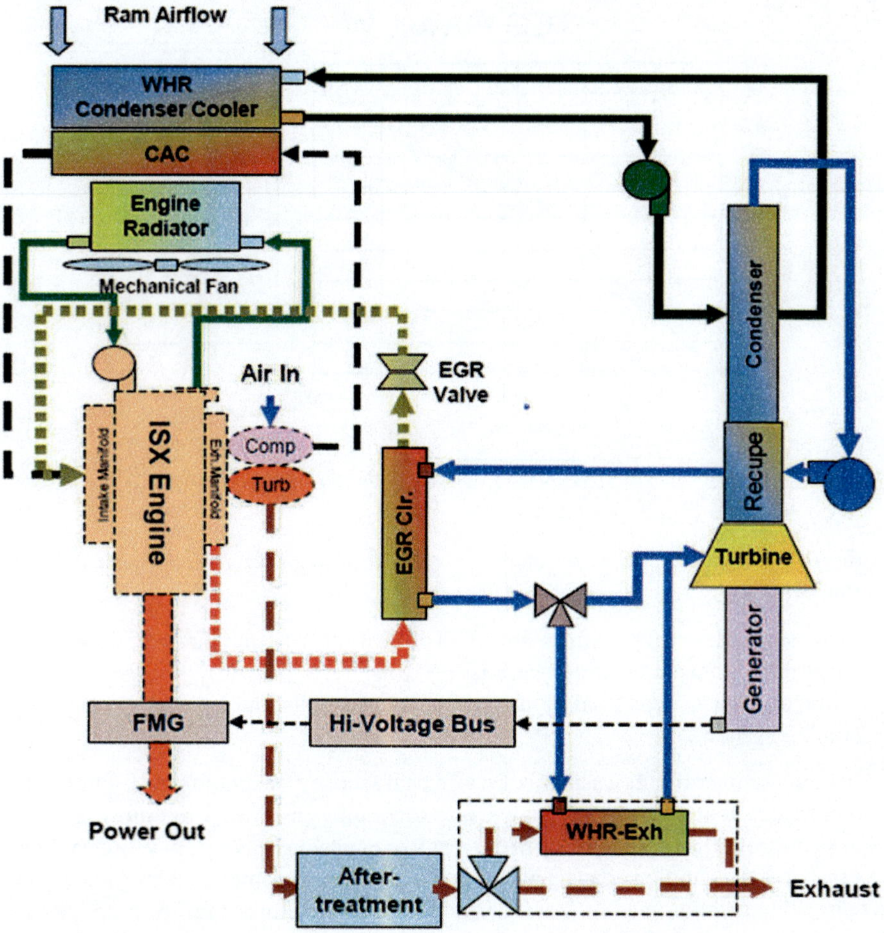

Fig. 2.40 Proposed schematic of Organic Rankine Cycle used in a heavy-duty truck [76]

The use of a thermoelectric generator (TEG) to generate electric power by utilizing temperature gradients within the engine and aftertreatment components has been explored but is yet to be used on a commercial scale. This power is used to supply electricity to components in the system. Kushch et al. [77] describe a system that was developed for heavy-duty diesel trucks. It is comprised of 72 thermoelectric modules, capable of producing 1 kW of electrical power at 12 V DC during nominal engine operation. The device utilized an engine coolant to cool the cold side of the modules, while the hot side of the modules was placed on the surface of a heat exchanger in the exhaust. The system output was found to be strongly dependent on engine load and less on engine speed as shown in Fig. 2.41.

Research into suitable thermoelectric materials and optimal heat exchanger designs are the main areas of focus in this technology. According to LaGrandeur

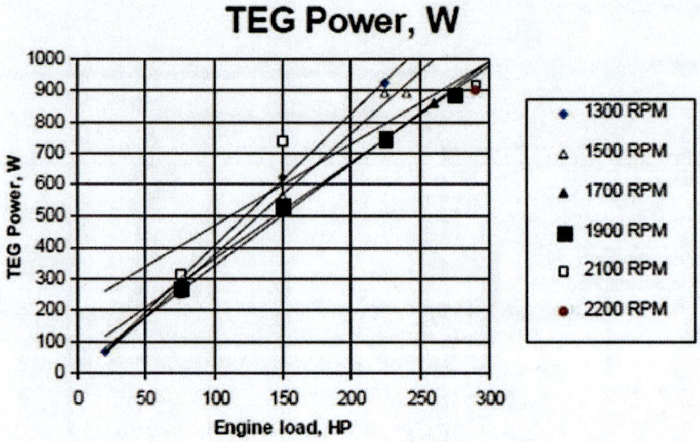

Fig. 2.41 Thermoelectric power output as a function of the engine load and speed [77]

et al. [78], thermoelectric waste heat recovery from diesel engines is most focused on the following:

- Increasing performance from lower (100–250 °C) to medium (250–500 °C) temperature thermoelectric materials.
- Extracting waste heat from more components than just the exhaust system (e.g., cooling system).

In their comparative assessment of waste heat recovery technologies, Eichler et al. [79] conclude that Organic Rankine Cycle technology is one of the most efficient ways to improve the fuel economy of current low emission heavy-duty vehicles. They assess that this technology can bring about 5% BTE improvement, making it superior to competing technologies such as Thermoelectric generators or turbo compounding that give only $\sim < 3\%$ BTE.

2.7.1.1 Impact of DEF Dosing

In urea SCR technology, water vaporization (Eq. 2.8) and thermolysis (Eq. 2.9) processes are endothermic, while the hydrolysis reaction (Eq. 2.10) is exothermic. Energy extracted for water vaporization and thermolysis can bring down the local gas temperature, while spray impingement on pipe or mixer surfaces can reduce wall temperatures and form solid deposits. It is also important to ensure good thermal insulation on mixers to prevent significant heat loss.

Koebel and Strutz [80] presented an energy analysis of the thermal decomposition of solid urea and urea solutions applicable to automotive $DeNO_x$ systems. Key findings from their analysis are summarized in Table 2.5.

It is seen that using 32.5% solution of urea in water raises the heat requirement for thermolysis by a factor of about 2.7 compared to dry solid urea due to the need

Table 2.5 Heat required to decompose urea (kJ/mol) [80]

	Thermolysis only (final temperature = 500 K)	Thermolysis only (final temperature = 600 K)	Thermohydrolysis (final temperature = 500 K)	Thermohydrolysis (final temperature = 600 K)
Urea (solid)	203.4	213.4	106.5	116.2
50% urea	360	382	263.1	284.8
32.5% urea	541.5	576.3	444.6	479.1

for evaporating water. However, the heat requirement for combined thermolysis + hydrolysis is smaller than that for thermolysis alone, since hydrolysis is exothermic. The authors concluded that gas-phase temperature drop can be modest (10–20 K). However, in modern systems that desire closer control of dosing, localized gas cooling can have an impact on the accuracy of exhaust gas temperature sensor measurements used to control dosing.

When spray impinges on a solid surface that is maintained below the film boiling temperature of the urea-water solution deposits can form. The energy extracted for water vaporization and thermolysis may be drawn from the surface, cooling it down further. Due care must be exercised to ensure the desired outcome of spray impingement on walls since solid deposits can form on walls that can be easily cooled. It must however be noted that there is no consensus on the value of this critical temperature due to the complexity of spray-wall interaction processes. Though wall temperature significantly influences the outcome of the drop-wall impingement, the influence of factors including impact velocity, droplet size, droplet temperature, physical properties of the liquid, and wall roughness are also significant. Munnannur et al. [81] conclude that flow recirculation, thermal gradients, and spray impingement on solid walls have major effects on deposit formation propensity. They found that:

- Recirculation zones in the direct path of the spray should be avoided to prevent the trapping of small particles. This consideration is especially important while using a spray with fine droplets and near low-temperature surfaces.
- In cases where an actively cooled injector must be used, design options that can reduce the thermal gradients between the cooled surface and gas, but without exceeding the temperature limits specified on the injector tip need to be used. The use of a thermal insulator was found to be beneficial in this respect.
- Direct impingement of spray on solid surfaces maintained below a critical temperature is not desirable. This calls for effective insulation of decomposition reactors and pipes that are directly exposed to varying ambient conditions.

Since deposits could be found at several locations including the injector tip, pipe walls, and mixers, the design of the aftertreatment system should ensure that high-temperature exhaust can reach all locations with deposit accumulation.

Schiller et al. [82] proposed a non-dimensional Excess Energy Ratio (EER) to identify operating conditions that would most likely produce deposits. EER is calculated as the ratio of energy available in the exhaust to the energy required to evaporate water from an initial temperature of 70 °C, and a low EER indicates a higher risk of deposits. The equation did not include the energy required to decompose urea content and additional terms are needed to account for it. For example, Southwest Research Institute (SWRI) proposed a corrected equation as follows [83]:

$$EER = \frac{m_{exh} C_{p_{exh}} (T_{exh} - 135)}{m_w C_{p_w} \Delta_1 + m_w h_{fg_w} + m_s C_{p_s} \Delta_2 + m_u C_{p_u} \Delta_2 + m_u h_{fg_u}} \quad (2.11)$$

where

u urea
s steam
w water
Δ_1 (100 – 70)
Δ_2 (135 – 70).

However, such lumped energy calculations do not account for local cooling effects. Detailed 3D simulations or physical tests may still be needed to optimize designs. Due to packaging constraints, efficiency requirements, and operating conditions, the formation of solid deposits is inevitable in certain applications and the use of control-based strategies that limit dosing or initiate deposit removal through one of the thermal management strategies discussed herein will be needed. The typical control strategies for deposit removal involve:

- Use of impingement tables to reduce the amount of DEF injected at low temperatures, and use of a controller to initiate DEF injection only at a specified temperature.
- Increasing the exhaust gas temperature to high values (500–600 °C) for a predetermined duration to help decompose the accumulated solid deposits.

2.7.2 *Maximum Exhaust Temperature Threshold*

As mentioned, action may be taken to lower exhaust temperatures under some operating conditions. Examples of these include when:

- protection of aftertreatment components is required
- safety concerns require limiting exhaust temperatures
- a lower exhaust temperature could improve tailpipe emissions
- sublimation of certain transition metals in aftertreatment catalysts need to be minimized.

As discussed in the Need for Aftertreatment Thermal Management section, one common type of catalyst used in diesel exhaust aftertreatment systems is vanadium-based SCR. These catalysts contain both vanadium and tungsten oxides, which become volatile at elevated temperatures. Figure 2.42 shows reactor-based vanadium

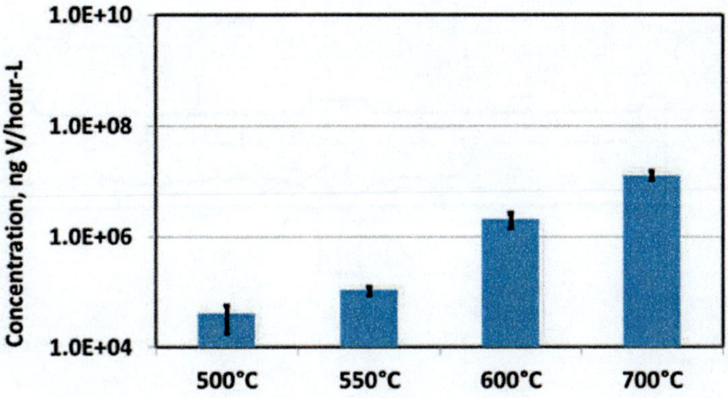

Fig. 2.42 Reactor-based release of vanadium as a function of operating temperature for a state-of-the-art V-SCR catalyst

release from a state-of-the-art vanadium SCR catalyst. The "L" unit in the denominator corresponds to the liters of vanadium SCR, and the "hr" corresponds to the duration the V-SCR is maintained at a given temperature. Vanadium volatility increases exponentially as the temperature increases from 500 to 700 °C. And, while it's not shown here, the same is true of tungsten oxides. Liu et al. have previously reported emissions of both vanadium and tungsten from state-of-the-art V-SCR catalysts as a function of exhaust temperature and gas composition [84]. The authors have shown that the release of vanadium and tungsten oxides is sensitive to water vapor concentration, in addition to exhaust temperature. The release of these metals can be harmful to both human health and the environment and guidance and regulations exist in the US, China, and Japan to prevent the release of vanadium from diesel aftertreatment systems. Furthermore, the US National Institute for Occupational Safety and Health (NIOSH) has recommended an exposure limit of 0.05 mg V/m^3 (15-min ceiling) for vanadium oxide(s) dust [85]. To prevent the release of vanadium and tungsten, it is necessary to limit the maximum temperature to which V-SCR catalysts are exposed. Another reason it may be advantageous to limit maximum exhaust temperatures can be seen in Fig. 2.2. It is clear that NO_x conversion reaches a maximum between 225 and 400 °C, and that conversion gradually decreases as the temperature increases beyond 400 °C. This trend applies to all SCR technologies on the market today. Thus, lower tailpipe NO_x emissions can be achieved by limiting the maximum exhaust gas temperature to 400 °C.

Safety concerns are another potential reason that a maximum exhaust temperature threshold may be required. US regulation 40 CFR § 86.1810 states that any device installed in a new motor vehicle to meet emissions regulations "shall not in its operation, function or malfunction result in any unsafe condition endangering the vehicle, its occupants, or persons or property near the vehicle." High exhaust temperatures can pose a safety risk to bystanders and may also pose a fire hazard to property close to a vehicle. One example of a potential fire hazard is non-road equipment operating over dry vegetation which may come into contact with hot exhaust gas or

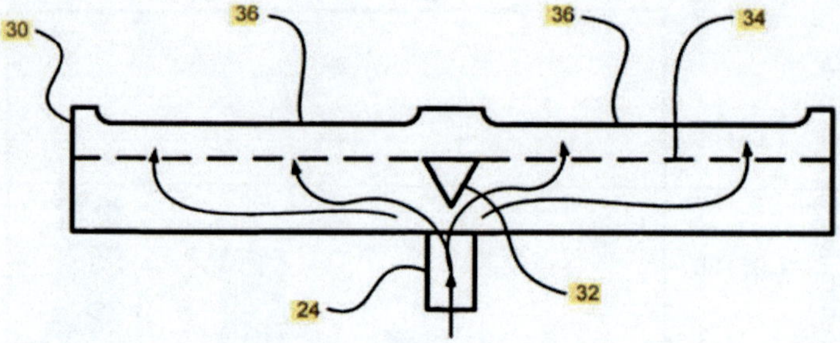

Fig. 2.43 Example of an exhaust diffuser. Arrows indicate the direction of exhaust flow [US8056328B2]

hot exhaust aftertreatment components located underneath a vehicle. This problem must be taken into account while designing the layout. Four methods are generally used to limit maximum exhaust gas temperatures. The first is similar to *Active Duty Cycle Adjustment*, previously discussed as a method to increase exhaust temperatures. Similarly, an operating condition (speed, fueling rate, etc.) can be selected which provides the required torque and also limits the exhaust gas temperature. The second to limit the temperature of diesel exhaust is with an engine derating. When derating, the maximum power output of the diesel engine is limited, reducing fueling and heat generation due to combustion. However, a derating has a significant impact on the vehicle mission, and, depending on its severity, can be mission disabling. For example, a semi-truck that is derated 80% will only have 20% of its maximum horsepower available and will no longer be capable of cruising at interstate speeds. For this reason, the derating strategy should be used as a last resort. The third method is via Waste heat recovery, discussed in the previous section. The WHR heat exchanger removes heat from the exhaust gas, thereby limiting the maximum temperature. Especially in applications with average exhaust gas temperatures above 400 °C, WHR could be a viable strategy for reducing exhaust gas temperatures. The final method that can be used to limit the temperature of exhaust gases leaving the aftertreatment system is an exhaust diffuser. US Patents US8056328B2, US7779961B2, US7604093B2 describe in detail potential exhaust diffuser designs. A diffuser mixes exhaust gas with ambient air and releases them from the exhaust system over a larger area than that of the exhaust pipe. This has the effect of reducing the temperature of gas emitted from the exhaust aftertreatment system. An image of a patented diffuser is shown in Fig. 2.43.

2.7.3 *Hybrids and Alternative Propulsion Systems*

Primary alternatives to fully diesel-based propulsion include hybrid electric vehicles, electric vehicles, and fuel cell electric vehicles. Diesel hybrids use a diesel engine with an electric motor and generator system and battery, in series or parallel configuration. The diesel engine increases the driving range, and electrical components increase efficiency and fuel economy by regenerating energy during vehicle braking and storing excess energy from the engine in a battery while coasting [86]. Based on the level of electric power available to propel the engine independently of diesel power, the hybrid vehicle could be classified as a micro, mild, or full hybrid. Full electric vehicles are also under development, and they have currently a range of up to 500 miles [87, 88]. Fuel cell electric vehicles, generally powered by either natural gas or pure hydrogen, do not produce tailpipe emissions, but the technology and the infrastructure to fuel them are in the initial stages of implementation. In these alternative technologies, thermal management should also aim to regulate temperatures of battery packs, electrical machines, or fuel cells for ensuring efficiency, longevity, and safety.

Barnitt [89] compared CNG, Gen I, and Gen II hybrid-electric propulsion systems used in New York City transit buses based on fuel economy, maintenance and operating costs per mile, and reliability. Conventional diesel was used as the baseline. The electric drive motor drove the vehicle and captured energy during regenerative braking by acting as a generator. Batteries supplied additional power during acceleration and hill climbing and stored energy recovered during regenerative braking and idling. The Gen I hybrid buses provided a much higher fuel economy than CNG and conventional diesel buses. The average fuel economy for the Gen II hybrid buses was lower than the Gen I hybrid buses, but propulsion-related maintenance costs were lower. Proust and Surcel [90] evaluated the reduction in fuel consumption and greenhouse gas emissions from three trucks equipped with hybrid technology. Two diesel-electric hybrid pick-up and delivery trucks and one diesel-electric hybrid utility truck equipped with an electric-driven PTO (power take-off) system were used in the evaluation. Results of the tests showed a reduction in fuel consumption and annual GHG emissions for hybrid trucks. The authors also provided some best practices in driving that include maximizing regenerative braking, accelerating moderately and progressively for speeds below 30 km/h, and avoiding transmission shifting at high engine speeds. Duty cycles and driving habits significantly influence the benefits from these vehicles, and naturally, these have a significant impact on the thermal management needs as well.

As Bennion and Thornton [91] point out, the subsystems for hybrid propulsion technologies will have an increased number of components, in addition to the exhaust aftertreatment system, that requires active thermal management. These components will have differing needs such as requiring heat for operation, cooling to reject heat, or operation within the specified temperature. Also, the reduction in waste heat can have negative impacts on functions such as cabin heating that use the recovered waste heat. Tao et al. [92] evaluated the performance of a computer-controlled thermal management system intended for a hybridized mid-size truck. They used thermal models for the heat-generating components and designed an optimal controller for ideal battery cooling and a model predictive controller (MPC) for the air conditioning system compressor pump operation. A nonlinear control strategy was introduced for the engine coolant temperature stabilization while minimizing the system power

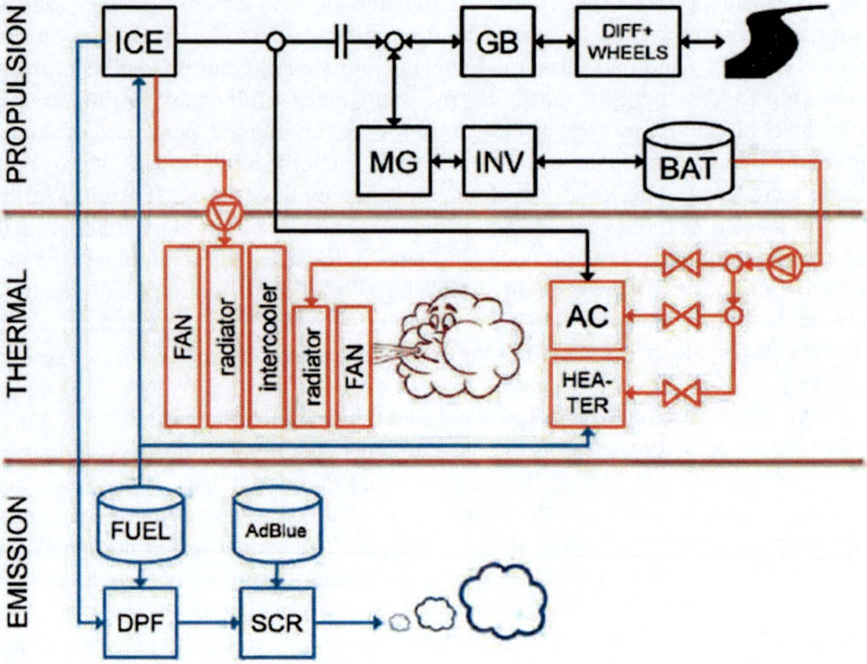

Fig. 2.44 Schematic representation of components and energy domains in a Euro-VI parallel HEV truck [93]

consumption. Van Reeven et al. [93] demonstrated the benefits of an Integrated Energy Management System strategy in a parallel hybrid electric truck for long-haul applications. A schematic in Fig. 2.44 shows the components subdivided into three energy domains. The control strategy integrates strategies for the power split between the combustion engine and the motor/generator, the thermal control of the battery, and that of the engine. These authors demonstrated that HEVs with an optimal thermal management control framework can enable 4–7% fuel savings for heavy-duty long-haul applications.

2.8 Summary

This chapter has described the thermal management of diesel exhaust aftertreatment. It began with an overview of the reasons why thermal management is a critical part of today's advanced diesel aftertreatment systems. These include exhaust emissions regulations, catalyst performance limitations related to exhaust temperatures, aftertreatment recovery, and DEF dosing and SCR solid deposit removal. Then, the suite of existing and promising strategies and technologies available to implement thermal management were described. Special attention was paid to the implications of the technologies on exhaust heat, emissions, and fuel efficiency. The technologies were divided into the following major categories:

- Engine-based Thermal Management
- Aftertreatment Fuel Introduction
- Heat Generation
- Aftertreatment Heat Retention
- Exhaust Heat Management.

In the section on engine-based thermal management, various thermal management strategies that primarily utilize the combustion of fuel in the engine to generate additional heat for the aftertreatment system were discussed. These strategies, specifically in-cylinder post-injection of fuel and exhaust restriction or intake throttling, are already in widespread use today to help regulate exhaust temperatures in diesel aftertreatment systems. In the section titled aftertreatment fuel introduction, strategies whereby diesel fuel is injected into the exhaust gases upstream of an oxidation catalyst, typically a DOC, were described in detail. The subsections detailed the practical challenges with this method of thermal management, namely, DOC light-off and quenching, DOC degradation, DPF temperature gradients during active regeneration, and hydrocarbon injector selection. In the section on heat generation, additional methods of heat generation, not involving fuel introduction in the cylinder or the aftertreatment system, were surveyed. These included uses of a dedicated burner device, electric heater, plasma burner, microwave heater, and some combinations of these. As discussed, these strategies are still under development and have not seen widespread use in diesel exhaust aftertreatment applications as of publication. The section on aftertreatment heat retention brought to light the methods used to retain the heat of exhaust gases as an important strategy to achieve the desired temperatures at the catalysts. It is likely that close-coupled and pre-turbo aftertreatment catalyst designs, one of the methods discussed in this section, will receive additional attention in the future since this is a straightforward way to increase catalyst temperature. Finally, exhaust heat management considerations including waste heat recovery, limiting maximum exhaust temperatures, and hybrid and other advanced propulsion systems were discussed. As regulations continue to become more stringent with future ultralow NO_x and PM/PN standards, thermal management of diesel exhaust aftertreatment systems will remain an important topic for the foreseeable future. Also, as discussed in the hybrids and alternative propulsion systems section, thermal management now must be integrated into system-level thermal management strategies for hybrid applications, increasing system complexity and requiring advanced integration and controls, bringing in rich possibilities for further research and development.

References

1. Greenhouse gas emissions and fuel efficiency standards for medium- and heavy-duty engines and vehicles—phase 2. US C.F.R. 40 & 49 (2016)
2. Nakano K, Okano H, Inoue K, Obuchi A (2018) Study on the prevention of face-plugging of diesel oxidation catalyst (DOC). SAE technical paper, vols. 2018-32-0069
3. Ottinger N, Xi Y, Foley B, Mnichowicz B, Liu J (2013) Hydrocarbon interactions with vanadium-based SCR catalysts—a mechanistic study. In: CLEERS Presentation
4. Wall JC, Hoekman SK (1984) Fuel composition effects on heavy-duty diesel particulate emissions. SAE technical paper, vol 841364
5. Hamzehlouyan T, Sampara C, Li J, Kumar A, Epling W (2014) Experimental and kinetic study of SO_2 oxidation on a Pt/Al_2O_3 catalyst. Appl Catal B 152–153:108–116
6. Xi Y, Ottinger N, Keturakis C, Liu ZG (2020) A case study of a Cu-SSZ-13 SCR catalyst poisoned by real-world high sulfur diesel fuel. SAE technical paper, vols 2020-01-1319
7. Lefebvre A (1989) Atomization and sprays. Taylor and Francis, Boca Raton, FL
8. Schaber P, Colson J, Higgins S, Thielen D, Anspach B, Brauer J (2004) Thermal decomposition (pyrolysis) of urea in an open reaction vessel. Thermochim Acta 424:131–142
9. Brack W, Heine B, Birkhold F, Kruse M (2014) Kinetic modeling of urea decomposition based on systematic thermogravimetric analyses of urea and its most important by-products. Chem Eng Sci 106:1–8
10. Zhang H, Xi Y, Su C, Liu Z (2017) Lab study of urea deposit formation and chemical transformation process of diesel aftertreatment system. SAE technical paper, vols 2017-01-0915
11. Meda L, Shu Y, Romzek M (2009) Heavy duty diesel after-treatment system analysis based design: fluid, thermal, and structural considerations. SAE technical paper, vols 2009-01-0624
12. Tan P, Duan L, Li E, Hu Z, Lou D (2020) Experimental study on thermal management strategy of the exhaust gas of a heavy-duty diesel engine based on in-cylinder injection parameters. SAE technical paper, vols 2020-01-0621
13. Parks J, Huff S, Kass M, Storey J (2007) Characterization of in-cylinder techniques for thermal management of diesel aftertreatment. SAE technical paper, vols 2007-01-3997
14. Iorio SD, Beatrice CGC, Giacomo ND, Napolitano P, Vassallo A (2011) Analysis of particle mass and size emissions from a catalyzed diesel particulate filter during regeneration by means of actual injection strategies in light duty engines, vols 2011-24-0210
15. Giechaskiel B, Munoz-Beuno R, Rubino L, Manfredi U, Dilara P, Santi GD, Andersson J (2007) Particle measurement programme (PMP): particle size and number emissions before, during and after regeneration events of a Euro 4 DPF equipped light-duty diesel vehicle. SAE technical paper, vols 2007-01-1944
16. Ramesh AK, Gosala DB, Allen C, Joshi M, McCarthy J Jr, Farrell L, Koeberlein ED, Shaver G (2018) Cylinder deactivation for increased engine efficiency and aftertreatment thermal management in diesel engines. SAE technical paper, vols 2018-01-0384
17. Morris A, McCarthy J (2020) The effect of heavy-duty diesel cylinder deactivation on exhaust temperature, fuel consumption, and turbocharger performance up to 3 bar BMEP. SAE technical paper, vols 2020-01-1407
18. Roberts L, Magee M, Shaver G, Garg A, McCarthy J, Koeberlien E, Holloway E, Shute R, Koeberlein D, Nielsen D (2014) Modeling the impact of early exhaust valve opening on exhaust aftertreatment thermal management and efficiency for compression ignition engines. Int J Engine Res 16(6):773–794
19. Ozel T, Hall MJ, Matthews R (2018) Increasing exhaust temperature of an idling light-duty diesel engine through post-injection and intake throttling. SAE technical paper, vols 2018-01-0223
20. Wall JC (2011) Engine technologies for the future. In: University of Wisconsin Engine Research Center symposium, Madison, WI, USA
21. Czarnowski R, Joergl V, Weber O, Shutty J, Keller P (2008) Can future emissions limits be met with a hybrid EGR system alone. In: DEER 2008, Dearborn, Michigan, USA

22. Yuksel A (2020) SCR versus EGR: what's the right choice for the rail industry? Cummins, 27 Aug 2019 [Online]. Available: https://www.cummins.com/news/2019/08/27/scr-versus-egr-whats-right-choice-rail-industry. Accessed 1 Aug 2020
23. Maier J (2003) Variable charge motion for 2007–2010 heavy-duty diesel engines. In: Deer conference, Newport, Rhode Island, USA
24. Reitz R, Hanson R, Splitter D, Kokjohn S (2009) High-efficiency, ultra-low emission combustion in a heavy-duty engine via fuel reactivity control. In: DEER conference, Dearborn, Michigan
25. Asad U, Divekar P, Zheng M, Tjong J (2013) Low-temperature combustion strategies for compression-ignition engines: operability limits and challenges. SAE technical paper, vols 2013-01-0283
26. Stanton D (2013) Systematic development of highly efficient and clean engines to meet future commercial vehicle greenhouse gas regulations. SAE Int J Engines 6(3):1395–1480
27. Dev S, Chaudhari HB, Gothekar S, Juttu S, Walke NH, Marathe NV (2017) Review on advanced low-temperature combustion approach for BS-VI. No. 2017-26-0042. SAE technical paper
28. Korakianitis T, Namasivayam AM, Crookes RJ (2011) Natural-gas fueled spark-ignition (SI) and compression-ignition (CI) engine performance and emissions. Prog Energy Combust Sci 37(1):89–112
29. Königsson F, Stalhammar P, Angstrom H-E (2011) Characterization and potential of dual-fuel combustion in a modern diesel engine. No. 2011-01-2223. SAE technical paper
30. May I, Pedrozo V, Zhao H, Cairns A, Whelan S, Wong H, Bennicke P (2016) Characterization and potential of premixed dual-fuel combustion in a heavy-duty natural gas/diesel engine. No. 2016-01-0790. SAE technical paper
31. Harrington J, Munshi S, Nedelcu C, Ouellette P, Thompson J, Whitfield S (2002) Direct injection of natural gas in a heavy-duty diesel engine. No. 2002-01-1630. SAE technical paper
32. Joshi A, Chatterjee S (2006) Development and field validation of an actively regenerating DPF system for retrofit applications. In: DEER Poster presentation
33. Chevron Corp. (2007). [Online]. Available: https://www.chevron.com/-/media/chevron/operations/documents/diesel-fuel-tech-review.pdf
34. Kumar A, Zokoe R, Joshi S, Kamasamudram K, Yezerets A (2018) Reactor system with diesel injection capability for DOC evaluations. SAE technical paper, vols 2018-01-0647
35. Kozlov A, Harris T, Salyards C (2010) Engine test for DOC quenching in DOC-DPF system for non-road applications. SAE Int J Fuels Lubr 3(1):397–405
36. Shakya BM, Sukumar B, Jesus Y, Markatou P (2015) The effect of Pt: Pd ratio on heavy-duty diesel oxidation catalyst performance: an experimental and modeling study. SAE Int J Engines 8(3):1271–1282
37. Gurupatham A, He Y (2008) Architecture design and analysis of diesel engine exhaust aftertreatment system and comparative study with close-coupled DOC-DPF system. SAE Int J Fuels Lubr 1(1):1387–1396
38. Green Car Congress (2017). [Online]. Available: https://www.greencarcongress.com/2017/05/20170508-iav.html
39. Cavatiao G, Jen H-W, Dobson DA, Warner JR (2009) Laboratory study to determine impact of Na and K exposure on the durability of DOC and SCR catalyst formulations. SAE technical paper, vols 2009-01-2823
40. Tong D, Zhang J, Wang G, Yang B, Cai K, Liu S, Abdalla A, Shuai S-J (2018) Experimental study and numerical interpretation on the temperature field of DPF during active regeneration with hydrocarbon injection. SAE technical paper, vols 2018-01-1257
41. Frobert A, Creff Y, Lepreux O, Schmidt L, Raux S (2009) Generating thermal conditions to regenerate a DPF: impact of the reductant on the performances of diesel oxidation catalysts. SAE technical paper, vols 2009-01-1085
42. Zelenka P, Telford C, Pye D, Birkby N (2002) Development of a full-flow burner DPF system for heavy-duty diesel engines. SAE technical paper, vols 2002-01-2787
43. Kimura M, Muramatsu T, Kunishima E, Namima J, Crawley W, Parris T (2011) Development of the burner systems for EPA2010 medium duty diesel vehicles. SAE technical paper, vols 2011-01-0295

44. Ramsbottom M, Birkby N, Khadiya N, Beesley S (2011) Development of a thermal enhancer™ for combined partial range burning and hydrocarbon dosing on medium and heavy duty engine applications. SAE technical paper, pp 2011-01-0298
45. Baier B, Schrewe K, Maurer B (2011) Two-stage electro-thermal supported HC (hydro carbon) conversion. SAE Int J Engines 4(1):508–514
46. Culbertson D, Khair M, Zha Y, Dlestelmeler J (2018) Exhaust heating system performance for boosting SCR low-temperature efficiency. SAE technical paper, vols 2018-01-1428
47. Pfahl U, Schatz A, Konieczny R (2012) Advanced exhaust gas thermal management for lowest tailpipe emissions—combining low emission engine and electrically heated catalyst. SAE technical paper, vols 2012-01-1090
48. Sharp CA, Webb CC, Neely GD, Smith I (2018) Evaluating technologies and methods to lower nitrogen oxide emissions from heavy-duty vehicles. Final report, Southwest Research Institute
49. Ning Z, He Y (1999) Experimental study on microwave regeneration characteristics of diesel particulate after-treatment system. SAE technical paper, vols 1999-01-1470
50. Penetrante BM, Brusasco RM, Merritt BT, Pitz WJ, Vogtlin GE (1999) Feasibility of plasma aftertreatment for simultaneous control of NOx and particulates. SAE technical paper, vols 1999-01-3637
51. Pyun SH, Lee DH, Kim KT (2014) Development of a unique plasma burner system for emission reduction during cold start of diesel engines. SAE technical paper, vols 2014-01-1490
52. Lee DH, Kim K-T, Cha MS, Lee JO, Song Y-H, Cho H, Kim Y-S, Song Y, Jee T (2009) Active regenerative DPF using a plasma assisted burner. SAE technical paper, vols 2009-01-1926
53. Robertson G (1979) A study of thermal energy conservation in exhaust pipes. SAE technical paper, vol 790307
54. De Sousa E (1993) Dual wall energy conserving pipes for internal combustion engines. SAE technical paper, vol 931095
55. Lee K-H, Shim C-Y, Kim W-B (2000) Development of dual wall air gap exhaust system. SAE technical paper, pp 2000-01-0205
56. Wurzenberger J, Wanker R, Schüßler M (2008) Simulation of exhaust gas aftertreatment systems—thermal behavior during different operating conditions. SAE technical paper, vols 2008-01-0865
57. Hamedi M, Tsolakis A, Herreros J (2014) Thermal performance of diesel aftertreatment: material and insulation CFD analysis. SAE technical paper, vols 2014-01-2818
58. Su C, Brault J, Munnannur A, Liu ZG, Harinath MSA, Dunnuck D, Federle K (2019) Model-based approaches in developing an advanced aftertreatment system: an overview. SAE Int J Adv Curr Pract Mobility 1(1):201–214
59. Host R, Moilanen P, Fried M, Bogi B (2017) Exhaust system thermal management: a process to optimize exhaust enthalpy for cold start emissions reduction. SAE technical paper, no 2017-01-0141
60. Joergl V, Keller P, Weber O, Mueller-Haas K, Konieczny R (2008) Influence of pre turbo catalyst design on diesel engine performance, emissions and fuel economy. SAE Int J Fuels Lubr 1(1):82–95
61. Subramaniam M, Hayes C, Tomazic D, Downey M, Bruestle C (2011) Pre-turbo aftertreatment position for large bore diesel engines—compact & cost-effective aftertreatment with a fuel consumption advantage. SAE Int J Engines 4(1):106–116
62. Watson M, Gonzalez J (2007) Engineered mat technology for high temperature close coupled applications. SAE technical paper, vols 2007-01-1077
63. Benson D, Potter T, Tracy C (1994) Design of a variable-conductance vacuum insulation. SAE technical paper, vol 940315
64. Burch S, Keyser M, Potter T, Benson D (1994) Thermal analysis and testing of a vacuum insulated catalytic converter. SAE technical paper, vol 941998
65. Burch S, Biel J (1999) ULEV and "Off-Cycle" emissions benefits of a vacuum-insulated catalytic converter. SAE technical paper, vols 1999-01-0461
66. Gaiser G, Seethaler C (2007) Zero-delay light-off—a new cold-start concept with a latent heat storage integrated into a catalyst substrate. SAE technical paper, vols 2007-01-1074

67. Hamedi M, Doustdara O, Tsolakisa A, Hartland J (2019) Thermal energy storage system for efficient diesel exhaust aftertreatment at low temperatures. Appl Energy 2019:874–887
68. Jankowski N, McCluskey P (2014) A review of phase change materials for vehicle component thermal buffering. Appl Energy 113:1525–1561
69. Leising C, Purohit G, DeGrey S, Finegold J (1978) Waste heat recovery in truck engines. SAE technical paper, vol 780686
70. Tennant D, Walsham B (1989) The turbocompound diesel engine. SAE technical paper, p 890647
71. Hopmann U (2004) Diesel engine waste heat recovery utilizing electric turbocompound technology. In: DEER Conference
72. Vuk C (2006) Electric turbo compounding: a technology who's time has come. In: DEER Conference
73. Hountalas D, Katsanos C, Lamaris V (2007) Recovering energy from the diesel engine exhaust using mechanical and electrical turbocompounding. SAE technical paper, vols 2007-01-1563
74. Nelson C (2006) Exhaust energy recovery. In: DEER Conference
75. Teng H (2007) Waste heat recovery of heavy-duty diesel engines by organic rankine cycle part i: hybrid energy system of diesel and rankine engines. SAE technical paper, vols 2007-01-0537
76. Nelson C (2010) Exhaust energy recovery. In: US Dept of Energy Annual Merit Review
77. Kushch A, Bass JC, Ghamaty S, Elsner NB (2001) Thermoelectric development at Hi-Z technology. In: Proceedings ICT2001, Twentieth international conference on thermoelectrics (ICT)
78. LaGrandeur J, Crane D, Eder A (2005) Vehicle fuel economy improvement through thermoelectric waste heat recovery. In: DEER Conference
79. Eichler K, Jeihouni Y, Ritterskamp C (2015) Fuel economy benefits for commercial diesel engines with waste heat recovery. SAE Int J Commer Veh 8(2):491–505
80. Koebel M, Strutz EO (2003) Thermal and hydrolytic decomposition of urea for automotive selective catalytic reduction systems: thermochemical and practical aspects. Ind Eng Chem Res 42(10):2093–2100
81. Munnannur A, Chiruta M, Liu Z (2012) Thermal and fluid dynamic considerations in aftertreatment system design for SCR solid deposit mitigation. SAE technical paper, vols 2012-01-1287
82. Schiller S, Brandl M, Hoppenstedt B, De Rudder K (2015) Wire mesh mixer optimization for DEF deposit prevention. SAE technical paper, vols 2015-01-0989
83. Southwest Research Institute (2016) Advanced combustion catalyst & aftertreatment technologies monthly progress report # 23. Southwest Research Institute
84. Liu ZG, Ottinger NA, Cremeens CM (2015) Vanadium and tungsten release from V-based selective catalytic reduction diesel aftertreatment. Atmos Environ 104:154–161
85. NIOSH (2020) Pocket guide to chemical hazards: vanadium dust. [Online]. Available: https://www.cdc.gov/niosh/npg/npgd0653.html. Accessed 2020
86. Chan C (2007) The state of the art of electric, hybrid, and fuel cell vehicles. Proc IEEE 95(4):704–718
87. Trend M (2017). [Online]. https://www.motortrend.com/news/first-electric-semi-truck-cummins-not-tesla/
88. Lambert F (2020). [Online]. https://electrek.co/2020/07/23/tesla-semi-electric-truck-production-texas/
89. Barnitt R (2008) In-use performance comparison of hybrid electric, CNG, and diesel buses at New York City transit. SAE technical paper, vols 2008-01-1556
90. Proust A, Surcel M (2012) Evaluation of class 7 diesel-electric hybrid trucks. SAE technical paper, vols 2012-01-1987

91. Bennion K, Thornton M (2010) Integrated vehicle thermal management for advanced vehicle propulsion technologies. SAE technical paper, vols 2010-01-0836
92. Tao X, Zhou K, Ivanco A, Wagner J (2015) A hybrid electric vehicle thermal management system—nonlinear controller design. SAE technical paper, vols 2015-01-1710
93. van Reeven V, Huisman R, Kessel J, Pham T (2013) Integrating energy and thermal management of hybrid trucks. In: Proceedings of the 12th international conference commercial vehicles truck, bus, van, trailer, 5–6 June 2013, Celle, Germany

Part II
Thermal Management Through Turbocharging and Insulation Between Aftertreatment Systems and the Engine

Chapter 3
Models for Instantaneous Heat Transfer in Engines and the Manifolds for 1-D Thermodynamic Engine Simulation

P. A. Lakshminarayanan, J. Galindo, J. M. Luján, J. R. Serrano, V. Dolz, P. Piqueras, and J. Gómez

Definitions, Acronyms, Abbreviation

Greek

α	Crank angle degree
α	Diffusivity coefficient defined as $\alpha = K/(\rho C_S)$
γ	Specific heat ratio
ε	Emissivity
μ	Fluid viscosity
μ_f	Fluid viscosity at the fluid temperature
μ_w	Fluid viscosity at the wall temperature
ρ	Density
σ	Boltzmann's radiation constant

Substantial portion of the chapter is taken from SAE2016-01-0670 with permission, Order Number: 1072610, 25 Oct 2020, Copyright Clearance Center.

P. A. Lakshminarayanan (✉)
Indian Institute of Technology Kanpur, Kalyanpur, Kanpur 208016, India

J. Galindo · J. M. Luján · J. R. Serrano · V. Dolz · P. Piqueras · J. Gómez
Universitat Politécnica de Valencia. CMT-Motores Térmicos, Camino de Vera, s/n., 46022 Valencia, Spain

P. A. Lakshminarayanan and A. K. Agarwal (eds.), *Handbook of Thermal Management of Engines*, Energy, Environment, and Sustainability,
https://doi.org/10.1007/978-981-16-8570-5_3

Acronyms

Bi	Biot number defined as $Bi = h \cdot \Delta x/k$
bmep	Brake mean effective pressure
BSFC	Brake specific fuel consumption
c	Speed
C_p	Specific heat at constant pressure
d	Duct diameter
d_c	Cylinder diameter
DPF	Diesel particle filter
Fo	Fourier number defined as $Fo = \alpha \cdot \Delta t/(\Delta x)^2$
h	Heat transfer coefficient
h_i	Inside heat transfer coefficient
$h_{i\text{-}2}$	Inside plus wall 1–2 heat transfer coefficient
$h_{i\text{-}3}$	Inside plus wall 1–2–3 heat transfer coefficient
h_o	Outside heat transfer coefficient
$h_{o\text{-}1}$	Outside plus wall 1–2–3 heat transfer coefficient
$h_{o\text{-}2}$	Outside plus wall 2–3 heat transfer coefficient
h_{or}	Outside radiation heat transfer coefficient
h_s	Enthalpy in isentropic conditions
HSDI	High-speed direct injection
IVC	Intake valve closing
IVO	Intake valve opening
K, k	Thermal conductivity
L	Coolant tube length
M_0	Torque at maximum engine power
N	Engine speed
Nu	Nusselt number
OEM	Original exhaust manifold
p	Pressure
$p2'$	Intake manifold pressure
PEM	Pulse exhaust manifold
PMEP, pmep	Pumping mean effective pressure
Pr	Prandtl number
Re	Reynolds number
S	Stroke
SEM	Synthesis exhaust manifold
T	Temperature
T	Torque
t	Time
T_3	Inlet turbine temperature
T_4	Outlet turbine temperature
t_c	Engine cycle duration
T_i	Fluid temperature inside ducts and cylinders

TIP	Turbine isentropic power
T_o	Fluid temperature outside ducts and cylinders
T_{w1}	Wall temperature at the inner calculation node
T_{w2}	Wall temperature at the central calculation node
T_{w3}	Wall temperature at the outer calculation node
u	Fluid velocity
VGT	Variable geometry turbine
ω	Engine rotating speed
X	Distance to the exhaust valve
Z	Number of cylinders
$\dot{m}$	Mass flow of fluid.
$\dot{W}$	Power

3.1 Introduction

The performance of after-treatment devices is dependent on the operating temperature apart from the control strategies [1]. A diesel particulate filter (DPF) reaches the targeted efficiency and completes regeneration [2] at around 600 °C [3]. While exploring the potential solutions to increase the temperature (T4) at the inlet to the after-treatment, the penalty on fuel consumption must be considered. BSFC may be meaningfully improved, at the same time decreasing T4 with suitable valve timing or by an integrated exhaust manifold [4] in the cylinder head for a diesel engine [5] or by the correct position DPF and DOC [6]. The turbo lag is caused by mechanical, fluid-dynamic, and thermal phenomena [7]. The heat transfer and the discrete spatial distribution of temperature in four zones, namely the intake line, the cylinders, the exhaust ports, and the exhaust line is estimated in conjunction with the Thermodynamics Model of the Engine [8–13].

3.2 Model for Heat Transfer from Turbocharger

Turbine case, bearing housing (H_2), and compressor case are each represented by one node [14] with two additional nodes for the bearing housing considering high temperature-gradients due to active cooling by oil/water (H_1, H_3), Fig. 3.1. All the resistances are convective to different fluids namely, oil, coolant, and air.

All temperatures are mass averaged resulting in a more accurate temperature prediction [15] than if time-averaged. The engine outlined in Table 3.1 is simulated using the commercial software GT Power [8], at six operating points (Table 3.2) at points, #A, #B, #C (low load), #D, #E, and #F (high load).

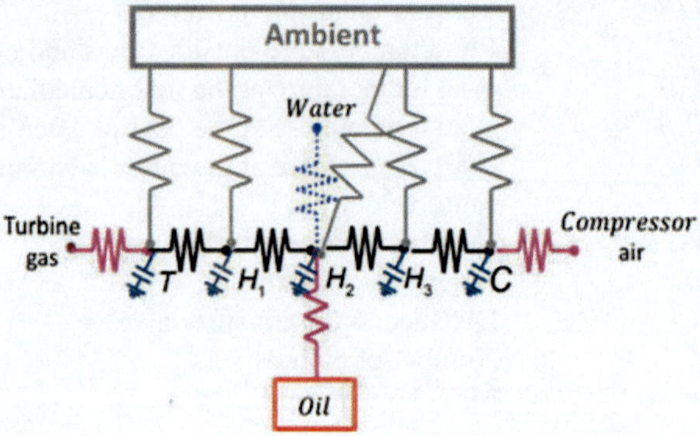

Fig. 3.1 Lumped turbocharger model for heat transfer

Table 3.1 Engine specification

Type	Turbocharged HSDI diesel
Displacement	1997 cm^3
Bore	85 mm
Stroke	88 mm
Number of cylinders	4 inline
Number of intake valves	2 per cylinder
Number of exhaust valves	2 per cylinder
Turbocharger model	Variable geometry turbocharger (VGT)
Compression ratio	15.5:1
Maximum power at speed	120 kW at 3750 rpm
Maximum torque at speed	340 Nm at 2000 rpm
EGR type	Cooled, high pressure with an intake throttle

Table 3.2 Engine points simulated in GT-power

Point ID	Speed (rpm)	Load (%)
#A	1500	0
#B	1500	25
#C	2000	25
#D	1250	75
#E	2000	75
#F	3500	100

3.3 Models for Heat Transfer: Ports, Intake and Exhaust Lines, Cylinders and Pistons

The main inputs to the model are the properties of the wall materials and the coolant such as density, specific heat, thickness, thermal conductivity, and viscosity. The four different areas are treated as follows.

Intake line

The heat transfer correlations [16, 17] for non-pulsating flows in conduits and the Hilpert correlation [18] for the outer surface of horizontal pipes are applied. However, the intake line does not influence thermal performance.

Engine cylinder

Three inner walls namely, the cylinder head, the liner, and the piston wall have different temperatures but the same internal heat transfer coefficient given by Woschni [19] and the outer wall by Cipolla coefficient [20] without considering nucleate boiling for a coolant flow of 0.8 l/min per kW of maximum engine power, Eq. (3.1).

$$h_0 = \left[0.063 + 0.733\left(\frac{ZL}{NM_0}\right)^{0.25}\right]\left(\frac{NM_0}{Z}\right)^{0.8} L^{-1.8} \tag{3.1}$$

where

Z Number of cylinders
M_0 Torque at maximum engine power
L Coolant tube length
h_o Outside heat transfer coefficient.

L is $1.1 \times d_c$ (cylinder diameter) in the cylinder and $L = Stroke$ in the liner cooling ducts. The resistance to heat between the gas and the water is the length from the centre of the piston crown to the position of the oil ring for conduction and the liner area for convection on the waterside (Fig. 3.2). The piston height is approximated to the diameter and the oil ring. Cooling by the lubricant is neglected.

Exhaust ports

Sieder-Tate correlation modified for the shape and turbulence (Eq. (3.2)) [21–27] is applied.

$$Nu = 0.10Re^{0.8}Pr^{0.33}\left(\frac{\mu_f}{\mu_w}\right)^{0.14} \tag{3.2}$$

where

Re Reynolds number
Pr Prandtl number

Fig. 3.2 Piston wall thickness scheme

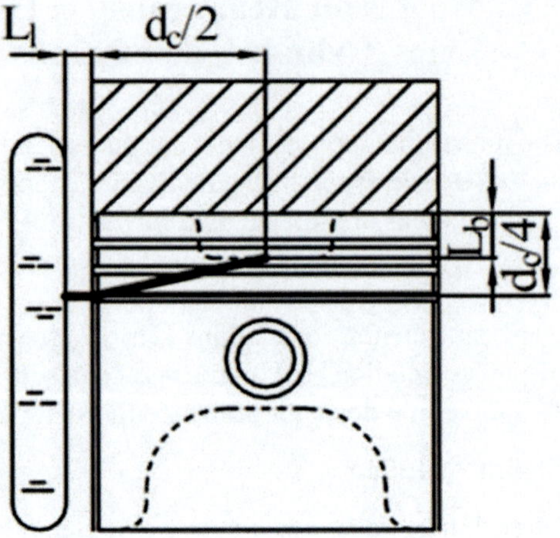

Nu Nusselt number
μ_f Fluid viscosity at bulk fluid temperature
μ_w Fluid viscosity at walltemperature.

Rearranging Eq. (3.2),

$$h_i = 6.8 \times 10^{-3} \frac{\rho^{0.8} u^{0.8} T_i^{0.3} C_p}{d^{0.2} (T_i + 110.4)^{0.2}} \left(\frac{\mu_f}{\mu_w} \right)^{0.14} \tag{3.3}$$

where

d diameter of the duct
h_i Inside heat transfer coefficient
T_i Fluid temperature inside ducts and cylinders
C_p Specific heat at constant pressure.

T_i from Sutherland [17] and u from the 1-D gas-dynamic model are obtained iteratively at every time step. Similarly, the temperature of the outer surface of the exhaust port is iterated [20, 23].

Exhaust line

A Nusselt versus Reynolds correlation by Reyes [28] is used to characterize the heat transfer, highly enhanced by the turbulence generated by the pulsation (Eq. 3.4) which decays with time and distance from the exhaust valve (Eq. 3.5).

$$Nu = 1.6 \overline{Re_\xi}^{0.4} C_{pos} \tag{3.4}$$

$$C_{pos} = 1 + 3e^{-\frac{1}{4}\frac{X}{d}} \tag{3.5}$$

where

Nu	Nusselt number
d	diameter of the duct
$\overline{u_\xi}$	weighted average of velocity of previous calculation instants [29]
Re_ξ	Reynolds number calculated with the velocity, and

C_{pos} is the turbulence coefficient that accounts for the distance between the duct and the exhaust valve (X). Rearranging Eq. (3.4) [23–27],

$$h_i = 7.2094 \times 10^{-4} \frac{\rho^{0.4}\overline{u_\xi}^{0.4} C_p T_i^{0.9}}{d^{0.6}(T_i + 110.4)^{0.6}} C_{pos} \tag{3.6}$$

where

d	diameter of the duct.
h_i	Inside heat transfer coefficient
T_i	Fluid temperature inside ducts and cylinders
ρ	fluid density.

The heat transfer from the outer surface, h_{or} is augmented by the radiation heat transfer coefficient [18] Eq. (3.7).

$$h_{or} = \frac{\varepsilon\sigma\left(T_{w3}^4 - T_0^4\right)}{T_{w3} - T_0} \tag{3.7}$$

where

Boltzmann's radiation constant, $\sigma = 5.67 \times 10^{-8}\mathrm{W}/\left(\mathrm{m}^2\mathrm{K}^4\right)$ and emissivity, $\varepsilon = 0.5$.

3.3.1 *Calculation of Wall Temperature*

Equation (3.8) of conductive heat transfer is solved by an explicit finite difference method of three nodes (left Fig. 3.3) with thermal inertia or without.

$$\frac{1}{\alpha}\frac{\partial T_w}{\partial t} = \frac{\partial^2 T_w}{\partial x^2} \tag{3.8}$$

where

T_w = wall temperature.

t = time

Diffusivity coefficient, $\alpha = K/(\rho C_S)$

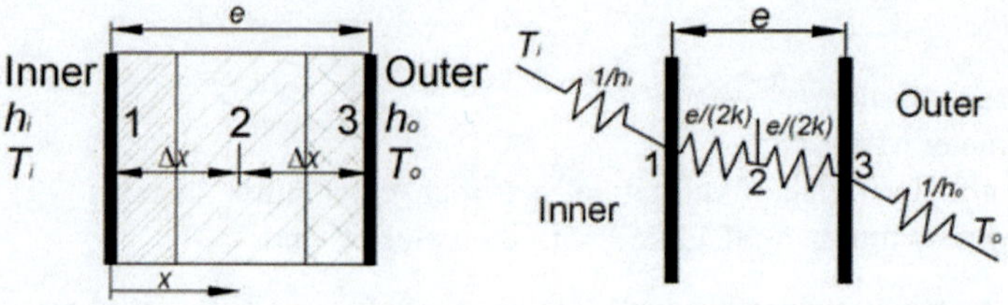

Fig. 3.3 (left) Scheme of wall mesh for temperature calculation with thermal inertia model and (right) scheme of thermal resistance for wall temperature calculation without thermal inertia model

Thermal inertia

The central difference approximations to the spatial derivatives and the forward difference to the time derivative [30] are applied to an element (left Fig. 3.3) with the initial condition.

$$T_w = T_{w0,t} = 0 \tag{3.9}$$

The equation is solved at time step ($p + 1$), using the explicit method by substituting explicit Eq. (3.9) into Eq. (3.8) at a node, n.

$$T_{w2}^{p+1} = F_0\left(T_{w3}^{p} + T_{w1}^{p}\right) + (1 - 2F_0)T_{w2}^{p} \tag{3.10}$$

where F_o is the Fourier number. The stability criterion is $F_o \leq 0.5$. Assuming convection transfer from an adjoining fluid, inside and outside the tube, and no heat generation,

$$T_{w1} = 2F_0\left(T_{w2}^{p} + Bi_i T_i\right) + (1 - 2F_0 - 2Bi_i F_0)T_{wi}^{p} \tag{3.11}$$

$$T_{w3}^{p+1} = 2F_0\left(T_{w2}^{p} + Bi_0 T_0\right) + (1 - 2F_0 - 2Bi_0 F_0)T_{w3}^{p} \tag{3.12}$$

where Bi is the Biot number and the stability criterion is $F_0(1 + Bi) \leq 0.5$. This model is too slow to calculate for engine steady conditions for a general initial value, because of thermal inertia. Therefore, to accelerate the convergence a strategy is established.

Strategy for fast convergence

Assuming steady heat flow and using the thermal resistance scheme (right, Fig. 3.3), the cycle average is used in Eq. (3.13) at the three nodes [29].

$$0 = h_{0-x}(T_0 - T_{wx}) + h_{i-x}(T_i - T_{wx}) \tag{3.13}$$

where

h	heat transfer coefficient
T	temperature
"x"	the node number and

$$h_{i-1} = h_i, h_{i-2} = \left(\frac{1}{h_i} + \frac{e}{2K}\right)^{-1}, h_{i-3} = \left(\frac{1}{h_i} + \frac{e}{K}\right)^{-1} \tag{3.14a}$$

$$h_{0-3} = h_0, h_{0-2} = \left(\frac{1}{h_0} + \frac{e}{2K}\right)^{-1}, h_{0-1} = \left(\frac{1}{h_0} + \frac{e}{K}\right)^{-1} \tag{3.14b}$$

Integrating the Eq. (3.13) for the complete engine cycle, (t_c) with h_o constant and assuming small variations of T_{w1}, $T_{w2,}$ and T_{w3},

$$\int_0^{t_c} h_{i-x(t)} T_{i(t)} dt - T_{wx} \int_0^{t_c} h_{i-x(t)} dt = h_{0-x} T_{wx} t_c - h_{0-x} T_0 t_c \tag{3.15}$$

Rearranging and simplifying Eq. (3.15), T_{w1}, $T_{w2,}$ and T_{w3} can be calculated as a function of $h_{i\text{-}2}$, $h_{i\text{-}3}$, $h_{o\text{-}2}$, $h_{o\text{-}1}$, h_i, Ti, h_o, $T_{o,}$ and t_c.

$$T_{w3(d,j)} = \frac{t_c h_0 T_0 + \sum_{k=1}^{n} h_{i-3(d,j,k)} T_{i(d,j.k)} \Delta t_{(k)}}{t_c h_0 + \sum_{k=1}^{n} h_{i-3(d,j,k)} \Delta t_{(k)}};$$
$$T_{w2(d,j)} = \frac{t_c h_{0-2} T_0 + \sum_{k=1}^{n} h_{i-2(d,j,k)} T_{i(d,j,k)} \Delta t_k}{t_c h_{0-2} + \sum_{k=1}^{n} h_{i-2(d,j,k)} \Delta t_{(k)}};$$
$$T_{w1(d,j)} = \frac{t_c h_{0-1} T_0 + \sum_{k=1}^{n} h_{i(d.j.k)} T_{i(d,j,k)} \Delta t_{(k)}}{t_c h_{0-1} + \sum_{k=1}^{n} h_{i(d,j,k)} \Delta t_{(k)}} \tag{3.16}$$

where

t	*time*
t_c	*cycle time*

and subscripts.

d and j	represent the duct and element numbers, and
k	represents the time step used to integrate the heat flow during the whole engine cycle.

Discussion about the Sensitivity of the Model

A single-cylinder engine (at full load, 2000 rpm steady operation) is used for demonstrating the sensitivity of the model. The iteration begins from an initial guess of temperature. The cycle-averaged temperature of the cylinder walls is shown on the

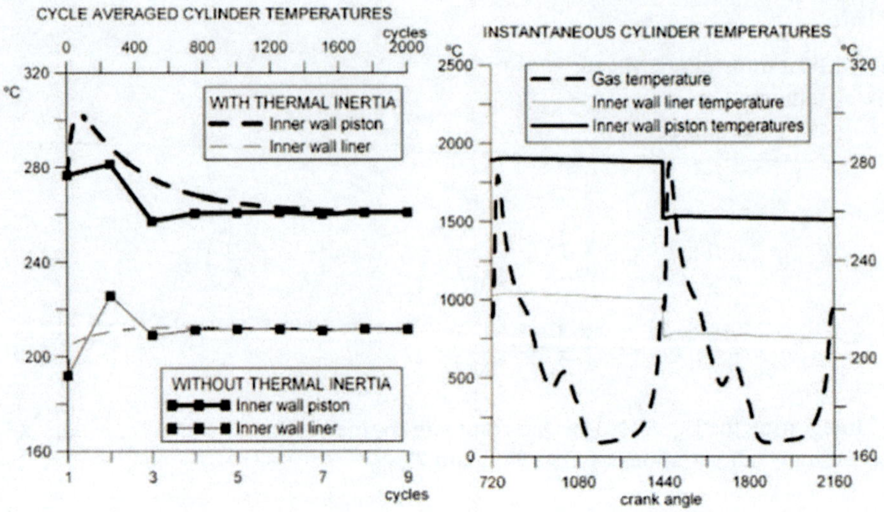

Fig. 3.4 Average (left) and instantaneous (right) values of wall temperatures calculated with and without thermal inertia

left of Fig. 3.4. With the *x*-scale in cycles, the dashed lines correspond to the computation up to 2000 cycles (120 s), considering thermal inertia by normal strategy. In the same graph, with the *x*-scale in seconds, the improved convergence of the temperature of the walls in nine engine cycles (0.54 s), by the new strategy.

The right of Fig. 3.4 shows the details of the accelerated strategy. The gas temperature varies widely (left *y*-axis) while the wall temperature (right *y*-axis) is almost constant due to thermal inertia. At the end of the cycle, the wall temperature is updated according to the strategy, Eqs. (3.11, 3.12).

The turbocharger speed and piston wall temperatures and in turn the turbocharger energy were insensitive to the variation of the most uncertain parameter namely the piston wall thickness by $\pm$ 40%.

3.3.2 Validation of Heat Transfer Model

The heat transfer model was validated for a full load transient test [7, 26, 31], at 1500 rpm, Table 3.3.

The rotor acceleration is caused by the positive difference between the turbine and compressor effective powers [7, 24]. The surplus torque derived from this excess power contributes to the acceleration of the rotating masses, as described by Newton's second law of motion; the acceleration indirectly validates the heat transfer model. Also, airflow, turbine inlet pressure, and turbocharger speed which are closely related to the turbine power agree well with the model. Similarly, the injected fuel and the wall temperature of the turbine inlet duct, the hottest point in the exhaust manifold,

Table 3.3 Experimental engine [32]

Parameter	
Speed, rpm	1500
Aspiration	Turbocharged
Number of cylinders	4
Swept volume, cc	1900

and the exhaust gas temperature which is linearly related to the air–fuel ratio validated the turbocharger energy balance.

3.4 Parametric Studies

First, the engine heat release model was successfully adjusted with experimental data to have a reliable predictive model, Fig. 3.5.

In all parametric studies, some variables have been controlled constant, by PIDs: (a) Torque by the injected fuel, (b) the intake manifold pressure by the VGT opening, (c) the oxygen concentration in the intake manifold after the mixing between fresh air and (d) high-pressure EGR by the EGR valve opening. If the EGR valve is fully open and the oxygen concentration is still high, an intake throttle upstream of the EGR junction reduces the air charge.

To obtain higher T4 with a low BSFC penalty parametric studies were done for the exhaust manifold layout, from the exhaust valve (station 1) to the turbine inlet (station 4), Fig. 3.6. The temperature profiles of the exhaust line for the six operation points are shown in Fig. 3.7. Figure 3.7 for the stations, and an additional station 5, the turbine outlet. The temperature drop across the exhaust port is the highest and hence, the studies were focused on this region.

The next highest drop across the turbine (between 4 and 5) is due to the expansion work and by heat transfer. The latter can be reduced, for example by thermal insulation.

3.4.1 Length and Distribution of Exhaust Ports

The first section of the exhaust ports is formed by two conduits from the respective exhaust valves up to their junction, 1–2, Fig. 3.6. The second section links the junction to the exhaust manifold inlet, 2–3. The first parametric study is focused on length reduction, Table 3.4. Some of them are non-feasible from the design point of view.

Figure 3.8 shows the change in temperature at the inlet to the after-treatment (T4) and BSFC change for baseline data (Fig. 3.7) and other operating cases (Table 3.4).

$$\Delta T_4 = T_{4\,case\#i} - T_{4\,baseline} \tag{3.17}$$

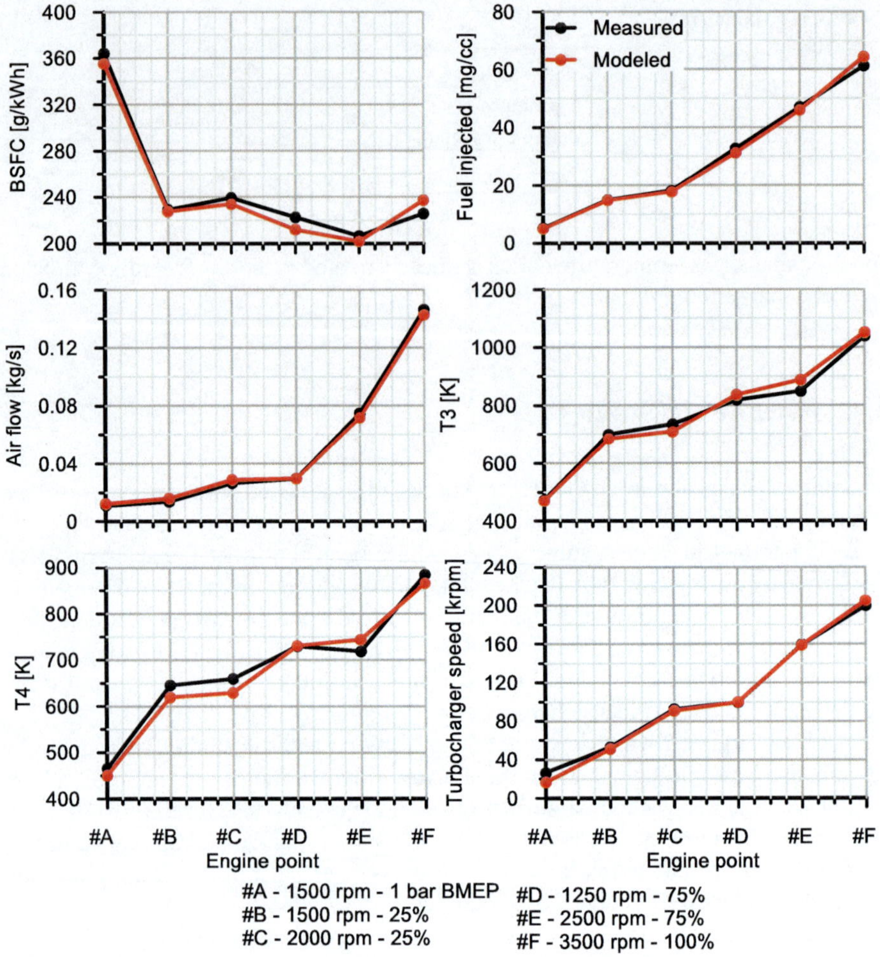

Fig. 3.5 Engine measured and modeled results comparison for the six operation points

$$\Delta bsfc[\%] = \frac{bsfc_{case\,\#i} - bsfc_{baseline}}{bsfc_{baseline}} \times 100 \tag{3.18}$$

The reduction of exhaust port length increases T4 linearly. T4 is less sensitive to the second section since the cross-section is higher and the instantaneous velocity is lower; hence, the heat transfer coefficient is lower. Also, the external surface area is higher in the first section, contributing to higher heat losses than in the common port. A reduction by 50 mm of the first section raises T4 by 20 °C at low loads. It is 30–40 °C in the medium and high loads, underpinning the importance of optimizing the exhaust ports design for fluid mechanics and heat transfer.

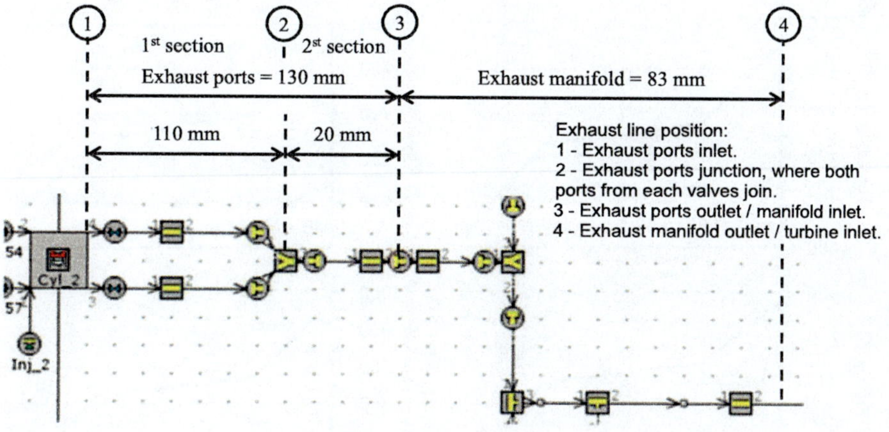

Fig. 3.6 Scheme of the baseline exhaust, lengths of exhaust ports, and distribution

Along with T4, fuel economy improves at high loads, because the turbine inlet temperature increases with reduced PMEP by higher VGT opening, and heat losses, Fig. 3.8d. On the contrary, in the partial load region, Fig. 3.8c, BSFC reduces or increases depending on the particular operating point; the benefits in pumping losses by VGT opening are lost by the need to close the EGR valve or intake throttle to maintain the air–fuel ratio, Fig. 3.9a; when the EGR valve opens maximum (cases 2, 3, and 4 in #C), the intake must be throttled in the manifold, Fig. 3.9b, to generate enough pressure difference to flow the requisite EGR.

3.4.2 *Valves and Ports Diameter*

Due to limited space in the cylinder head, any increase in the diameter of exhaust valves reduces the size of intake valves and the total area of the valves remains nearly the same (Fig. 3.10). The section next to the valves has the same diameter as the corresponding valve; the outlet is larger achieving a tapered geometry.

The ratio of intake to exhaust valve area is selected as a parameter to define the three cases, Table 3.5:

- Case #1 is the baseline, characterized by a higher intake area than the exhaust.
- Case #2, the valve diameters of the intake and exhaust are equal
- Case, #3 represents a design where the valve diameter of the exhaust is higher than that of the intake; the area ratio of intake to exhaust valve is the inverse of case #1.

The reduction of the intake valve decreases volumetric efficiency. To offset this loss, a control strategy of constant boost pressure is considered at high loads #D, #E, and #F, Fig. 3.11; thus the fresh air flow is maintained constant. Figure 3.11a, b

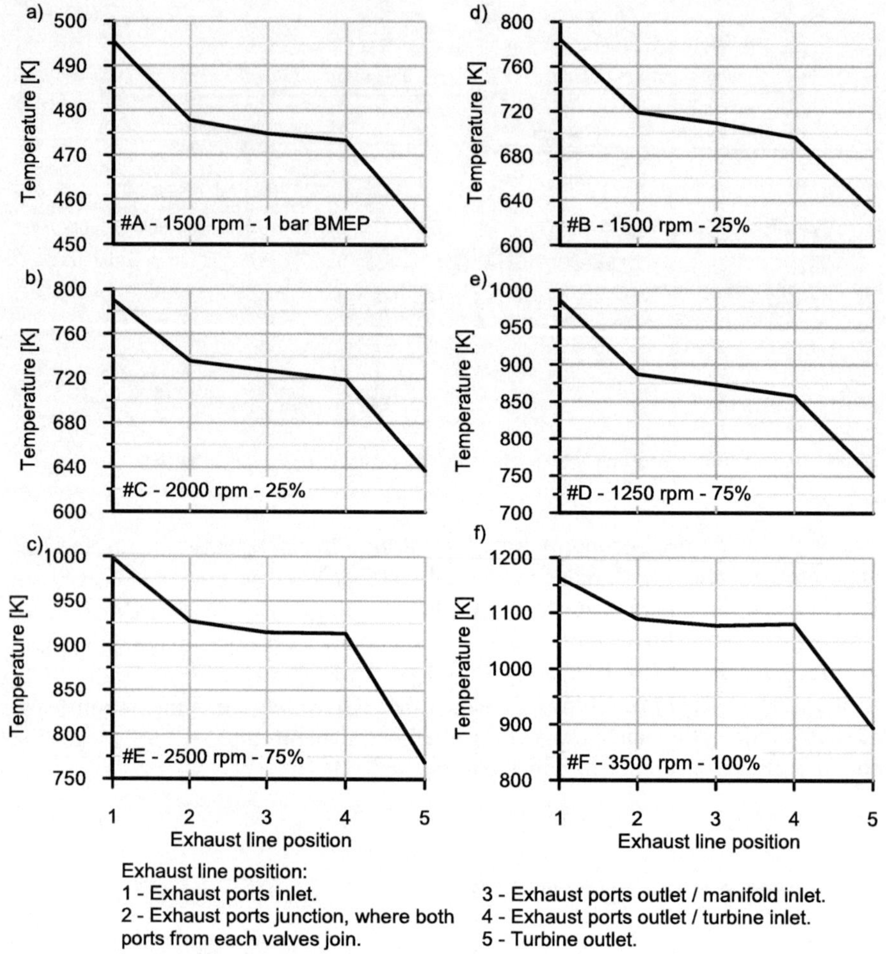

Fig. 3.7 Temperature variation across the exhaust line in the baseline configuration for the six operation points

Table 3.4 Exhaust port length of each case of the parametric study

Case	L total	First section pipe, mm		Second section pipe, mm	
		L	D in/out	L	D in/out
#1 (baseline)	130	110	22/25.5	20	41.5/41.5
#2	80	60	22/25.5	20	41.5/41.5
#3	40	20	22/25.5	20	41.5/41.5
#4	20	20	22/25.5	0	– /–

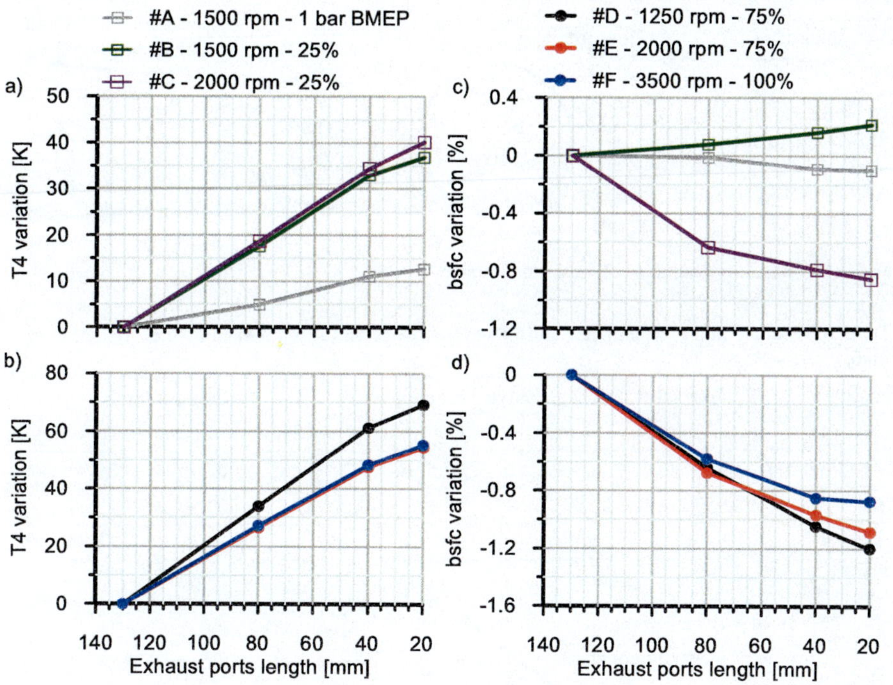

Fig. 3.8 Aftertreatment inlet temperature and BSFC variations at low and high load operating points as a function of the exhaust ports length

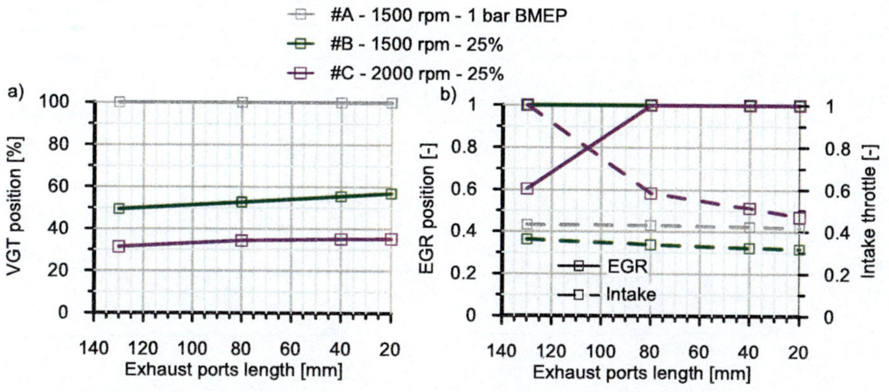

Fig. 3.9 VGT and EGR valve position at low load operating points as a function of the exhaust ports length

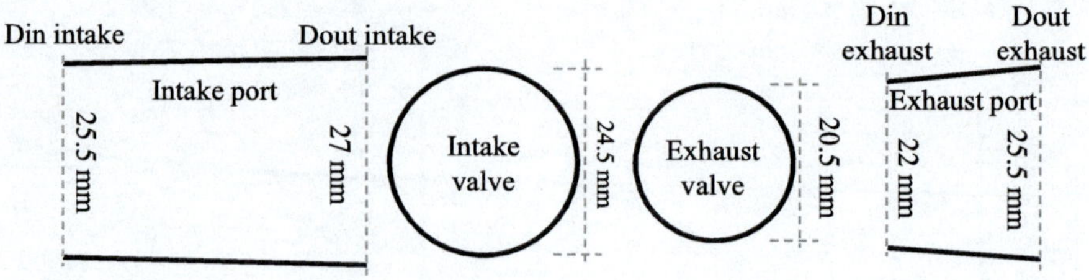

Fig. 3.10 Baseline geometry of valves and ports

Table 3.5 Intake and exhaust ports and valves geometry variation in each case of the parametric study

Case	Intake valve	Exhaust valve	Intake ports		Exhaust ports	
	Dia (mm)	Dia (mm)	Din (mm)	Dout (mm)	Din (mm)	Dout (mm)
#1 (baseline)	24.5	20.5	25.5	27	22	25
#2	22.59	22.59	21.34	22.59	22.59	25.67
#3	20.5	24.5	19.36	20.5	24.5	27.84

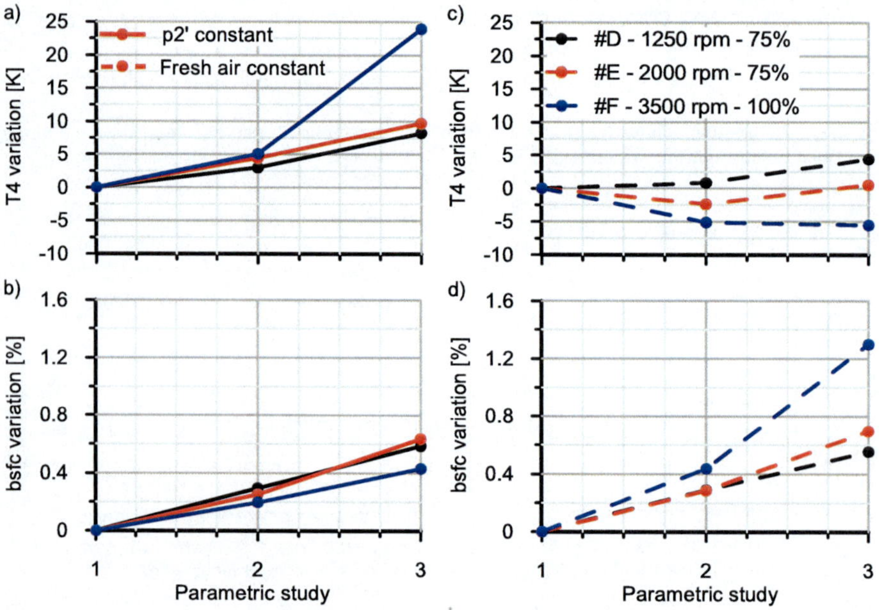

Fig. 3.11 Variations of after-treatment temperature and BSFC at high loads as a function of the valve diameters and the control strategy

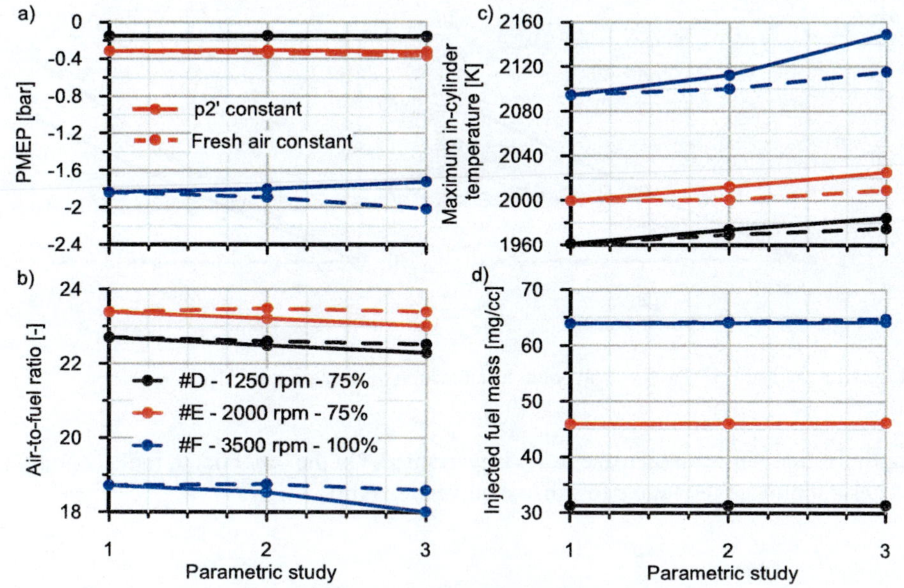

Fig. 3.12 PMEP, air–fuel ratio, maximum cylinder temperature, and injected fuel at high loads as functions of the valve diameter and the boost control strategy

refer to constant p2′ strategy, and (c) and (d) to constant fresh air control. The study of (a) and (c) concludes that the control of boost is more beneficial in raising T4 by 10 K than the constant mass flow strategy, increasingly so as the intake to exhaust valve area ratio decreases. Yet, it pays a penalty in fuel consumption caused by the decrease in the intake valve area. The penalty trend is similar for both strategies.

The higher fuel penalty in fresh mass flow control is because the volumetric efficiency reduces. For example #F, the VGT is closed to increase the boost for keeping the airflow constant; consequently, the turbine inlet pressure increases, and PMEP increases, as shown by the blue dashed lines, Fig. 3.12a. Moreover, the air–fuel ratio decreases with a reduction in volumetric efficiency, Fig. 3.12b. It decreases even when fresh mass flow is controlled and the decrease is more pronounced in the case of boost pressure control. As a result, the maximum cylinder temperature increases, and the indicated efficiency decreases, Fig. 3.12c, leading to the BSFC penalty in the case of boost pressure control. Though the fuel economy is less affected when fresh mass flow control is imposed, the final effect is higher as PMEP increases as discussed earlier.

Since the injected fuel mass is varied to keep the torque constant, it is higher in the case of constant fresh air control strategy, Fig. 3.12d. Since boost pressure control is better, only this strategy is applied at low load points. Variation in T4 is low, Fig. 3.13a. The BSFC penalty is nearly 1% and higher than at high load, Fig. 3.13b. Therefore, efforts are to be concentrated on valve and port geometry

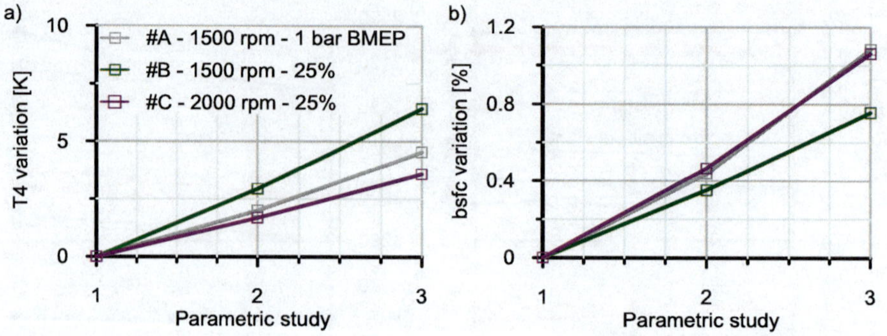

Fig. 3.13 T4 and BSFC at low load points as a function of diameters of valves

optimization while an increase in T4 is warranted. On the other hand, fuel economy criteria influence the intake to exhaust valve area ratio.

3.4.3 Exhaust Valve Timing

The exhaust cam can be phased with respect to the baseline 15° to just avoid interfering with the piston. The fuel economy and volumetric efficiency are the highest at baseline timing, except at low load #A, Figs. 3.14 and 3.15; PMEP is also in the minimum for all operating points except #F. BSFC increase is not very high when the exhaust valve opening is advanced.

Retarding of the exhaust valve timing (Fig. 3.14) increases BSFC without affecting the gas temperature. When advanced (a) the valve overlap reduces and (b) the gas temperature exiting the cylinder increases with a slight worsening of the fuel economy. Consequently, the induced exhaust by the fresh air is less efficient. This leads to an increase of the burned gases fraction inside the cylinder at the combustion start and the decrease of volumetric and indicated efficiencies, Fig. 3.15. T4 increases by about 15 K at low loads. At high loads, a higher temperature rise of 20–60 K is obtained.

3.4.4 Multi-step Valve Opening

Multi-step valve opening comprises a second valve opening [33] to allow internal EGR. It affects T4 and BSFC. Two different strategies are considered:

- The intake pre-opening strategy is to be applied when the cylinder pressure is higher than the intake pressure during the exhaust phase. A part of the burned

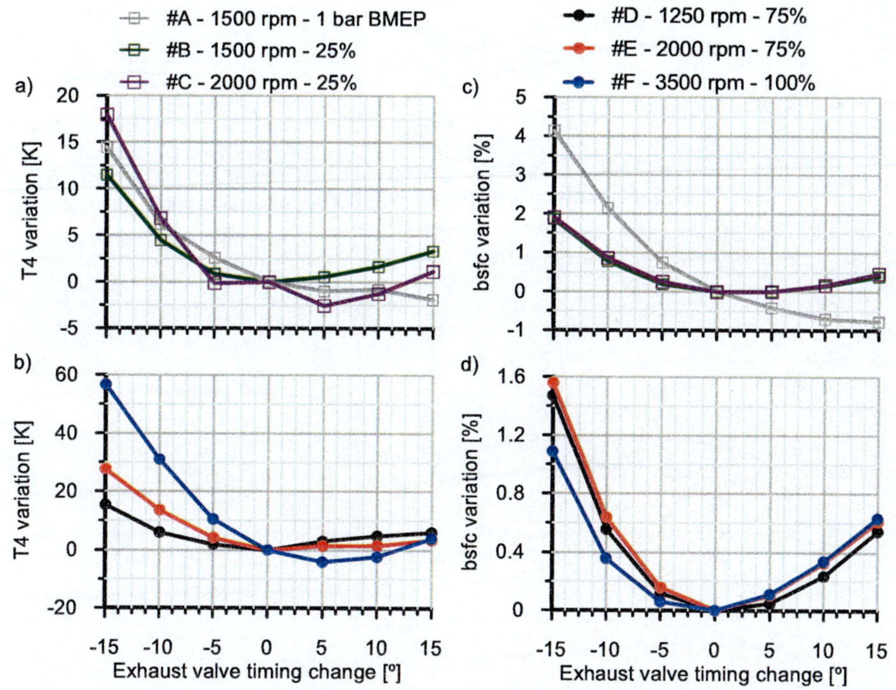

Fig. 3.14 T4 and BSFC at low and high loads as a function of the exhaust valve advance with reference to the baseline

gases moves back towards the intake port. Blue areas identify the range in which intake pre-opening strategy is applied.

- The exhaust post-opening strategy aims at an increase of the internal EGR by making a part of the exhaust gases go back to the cylinder during the intake. Red areas identify the suitable range where the cylinder is at a lower pressure than the exhaust e during the intake.

The proposed valve lift profiles fulfilling these constraints are shown in Fig. 3.16:

- Intake pre-opening window of 60°crank with a maximum lift of 0.42 mm
- The exhaust post-opening window is 100°crank with a maximum lift of 0.80 mm

Four cases have been considered, Figs. 3.16, 3.17 and 3.18.

- case #1: Baseline
- case #2: only intake pre-opening
- case #3: only exhaust post-opening
- case #4: a combination of both strategies

The cylinder temperature increases as a result of the high internal EGR and the decreased mass flow, Fig. 3.17.

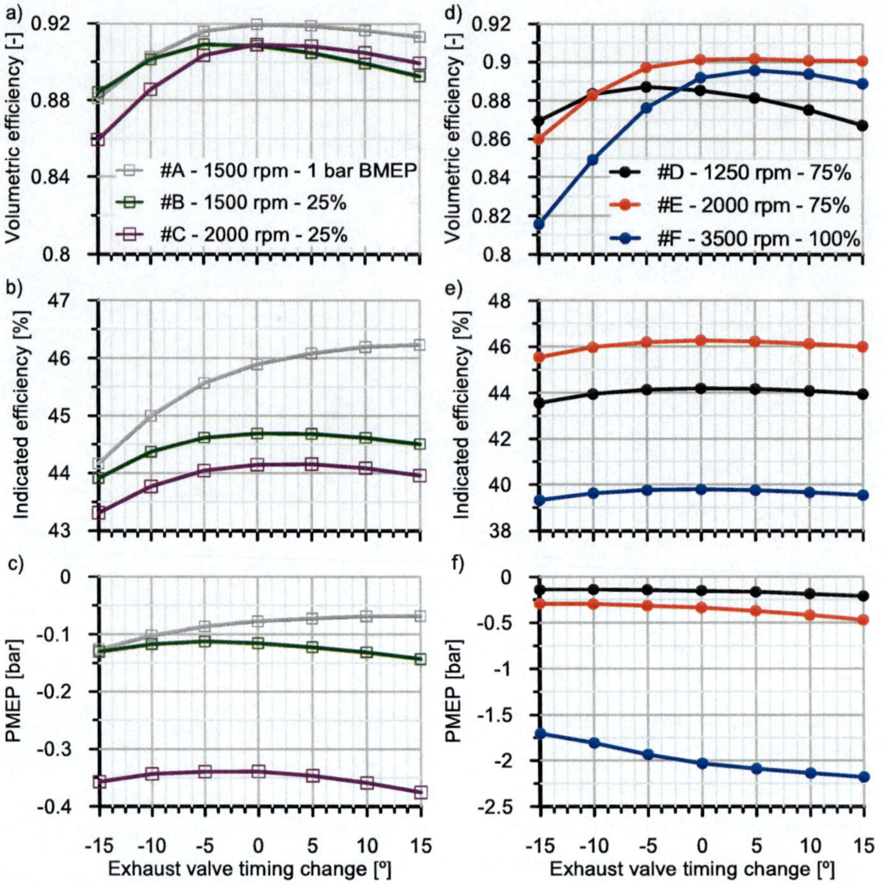

Fig. 3.15 Volumetric efficiency indicated efficiency and PMEP at low and high loads as a function of the exhaust valve opening with respect to baseline

Fig. 3.16 Proposed valve lift profiles for intake pre-opening and exhaust post-opening application

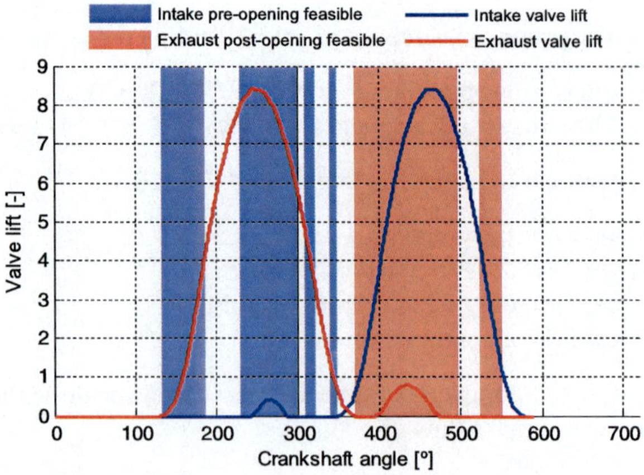

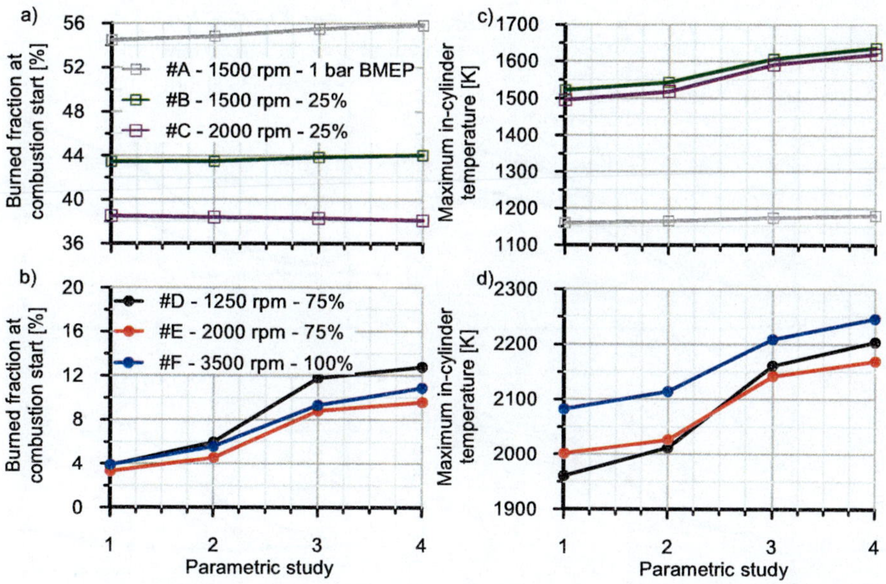

Fig. 3.17 Multi-step valve opening strategy: burned fraction at start of combustion and maximum cylinder temperature at low and high loads

BSFC increases at all loads and speeds, Fig. 3.18, especially at medium speeds, as the engine load increases. T4 improves when the exhaust valve is opened for the second time, Fig. 3.18a, b. At low loads, it is 25–50 °C depending on its combination with the intake valve pre-opening and at high loads by 60–90 °C.

On the contrary, at very low loads, BSFC improves slightly with a marginal increase in T4. At high speeds and full load (#F), the increase in the turbine inlet temperature offsets the decrease in mass flow and resulting in increased VGT opening and consequent reduction in pumping losses. As a result, T4 increases beneficially with an insignificant increase in BSFC.

3.5 Conclusions

Suitable convective correlations are used for the heat transfer coefficient for the different ducts and cylinders of the engine. An important contribution of this work is the calculation of the wall temperatures by a 3-nodes finite-difference scheme accounting for thermal inertia. A strategy using a cycle-averaged thermal balance is devised to accelerate the convergence of steady-state simulation.

Increasing T4 is important for the catalytic reactors to quickly reach target conversion efficiency and reduce the frequency of active regeneration cycles. The studies confirm the potential to widen the operating range in which catalyst reactors perform

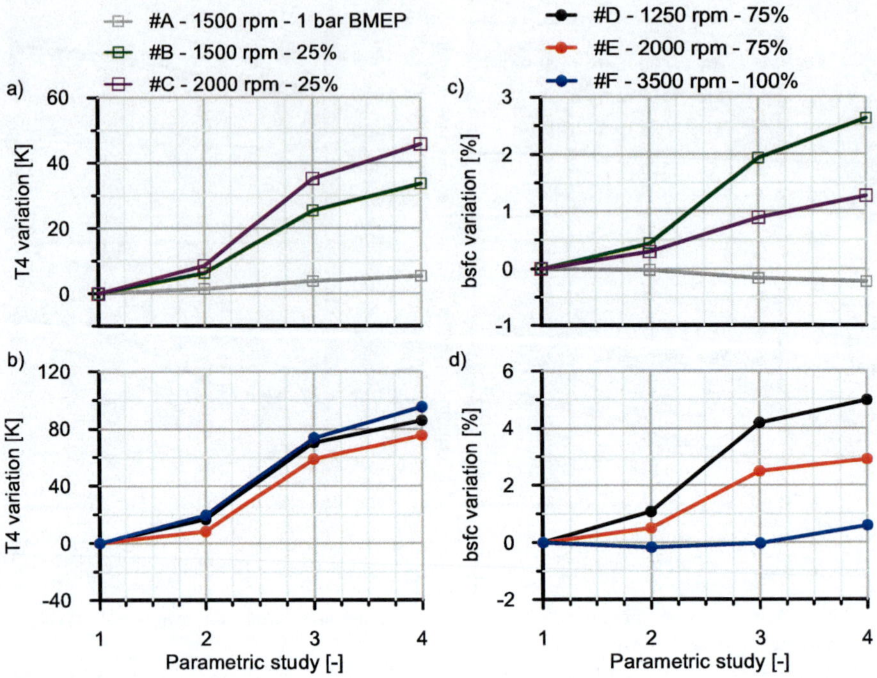

Fig. 3.18 Multi-step valve opening strategy: turbine outlet temperature and BSFC at low and high loads

and regenerate passively in a steady state. Also, the studies endorse an increase in T4 without any serious negative effect on BSFC. Thus, strategies for reducing the time for light-off of the different exhaust after-treatment systems can be designed.

Figure 3.19 summarizes the parametric studies and operation points with respect to the baseline [34]. The top left quadrant is interesting to the engine designers where T4 increases and BSFC decreases; exhaust length studies are in this quadrant. The

Fig. 3.19 BSFC increase with respect to T4 increase for all parametric studies and all operation points simulated

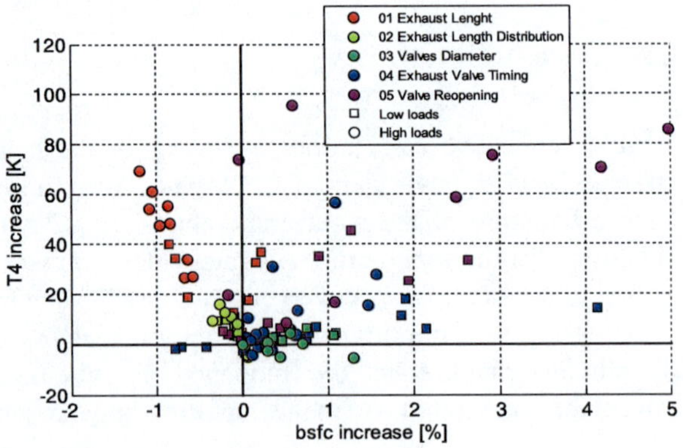

rest of the studies are mostly in the top right quadrant, where both T4 and BSFC increase; here, the design and operation parameters must be carefully considered.

Changing the valve diameter has a very small influence on T4 but greatly affects the BSFC compared with the other studies. Valve reopening and valve timing studies are mainly in the top right quadrant, but the dispersion of the points is very large: points without BSFC penalty but with significant T4 increase and points without T4 increase but remarkable BSFC penalty.

Finally, exhaust length studies show a linear increase in T4 and a decrease in BSFC as exhaust length is reduced and the common branch is longer. Exhaust ports have to be as short as possible and with a small perimeter to reduce heat transfer losses but with a large area to reduce velocity; therefore, the best architecture is with a large single exhaust valve! Of course, these design changes are limited by other mechanical concerns.

Modification of ports increases T4 at rated power accompanied by excessive T3. Therefore, active strategies to cool the ports are needed at rated power. However, active control strategies such as intake and exhaust valve-timings to increase T4 can be selectively performed whenever higher T4 is needed, avoiding high T3 at rated power.

Incidentally, the increase in T4 by about 50 °C contributes to a marginal increase in temperature of the oil that comes in contact with the small area of the bearing housing of the turbocharger.

3.6 Appendix 1: Potential of different designs of exhaust manifolds in saving thermal Energy for transient Performance

The boost pressure, airflow, and turbine inlet pressure develop fast with the insulated dual-walled manifold with its internal walls (1-mm of iron) of low thermal inertia material, an air gap, and external insulation [14, 35], Fig. 3.20. This manifold is compared with another without any insulation and with walls (2 mm of stainless steel) of 13-fold higher thermal inertia. The energy delivered during the first two seconds, however, increased more only by1.7% for the "low thermal inertia" configuration than the "high thermal inertia" case. A third possibility of insulating only the straight part of the ports by inserting a duct of iron (1 mm thick) inside the straight part of the exhaust ports was also modelled (Fig. 3.21). For reference, a photograph of the original exhaust manifold (OEM) of cast iron with a wall thickness of about 6 mm is shown in Fig. 3.22.

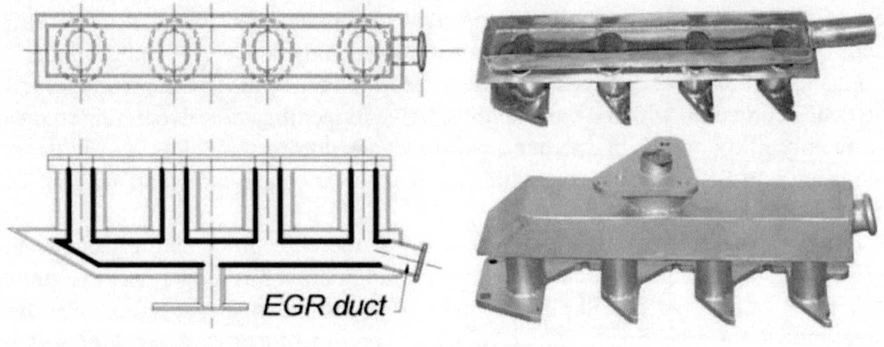

Fig. 3.20 Exhaust manifold scheme with dual walled air gap and two steps of the manufacturing procedure

Fig. 3.21 Scheme of the port with insertion for thermal insulation

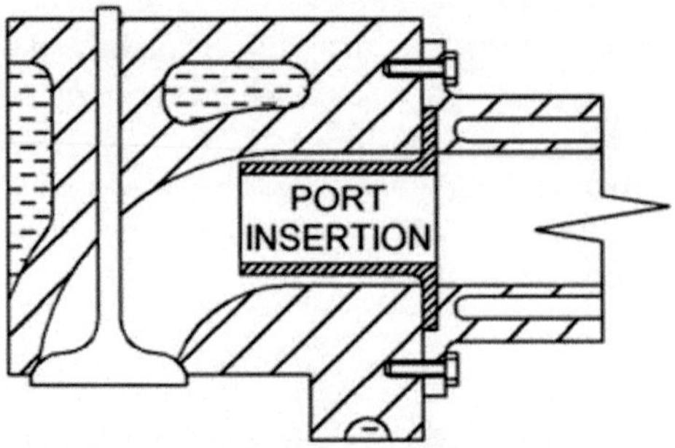

Fig. 3.22 Original grey cast iron exhaust manifold

3.7 Synthesis of Different Designs

To optimally improve engine transient response at low engine speeds, a synthesized exhaust manifold (SEM) is proposed as shown in Fig. 3.23, with the theoretical and experimental learnings from the work on PEM, insulation, and thermal inertia [35]:

- with a directional shape,
- reduced interference of pulses by providing a reed at the manifold outlet up to the turbine scroll
- reduced length and volume,

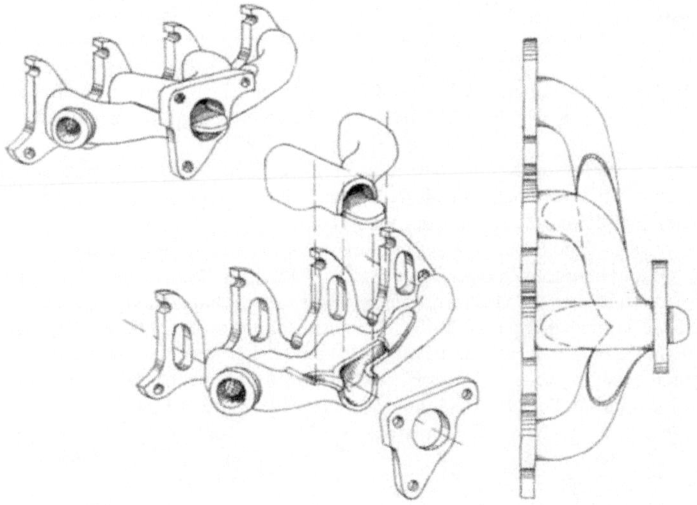

Fig. 3.23 Synthesis exhaust manifold design

- dual walled air gap and thermal inertia of the internal wall 88.3% lower than grey cast iron

The computed interference between pressure pulses reduces in SEM and the volumetric efficiency and airflow are higher by 1% than in the case of the OEM. Also, the temperatures at the turbine inlet and outlet increase by 50 °C, at a steady state. Moreover, the reduced volume of the SEM design helps maintain the kinetic energy, which together with the increase in temperature increases the total energy and consequently the boost. An increase in the airflow is observed during the transient and at steady conditions. Hand in hand, the brake mean effective pressure (bmep) or the engine torque improves.

The maximum torque, 218 Nm (14.05 bar bmep) by the SEM is 6.6% higher than by the OEM. Using the SEM, the time taken to reach 90% of the maximum torque is 1.62 s; it is 14% lower than with the OEM. The energy delivered, till t90%, by the engine increases only by 1.3%.

Acknowledgements The work has been partially supported by the Spanish Ministry of Economy and Competitiveness through grant number TRA2013-40853-R. The authors wish to thank Antonio Peris and José Gálvez for their support in the engine testing and workshop activities.

The authors acknowledge with gratitude the copyright permission from the SAE to use the authors' paper, under license 1072610-1 dated 25 Oct 2020.

References

1. Chi J (2009) Control challenges for optimal NOx conversion efficiency from SCR after-treatment systems. SAE Technical Paper 2009-01-0905. https://doi.org/10.4271/2009-01-0905
2. Stenning L (2010) Strategies for achieving pre-DPF regeneration temperatures using in cylinder post injection on a common rail diesel engine with EGR, DOC, and intake throttle. SAE Technical Paper 2010-36-0306. https://doi.org/10.4271/2010-36-0306
3. Guan B, Zhan R, Lin H, Huang Z (2015) Review of the state-of-the-art of exhaust particulate filter technology in internal combustion engines. J Environ Manage 154:225–258
4. Deng J, Stobart R (2009) BSFC investigation using variable valve timing in a heavy-duty diesel engine. SAE Technical Paper 2009-01-1525. https://doi.org/10.4271/2009-01-1525
5. D'Ambrosio S, Ferrari A, Spessa E, Magro L et al (2013) Impact on performance, emissions and thermal behavior of a new integrated exhaust manifold cylinder head euro 6 diesel engine. SAE Int J Engines 6(3):1814–1833. https://doi.org/10.4271/2013-24-0128
6. Payri F, Serrano J, Piqueras P, García-Afonso O (2011) Performance analysis of a turbocharged heavy-duty diesel engine with a pre-turbo diesel particulate filter configuration. SAE Int J Engines 4(2):2559–2575. https://doi.org/10.4271/2011-37-0004
7. Benajes J, Luján JM, Bermúdez V, Serrano JR (2002) Modelling of turbocharged diesel engines in transient operation. Part 1: insight into the relevant physical phenomena. Proc IMechE Vol 216 Part D J Autom Eng 431–441:D06401
8. GAMMA Technologies. http://www.gtisoft.com
9. OpenWAM webpage, CMT-Motores Térmicos, Universitat Politècnica de València. www.openwam.org
10. Galindo J, Serrano JR, Arnau FJ, Piqueras P (2008) Description and analysis of a one-dimensional gas-dynamic model with independent time discretization. In: Proceedings of the spring technical conference of the ASME internal combustion engine division, pp 187–197
11. Galindo J, Serrano JR, Arnau FJ, Piqueras P (2009) Description of a semi-independent time discretization methodology for a one-dimensional gas dynamics model. J Eng Gas Turbines Power 131(3):034504
12. Serrano JR, Olmeda P, Arnau FJ, Dombrovsky A et al (2015) Analysis and methodology to characterize heat transfer phenomena in automotive turbochargers. J Eng Gas Turbines Power 137(2):021901
13. Serrano JR, Olmeda P, Arnau FJ, Dombrovsky A (2015) Turbocharger heat transfer and mechanical losses influence in predicting engines performance by using one-dimensional simulation codes. Energy 86:204–218
14. Baines N, Wygant KD, Dris A (2010) The analysis of heat transfer in automotive turbochargers. J Eng Gas Turbines Power 132(4):042301. https://doi.org/10.1115/1.3204586
15. Caton J (1982) Comparisons of thermocouple, time-averaged and mass-averaged exhaust gas temperatures for a spark-ignited engine. SAE Technical Paper 820050. https://doi.org/10.4271/820050
16. McAdams WH (1954) Heat transmission. McGraw-Hill
17. Chapman AJ (1960) Heat transfer. Macmillan Publishing Company (New York)
18. Hilpert R (1933) Wärmeabgabe von geheizen Drahten und Rohren, Forsch. Geb. Ingenieur-wes, vol 4, p 220
19. Woschni G (1967) A universally applicable equation for the instantaneous heat transfer coefficient in the internal combustion engine. SAE paper 670931
20. Cipolla G (1988) Heat transfer correlations applicable to the analysis of internal combustion engine head cooling. Heat and mass transfer in gasoline and diesel engines. Proceedings of the international center for heat and mass transfer. Afgan and Spalding pp. 373–396
21. Rohsenow WM, Harnett JP, Ganic EN (1985) Handbook of heat transfer fundamentals. McGraw Hill
22. Santos R (1999) A study on the available energy in the exhaust gases of diesel engines. Ph.D. Thesis. Text in Spanish. Universidad Politécnica de Valencia, Spain

23. Luján JM, Serrano JR, Arnau F, Dolz V (2003) Heat transfer model to calculate turbo-charged HSDI diesel engines performance. SAE paper 2003-01-1066
24. Payri F, Galindo J, Serrano JR (2002) Variable geometry turbine modelling and control for turbocharged diesel engine transient operation. Thermo- and Fluid-dynamic processes in diesel engines, pp 189–210. Springer-Verlag. ISBN 3-540-42665-5
25. Payri F, Reyes E, Galindo J (2001) Analysis and modelling of the fluid-dynamic effects in branched exhaust junctions of I.C.E. Int J Gas Turbine Power Trans ASME 123(1):197–203
26. Payri F, Reyes E, Serrano JR (2000) A model for load transients of turbocharged diesel engines. 1999 SAE Trans J Engines 108-3:363–375
27. Benajes J, Luján JM, Serrano JR (2000) Predictive modelling study of the transient load response in a heavy-duty turbocharged diesel engine. SAE paper 2000-01-0583
28. Reyes M (1994) Modelo de transferencia de calor para colectores de escape de motores alternativos. Ph.D. Thesis. Text in Spanish. Universidad Politécnica de Valencia, Spain
29. Benajes J, Torregrosa A, Reyes M (1991) Heat transfer model for I.C. engine exhaust manifolds. Proc EUROTHERM 15:189–194
30. Incropera FP, David PD (1996) Fundamentals of heat and mass transfer, 4th edn. Wiley, New York. ISBN 0471304603
31. Galindo J, Luján JM, Serrano JR, Hernández L (2005) Combustion simulation of turbocharged HSDI diesel engines during transient operation using neural networks. Appl Therm Eng 25:877–898
32. Galindo J, Luján JM, Serrano JR, Dolz V, Guilain S (2004) Design of an exhaust manifold to improve transient performance of a high-speed turbocharged diesel engine. Exp Thermal Fluid Sci 28:863–875
33. Benajes J, Reyes E, Luján JM (1996) Modelling study of the scavenging process in a turbocharged diesel engine with modified valve operation. Proc Inst Mech Eng C J Mech Eng Sci 210:383–393
34. Serrano JR, Piqueras P, Navarro R, Gómez J, Michel M, Thomas B (2016) Modelling analysis of aftertreatment inlet temperature dependence on exhaust valve and ports design parameters. No. 2016-01-0670. SAE Technical Paper
35. Galindo J, Luján JM, Serrano JR, Dolz V, Guilain S (2004) Design of an exhaust manifold to improve transient performance of a high-speed turbocharged diesel engine. Exper Thermal Fluid Sci 28(8):863–875

Chapter 4
Variable Geometry Turbocharger Technologies for Exhaust Energy Recovery and Boosting

D. A. Subramani and K. Ramesh

4.1 Supercharging and Turbocharging

Supercharging is the process of supplying the air into an engine with a higher density than the atmosphere. This increases the amount of air supply to the engine and allows an increased amount of fuel supply to burn, and increases the engine power output. The air with higher pressure is generated with the help of the air compressor, and the same is driven by using engine power output from the crankshaft. For the compressor to deliver the air at higher pressure, about 4 bar (a) for a typical high power intensity diesel engine at rated power, the compressor's power is about 40–50% of the power developed by the engine. This is the loss of fuel energy and economically not viable for conventional engine applications. An alternate way of delivering the air to the engine with an air compressor driven by a turbine using the engine's exhaust gas is called turbocharging.

Turbocharging is a specific method of supercharging, uses the energy from the hot exhaust gas from the engine that would normally go to waste to drive the supercharging compressor. So, the power required by the turbocharger compressor is not extracted from the engine power output. The benefit is obtained without loss of efficiency. Note that the method of turbocharging for the engines with different fuels (diesel, gasoline, and gas) remains the same except for the level of compressed air pressure level and mass flow rate. The compressor outlet pressure is also called boost pressure, and the level of boost pressure is normally less with the gasoline engine and the engines operating with the gases considering the knocking limitation. For diesel engine applications, the level of boost pressure is high. For higher engine power output, a higher level of airflow rate at higher pressure is required.

D. A. Subramani · K. Ramesh (✉)
Turbo Energy Private Limited (TEL), Chennai 600028, India
e-mail: ramesh.k@turboenergy.co.in

P. A. Lakshminarayanan and A. K. Agarwal (eds.), *Handbook of Thermal Management of Engines*, Energy, Environment, and Sustainability,
https://doi.org/10.1007/978-981-16-8570-5_4

First attempts to use exhaust gas energy to drive a supercharging compressor were proposed by Büchi [1]. The typical arrangement of the turbocharged engine system is shown in Fig. 4.1. The exhaust gas from the engine drives the turbine, and the compressor installed at the other end of the turbine compresses the air from atmospheric condition to the required level of pressure. The charge air cooler (CAC) is used to cool down the compressed air and to increase the air density before entering the engine intake manifold.

Based on the nature of energy recovery, turbocharging is classified into two categories: constant pressure turbocharging and pulse system turbocharging. In constant pressure turbocharging, the exhaust gas from all the cylinders is connected to a common large exhaust manifold. The gas supplied to the turbine is nearly at a steady-state condition, and the turbine efficiency is maximum. The arrangement of the exhaust manifold and turbine is simple. However, the kinetic energy of the exhaust leaving the exhaust ports or valves is not fully utilized. In pulse turbocharging, the pulsating energy is transferred to the turbine through a narrow pipe. The optimization of the exhaust piping system helps in improving the utilization of pulse energy. Based

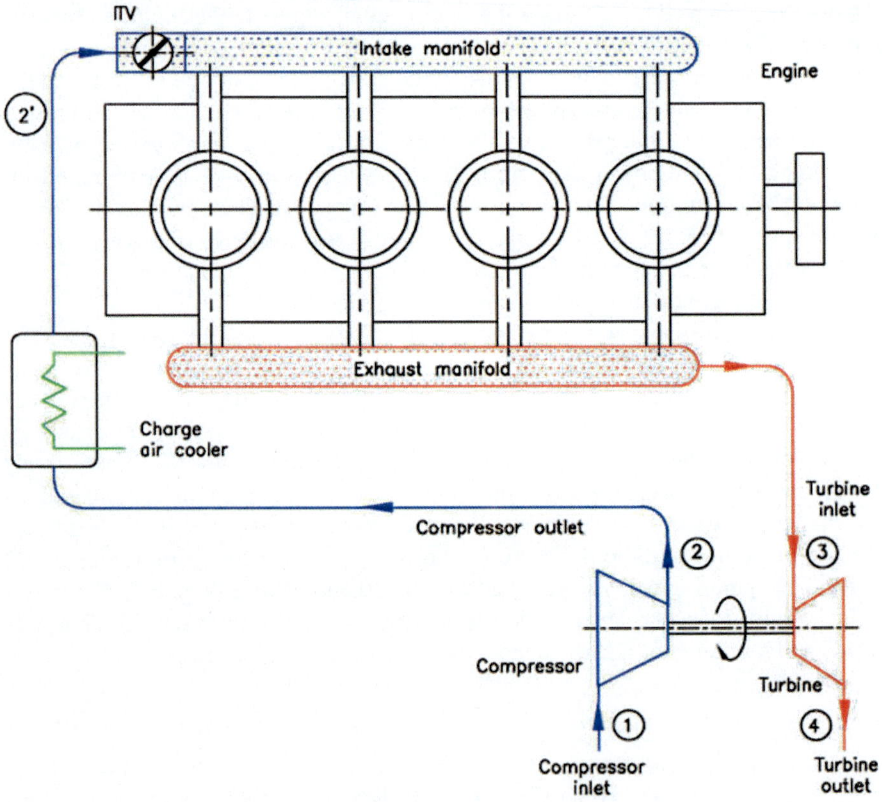

Fig. 4.1 Arrangement of the turbocharged engine system

on the nature of the engine arrangement and airflow requirements, the turbocharging system is suitably selected.

4.2 Types of Turbocharger

Based on the gas entry to the turbine, the turbocharger (TC) is broadly classified into two categories. One is the fixed turbine geometry (FTG), and another one is the variable turbine geometry (VTG). Figure 4.2 shows the TC configurations with FTG and VTG. In FTG, the exhaust gas from the engine is directly supplied to the turbine wheel. The flow angle at the entry to the turbine wheel is based on the shape of the turbine housing volute. The flow angle varies with mass flow rate and leads to incidence loss at off-design conditions, leading to a reduction of turbine efficiency. The selection of the turbine volute size (flow area) defines the turbine flow capacity. To achieve higher engine torque at lower engine speeds, a smaller size turbine is used; however, this leads to a high exhaust gas back pressure at a higher mass flow rate (at higher engine speeds) as the swallowing capacity of the turbine becomes insufficient. Consequently, there is a power loss and an increase in engine brake-specific fuel consumption (BSFC). To limit the exhaust gas pressure, the excess gas is bypassed through the waste-gate system. A TC with FTG is capable of producing a medium level of boost pressure and mass flow rate and is used for medium-duty engine applications. As the flexibility in altering the turbine flow area is not available, the turbine and compressor must be carefully matched to handle the air and exhaust gas flow.

In VTG, the exhaust gas is supplied to the turbine wheel through a set of nozzle vanes, which is arranged circumferentially upstream of the turbine wheel. Further, the flow velocity is varied through the nozzle vanes. The flow angle at the entry to the turbine wheel is based on the opening position of the nozzle vanes. At lower engine speeds and loads, the nozzle vanes are set at the minimum opening position, the flow velocity accelerates, and higher rotor turbine speed is achieved. Hence, the higher airflow from the compressor enables higher fuel injection to enhance the engine torque and acceleration. At higher engine speeds and loads, the exhaust mass flow is high, and the opening area of the nozzle vanes is increased to limit the turbocharger speed and hence the airflow and the boost pressure. The optimization of nozzle vane opening over the engine operating range ensures improved engine performance and BSFC.

With varying the turbine flow area using the nozzle vanes, the matching of the turbine and compressor becomes highly flexible. Further, while downsizing and down-speeding an engine, the use of VTG helps in managing the airflow rate effectively and contributes to the reduction of CO_2. A TC with VTG is capable of producing a higher level of boost pressure and mass flow rate, and the same can be used where the fast transient response of a vehicle is required. For further higher

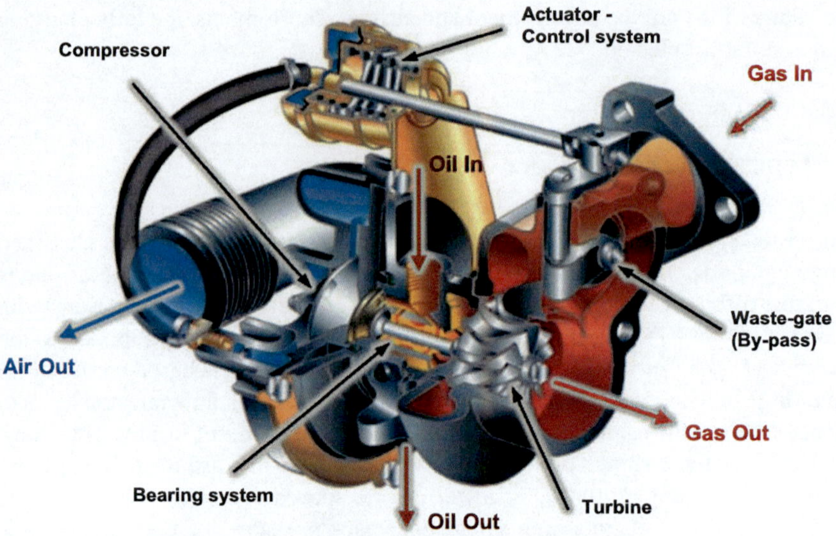

a. TC with fixed turbine geometry (FTG)

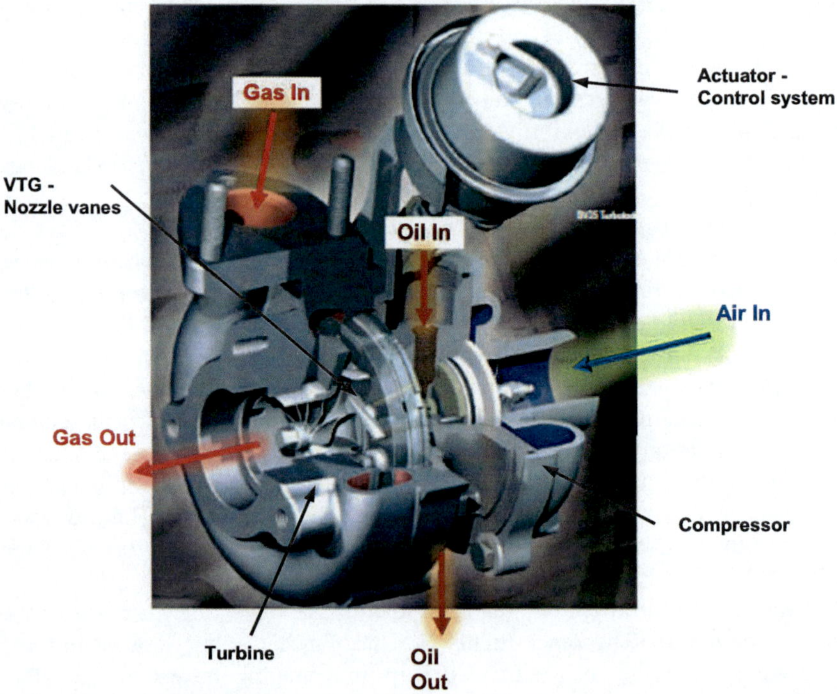

b. TC with variable turbine geometry (VTG) (TurboNews, BorgWarner Turbo Systems, 2005)

Fig. 4.2 TC configurations [2]

torque at a lower speed, the multiple-stage TC is used. The use of VTG and a multiple-stage TC helps in regulating the exhaust gas recirculation for controlling oxides of nitrogen (NO_x) emission.

4.3 Turbocharger Construction

The sectional view of a typical turbocharger is shown in Fig. 4.3. The turbocharger consists of three major components: a compressor, a turbine, and a bearing system. Though other types of compressors and turbines are available, centrifugal compressors and radial turbines are widely used for TC applications. The rotating shaft is attached to the turbine wheel as an integral part of the turbine wheel sub-assembly. The compressor wheel is mounted at the other end of the sub-assembly forming the rotor assembly. A journal bearing system supports the rotor.

The turbine and compressor casings are attached to the bearing housing. The oil is supplied to the bearing system for lubricating the rotor assembly and for carrying away the excess heat in the system. Besides, the oil flow damps vibrations, provides stability to the rotor, and reduces the turbocharger noise. For turbochargers with a higher duty cycle and gasoline engine applications, bearings are water-cooled to limit the operating temperature and avoid oil coking. To prevent the seepage of air from

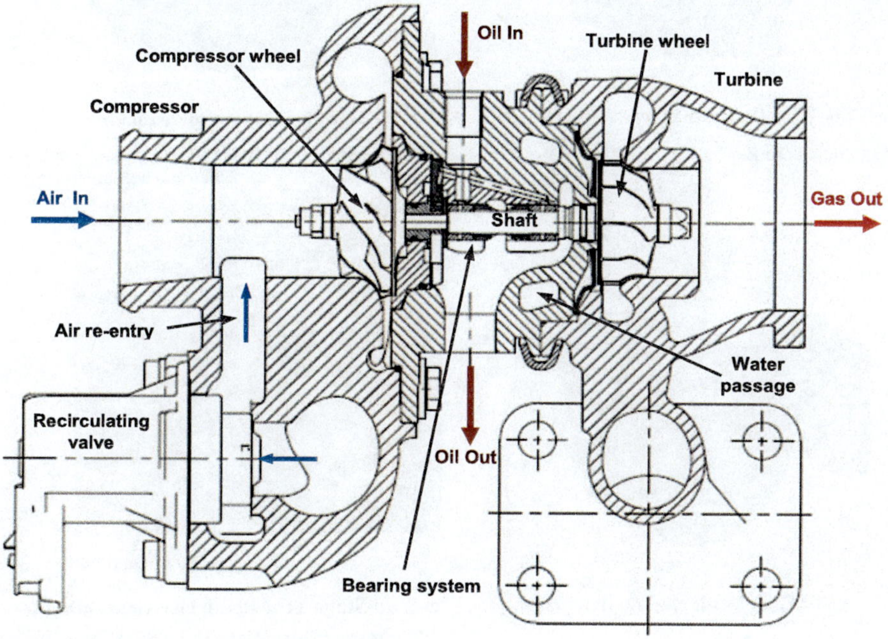

Fig. 4.3 A sectional view of the turbocharger

the compressor and gas from the turbine into the bearings, sealing rings are used on the turbine and compressor sides.

For TC with FTG, usually, a self-regulated pressure type actuator sensing compressor outlet pressure is used for actuating the waste-gate valve. However, for better control of the emission, a vacuum-type actuator with position sensor feedback or electrical actuators is also used with either FTG or VTG. Figure 4.4 shows FTG, VTG, and two-stage TC configurations with different actuation systems. Figure 4.5 shows the nozzle vanes actuating mechanism for VTG. The use of an electrical actuator helps in the rapid opening and closing of the WG valve and nozzle vanes to achieve the required turbine speed and hence the airflow. Also, it helps in raising the exhaust gas pressure before the turbine to support the EGR operations and raising the gas temperature after the turbine for thermal management of the exhaust after-treatment system.

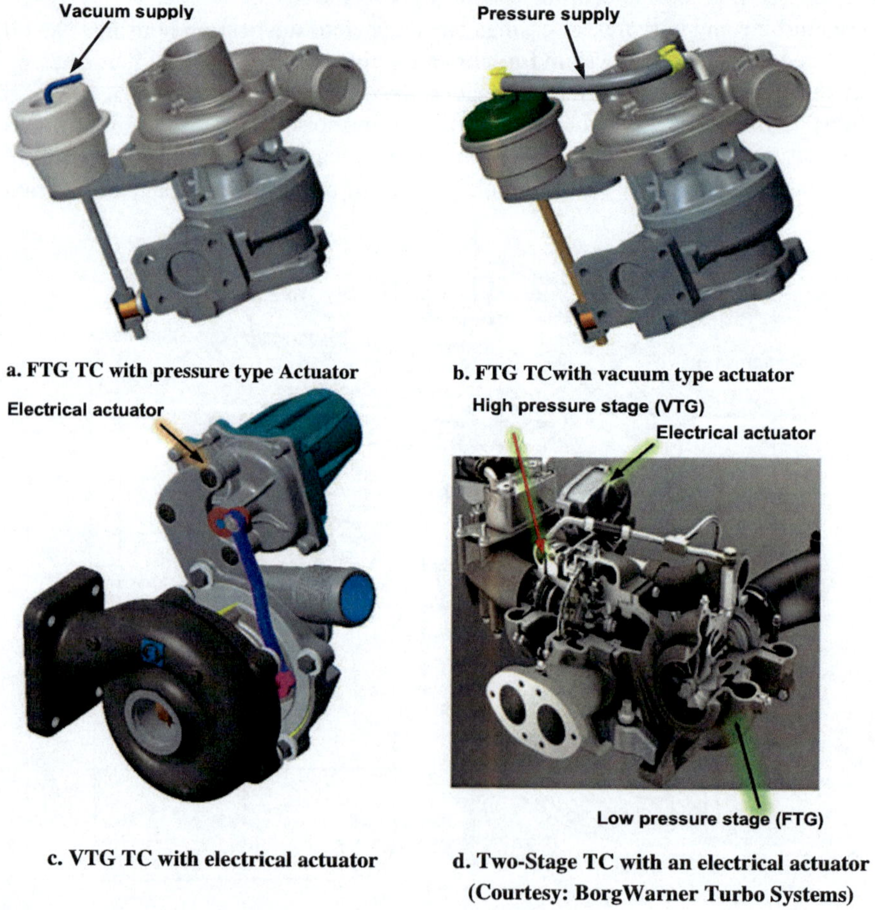

Fig. 4.4 Turbochargers with different actuation systems

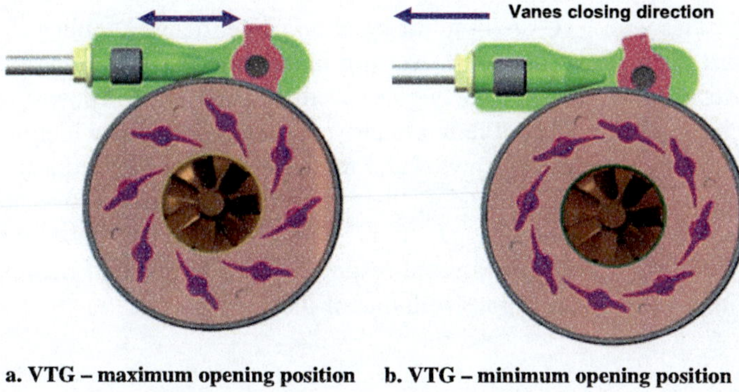

Fig. 4.5 VTG—Nozzle vanes actuating mechanism

The performance characteristic of the TC depends on the efficiency of the compressor, turbine, and bearing system. To have improved overall efficiency, the compressor and turbine specifications are optimized based on the range of flow and pressure for the intended engine operating regime. The selection of appropriate bearing journals plays a major role in the reduction of rotor friction and improvement in transient response.

4.4 Engine and Turbocharger: Performance Characteristics

The engine power output is proportional to the size of the engine. Further, for the given cylinder capacity, the power output is based on the amount of fuel being burned. Today, the power to weight ratio of automotive engines is increased manifold, and the engines are downsized 40–50% based on the type of fuel being used, refer to [3, 4]. The specific power output for the typical diesel engine is up to 80 kW/L, and for the gasoline engine is around 95 kW/L. The net increase in power is proportional to the fuel being burned with sufficient air. The advanced fuel injection equipment with common rail and electronic controls aids (1) in injecting the fuel in multiple shots at high pressure enabling fine atomization of fuel at rate and duration determined at will (2) to raise the torque/brake mean effective pressure (BMEP) over a wide range of speeds. To cope with the fuel supply rate, air should be supplied to the engine. For the given engine cylinder capacity, the power can be increased by turbocharging, nearly double that of a naturally aspirated engine. Also, the mechanical efficiency of the engine improves, accompanied by a reduction in fuel consumption and exhaust emissions. The selection of turbochargers plays a major role in achieving the desired engine torque, rated power, BSFC, and reduction of emissions. Also, the transient response of the engine depends on the rate of increase in torque which in turn depends on the rate of increase in the air supply by the TC.

Whilst the TCs with FTG are used for medium power density applications, those with VTG are used for applications desiring high torque at lower engine speeds. Further, for very high torque at lower engine speeds and higher rated power, the multi-stage turbocharger is preferred. Predominantly, two-stage TCs are widely used. The application of a three-stage TC is, however, limited. Figure 4.6 shows the comparison of different turbocharger technologies, the torque rise, and the relative merits and constraints. The VTG and two-stage TC provide progressively higher torque and power than the FTG. Also, the transient response and emissions of the engine and the vehicle improve commensurate with that of the TC.

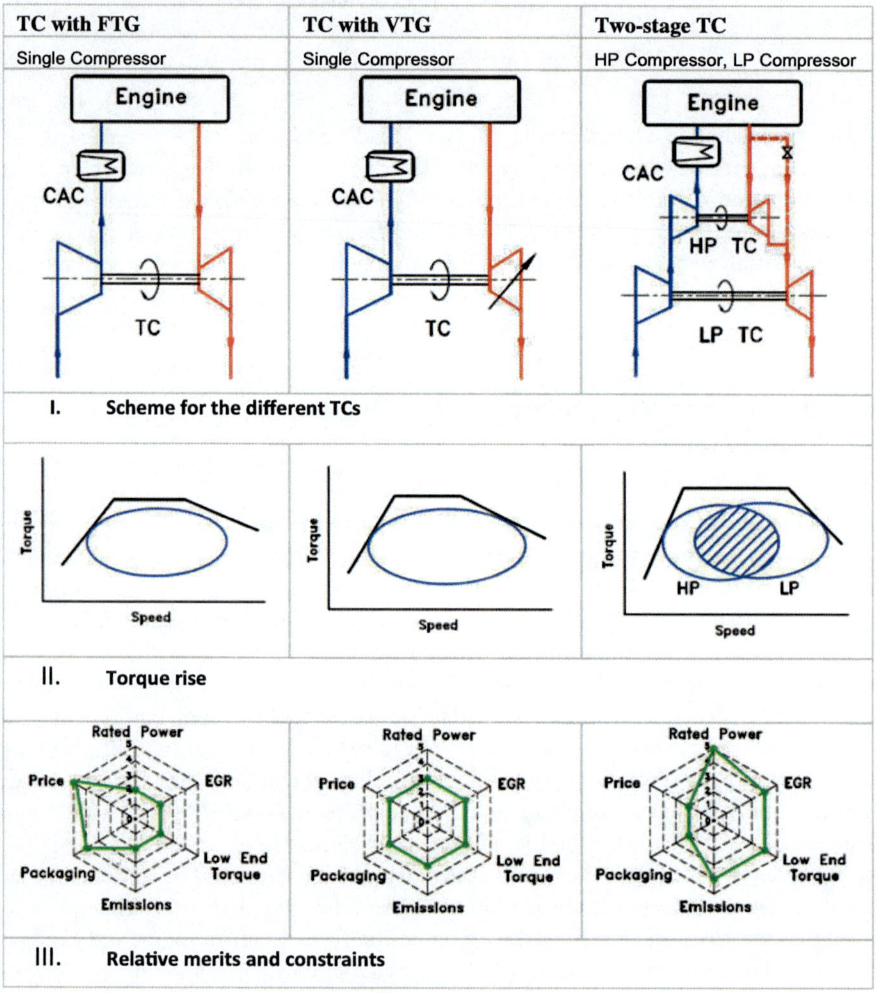

Fig. 4.6 Comparison of different turbocharger technologies

The two-stage TC includes the combination of two different TCs with a smaller size turbine and compressor for high-pressure stage TC and a larger size turbine and compressor for low-pressure stage TC. For improved transient response, the VTG is used in the high-pressure stage. The two-stage TC is complex and requires additional control systems for maneuvering the turbochargers switching over from higher pressure (HP) to low pressure (LP) and vice versa, depending on the engine speeds and loads. Therefore a suitable TC must be selected considering the performance requirements.

4.4.1 Engine Torque and Brake Mean Effective Pressure

Figure 4.7 shows the brake mean effective pressure (BEMP) characteristics for a typical 1.5 L diesel engine with different turbochargers. For an engine with FTG, the maximum BMEP is around 22 bar, whereas, for the engine with VTG, the maximum BMEP is around 25 bar. With a two-stage TC, the BMEP is around 30 bar. From minimum full load point 'A', the rate of increase of BMEP is higher with VTG and two-stage TC compared to FTG and the maximum BMEP is achieved at lower engine speeds. Normally, the peak BMEP is maintained over the speed range (B–B′). This range is progressively higher for the engine with TC having VTG and two-stage TC compared to the TC with FTG. The rate of increase in BMEP with speed improves the engine transient response.

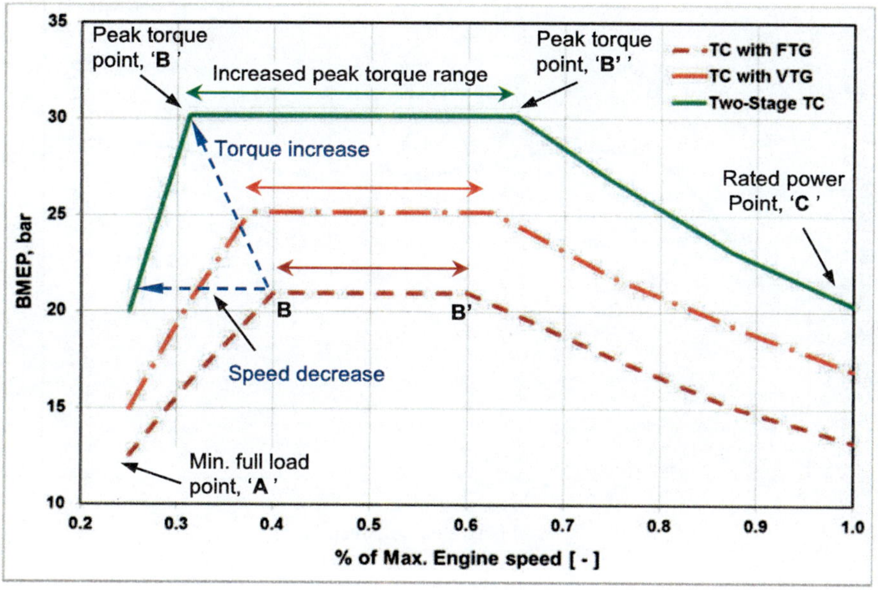

Fig. 4.7 BMEP characteristics for diesel engine

For TC with FTG, the low-end torque can further be increased with the use of a smaller size turbine. However, at speeds higher than peak torque speed, the engine backpressure increases due to small turbine swallowing capacity and leads to an increase in BSFC. Typically, the exhaust gas by-pass is less than 35%. Higher bypassing is avoided since the turbine efficiency drops steeply; what is more, providing a larger area for the by-pass port is mechanically limited. With the increase in engine power, the peak firing pressure increases. It can be as high as 175–250 bar, depending on the BMEP. Further, the temperature of the exhaust gas increases to 850 °C with a boost pressure of about 3.5 bar (absolute) for diesel applications. For gasoline application, the exhaust gas temperature can go up to 1050 °C. To withstand the higher pressure and temperature and without any physical deformation, the material for the engine parts and the turbocharger-turbine parts are selected.

4.4.2 Engine Air Mass Flow Rate

The airflow to the engine is described using the excess air factor (λ).

$$Excess\,air\;factor,\;Lambda\,(\lambda) = \frac{Actual\,air\;fuel\,ratio}{Theoretical\,air\;fuel\,ratio} \tag{4.1}$$

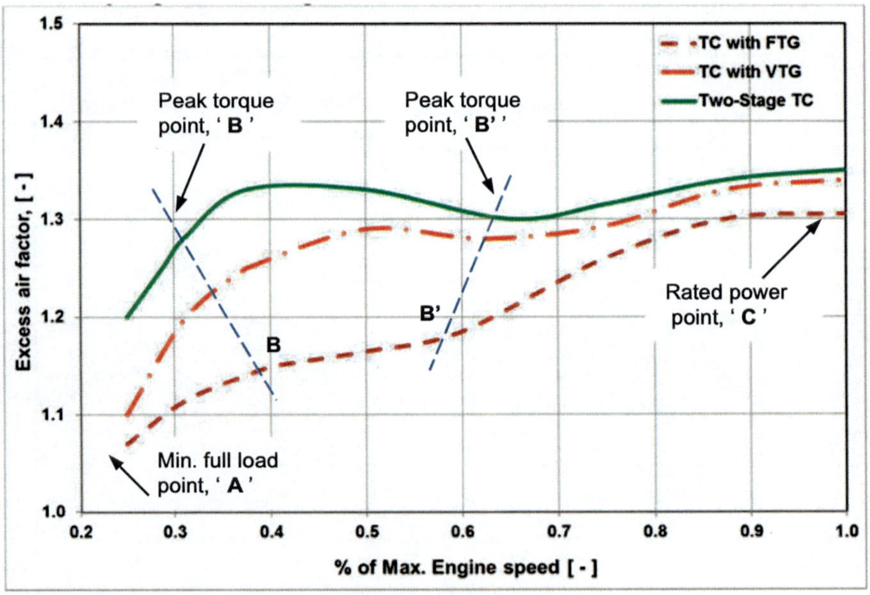

Fig. 4.8 Excess air factor for diesel engine with different turbochargers

Figure 4.8 shows the trend of the excess air factor (λ) for an engine with different turbochargers. At minimum full load point (A), for FTG, the lambda is around 1.07; it is nearly 1.1 and 1.2, respectively, for the VTG and two-stage TC. At peak torque point (B), the lambda is around 1.15 for FTG, and it is around 1.24 and 1.3, respectively for VTG and two-stage TC. To limit the oxides of nitrogen (NO_x), lambda is maintained to about 1.2–1.30 at the mid-speeds (B′). Beyond the peak torque region, the lambda rises gradually up to B′ and rated speed C′ for FTG; on the other hand, for the VTG and two-stage TC, the lambda gradually reduces up to B′ and then increases up to the C′. At high speeds, the airflow is limited by considering the trade-off between smoke and NO_x and the margins for the maximum turbocharger speed and compressor choke.

4.4.3 *Brake Specific Fuel Consumption*

Figure 4.9 shows the trend of the brake-specific fuel consumption (BSFC) of an engine with different turbochargers. Normally, an engine is tuned to have the minimum BSFC in the peak torque region. At lower engine speeds and rated power, the BSFC is higher. With the VTG and two-stage TC, it is lower than with FTG. The difference is around 5–10 g/kW-h (about 3–5%). Fine-tuning the opening of nozzle vanes of VTG or optimum sharing of high-pressure and low-pressure TC stages improves the overall turbocharger efficiency at any given operating point (speed and

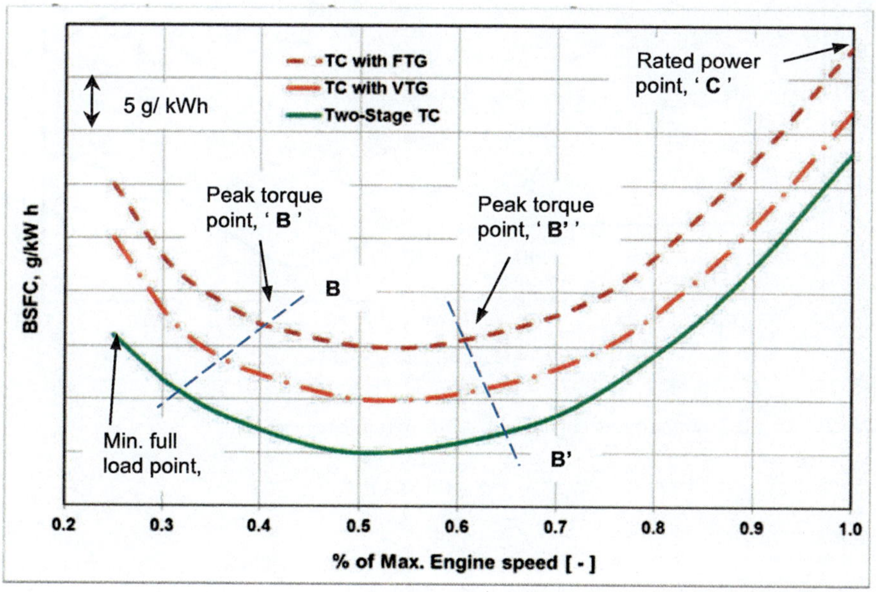

Fig. 4.9 BSFC for diesel engine with different turbochargers

load). Thus, optimum air delivery and boost pressure are achieved accompanied by minimum backpressure and hence improved BSFC.

4.4.4 Smoke Emission

As a result of higher air availability with VTG and two-stage TCs, the combustion of fuel is better, and it contributes to the reduction in smoke. Figure 4.10 shows the trend of the smoke from an engine with different turbocharger configurations. At lower engine speeds, the smoke from an engine with VTG is 0.3 FSN lower than with FTG; at peak torque region (higher engine speeds), the smoke is 0.5 FSN lower. At lower engine speeds, with two-stage TC, the smoke is about 1.7 FSN lower than with FTG. The smoke level is lower with VTG and two-stage TC compared to FTG. The reduction of smoke contributes to the decrease in particulate matter and particulate number.

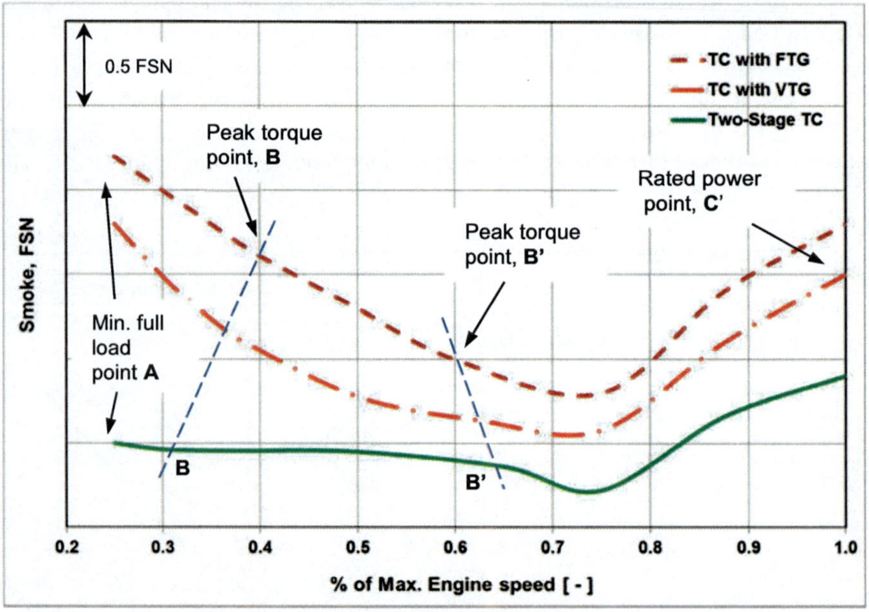

Fig. 4.10 Smoke level for diesel engine with different turbochargers

4.4.5 Compressor Characteristics

The characteristics of the compressor are described by reduced volume flow rate, pressure ratio, and efficiency; the first two terms are defined as follows.

$$Reduced\ volume\ flow\ rate = \frac{m_c\sqrt{\frac{T_1}{T_0}}}{\rho_0\left(\frac{P_1}{P_0}\right)} \tag{4.2}$$

$$Pressure\ ratio = \frac{P_{02}}{P_{01}} \tag{4.3}$$

where,

m_C air mass flow rate through the compressor, kg/s.
P_0 reference ambient pressure = 1000 mbar.
P_1 pressure at the compressor inlet, bar (abs).
P_{01} total pressure at the compressor inlet, bar (abs).
P_{02} total pressure at the compressor outlet, bar (abs).
T_0 reference ambient temperature = 298 K.
T_1 the temperature of the air at the compressor inlet, K
ρ_0 density of the air at reference ambient condition, kg/m^3.

The isentropic efficiency of the compressor plays a significant role in generating the boost pressure at a faster rate and is given by

$$\eta_c = \frac{T_{01}\left[\left(\frac{P_{02}}{P_{01}}\right)^{\frac{\gamma_a-1}{\gamma_a}} - 1\right]}{T_{02} - T_{01}} \tag{4.4}$$

where,

T_{01} the total temperature of the air at the compressor inlet, K
T_{02} the total temperature of the air at the compressor outlet, K

Figure 4.11 shows the compressor characteristics and the operating lines of an engine with FTG and VTG. Based on the airflow and the pressure ratio requirements, the size of the compressor is selected so that the engine operating regime lies within the maximum efficiency islands of the compressor, leaving sufficient margins for surge, choke, and maximum operating speed considering the safe operation of the compressor.

The rate of increase in pressure ratio of the engine operating lines is higher for the TC with VTG than with FTG. Also, for the entire operating range of the engine, the pressure ratio is higher with VTG. The part-load curves both at constant speeds and constant loads also lie in the efficient operating zone of the compressor. Beyond

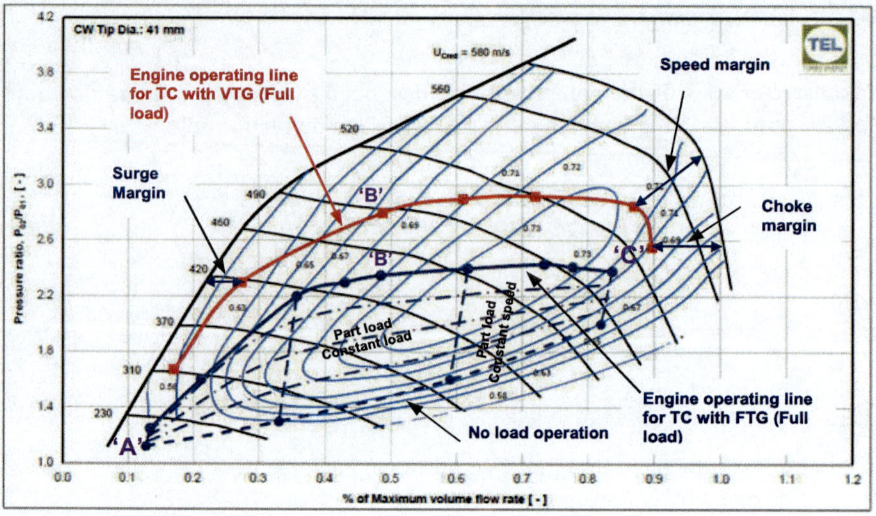

Fig. 4.11 Compressor characteristics

50% of the speed and load, the compressor efficiency is high. If the engine operates predominantly in this zone, then the cycle efficiency of the engine is high.

4.4.6 Turbine Characteristics

The turbine performance is defined using different curves showing the relationship between the expansion ratio, turbine mass flow parameter (non-dimensional mass flow rate MFP), and mechanical-cum-turbine efficiency (η_{M*T}).

$$Expansion\,ratio = \frac{P_{03}}{P_{4s}} \tag{4.5}$$

$$Turbine\,mass\ flow\ parameter,\ MFP = \frac{m_T\sqrt{T_{03}}}{P_{03}} \tag{4.6}$$

Mechanical-cum-turbine efficiency,

$$\eta_{M*T} = \frac{m_c C_{pa}(T_{02} - T_{01})}{m_T C_{pg} T_{03}\left[1 - \left(\frac{P_{4s}}{P_{03}}\right)^{\frac{\gamma_g - 1}{\gamma_g}}\right]} \tag{4.7}$$

where,

m_T gas mass flow rate through the turbine, kg/s.
P_{03} total pressure at turbine inlet, bar (a).
P_{4s} static pressure at turbine outlet, bar (a).
T_{03} the temperature of the gas at turbine inlet, K

Since the exhaust kinetic energy is not recovered, the static pressure is considered at the turbine outlet. A typical turbine performance map of the mass flow parameter and the expansion ratio is shown in Fig. 4.12 for both the FTG and VTG. The parameters are turbine speed-dependent.

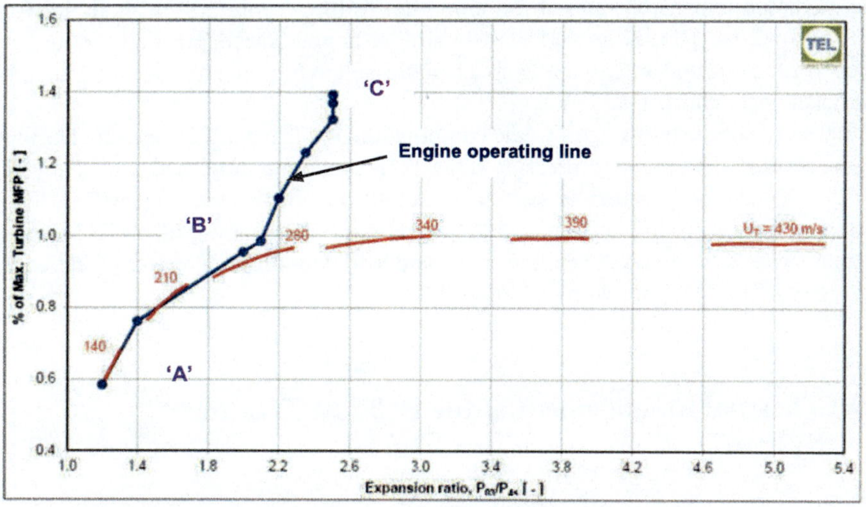

a. **Fixed turbine geometry**

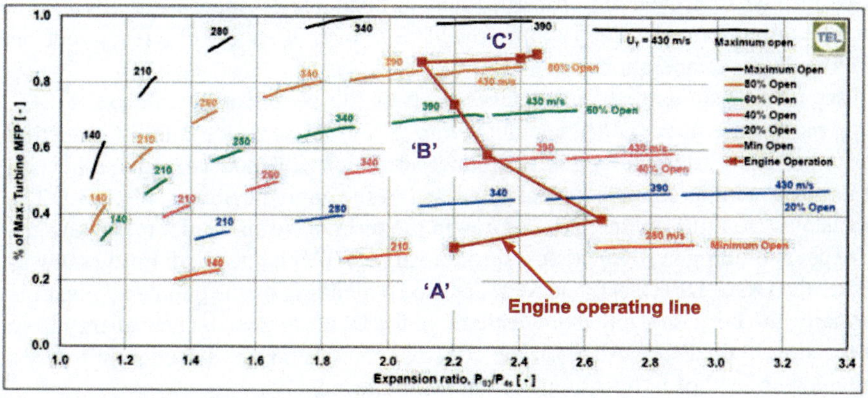

b. **Variable turbine geometry**

Fig. 4.12 Turbine mass flow parameter versus expansion ratio

For a given turbine housing cross-section, the MFP increases with expansion ratio and chokes beyond a certain expansion ratio, indicating the maximum flow handling capacity of the turbine. For VTG, choking occurs at every nozzle vane-opening beyond a certain expansion ratio. The maximum flow capacity is limited by the turbine wheel exducer area in the case of VTG and the turbine housing cross-section in the case of FTG. In the case of VTG, the opening of the nozzle vanes area increases the mass flow rate. In FTG, Fig. 4.12a, beyond the peak torque point, 'B', the exhaust gas is by-passed to limit the turbocharger speed and hence the airflow rate. In the case of VTG, Fig. 4.12b, the nozzle vanes opening area is reduced to increase the turbocharger speed at lower engine speeds, and hence the airflow rate. The increased expansion ratio at the minimum full load point 'A' is about 2.2 for VTG and 1.2 for FTG (Fig. 4.12a) for a similar engine speed. Beyond the torque point 'B', the nozzle vanes opening area is increased to limit the turbocharger speed, and hence the airflow rate.

In the case of the two-stage TC, the combination of FTG and VTG is used. During engine acceleration, the high-pressure stage TC is predominantly used, and at higher engine speed, the operation is switched over to the low-pressure stage TC. The switching over from higher pressure to low pressure and sharing of the TC stages are maneuvered based on the engine speed and load conditions. The rate of increase in pressure ratio is higher with two-stage TC.

4.5 Thermal Management of the Exhaust Treatment System

The combustion and engine power are improved by the advanced engine design and the cutting-edge fuel injection system. Also, the energy conversion in the innovative turbine improves the efficiency and turbocharger response, improving the overall engine performance. The improvement in overall engine efficiency reduces the exhaust gas temperature.

Various emissions-treatment devices such as diesel oxidation catalyst (DOC), diesel particulate filter (DPF), lean NO_x trap (LNT), or selective catalyst reduction (SCR) are used after the engine to reduce emissions in modern engines beyond the improvement in combustion achieved within the engine cylinder, Fig. 4.13. The exhaust gas from the engine passes through the turbine. At the outlet to the turbine, the exhaust braking system/exhaust throttle valve (ETV) is installed. Further downstream, the DOC, DPF, LNT, and SCR are used depending on the engine application. Normally, the DPF is installed downstream of the DOC, to gain the heat energy from the reaction in the DOC. Also, for the effective conversion, the selection of catalyst with a lower level of light-off temperature is preferred.

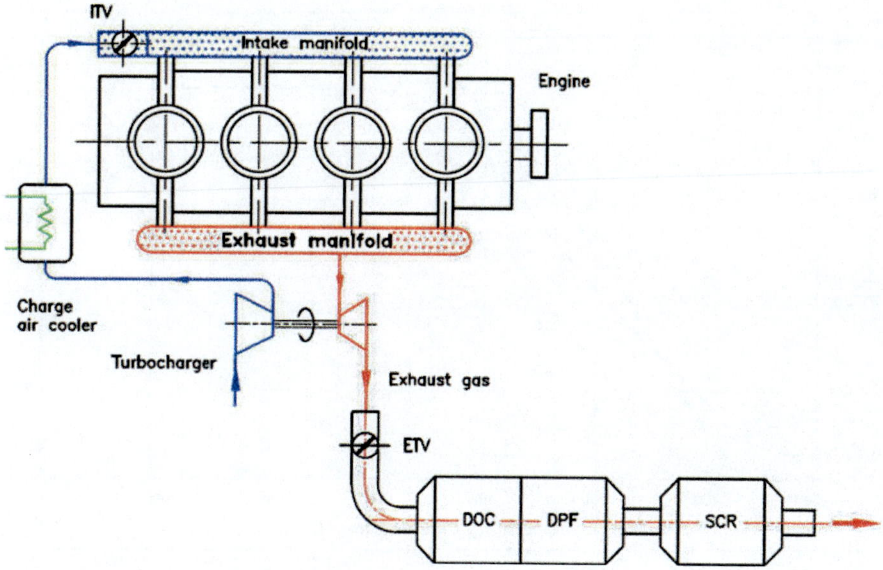

Fig. 4.13 Turbocharged engine with exhaust after-treatment system

4.5.1 Cold Start

The emission levels are higher during engine cold start, and warm-up conditions than at a steady-state since the catalysts after the engine have not reached the respective light-off temperatures. Further, during the winter, the duration of the warm-up cycle is still higher, considering the higher level of heat loss to the surroundings. Even after the warm-up, there is no relenting in applications like city transport buses/cars, airport shuttle services, and industrial-goods delivery vehicles, the engines normally operate at very low loads, and correspondingly the exhaust gas temperature is lower. Similarly, some applications involve frequent start and stop cycles. For every start, the catalyst needs a definite time to warm up.

During engine cold start, to initiate the catalyst reaction, the temperature of the exhaust gas and the surface of the catalyst should be sufficiently above the light-off temperature to trigger the chemical reactions, Fig. 4.14. At a lower temperature, the rate of conversion of emissions to harmless products is low, and significant amounts of CO, HC, and NO_x leave the tailpipe. Beyond the catalyst light-off temperature, which is 250–300 °C, the conversion efficiency is high. Therefore, the catalyst temperature must be raised as quickly as possible from the cold. Start of Figure.

At the minimum full load point, the typical exhaust gas pressure and the temperature at exhaust manifold (before TC, i.e., turbine inlet) are around 1.3–1.5 bar (abs) and 500–600 °C, respectively, for the engine with FTG. After the turbine, the pressure is nearly equal to the ambient, and the temperature is 400–450 °C. Further, the

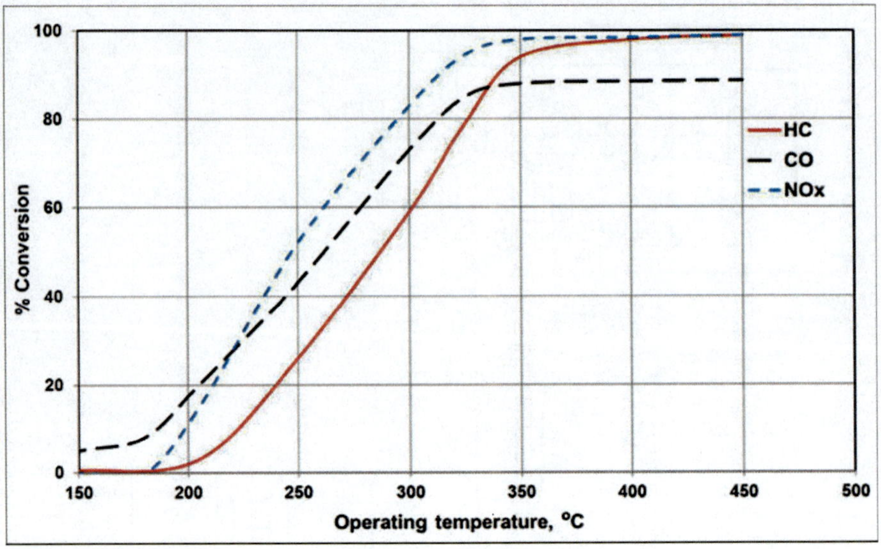

Fig. 4.14 Three-way catalyst performance over exhaust temperature (www.open.edu)

exhaust gas temperature is reduced due to expansion in the turbine outlet pipe before entry to the catalyst converter.

- During a cold start, the engine runs at no load, and the temperature of the exhaust gas is lower. The exhaust temperature is 300–350 °C.
- Additionally, during the engine warm-up cycle, the loading of the engine is limited by the peripheral system like lighting, air conditioning, and battery charging.
- Further, at lower engine speeds, the exhaust gas mass flow rate is also lower.
- With this typical boundary condition and generally with the large pipes and ducts after the turbine, further expansion of the exhaust gas reduces the gas temperature before the catalytic converter.
- The velocity of gases drops in the ducts and heat is lost due to higher resident time in the ducts.
- The gases transfer substantial heat to the metallic parts before they are warmed up. This constitutes a further drop in gas temperature.
- The duration of the warm-up cycle depends on the ambient temperature as well. For typical ambient temperature between − 15 and 10 °C, the warm-up period is 180–300 s. At higher ambient temperature, the warm-up cycle is relatively shorter.
- Consequently, the energy available in the exhaust gas is not sufficient to raise the catalyst above the light-off temperature fast. Thus the warm-up cycle becomes longer, and hence the emissions increase during cold start conditions.
- If the engine is used for hybrid vehicle applications, the engine load can be slightly increased to charge the battery, and hence the warm-up cycle duration can be reduced.

To reduce the emissions during cold start conditions, the combination of the following strategies can be followed.

- Optimization of combustion parameters for reduced engine-out emissions
- Reduction of emissions using exhaust after-treatment systems
- Enhancing the catalyst light-off using the thermal management systems
- Effective monitoring of the emissions and control of the various parameters using the sensors and engine calibration strategies

Late injection of fuel in the case of diesel engine and late ignition in the case of gasoline engine increases the temperature of the gas due to late burning of fuel in the power stroke. The fuel injection pressure, intake air pressure of a turbocharged engine, and an optimum combustion chamber help further increase the exhaust gas temperature. After combustion, the early opening of the exhaust valve and the fuel injection just before the catalytic converter increases the temperature of the exhaust gas. Also, the closing of the intake throttle valve (ITV) restricts the flow area supplying less air to the engine; it enriches the air–fuel mixture and the exhaust gas temperature increases. Similarly, the exhaust throttle valve (ETV) can be maintained to a nearly closed position to increase the exhaust gas pressure and temperature with the reduction of space velocity. Some of these options can be applied in combination. For example, the exhaust gas temperature can be increased during the warm-up cycle by grouping optimized valve timings and fuel injection timing combined with ITV and ETV. However, all these techniques reduce the engine power and efficiency and increase the BSFC. The heat transfer from the catalytic converter also must be reduced by with the external insulation and by suitably optimizing the shape and size.

The exhaust gas temperature at the inlet to the catalytic converter can be enhanced by optimizing the turbocharger-boosting system. It may be noted that the turbocharger itself cannot solve the challenges faced with temperature management, and it has to be done with a combination of engine, turbocharger, and exhaust after-treatment systems. The following options are considered regarding the turbocharger system.

1. Optimized turbine housing and wastegate port
2. Turbine by-pass for TC with FTG
3. Opening of the nozzle vanes for TC with VTG
4. Combination of TC with VTG and external by-pass
5. Inner thermal insulation of turbine housing
6. Outer thermal insulation of turbine housing and exhaust manifold
7. Air gap insulated exhaust manifold and turbine housing
8. Integration of turbine housing with exhaust manifold
9. Exhaust after-treatment system before the turbine

The first four options are active methods; the others are passive.

4.5.2 *Optimized Turbine Housing and Wastegate Port*

The expansion of gas in the turbine reduces the gas temperature at the outlet, the turbine acting as a thermal sink. During a cold start, the heat transfer rate is high, and it results in very low temperatures at the after-treatment systems. The optimization of flow paths in the turbine housing from the inlet to the exit and the wastegate port improves the gas temperature. Figure 4.15 shows the temperature distribution, and Fig. 4.16 compares the temperature rise from the cold start condition for basic and optimized flow paths. In the basic turbine, the by-pass flow is directed towards the turbine outlet wall, and the gases from the turbine outlet and the by-pass port are mixed too early, and as a result, the overall gas temperature at the inlet to the catalytic converter is less.

In the optimized turbine housing, the by-pass gas is directed towards the center of the outlet pipe, and its penetration in the mainstream flow from the turbine rotor is high. Hence, the overall gas temperature is high. Typically, the temperature of the gases is higher by 220 K in optimized turbine housing than in the basic housing.

The reduction in the heat sinking behavior of the housing enables earlier catalytic light-off during cold start conditions. Additionally, the optimized turbine housing improves the flow uniformity at the inlet to the catalytic converter.

Figure 4.17 shows the optimized wastegate port along with the flow profile. The wastegate port orientation is optimized for mainstream flow. This results in a significant reduction in exhaust backpressure before the turbine inlet. The flow from the

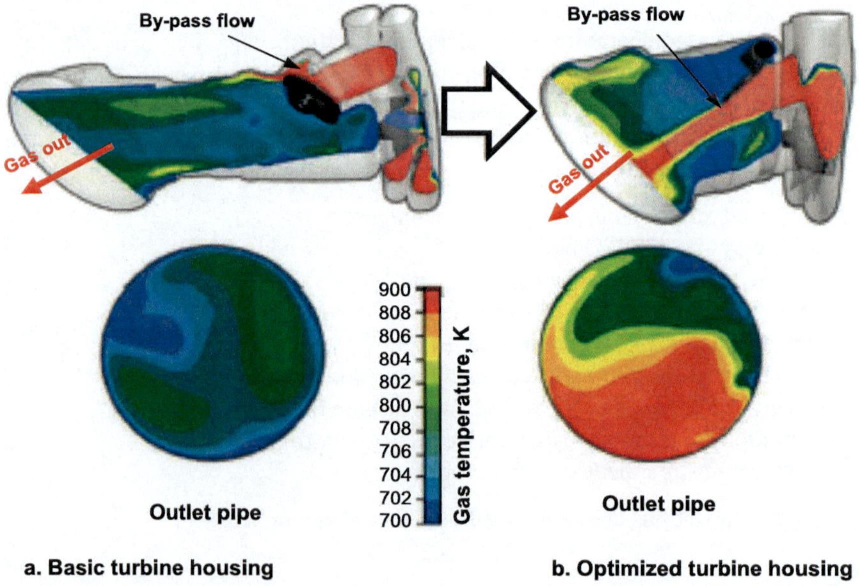

Fig. 4.15 Temperature distribution in the turbine housing (*Courtesy* BorgWarner Turbo Systems)

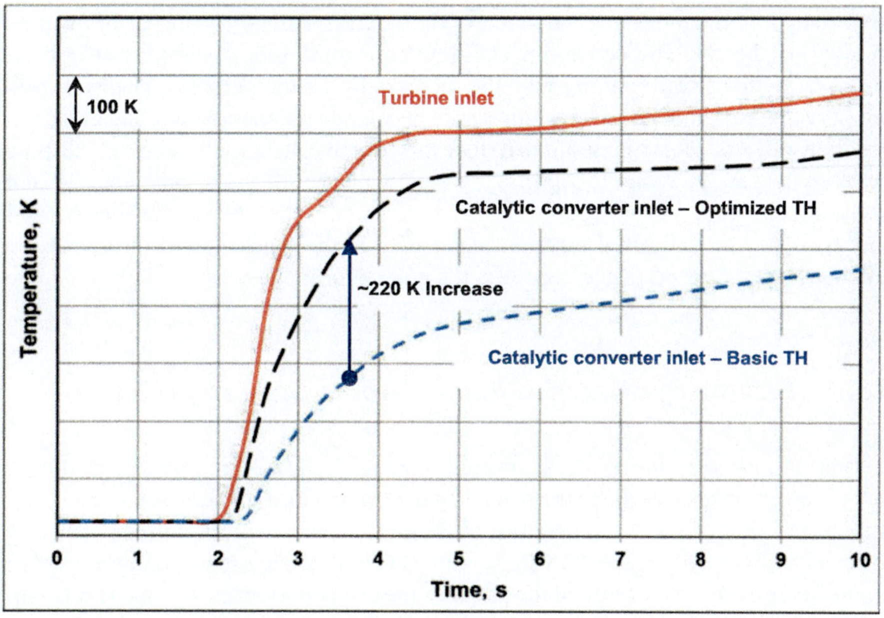

Fig. 4.16 Temperature rise during cold start condition

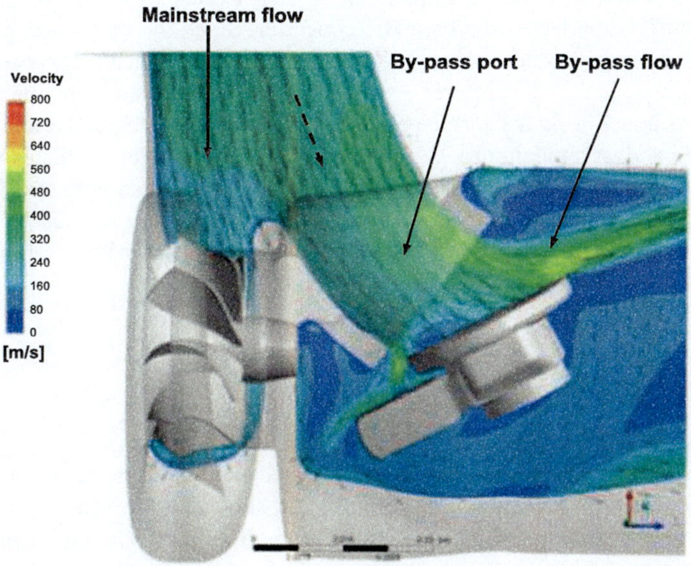

Fig. 4.17 Optimized turbine-waste gate port (*Courtesy* BorgWarner Turbo Systems)

wastegate is directed towards the center of the catalyst with minimum surface wetted by gases in the catalytic converter. The flow uniformity index is improved in front of the catalytic converter along with the increase in gas temperature. The improved pressure gradient and better flow uniformity accelerate the thermal loading and catalyst light-off. Also, it is important to reduce the effective distance between the turbine outlet to the catalytic inlet by optimizing the boundaries of the turbine housing and catalytic converter. A close-coupled catalytic converter is helpful to conserve thermal energy. Also, the optimized turbine shapes help in the quick rise of turbine speed and boost pressure and hence improves the engine torque.

4.5.3 *Turbine By-Pass with Fixed Turbine Geometry (FTG)*

During idling, the wastegate is normally closed; however, the speed of the turbocharger is very low since the exhaust gas flow and its temperature are low. With turbine reaction, the gas temperature at the turbine outlet drops further, delaying the catalyst light-off. Figure 4.18 shows the FTG with the wastegate either closed or open. To raise the temperature of the gas after the turbocharger, the turbine is substantially bypassed by the flow. Figure 4.19 shows the exhaust gas temperature profile for the typical TC with FTG with the wastegate in the open position. With an optimized by-pass path and maximum opening of the wastegate valve, the maximum hot gas flow is directly supplied to the catalytic converter. Since the by-passed gas is not passing over a large surface of the cold turbine housing, the heat loss is contained. The gas flow at high temperatures is directed towards the center of the catalytic converter without hitting the surface of the walls of the converter, and it allows a fast increase of catalyst temperature. With the wastegate partially open, the exhaust gas temperature at the turbine outlet is around 40–50 K higher than with the wastegate closed. Normally, during the warm-up cycle, the engine speed is increased to around 50–75% of the maximum speed of the engine, and the load is maintained low. With this, a further increase in exhaust gas temperature is achieved.

4.5.4 *Opening of the Nozzle Vanes of VTG*

In the case of VTG, the temperature of the gas at the outlet of the turbine can be varied by opening the nozzle vanes. Figure 4.20 shows nozzle vanes of a VTG at maximum and minimum open positions. With the minimum opening of the nozzle vanes, the expansion is high, and the turbine outlet temperature is low. With maximum opening/partial opening of the nozzle vanes based on the exhaust gas mass flow rate, the expansion of the gas is reduced, and the turbine outlet temperature is high. The higher gas temperature contributes to the heating of the catalytic converter. Thus, VTG provides flexibility to thermal management of the exhaust treatment devices and emission control during cold start.

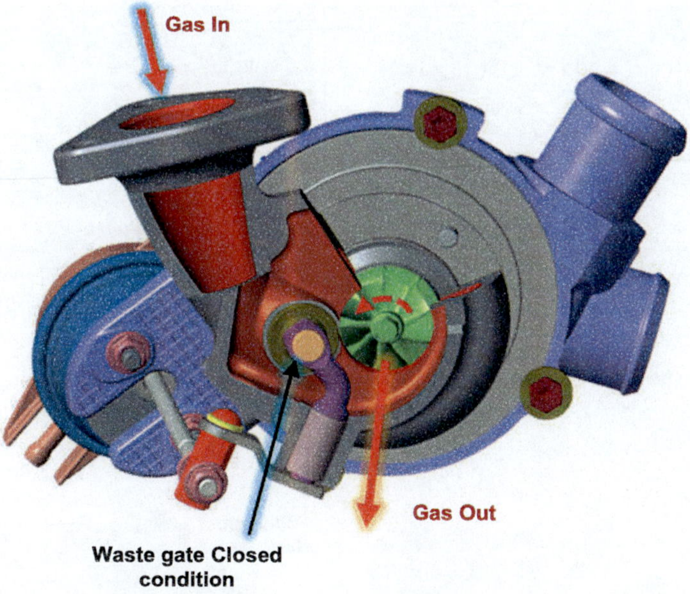

(a) Wastegate closed

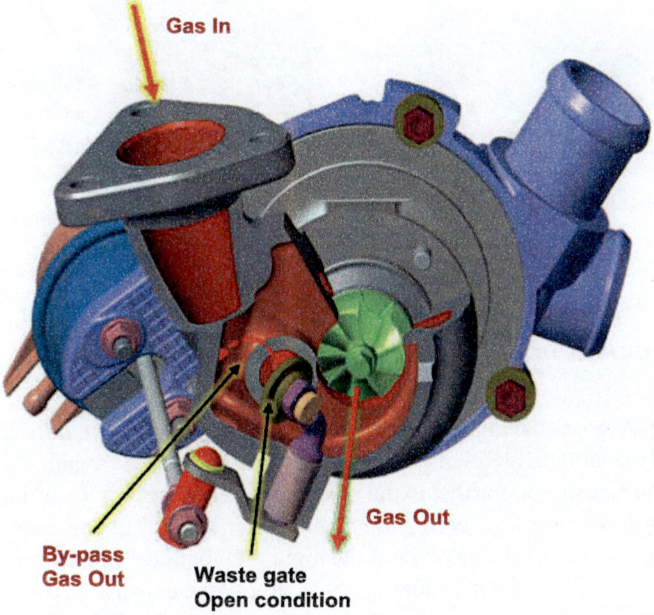

(b) Wastegate open

Fig. 4.18 TC with FTG—wastegate position

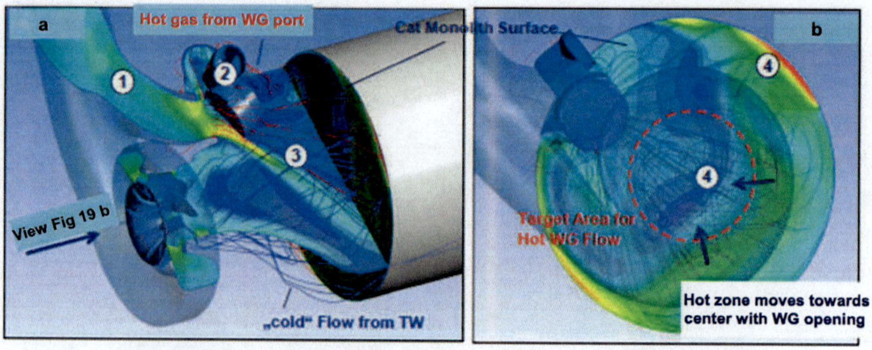

Fig. 4.19 Exhaust gas temperature profile at wastegate opening: 1 optimized flow path, 2 wastegate open, 3 hot gas flow, 4 gas flow towards the center of the catalytic converter (*Courtesy* BorgWarner Turbo Systems)

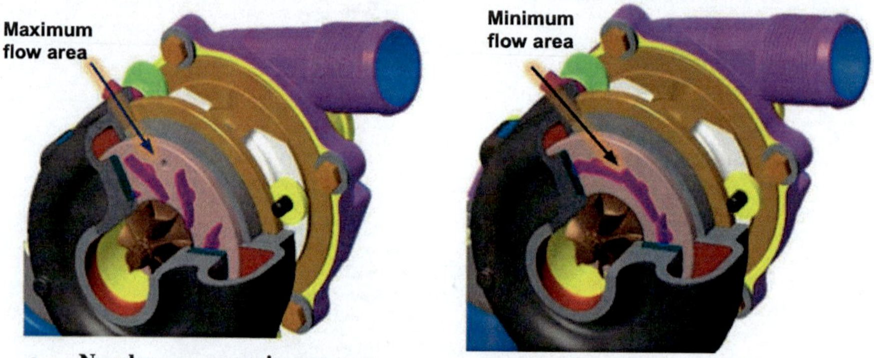

a. Nozzle vane – maximum open **b. Nozzle vane – minimum open**

Fig. 4.20 VTG—nozzle vane opening positions

4.5.5 Combination of VTG and External By-Pass

For engine applications having higher torque at lower engine speeds, normally, VTG is used. For thermal management, the vanes are partially open to gain the thermal energy at the turbine outlet. Further to the increase of heat energy at the turbine outlet, an external by-pass is installed before the turbine inlet, Fig. 4.21. The opening of the external by-pass increases the gas flow at the higher temperature at the turbine outlet. This configuration also helps in reducing considerably the exhaust gas backpressure at higher engine speeds in case of the high power intensity engines. The use of an external by-pass increases the exhaust gas temperature at the turbine outlet approximately by 25 K for the typical arrangement. However, it has a deleterious impact on torque and the specific fuel consumption by 2%. To compensate for the air loss due

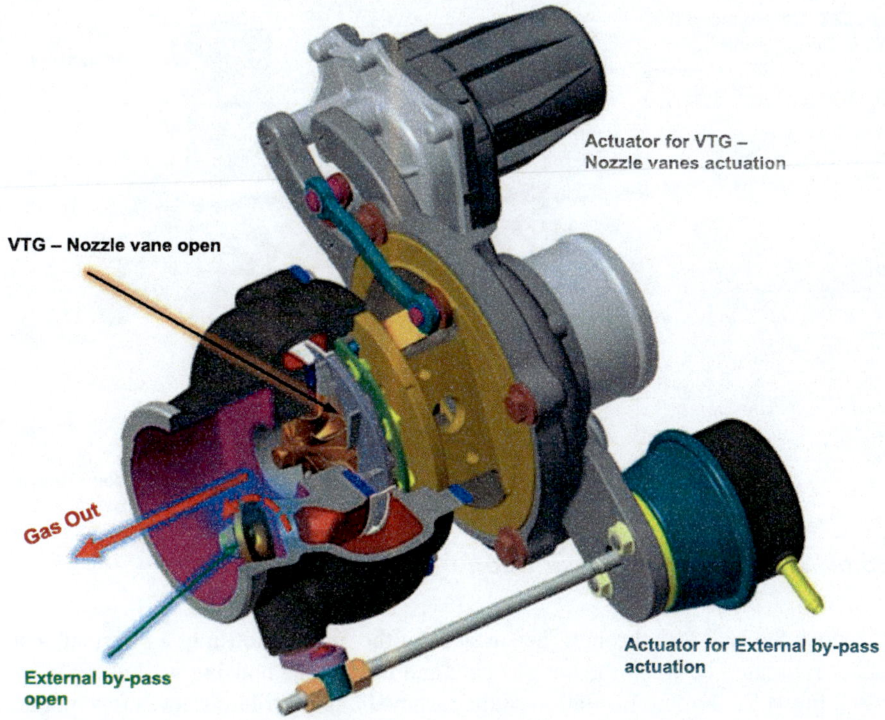

Fig. 4.21 Schematic of the VTG TC with external by-pass

to exhaust gas by-pass, the secondary air supply system like e-Booster can be used at lower engine speeds.

The increase in low-end torque further increases the exhaust gas temperature by ~ 30 K at the turbine outlet and improving the catalyst light-off. The result is a reduction of NO_x by 40% compared to the base VTG. When combined with an e-Booster, the opening of the VTG results in higher specific fuel consumption by 10–12% at lower engine speeds during the cold start warm-up cycle. However, the use of the e-Booster permits an increase in the torque at lower speeds and improves the engine transient response in addition to the thermal management for catalytic light-off.

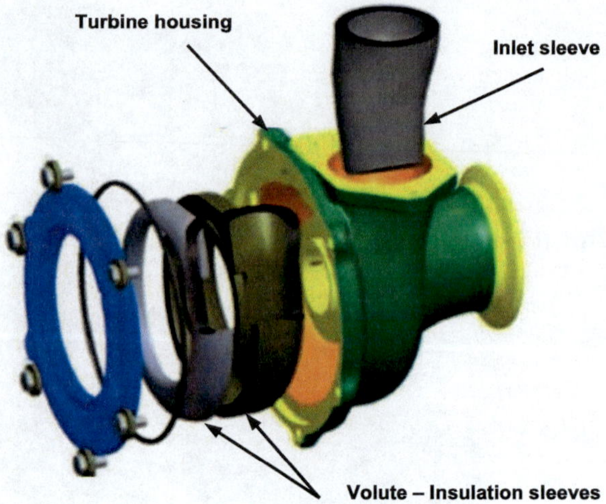

Fig. 4.22 Schematic of the turbine housing with inner thermal insulation (*Courtesy* BorgWarner Turbo Systems)

4.5.6 Inner Thermal Insulation of Turbine Housing

To reduce the heat loss through the surfaces of the turbine housing, a physical heat barrier is created between the hot gas path and the turbine housing wall using insulation, Fig. 4.22. For the insulation of the turbine housing volute sleeves forming an inner shell are inserted inside the housing. Considering the complexity, the volute, inlet, and outlet pipes are made of multiple pieces and joined suitably by welding. Figure 4.23 shows the tailpipe NO_x and the specific fuel consumption of an engine using a turbine housing with and without inner thermal insulation. For the typical diesel engine, the heat loss from the housing with inner insulation is only 30% of the turbine housing without insulation. This results in a 2.1% reduction in specific fuel consumption for the typical NO_x emissions corresponding to the Euro-6 limit [2].

4.5.7 Outer Thermal Insulation of Turbine Housing and Exhaust Manifold

The heat transfer can be further reduced by insulating the exhaust manifold and the turbine housing using thermal shields, Fig. 4.24. The insulation of the turbine housing reduces thermal conduction and heat radiation and results in a reduction of the cooldown of the exhaust after-treatment system. The inner skin of the insulation is normally made using stainless steel wire mesh. Based on the operating temperature, either silica-based glass fiber or calcium magnesium silicate wool is used on the mesh. The external surface of the skin is made using powders from lava rock. These heat shields are suitable for temperatures up to 1000 °C continuously, refer to www.elevatecars.com. While wrapping the manifold and turbine housing using a

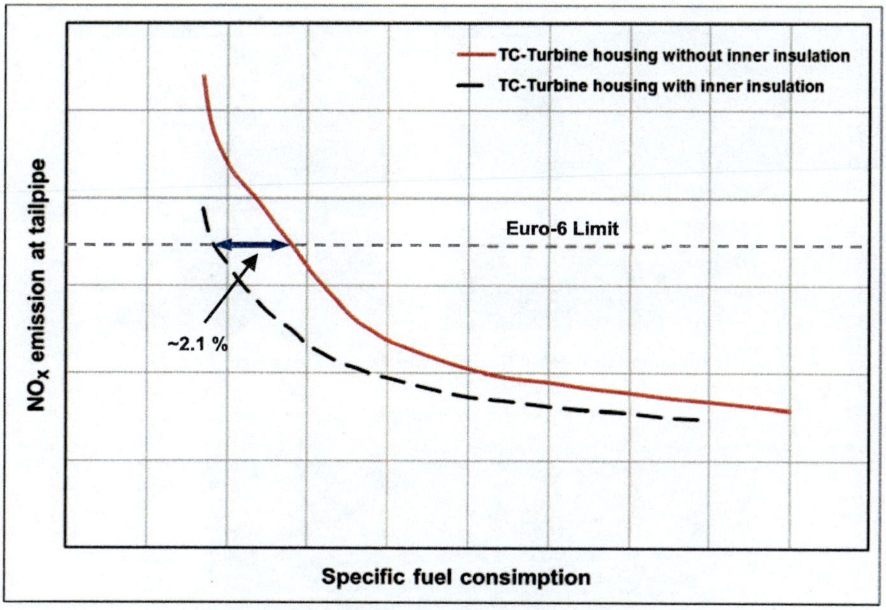

Fig. 4.23 NO_X and the specific fuel consumption with inner thermal insulation [5]

typical heat shield blanket improves the thermal management during cold start and warm-up conditions, the excessive heat during the full load and full speed and hot environment conditions must be managed by selecting suitable materials for the exhaust manifold and the turbine housing, and by maintaining a proper air gap between them. The use of a heat shield also acts as a sound absorbent. A sufficient level of flexibility must be ensured to avoid the straining of the heat shield at higher temperatures. Also, the insulation on the exhaust manifold and turbine housing acts as a thermal barrier to the surrounding components. An increase in temperature by about 12 °C is observed during a typical transient cycle, and the benefit increases with time, Fig. 4.25.

4.5.8 Air Gap Insulated Exhaust Manifold and Turbine Housing

Figure 4.26 shows the schematic of the integrated double-walled (with air gap) exhaust manifold and the turbine housing. This is also called air gap insulated turbine housing. Such housings are used for higher operating temperatures. The mass of the inner shell is less, and hence the heat transfer to the outer surface is low. Figure 4.27 shows the exhaust gas temperature rise with regular cast iron and air gap insulated manifold-turbine housing for a typical driving cycle. The use of an air gap insulated

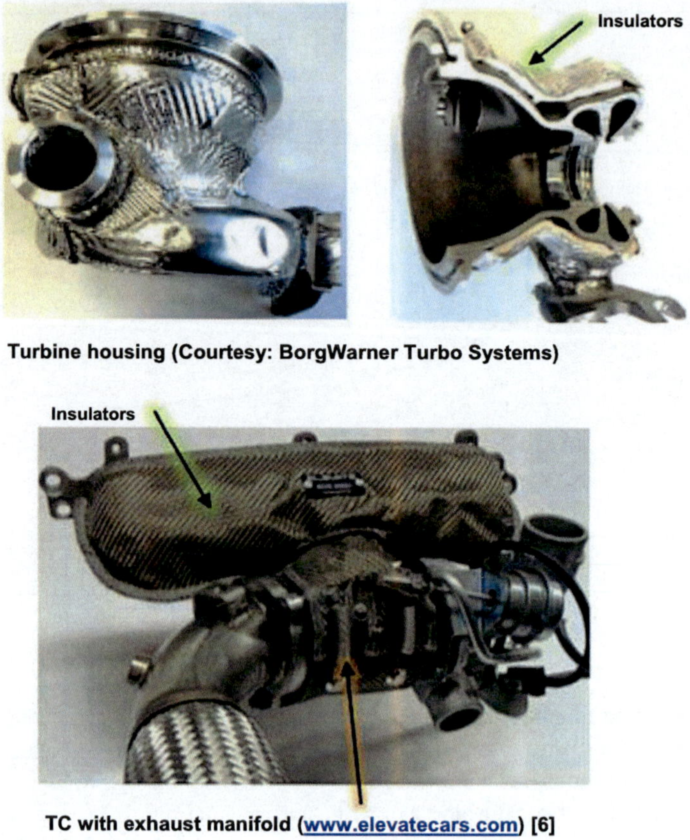

Fig. 4.24 Thermal insulation of exhaust manifold and turbine housing [6]

exhaust manifold and turbine housing reduces the heat loss by 45–55% compared to a conventional cast-iron manifold.[1] The exhaust gas temperature gain is around 25–30 °C at a given turbine inlet temperature. It is expected to reduce the HC, CO, and NO_x to about 20–50% compared to the conventional exhaust manifold and turbine housing of cast iron.

With this concept, the reduction in heat loss can be achieved along with the reduction of the engine weight. The shape of the inner shells in the exhaust manifold and turbine housing forming the gas passage is created by the pressing process. Based on the operating temperature, the basic inner wall thickness is around 1–1.5 mm made out of austenitic stainless steel material. Since the outer wall is experiencing lower temperatures compared to the inner wall, they can be of ferritic stainless steel material.

[1] Editor: More details of estimation is available in chapter 3.

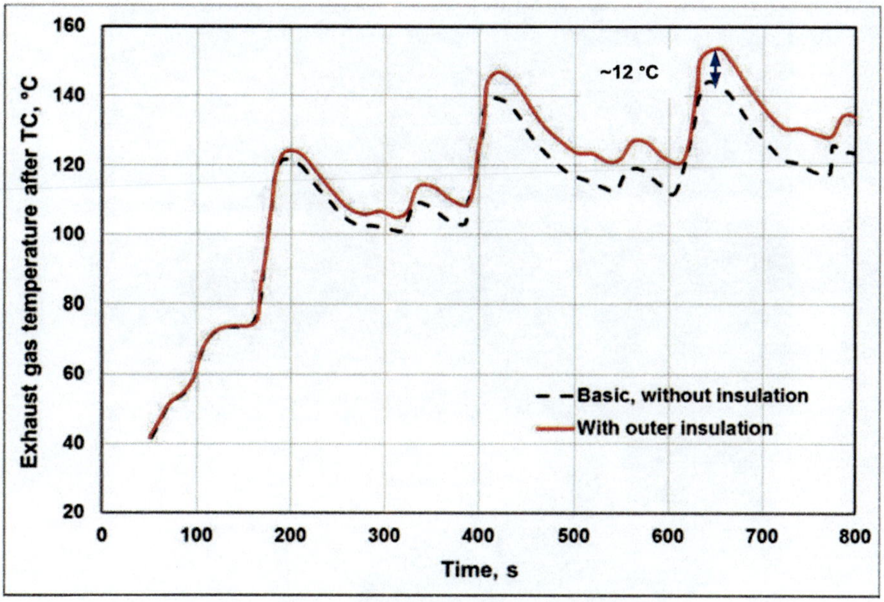

Fig. 4.25 Exhaust gas temperature rise with outer thermal insulation (*Courtesy* BorgWarner Turbo Systems)

4.5.9 Integration of Turbine Housing with Exhaust Manifold

Conventionally, the turbine housing is mounted on the exhaust manifold using bolted joints or using the V-bands depending on the geometry of the joining faces. There is a definite increase in length between the inlet to the exhaust manifold and the turbine wheel inlet, considering the convergence of passage at the exit of the exhaust manifold and inside the turbine housing. Due to the thickness of the flanges of the manifold and the turbine, the distance increases further.

Figure 4.28a shows the conventional arrangement of the individual turbine housing assembled on the engine exhaust manifold. Figure 4.28b shows the turbine housing and engine exhaust manifold integrated as a single cast part.

- This primarily
  - eliminates the flanges at the exhaust manifold and turbine housing mating surfaces
  - results in effective utilization of pulse energy in the exhaust gas, and
  - improves the transient response of the turbocharger and engine.

The distance '*a*' between the merging zone of the manifold and the center of the turbine is higher when individual parts are assembled than when integrated. The higher the distance, the higher are the gas travel path and heat transfer, and hence higher is the drop in gas temperature at the turbine outlet. The exhaust flow

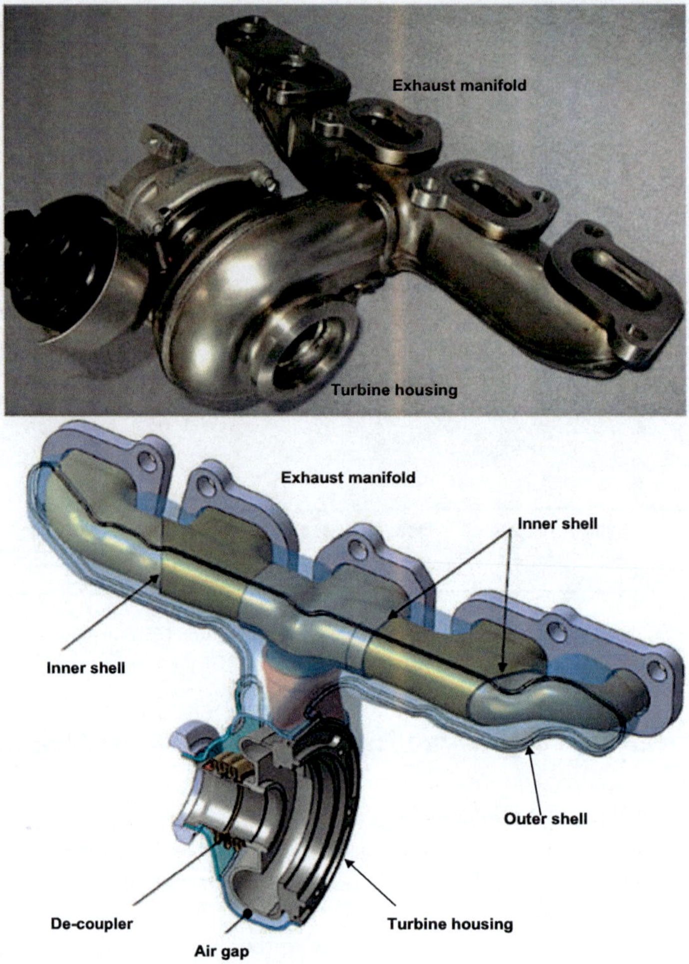

Fig. 4.26 Air gap insulated exhaust manifold and turbine housing (*Courtesy* BorgWarner Turbo Systems)

passages within the manifold can be further optimized while integrating with the turbine housing to reduce the flow path and conserve the heat energy.

4.5.10 Exhaust After-Treatment System Before the Turbine

Figure 4.29 shows the schematic of the exhaust after-treatment system installed before the turbocharger-turbine. This arrangement provides the heat energy to the catalytic converter at high temperatures, reduces the heat losses, and supports the

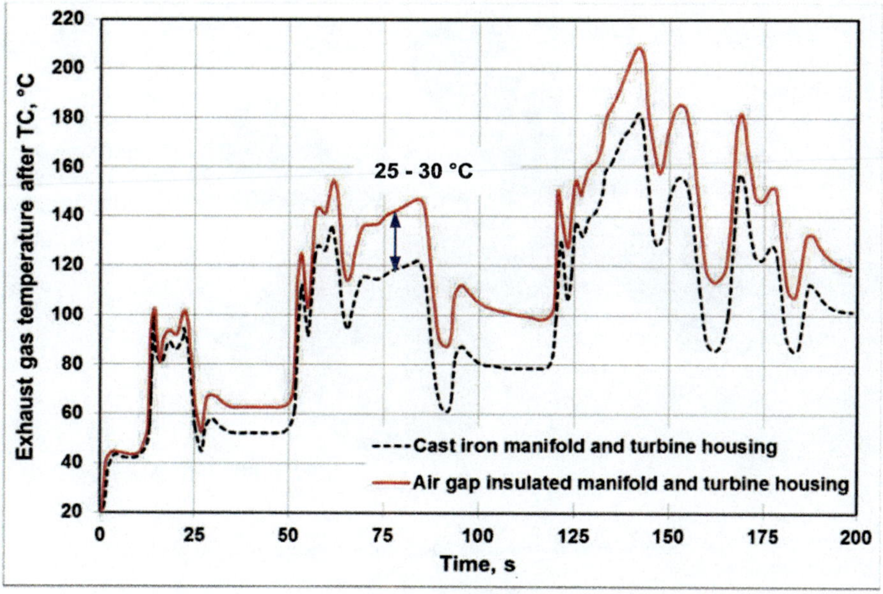

Fig. 4.27 Temperature rise with cast iron and air gap insulated turbine housings

faster catalytic light-off. This system is ideal for engine applications having a frequent start-stop cycle and cold ambient. However, there is a reduction in turbine power and hence the air supply to the engine by the TC. The position of the ETV should be fixed appropriately based on the application.

4.6 Summary

The salient performance parameters of a high-power intensity diesel engine and methods of achieving them with the help of a turbocharger are discussed. In particular, the benefits of a turbocharger with variable turbine geometry are discussed. The selection of a turbocharger for the right match with the engine plays a significant role in achieving good engine performance. Further, based on the frequent engine operating zone for the intended application, the optimization of variable turbine geometry (nozzle vanes opening) provides flexibility to the thermal management of the after-treatment systems and benefits in terms of the BMEP, transient response, BSFC, and emissions. Also, during cold engine staring, the optimization of the variable turbine geometry (nozzle vanes opening) along with its position feedback via engine electronics control unit supports quick heating and maintains the temperature of the exhaust treatment system throughout the engine operating cycle. Of the different methods available for thermal management, the most appropriate option must be selected for maximum effectiveness by considering the type of the engine, the turbocharger, and vehicle configurations as well as the application.

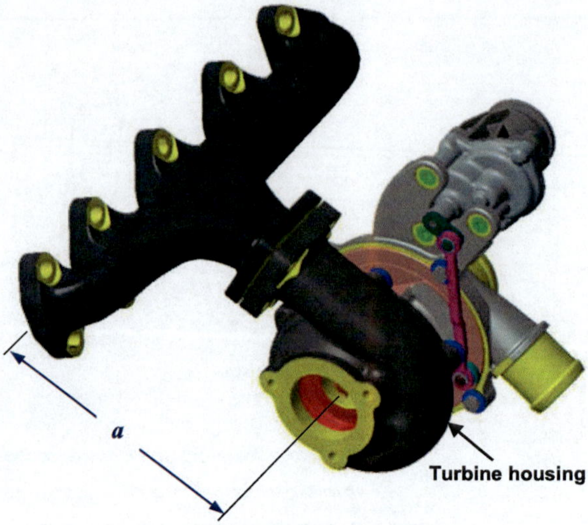

(a) **Engine exhaust-manifold and turbine housing (individual parts)**

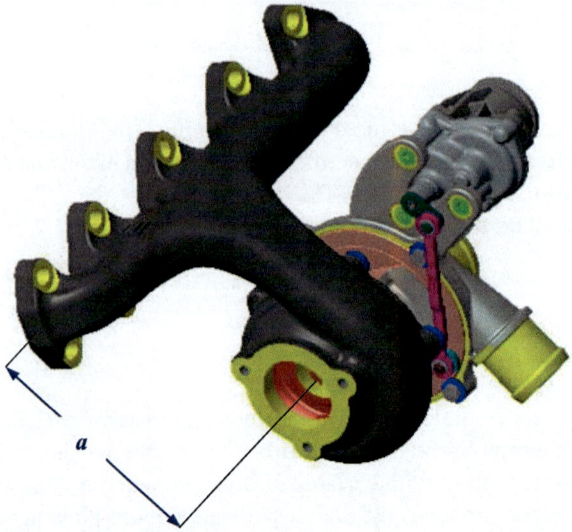

(b) TC-turbine housing integrated with engine exhaust manifold

Fig. 4.28 TC-turbine housing installation on the engine exhaust manifold

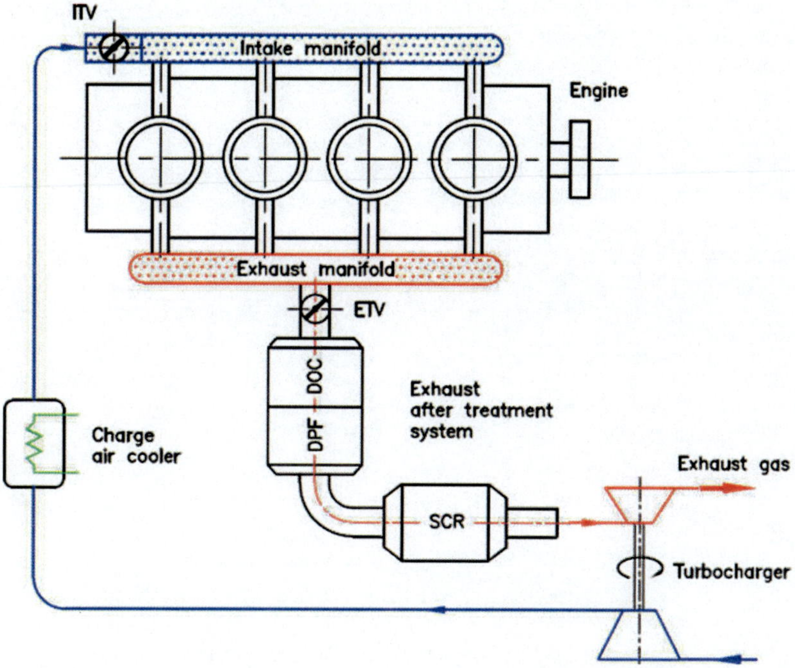

Fig. 4.29 Schematic of the exhaust after-treatment system before the turbine

Acknowledgements The authors would like to thank Turbo Energy Private Limited, Chennai, for permitting us to publish the chapter. The authors express their thanks to Courtesy: BorgWarner Turbo Systems Team (Germany) for supporting the technical contents. Thanks to Elevate Cars, Inc., California, USA for permitting to use of the images and technical data on the thermal heat shield. Also, thank our colleagues Mr. Tamilarasan, Mr. Chandrasekar, Mr. Sreedhar, and Mr. Karthi for their help during the chapter's preparation.

References

1. Büchi A (1905) Patent No. 204630. Highly supercharged compound engine. Imperial Patent Office of the German Reich
2. TurboNews (2005) The info magazine of borgwarner turbo systems. Issue. 01/05
3. Leduc P, Dubar B, Ranini A, Monnier G (2003) Downsizing of gasoline engine: an efficient way to reduce CO_2 emissions. Oil Gas Sci Technol Rev IFP 58(1):115–127
4. Kirwan JE, Shost M, Roth G, Zizelman J (2010) 3-cylinder turbocharged gasoline direct injection: a high-value solution for low CO_2 and NOx Emissions. SAE Int J Engines 3(1):355–371. No. 2010-01-0590
5. Burke R, Liu Y, Vijayakumar R, Dalby J (2019) Inner-insulated turbocharger technology to reduce emissions and fuel consumption from modern engines. SAE Technical Paper 2019-24-0184. https://doi.org/10.4271/2019-24-0184
6. www.elevatecars.com/elevate-volvo-t5-k04-k16-turbo-heat-shield-blanket.html

Chapter 5
Thermal Management Through Insulation Design—Passenger Car Platforms

Ravi Ranjan, Srinivas Gunti, and Parvej Khan

5.1 Introduction

Underhood is the engine compartment whereas the underbody is below the vehicle floor just above the ground. The underhood heat management is very important for a vehicle powered by an internal combustion engine. After the conversion of fuel energy to mechanical work, around 61% of the energy is lost in the form of heat energy, Fig. 5.1 [1]. The vehicle system has to manage the heat loss through conduction, convection, and radiation to the surroundings without injuring the general public and the precious equipment inside the vehicle or making the occupants uncomfortable. It becomes more critical to direct the heat transfer properly in the new age vehicles where packaging components under the hood are highly compact. Thus it is important to insulate the components from the hot spots to ensure functionality and any thermal hazards [2].

5.2 Thermal Insulation Design Consideration

A typical distribution heat loss inside the engine bay as shown in Fig. 5.2 is considered for the design of thermal insulation by giving suitable weightage [4].

R. Ranjan · S. Gunti (✉) · P. Khan
Mahindra & Mahindra Ltd., Mahindra Research Valley, Mahindra World City, Chengalpattu, Tamil Nadu 603004, India
e-mail: srinivas.gunti@mahindra.com

P. A. Lakshminarayanan and A. K. Agarwal (eds.), *Handbook of Thermal Management of Engines*, Energy, Environment, and Sustainability,
https://doi.org/10.1007/978-981-16-8570-5_5

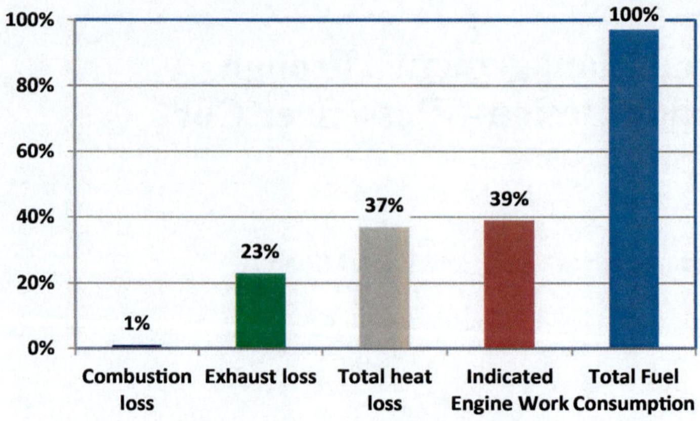

Fig. 5.1 Typical heat loss distribution for a combustion engine [3]

Fig. 5.2 Vehicle engine bay heat dissipation

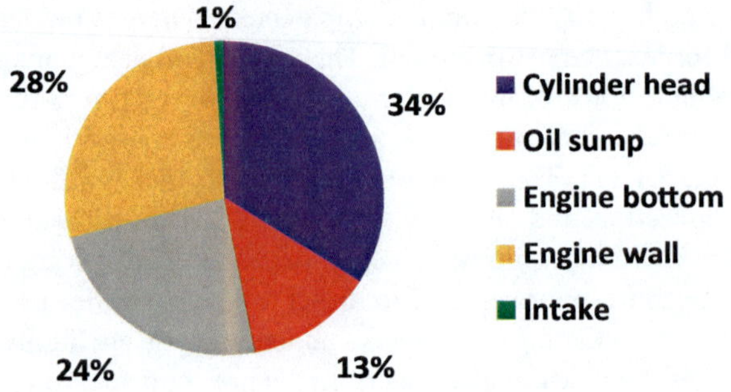

5.3 Insulation Applicability Criteria

Figure 5.3 refers to the application of a single, double shell, or sandwich-type heat shield based on the distance from the hotspots and their surface temperatures. However, as the packaging inside an engine compartment of a modern automobile is very compact and distances between the surrounding components are marginal, it is becoming increasingly difficult to place the parts at great distances to avoid heatshields. The hotspots like the catalytic converter and the turbocharger are placed tightly underhood and they need special attention for thermal insulation with the interface parts.

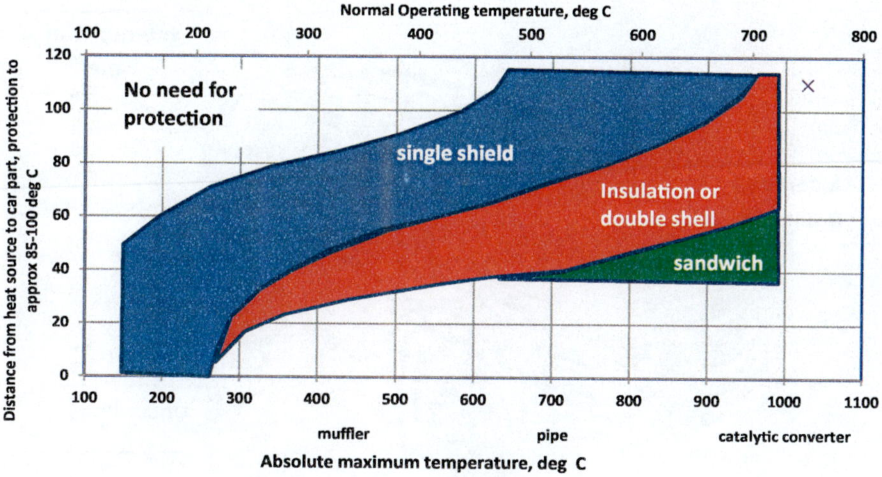

Fig. 5.3 Heat shield applicability

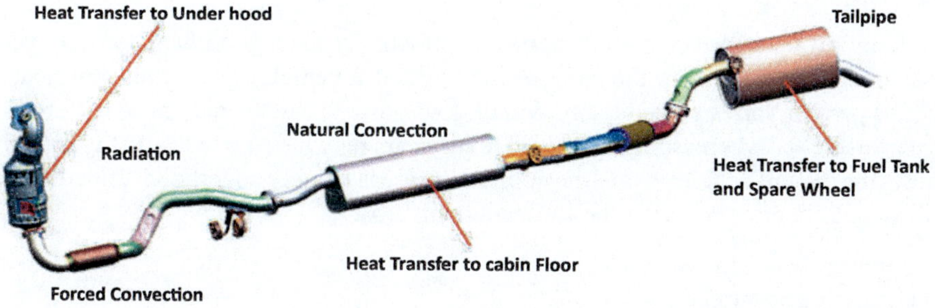

Fig. 5.4 Exhaust hot spot area

5.4 Modes of Heat Transfer in Typical Hot Spot Areas

Figure 5.4 shows the surface profiles of the engine exhaust system containing high temperature hot spots. Heat is transferred underhood from the close-coupled catalysts and the ducts essentially by radiation; it is transferred underbody predominantly by natural convection from the ducts, the silencer, and other catalysts. These parts as well as the affected areas need thermal insulation.

5.5 Heatshield Applications

The most important function of heatshields is to protect the sensitive vehicle components from the heat and to provide optimal thermal comfort for the occupants. Heat

Fig. 5.5 Vehicle level thermal insulation application

shielding products are custom designed to effectively manage radiant, convective, and conductive heat from the front to the back of a vehicle. The number of heat-shields per car varies considerably. Small, low-power vehicles may need as few as two heatshields, whereas high-powered luxury cars may have up to thirty. Figure 5.5 shows the overall vehicle insulation application areas namely underhood, underbody, and vehicle interior covering the following parts [5]:

- Exhaust manifold
- Catalytic converter
- Turbocharger
- Starter motor
- Wiring Harness (Engine compartment and After-treatment System)
- Rubber Hoses
- Fuel Lines

5.6 Types of Heat Shields Based on Design

Depending upon the thermal intensity of the sources various types of insulation (Table 5.1) are applied.

Single shell heatshields

Single shell heat shields are used for protection against heat sources of relatively low temperature, especially when sufficient space is available.

Table 5.1 Metal heat shield types and application area

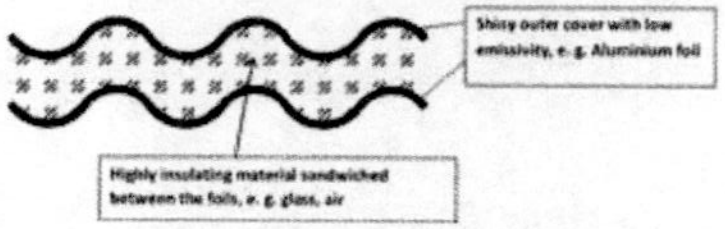

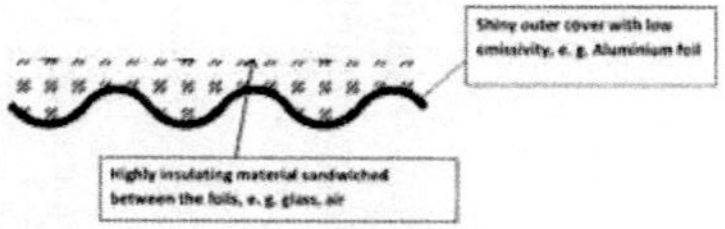

type	Style	Construction	Material	Emissivity New → oxidized	Application
1	Single wall embossed		Aluminum	0.1 → 0.3	Under hood/underbody
2	Single wall flat		Aluminum	0.1 → 0.3	Under hood/underbody
3	Dual wall embossed		Aluminum	0.1 → 0.3	Under hood/underbody
4	Dual wall with glass	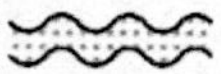	Aluminum-glass-aluminum	0.1 → 0.3	Under hood/underbody
5	AMS (all metal shield)	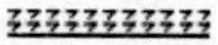	Aluminum-mesh-aluminum	0.1 → 0.3	Under hood/underbody
6	Stick-on heat shield		Aluminum-glass	0.3 → 0.6	Under hood/underbody
7	AMS stainless (all metal shield)		Stainless-steel mesh-stainless	0.15 → 0.65	Under hood/underbody/exhaust-direct mount
8	Single wall flat		Aluminized steel	0.3 → 0.6	Under hood/underbody
9	Dual wall (manifold)		Aluminum-clad steel-glass-aluminized steel	0.3 → 0.6	Exhaust/manifold-direct mount

Double shell heatshields

These heat shields made of two aluminum sheets are used for moderate temperatures and limited package restrictions.

Sandwich heatshields

For protection against the highest temperatures and in cases of severe space limitations, heat shields of sandwich-type are used. They usually consist of a single carrier sheet (0.3–1.0 mm thick aluminum), an insulating core, and a cover (0.03–0.1 mm aluminum foil or 0.2–0.5 mm aluminum sheet). The core can be built up of several layers of embossed aluminum foils (0.03–0.05 mm thick), which are sandwiched together. The core can also be an insulating mat of ceramic felt or glass fiber. Depending on the maximum recommended operating temperature, other types of fibers of plastic, cotton, wood, etc. can also be used.

5.7 Heat Shield Application: Area and Types

5.7.1 Engine Compartment Area

The engine compartment is the heart of the vehicle consisting of a power pack and other essential auxiliary systems like intake, engine cooling, air conditioning, and exhaust aftertreatment. For discussions, experimental data obtained during the development of a commercially successful vehicle are presented here.

1. **Intake system**

An intake system provides fresh and clean air to the engine for combustion. It plays an important role in emission control and performance, Fig. 5.6. The numbers in the legends indicate the surface temperatures of different parts of an optimally insulated engine. Higher intake air temperature leads to volumetric efficiency loss resulting in higher fuel consumption and lower power. It is a good practice to insulate the

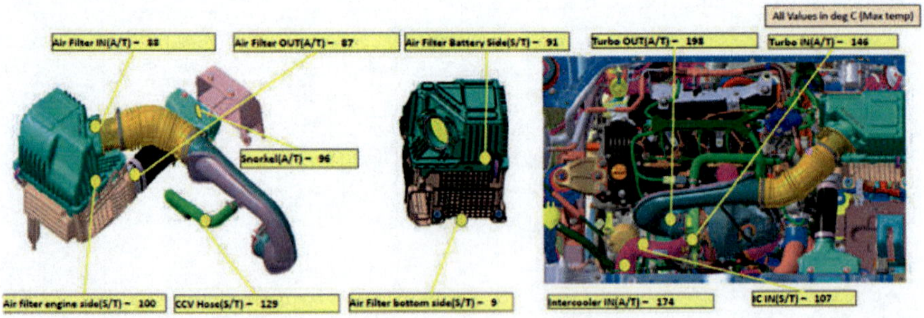

Fig. 5.6 Intake system

hose duct to reduce or stop the heat transfer from the engine compartment to the charge air. In the particular example, insulation was not needed because the surface temperatures and charge air temperatures were below the threshold limits set by the designer, usually 5–10 °C above the ambient. However, if the temperatures were found higher the recommended insulation would be:

Application area	Intake system duct, air filter box
Component material	Plastic (polypropylene talc 20)
Maximum recommended operating temperature	100 °C
Heatshield	Aluminium foil

2. **Engine top cover**

For aesthetics as well as to provide insulation from the sound, engine insulation covers are provided. The usual choice of the material is:

Application area	Engine cover
Component material	Plastic
Maximum recommended operating temperature	100 °C
Heatshield	Aluminium foil

The numbers flagged with the names of the parts show the surface temperatures of a well laid-out engine in °C, after optimal insulation [6] (Fig. 5.7).

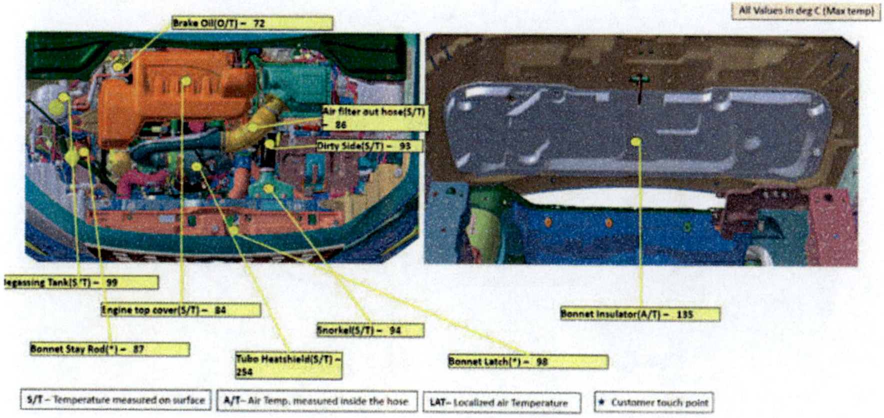

Fig. 5.7 Engine bay under the hood

3. Cooling and air-conditioning System

The engine cooling system consists of a cooling jacket to which heat from the engine combustion chamber is transferred such that the metals are within safe temperature limits. The surfaces of the cylinder head as well as the crankcase holding the jacket emit heat mainly by convection and partly by radiation to the surroundings. Further, the coolant in the jacket carries away the heat to the heat exchangers to reject the heat to the environment. In the case of automobiles, the environment is air and in marine applications, it is seawater. The hot air from the compressor of the turbocharger must be cooled to lower nitric oxide emissions and improve power output as well as fuel economy. Thus, an automobile cooling system (Fig. 5.8) consists of an intercooler, a radiator, a fan shroud, and a fan. The materials of the system must be protected from hotspots the catalytic converter radiating excessive heat energy. Insulation was not needed because the surface temperatures were below the threshold limits. However, if the temperatures were found higher the recommended insulation would be:

Application area	Radiator fan shroud heat shielding due to exposer to the catalytic converter, cooling hoses
Component material	Plastic (polyamide 66 glass-filled 30)
Maximum recommended operating temperature	120 °C
Heatshield	Single flat wall

The same material is used for insulation of the piping of the air-conditioning system to minimize the loss of cooling to the engine bay compartment. For Protecting the rubber hoses, aluminium sleeves (Spiral Multilayer) are used.

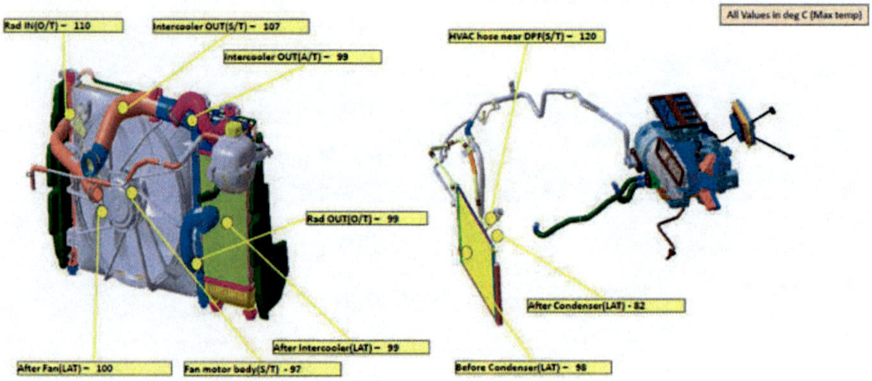

Fig. 5.8 Cooling and air conditioning system

4. Engine Compartment Firewall

The firewall in the engine bay works as a thermal barrier and restricts the heat transfer into the passenger cabin from the engine compartment. Further, it is important to provide the cabin with the right insulation for improving the performance of heating, ventilation, and air conditioning (HVAC) systems and reducing the power consumed by the auxiliary equipment. Figure 5.9 shows the application area with the numbers indicating the temperatures of different surfaces in °C, after optimally applying the heat insulation:

Application area	Engine compartment firewall
Component material	Paraformaldehyde + black nonwoven fabric + process fabric + ethylene vinyl acetate
Maximum recommended operating temperature	100 °C
Heatshield	Aluminium foil

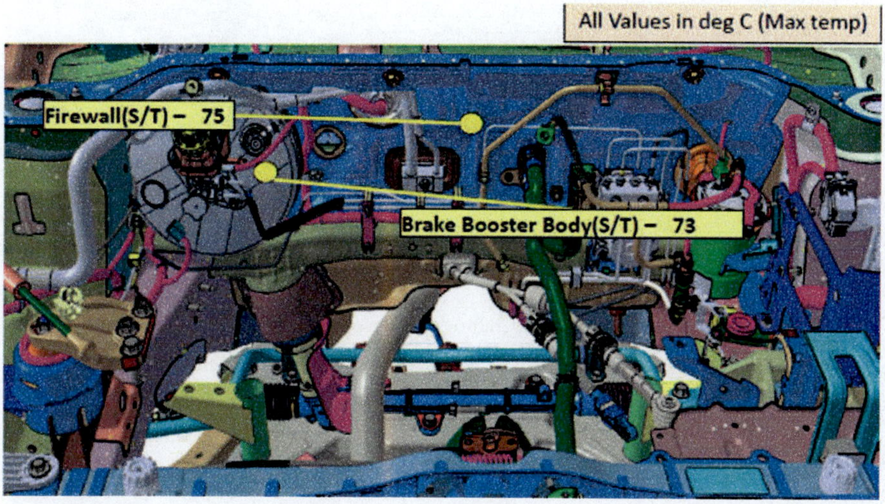

Fig. 5.9 Engine compartment firewall

5. **Aftertreatment system**

With increasingly stringent emission standards like Euro-6 or BS-VI have more of the after-treatment systems are kept close to the engine in the hot zone [7]. Temperatures of the surfaces of the catalysts can go up to 800 °C. The aftertreatment systems are insulated with the help of heat shields to maintain the temperatures above light off as well as protect the neighbouring equipment and the customer (Fig. 5.10).

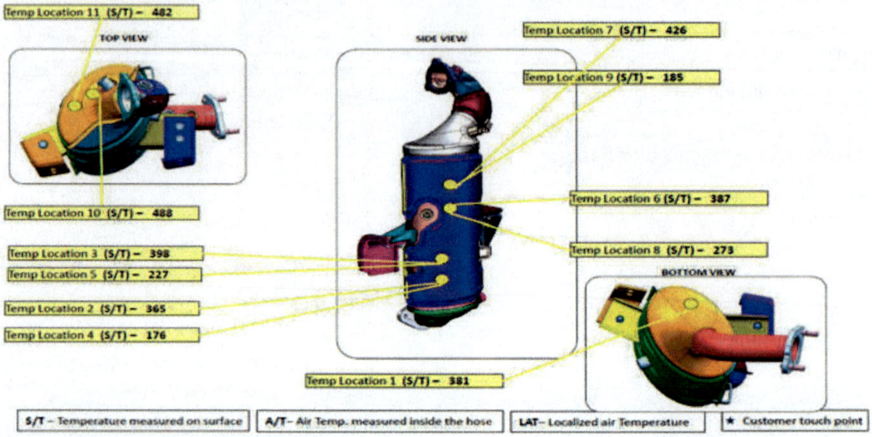

Fig. 5.10 Aftertreatment system

Application area	Catalytic converter
Component material	Steel
Maximum recommended operating temperature	800 °C
Heatshield	AMS stainless

6. **Exhaust system**

As the exhaust system carries about 23% of heat energy released from the fuel, it becomes very hot at full load. The system passes through the tunnel in the underbody area and is protected with heatshields (Fig. 5.11).

Application area	Underbody, fuel tank, spare wheel
Component material	Mild steel, plastic
Maximum recommended operating temperature	400–500 °C
Heatshield	AMS

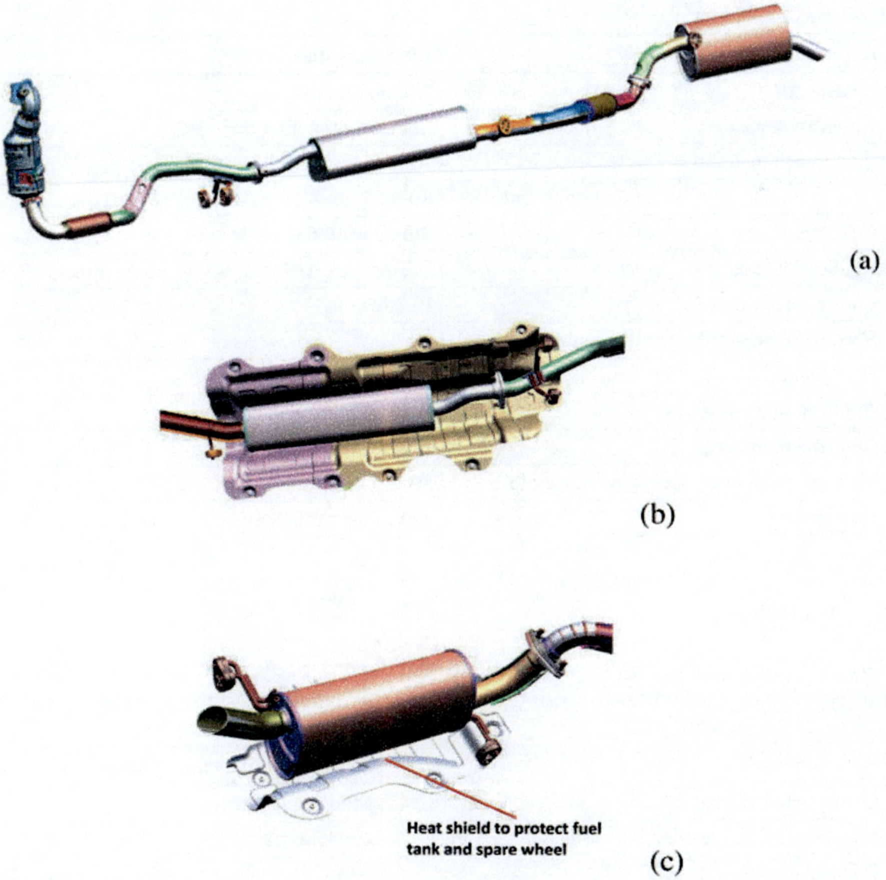

Fig. 5.11 **a** A full exhaust system: a close-coupled catalyst in the engine compartment, ducts, catalysts underbody, silencer, tailpipe, **b** details of the arrangement of heat shield for the catalyst underbody, **c** details of the arrangement of heat shield for the silencer underbody

7. Electrical Systems

Communication between various systems and subsystems takes place with the help of electric systems. Any fault in the electrical system disables the vehicle completely. Some of the systems like wiring harness, fuse box, ECU, battery pack, alternators are typical examples, Figs. 5.12, 5.13, 5.14, and 5.15. Here, the safe operating temperatures are highlighted in the legends for a well-optimized system for heat transfer using heatshields.

Application area	LV battery, fuse box
Component material	Plastic
Max. recommended operating temperature	100 °C

(continued)

(continued)

Application area	LV battery, fuse box
Heatshield	Aluminium foil
Application area	Starter motor, alternator, ECU
Component material	Aluminum
Max. recommended operating temperature	80 °C
Heatshield	Single wall, flat
Application area	Wiring harness in the engine compartment
Component material	Plastic
Max. recommended operating temperature	80 °C
Heatshield	Aluminium foil
Application area	Fuse box, headlamp
Component material	Plastic
Max. recommended operating temperature	80 °C
Heatshield	Aluminium foil

8. **Underbody**

The base for the passenger compartment is under the body, and the area should be protected from the radiation from the road and the exhaust system by heat shields:

Application area	Underbody, fuel tank, spare wheel
Component material	Mild steel, plastic
Maximum recommended operating temperature	400–500 °C
Heatshield	AMS

Figure 5.16 shows the temperatures at representative points after the ideal application of the heatshields under the body of a commercial successful car [3].

5.8 Protection Against Heat: "Hot-Spot"

Some examples of the heatshield application on the hot spots like (a) the catalytic converter and the after-treatment system, (b) underbody, and (c) fuel tank are shown in Fig. 5.17.

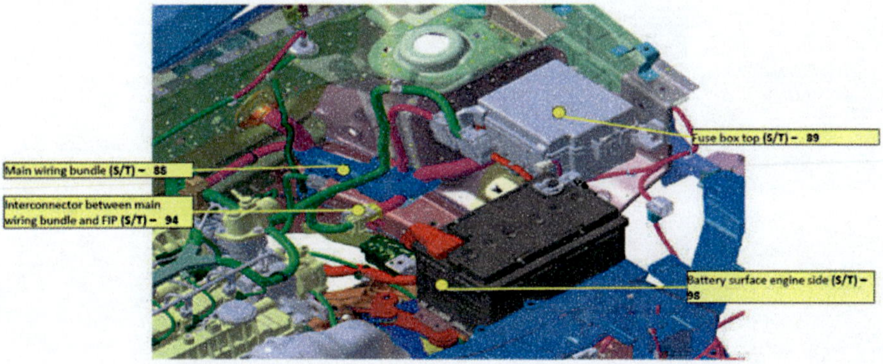

Fig. 5.12 Electric system-battery and fuse box

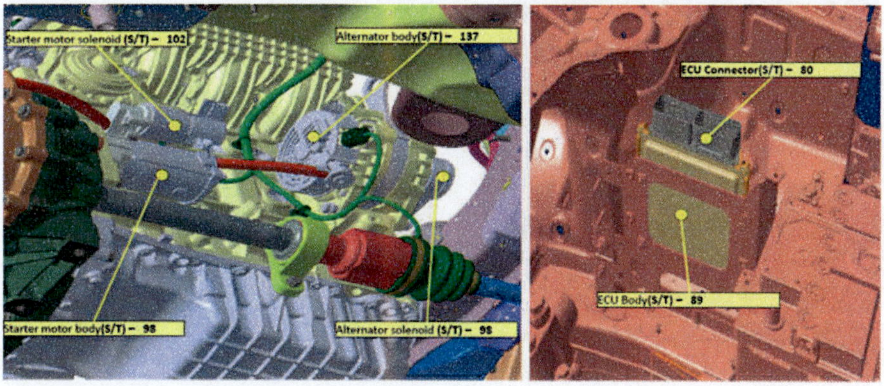

Fig. 5.13 Electric system: alternator and ECU

5.9 Thermal Mapping for Design Validation

5.9.1 Mapping Surfaces Using a Thermal Camera

A thermal imaging camera (Fig. 5.18) is useful in thermal mapping and evaluating the surface temperatures at the customer touchpoints (Table 5.2) and at the locations that affect the thermal performance of the vehicle and the fuel economy, mainly at kerb idle and hot shut-off conditions whice are severe in terms of underhood thermal management.

Thermal Imaging Camera

All objects emit radiation as a function of their temperatures which is a fraction of the black body radiation; the fraction is given by the emissivity of the surface. At lower temperatures say, less than 500 °C more of the emission is in the infrared region and

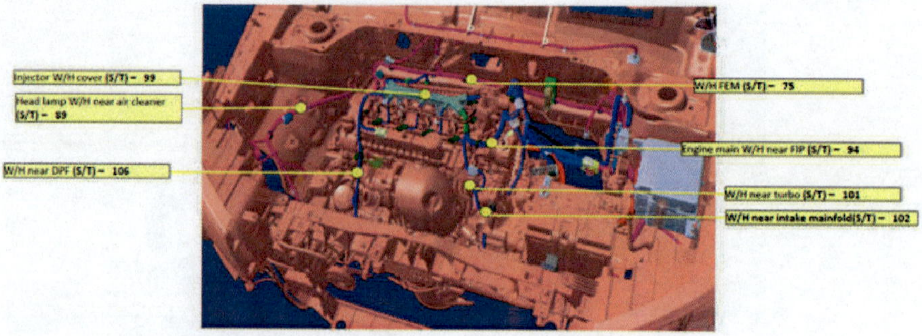

Fig. 5.14 Wiring harness

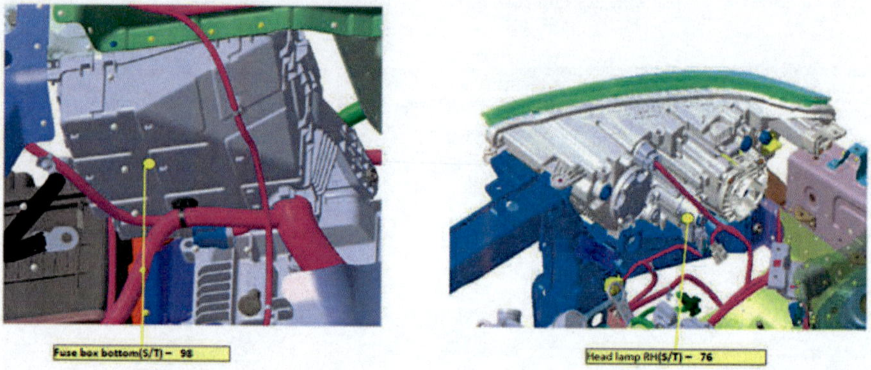

Fig. 5.15 Head lamp

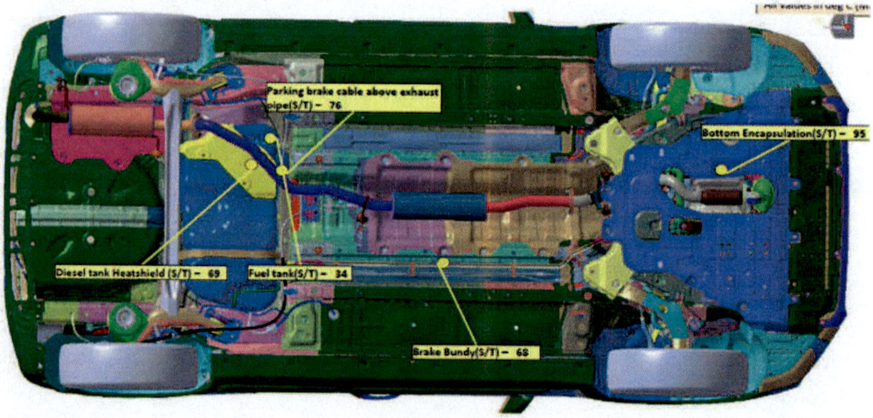

Fig. 5.16 Under body

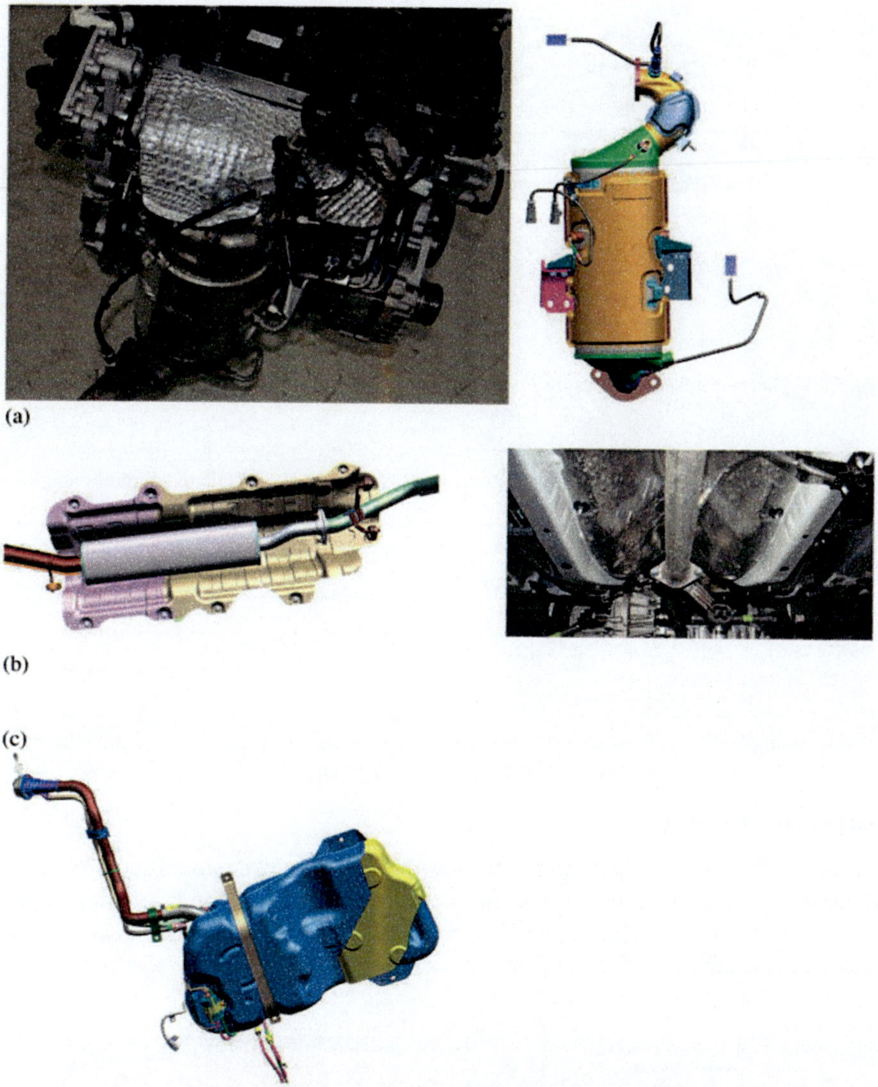

Fig. 5.17 Examples of application of heatshields at hot spots **a** catalytic converter-after-treatment system, **b** underbody, **c** fuel tank (yellow-colored sheet is the insulation)

shifts to the visible region at higher temperatures. The higher the temperature, the more is the infrared radiation emitted by a surface.

A thermal imager camera (also called an infrared camera or thermographic camera) forms an image using infrared radiation beyond the visible spectrum, Fig. 5.18. They operate at wavelengths as long as 14,000 nm (14 μm) and the lens of the camera is transparent to infrared light unlike glass used in ordinary cameras,

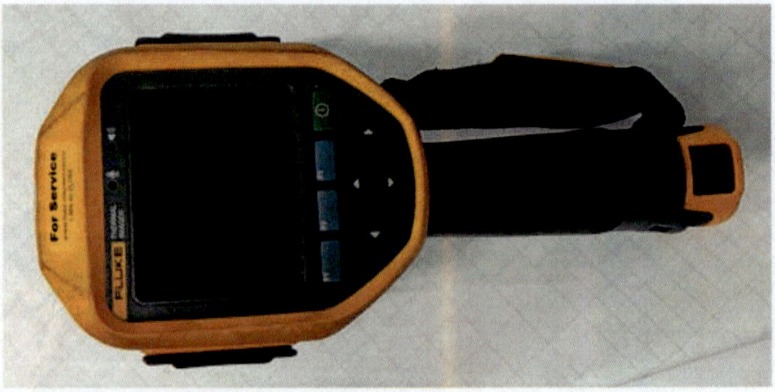

Fig. 5.18 Thermal imager camera by *Fluke*

Table 5.2 Customer touchpoints

	Customer touchpoint	Acceptable temperature range (°C)
1	Hood latch	50–70
2	Bonnet stay rod	50–70
3	Engine cover	50–70
4	Cabin floor area	35–40

which is opaque to infrared light. For the measurements of surface temperatures, a thermal imaging camera by FLUKE has been used [8].

Surface temperatures

Figures 5.19, 5.20, 5.21, 5.22 and 5.23 show the temperature contours at the locations identified as customer touchpoints after optimizing the insulation in relevant zones.

It is important to maintain the temperatures of surfaces near the intake system, the exhaust system, the firewall, and the cabin-floor area below the allowable limits

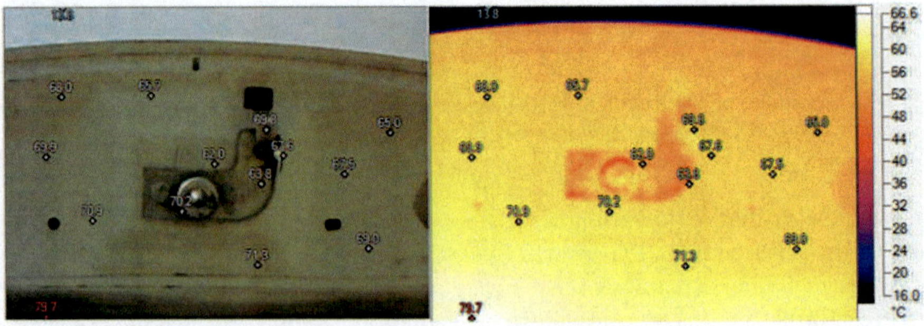

Fig. 5.19 Bonnet hood latch, mapping temperature distribution

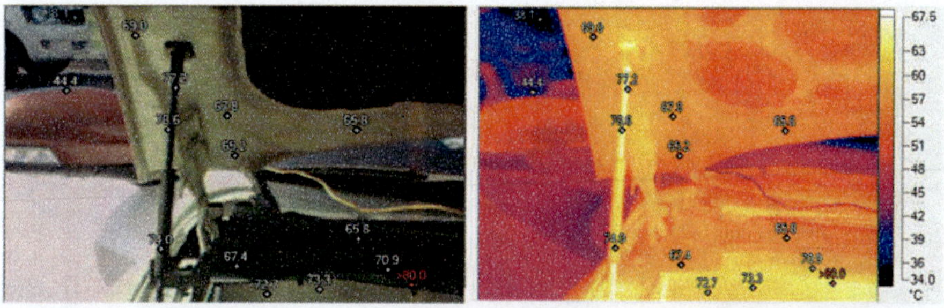

Fig. 5.20 Bonnet stay rod, mapping temperature distribution

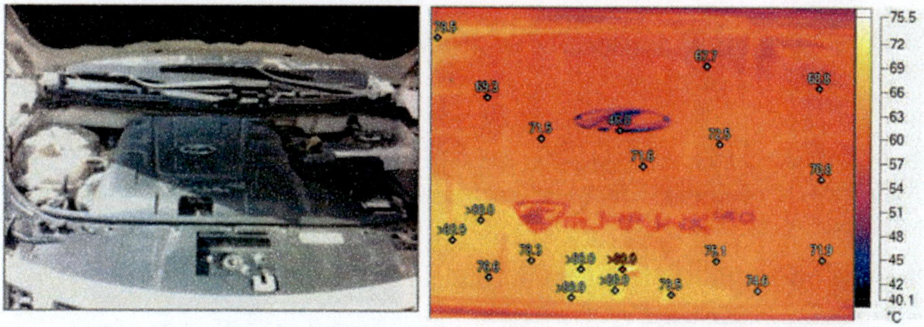

Fig. 5.21 Engine cover, mapping temperature distribution

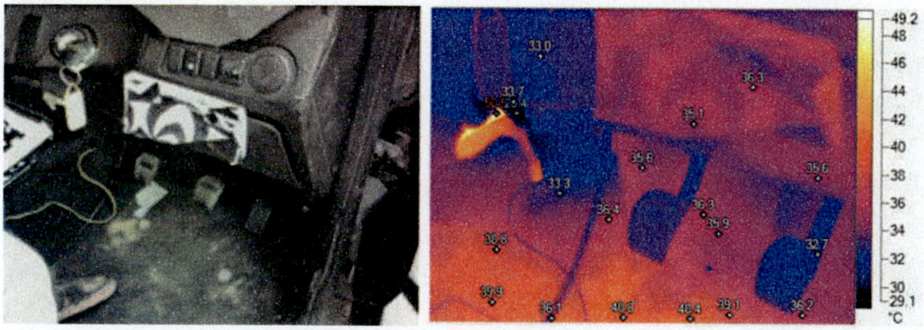

Fig. 5.22 Cabin floor area, mapping temperature distribution

to meet the thermal comfort and to achieve the best vehicle performance for fuel economy and emissions.

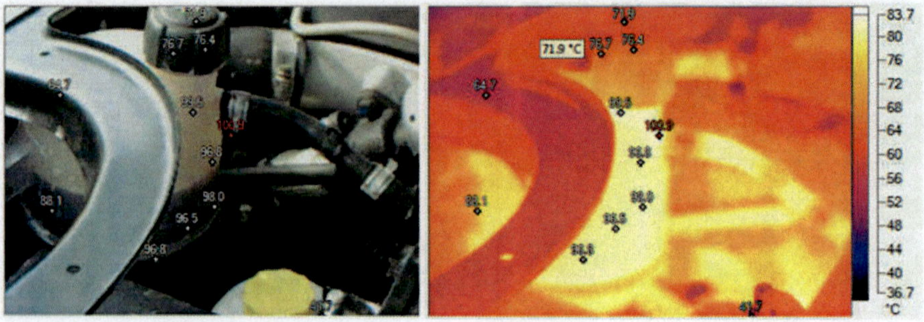

Fig. 5.23 Degassing tank, mapping temperature distribution

5.9.2 Mapping Temperature of Flows and Surfaces Using Thermocouples

The underhood heating happens after the vehicle undergoes high thermal loads after a sequence of operating conditions such as highway driving and grading in hot environments above 40 °C.

Temperatures of flows of coolant, oil, and air

Thermocouples are placed in the centre of flow conduits of oil, coolant, and air to measure the temperature of the fluids at the inlet and the outlet of the equipment or systems like coolers, radiators, the cylinder head, and the crankcase. When a thermocouple is used in the soft hose, to avoid spillage, the skin of the thermocouple is carefully removed without damaging the insulation of the wire and the thermocouple tip before inserting it in the hose, Fig. 5.24. For external flows over large areas, three or four thermocouples are kept in the flow path which helps in arriving at an average temperature of the flow, Fig. 5.25.

Thermocouples for surface temperature measurement

When applied on a surface, the tip of the thermocouple is covered with adhesive insulation to rule out convection and radiation from the surroundings to the tip, Figs. 5.26 and 5.27. To keep track of the temperatures of some typical surfaces, e. g., catalyst surface, it is often necessary to use a robust K-type thermocouple, Fig. 5.28. Sometimes it becomes tricky to measure the temperatures like that of the oil sump and it may be necessary to contrive special solutions; in Fig. 5.29, the thermocouple is inserted in the drilled bolt hole of the oil sump itself.

Fig. 5.24 Thermocouple installed at soft joints of the radiator header

Fig. 5.25 Measurement of temperature of airflow from the radiator using thermocouples

5.10 Assembly of Heatshields

The assembly technique must be chosen to satisfy the requirements of

- Stiffness and strength,
- Feasibility for disassembly, e. g., maintenance, repair and
- Lifetime performance, e.g., corrosion resistance.

Fig. 5.26 Thermocouple tip placed over the surface and the tip is covered with an adhesive insulation

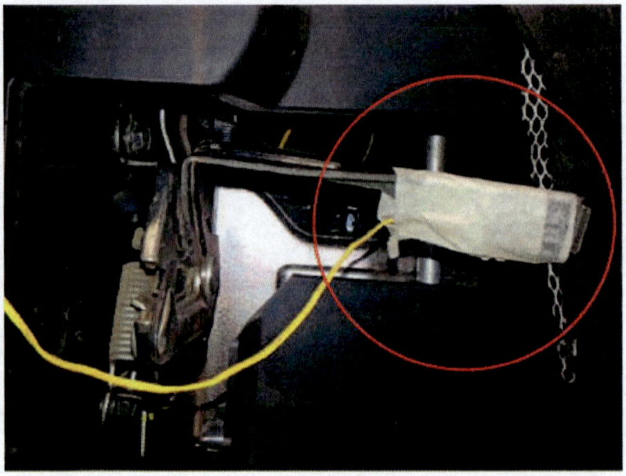

Fig. 5.27 Thermocouple tip placed over the metal surface and the tip is covered with an adhesive insulation

Rivets and screws which are electrolytically galvanized for protection against corrosion are often used to assemble the heatshields to the car body. Also, aluminum heat shields are commonly attached to the steel car body using nuts and sacrificial washers, Fig. 5.30.

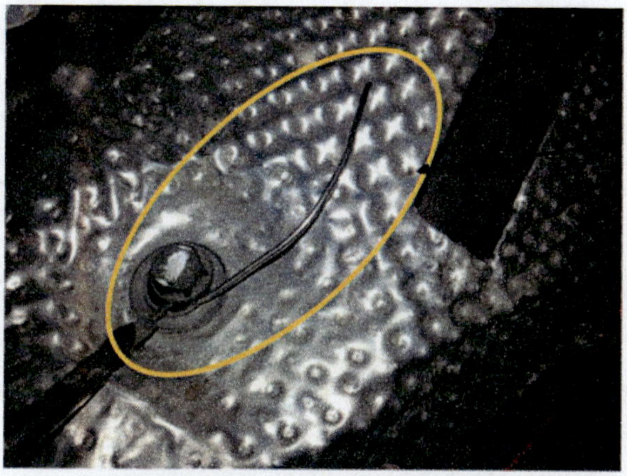

Fig. 5.28 K-type thermocouple fitted over the catalytic converter heat shield to measure the temperature

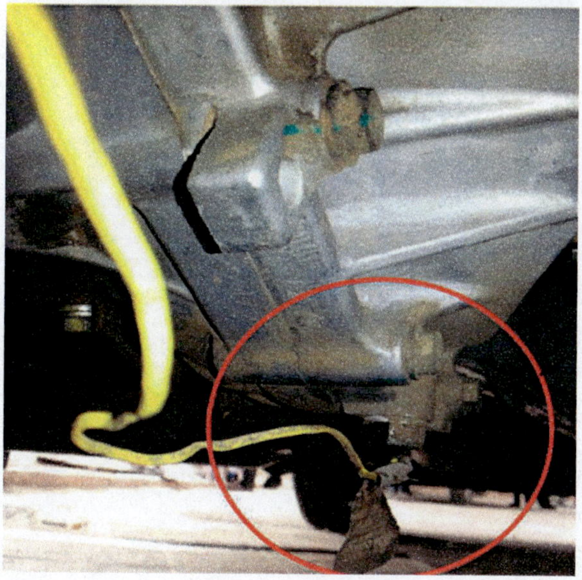

Fig. 5.29 For the measurement of the temperature of the oil sump, the thermocouple should be inserted in the drilled bolt of the oil sump itself

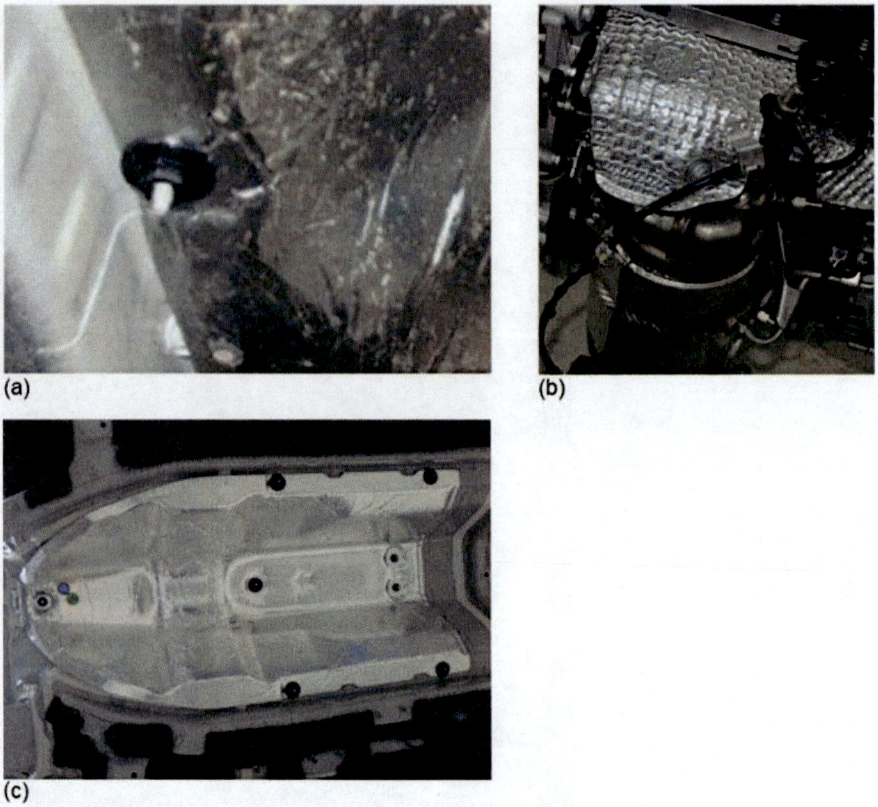

Fig. 5.30 Heatshield mounting scheme. **a** "Scrivet" type mounting. **b** Direct bolting the heatshield. **c** With nut and washer

Acknowledgements The authors of the chapter thank Mahindra & Mahindra Ltd. for supporting the publication in terms of using the photographs, images and terminology.

References

1. Roberts A, Brooks R, Shipway P (2014) Internal combustion engine cold-start efficiency: a review of the problem, causes and potential solutions. Energy Convers Manage 82:327–350
2. What is thermal insulation—thermal insulator—definition (thermal-engineering.org). https://www.thermal-engineering.org/what-is-thermal-insulation-thermal-insulator-definition/. Accessed on 6-6-2021
3. Heat shield materials, thermal barrier heat shield materials|ZircoFlex® from Zircotec. https://www.zircotec.com/heat-shield-materials/. Accessed on 6-6-2021
4. Ali A, Kamal M, Xianjun H, Turkson RF, Ezzat M (2016) An analytical study of tribological parameters between the piston ring and cylinder liner in internal combustion engines. Proc Inst Mech Eng Part K J Multi-body Dyn 230(4):329–349

5. Thermal insulation materials, technical characteristics and selection criteria (fao.org). http://www.fao.org/3/Y5013E/y5013e08.htm. Accessed on 6-6-2021
6. Victor Reinz® Sound Absorbing Heat Shield. https://www.reinz.com/OE/EN/PRODUCTS/Heat-Shields/Sound-Absorbing-Heat-Shield.aspx. Accessed on 6-6-2021
7. Heat Shields. http://www.fitwelgasket.com/heat_shields.html. Accessed on 6-6-2021
8. Fluke Ti480 PRO infrared cameralfluke. https://www.fluke.com/en-in/product/thermal-cameras/ti480-pro. Accessed on 6-6-2021

Part III
Techniques for Early Light-Off of Aftertreatment Systems

Chapter 6
Diesel Engine Throttling—The Classical Tool: To Adapt Exhaust Gas Temperature for Emission Control by Catalysts and Filters: From Its Beginning to the State of the Art in Euro 6/VI

A. Mayer, A. Amstutz, L. Guzzella, Y. Hohl, F. Jaussi, S. Kany, Chr. Lämmle, F. Legerer, Th. Lutz, P. Nöthiger, M. Wyser, H. Stieglbauer, and J. Czerwinski

---

A. Mayer (✉)
TTM, Fohrhölzlistrasse 14 b, CH 5443 Niederrohrdorf, Switzerland

A. Amstutz · L. Guzzella · Th. Lutz
ETH, Rämistrasse 101, CH 8092 Zürich, Switzerland
e-mail: amstutz@idsc.mavt.ethz.ch

L. Guzzella
e-mail: lguzzella@ethz.ch

Th. Lutz
e-mail: thomas.lutz@alumni.ethz.ch

Y. Hohl · F. Jaussi
Liebherr Machines Bulle SA, Rue Hans-Liebherr 7, CH 1903 Bulle, Switzerland
e-mail: yves.hohl@liebherr.com

F. Jaussi
e-mail: Francois.Jaussi@liebherr.com

S. Kany
TSH, Gewerbestrasse 7, D 79801 Hohentengen, Germany

Chr. Lämmle
Computer Flow Solutions, Technoparkstrasse 1, CH 8005 Zürich, Switzerland
e-mail: laemmle@combustion-flow-solutions.com

F. Legerer
VERT, Verification of Emission Reduction Technologies, c/o JCA Treuhand AG, Badener Strasse 9, CH-5200 Brugg, Switzerland
e-mail: ztdr.legerer@aon.at

P. A. Lakshminarayanan and A. K. Agarwal (eds.), *Handbook of Thermal Management of Engines*, Energy, Environment, and Sustainability,
https://doi.org/10.1007/978-981-16-8570-5_6

Acronyms

ADECS	Verified Diesel Emissions control Strategy
ARB	Air Resources Board
BMEP	Brake mean effective pressure
BSFC	Brake specific fuel consumption
CNC	Condensation nucleus counter
CO	Carbon monoxide
CO_2	Carbon dioxide
CRT	Continuous regeneration trap (JMC patent)
DC	Diffusion charger
DMA	Differential mobility analyzer
DPF	Diesel particulate filter
EC	Elemental carbon
ECU	Electronic control unit
EGR	Exhaust gas recirculation
EPA	US-environment protection agency
FBC	Fuel borne catalyst
FSN	Smoke number
HC	Hydrocarbons
HPL	High-pressure loop
LPL	Low-pressure loop
Nm	Nanometer
NO	Nitric oxide
NO_2	Nitrogen dioxide
NOx	Nitrogen oxide
O_2	Oxygen
PAS	Photoemission aerosol sensor
PASS	Photoacoustic soot sensor
PM	Particulate matter
ppm	Part per million
SCR	Selective catalytic reduction
SiC	Silicon carbide
SMPS	Scanning mobility particle sizer

P. Nöthiger
PNE, Mühletalweg 6, CH 4600 Olten, Switzerland
e-mail: pne@pop.agri.ch

M. Wyser
BUWAL, Bundesamt für Umwelt, Wald und Landschaft, CH-3003 Bern, Switzerland

H. Stieglbauer
STARFILTER, am Katzernbuckel 59, D 82377 Penzberg, Germany

J. Czerwinski
AFHB, Abgasprüfstelle der Fachhochschule Biel, Weiherweg 17a, CH-2562 Port, Switzerland

TAC	Toxic air contaminant
TC	Turbocharger
TRU	Transport refrigeration unit
VERT	Reduction of emissions of diesel in tunneling
VOCs	Volatile organic compounds
VTG	Variable turbine geometry

6.1 Part 1: Temperature Management by Throttling

6.1.1 Introduction

The trap could be prevented from getting clogged by burning the soot collected in it periodically; however, the combustion of soot takes place only when the temperature is in the range of 500–600 °C. The range of temperature is dependent upon the composition, morphology, and graphite structure of the soot, and the composition of the exhaust gas.

Such temperatures are seldom attained in supercharged diesel engines used in vehicles. Therefore, the ignition temperature of soot must be reduced, for example by some catalyst (coated or fuel borne); alternatively, the temperature of the exhaust gas must be increased. Catalysts lower the soot ignition temperatures may to about 250–350 °C and it is reasonable for most of the operational conditions. Even then, the engines in city drive may run for long at low loads, and the filter gets clogged resulting in high back pressure, loss of power seriously affecting the fuel economy accompanied by greater temperature gradients, and thermal stresses on structural parts. Further, the accumulated soot can burn uncontrolled at temperatures exceeding 1000 °C, putting the materials of the trap in danger.

6.1.2 Regeneration Requirements

When the backpressure sensor shows that the trap has reached the soot mass for starting regeneration, the exhaust gas temperature must be augmented. However, the exothermal reaction rate should be within limits, to avert overheating of the filter, by limiting the energy supply or the oxygen. Conversely, if the hard layers of soot formed by slow deposition (cake formation) or intermittent regenerations are to be burnt, its ignition may be improved by increasing the excess oxygen. The quick regeneration without damaging the filter is aimed at while optimizing the combustion. In other words, the combustion should be completed to maintain the average back pressure below the level that is considered as the serious external load on the engine. Also, during regeneration, no secondary emissions of toxic substances may be released. Finally, regeneration must be completed without the operator noticing any intervention by the apparent loss of power and torque.

6.1.3 Current Systems

Currently, the gases are heated by burners or electric heaters. The temperature increase by throttling was attempted originally by Pattas [1], by retaining the hot residual gas in the cylinder and reducing the supply of fresh air for combustion. The method has not been widely applied due to problems of design of the hot flaps, soiling of cylinders as well as high thermal loads.

For instance, Mayer [2] introduced intake throttling by controlled exhaust recycling of pressure waves in an open COMPREX supercharging cycle, control of which is extensively analyzed by Amstutz [3, 4]. This method is, easier to build, and control than throttling hot gases.

For now, intake throttling is widely applied for controlling the exhaust recirculation to achieve the required pressure difference between the exhaust and the intake [5].

6.1.4 Concept of Intake Throttling

Excess-air to the tune of $\lambda = 6$–8 at part loads of diesel engines is because the torque and power are controlled by fuel flow only without controlling the airflow, in turn contributing to the improved fuel economy of a diesel engine over an SI-engine, as the pumping and heat losses are lowered. However, low temperatures are not favoured from the viewpoint of the quality of combustion, and emissions of CO and hydrocarbons at light loads. What is more, regeneration is not easily started at low loads.

By reducing the airflow at low loads, the air–fuel ratio is brought close to the stoichiometric, and the exhaust temperature increases close to full load operation sufficient for regeneration. The expected increase in exhaust temperature is somewhat moderated by the increased heat flow through cylinder walls. This technique may also be used optimally for reducing emissions.

When sudden acceleration demands a quick change of load, the airflow must be immediately commensurate with minimal delay. Therefore, the response of the dynamic throttling control should be sufficient for the trouble-free operation of an engine. Concepts of control, suitable for solving this problem are described in detail in Refs. [2, 3].

While throttling a substantial depression is generated at the intake side, hence, there is increased fuel consumption as a consequence. However, the overall effect on fuel economy is negligible since the duration of throttling is limited to only ten minutes in eight hours of operation.

6.1.5 Computational Simulation

To evaluate throttling design parameters, simulation studies were conducted and verified by measurements at the Department of Internal Combustion Engines at the Federal Polytechnic Institute ETH, Zurich. See Table 6.1.

6.1.5.1 Computational Model

The model of the thermodynamic process is set up in the software, *GT-Power* where the intake and exhaust Flow phenomena are completely accounted for, and, hence the throttling at any point in the system is easily analyzed. The high-pressure process in the cylinder is considered by a simplified dual-zone combustion model with heat losses across the walls. The low-pressure part is employed using the one-dimensional equations of gas dynamics (Chap. 12). The turbine and the compressor characteristics of the turbocharger are implemented in the model.

The design alternatives are studied using a throttling flap installed at different positions before compressor, after the compressor, and before the turbine. By simultaneously changing the angle of the flap and fuel mass flow, it is possible to operate the engine at constant excess air of $\lambda = 2$ at various part loads, at constant engine speed.

The measured temperatures, pressures, cylinder pressure–time diagrams are compared with the model at a BMEP = 10.5 bar. The heat release of combustion at some points of reference derived from the indicated pressure serves as calibration input for validation. Data at lower loads are obtained by extrapolation.

Table 6.1 Principal features of a modern turbocharged diesel engine with a common rail injection system

Parameter		
Bore	122	mm
Stroke	142	mm
Displacement	6.64	L
Ratio of compression	17.2	–
Number of cylinders	4	
Number of valves per cylinder	2	
Nominal power at engine rpm	180/2100	kW
Maximum torque at engine rpm	1050/1150–1500	Nm
Supercharged	TC with IC	
Injection-system	Common-rail	
EGR	None	

6.1.5.2 Throttling Before Turbocompressor

A flap at the inlet of the compressor is the easiest to install. The load is reduced in steps from full load to low load (BMEP = 1 bar) at constant air excess and a speed of 1400 rpm. Figure 6.1 shows the exhaust temperature due to throttling.

All through the regular operation without throttling, λ increases from $\lambda = 2$ to 8 with decreasing load, which reduces the exhaust temperature from about 800 K to less than 450 K. However, throttling to a predetermined level of $\lambda = 2$, keeps the exhaust temperature a constant. At low loads, the exhaust temperatures increase impressively by about 300 °C (BMEP = 1 bar). The accompanying increase in heat transfer losses across the walls is shown in Fig. 6.2.

Throttling relatively increases the heat loss to the walls from 13% at full load to 17% at the extremely low load. Also, it increases the specific fuel consumption by up to 80 g/kWh due to increased pumping work, Fig. 6.3. The losses due to the differences become more prominent at lower loads. However, the losses are noticeable only for a short time, without affecting the overall performance.

In Fig. 6.4 pressures before and after turbocharger with and without throttling are presented as a function of load. With intake throttling, the pressure before the compressor is almost constant, i.e. more or less the ambient pressure, while the pressure between the throttle and compressor intake decreases substantially. To achieve the intended effect of reduced airflow, the pressure before the compressor has to drop to about 0.5 bar; thus, the pressure-ratio at the throttle reaches almost two and within

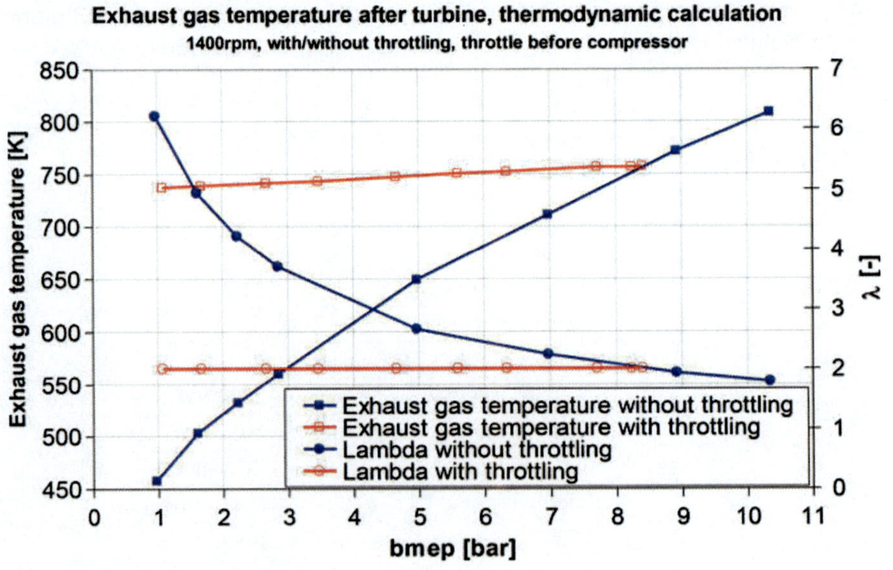

Fig. 6.1 Temperatures of exhaust gas with and without throttling for constant air excess before compressor at 1400 rpm (computational model)

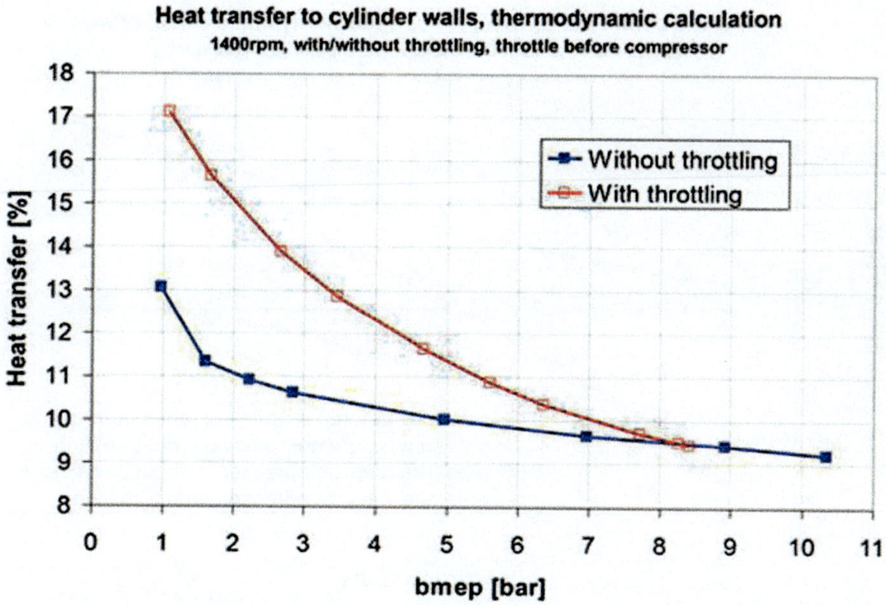

Fig. 6.2 Losses due to heat transfer through cylinder walls while throttling before compressor (computational model)

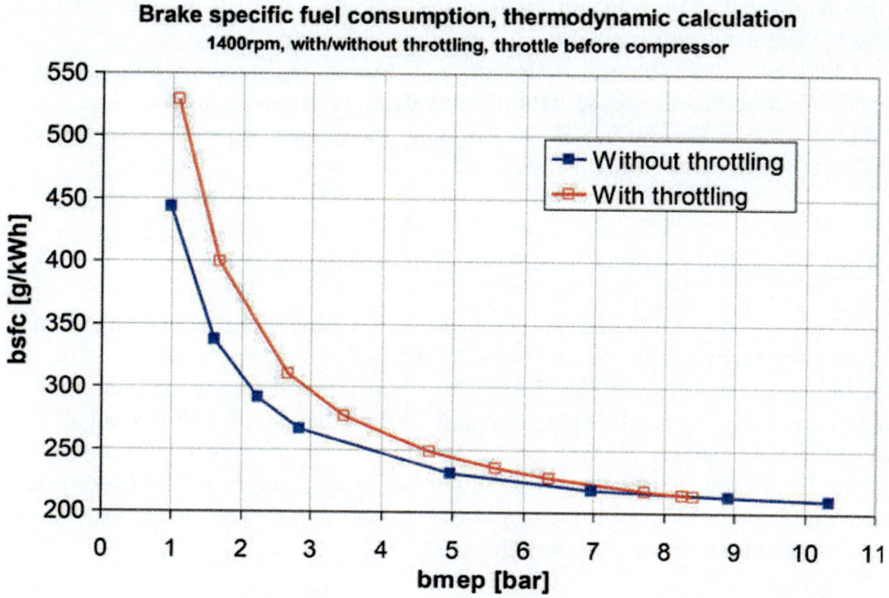

Fig. 6.3 Specific fuel consumption as a function of load with/without throttling (computational model)

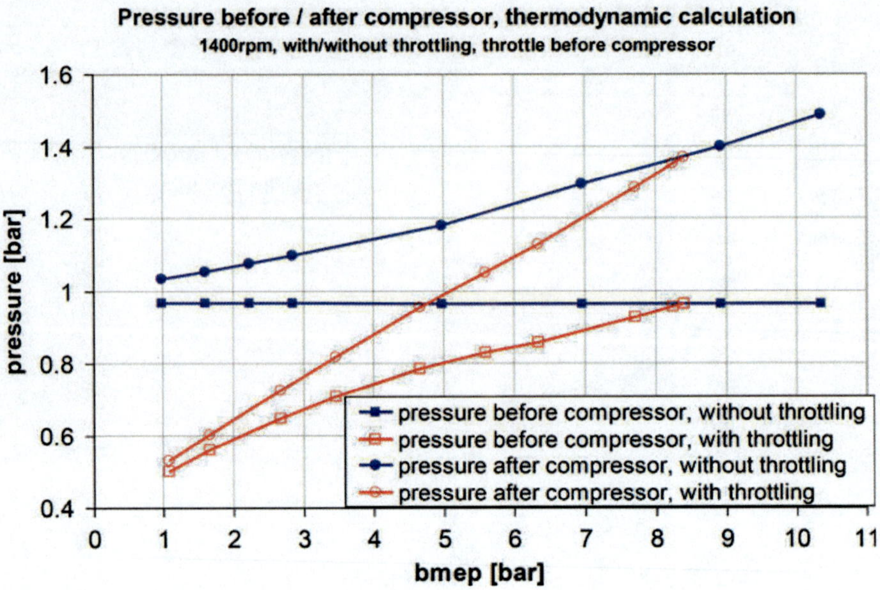

Fig. 6.4 Pressure before/after the compressor, with and without throttling (computational model)

the zone formed by the flap and the duct, the airflow reaches the velocity of sound. This hints at the necessary mechanical strength of the flap.

Additionally, the extraordinary low pressure extends to the rear of the compressor wheel affecting the oil-sealing with the attendant concern regarding the loss of oil. Another worry is the shifting of the operating line in the compressor map with the consequent risk of surging, Fig. 6.5.

6.1.5.3 Throttling After Compressor of a Turbocharger

Introduction of the throttle after the compressor preferably after the intercooler avoids the disadvantage of generating a severe depression, which causes oil loss. Also, the smaller and compact throttling flap enables easier installation in the manifold of compressed air. The results of the thermodynamic calculation of this configuration are plotted in Fig. 6.6.

The results are the same as for throttling before the compressor. At lower loads, the temperature after the turbine increases by about 300 K. The engine response is equivalent in both the throttle configurations.

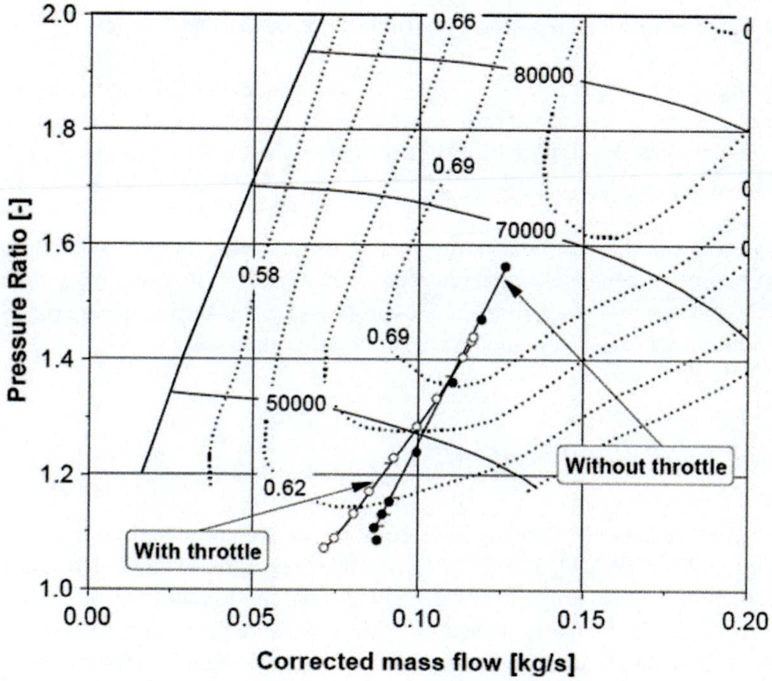

Fig. 6.5 Characteristics of the compressor of the turbocharger with both operating lines, with and without throttling; turbo rpm (solid line) and compression efficiency (dotted line) are plotted parameters

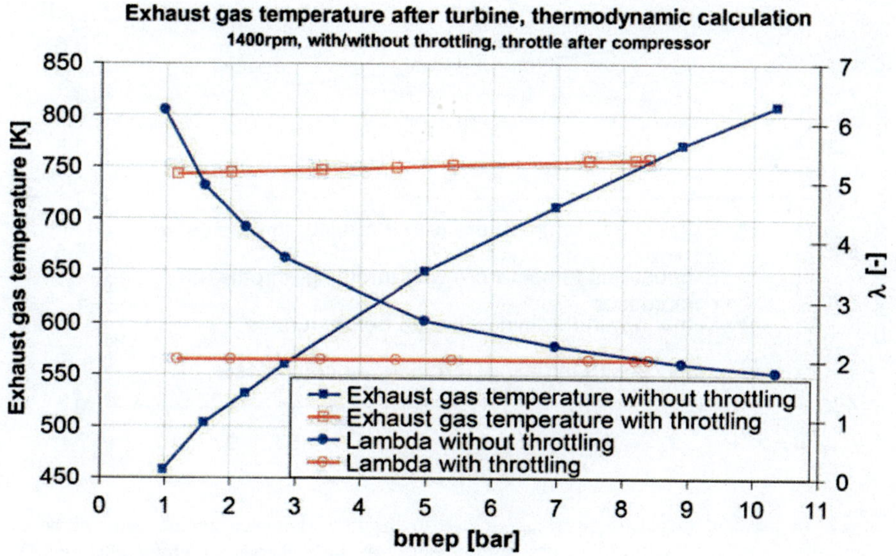

Fig. 6.6 Exhaust gas temperature (computational model)

6.1.5.4 Throttling Upstream of the Turbine of the Turbocharger

This configuration is advantageous but at the same time challenging because of the high temperatures in the case of throttle after the compressor, Fig. 6.7, though the fuel consumption increases significantly higher than by the throttling of the air, Fig. 6.8. The efficiency loss is mainly due to pumping losses, inferior combustion. The exhaust temperature rises mainly due to a drop in the air–fuel ratio.

Other than the difficulties regarding the design and installation of the flap, this method causes a completely different engine response to scavenging, combustion character, and residual gases compared to throttling the intake. Also, there is a serious risk of fouling the engine oil and hotter structural components.

6.1.6 Experimental Verification

Experimental results of throttling after compressor are compared with the simulation, Figs. 6.9, 6.10, 6.11 and 6.12. Generally, they agree well. Below BMEP = 4, extremely low mass flow did not achieve the required pressure ratio, and hence, λ rose to 2.4 experimentally, instead of the desired 2. Consequently, the exhaust gas temperature decreases, and hence, the calculation deviates from the experiment,

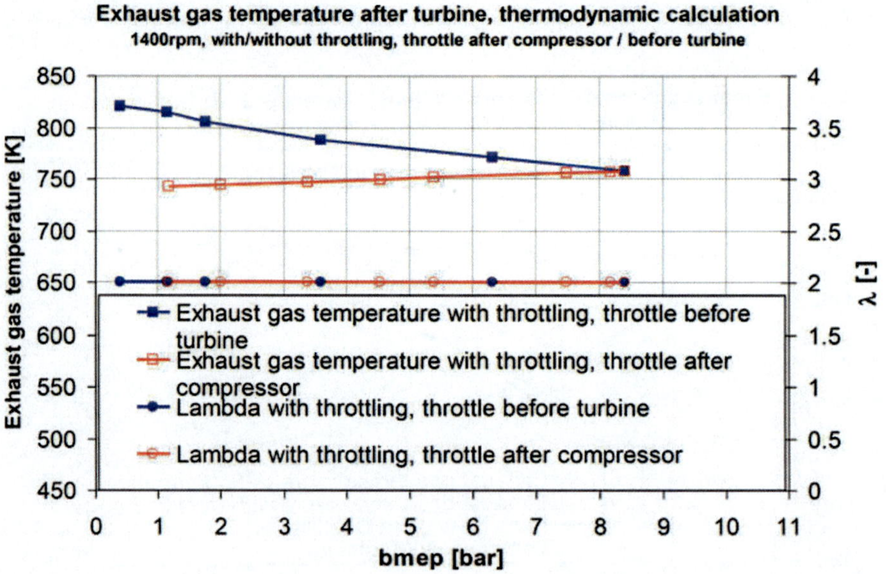

Fig. 6.7 Increase of exhaust-temperature **a** due to throttling between engine and turbine of turbocharger, and **b** due to throttling after compressor of the turbocharger (computational model)

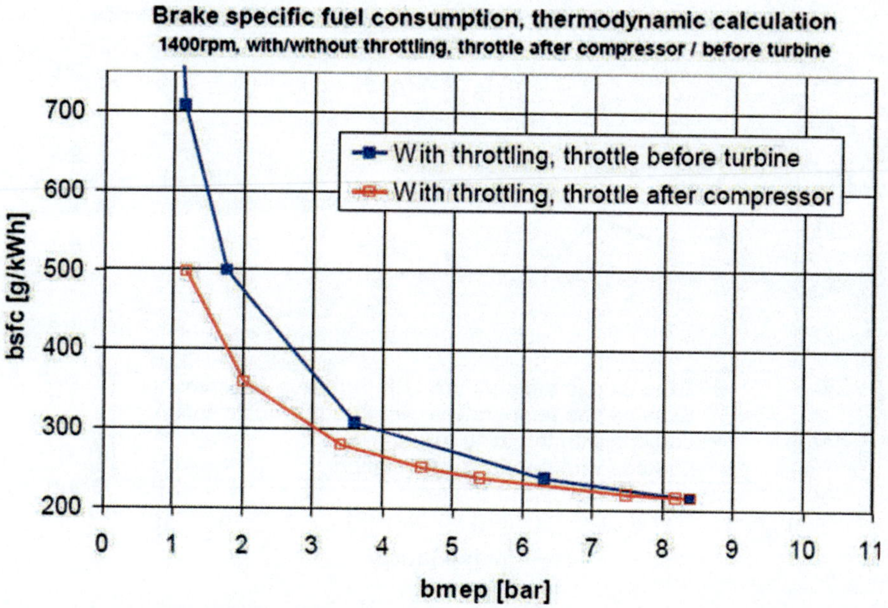

Fig. 6.8 Comparison of increased specific fuel consumption due to **a** throttling before the turbine and **b** after the compressor (computational model)

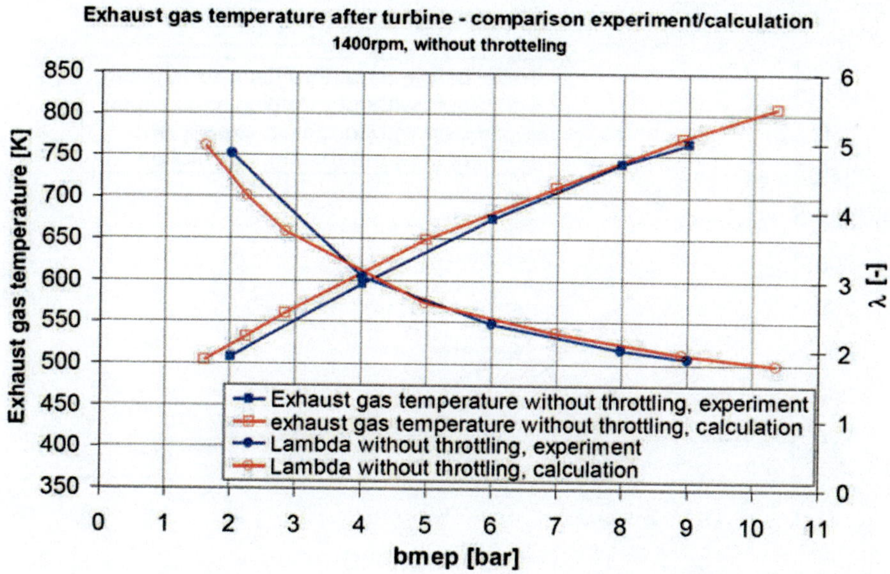

Fig. 6.9 Exhaust gas temperature after the turbine without throttling; comparison of experiment and computational simulation

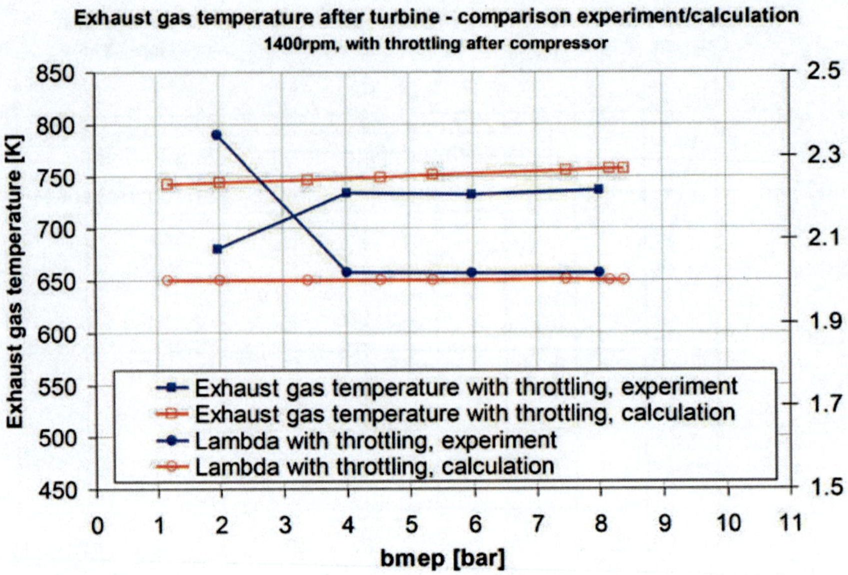

Fig. 6.10 Exhaust gas temperature after turbine with throttling; comparison of experiment and computational simulation

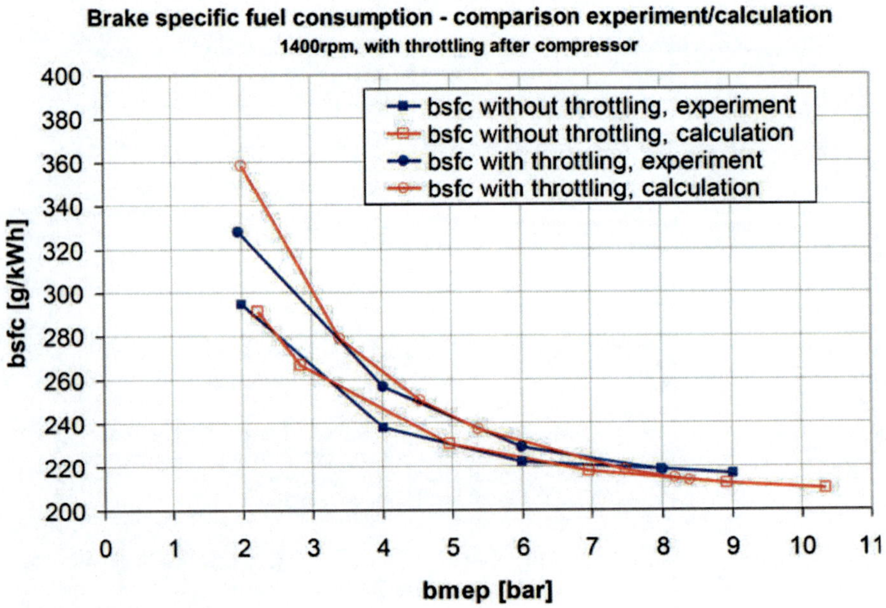

Fig. 6.11 Specific fuel consumption with throttling after compressor; comparison of experiment and computational simulation

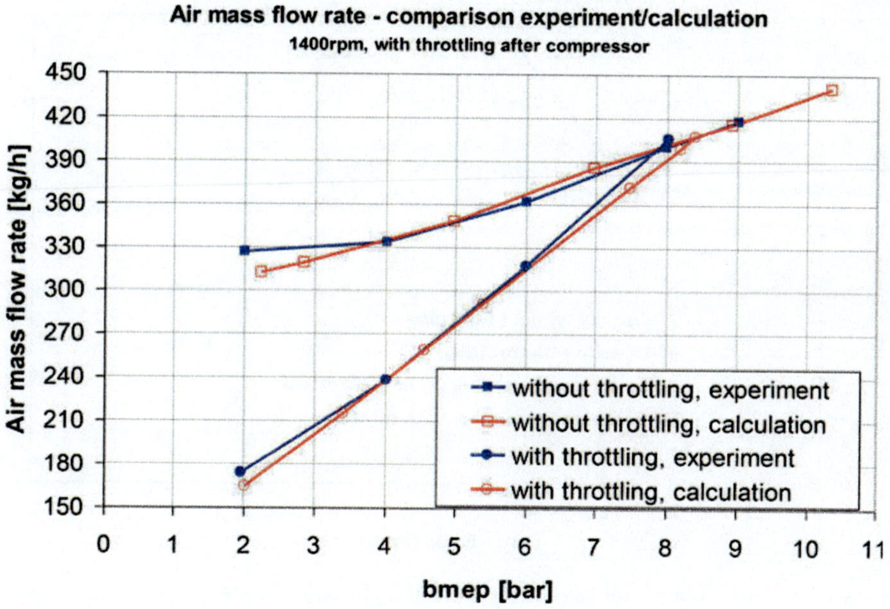

Fig. 6.12 Air mass flow with and without throttling; comparison of experiment and computational simulation

Figs. 6.9 and 6.10. The throttle must be mechanically robust and quite accurate in shape.

The engine experiments are also used to determine the effects of throttling on emissions and noise, Figs. 6.13, 6.14 and 6.15.

Smoke, NOx, and noise of combustion are inferior due to throttling. Higher temperatures bring about the formation of more nitrogen oxides. A higher gradient of heat release leads inevitably to higher noise.

6.1.7 Concepts of Control

Several strategies are available for manual or automatic control of flap positions.

- During stationary operation, it is possible just to turn the flap manually to a fixed end-position and switch it back to the neutral position after the back-pressure is relieved. Such solutions already exist for retrofit in off-road vehicles.
- In computer-controlled engines, the flap angle is programmed as a function of engine speed and load. The actuator must change between extreme positions within a tenth of a second.

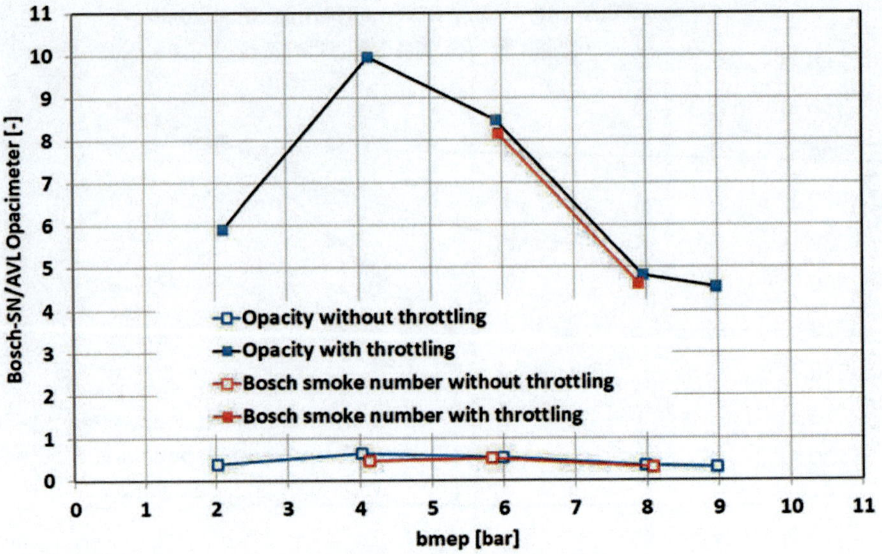

Fig. 6.13 Bosch smoke number and opacity with and without throttling; note the low end of load: due to throttling there is not sufficient mass flow for measuring smoke

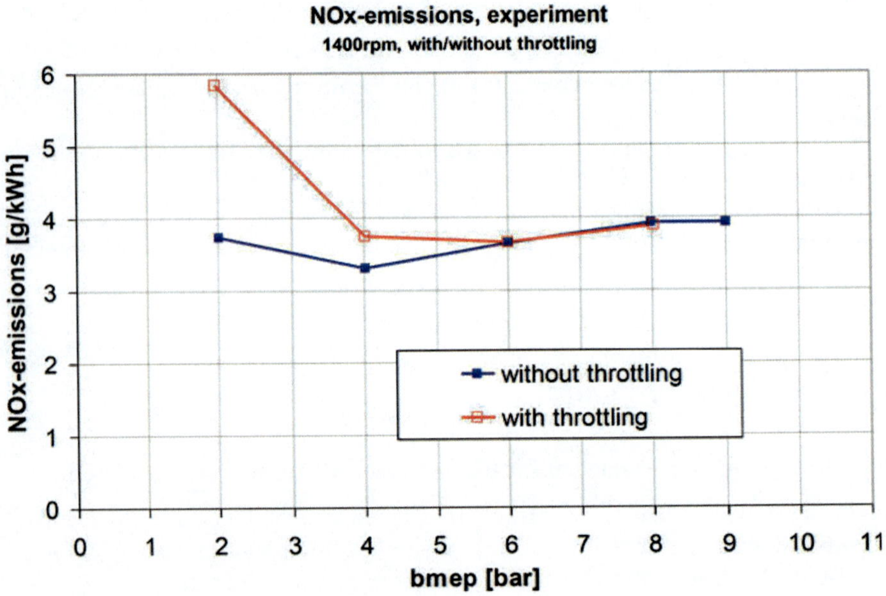

Fig. 6.14 NOx emissions with and without throttling

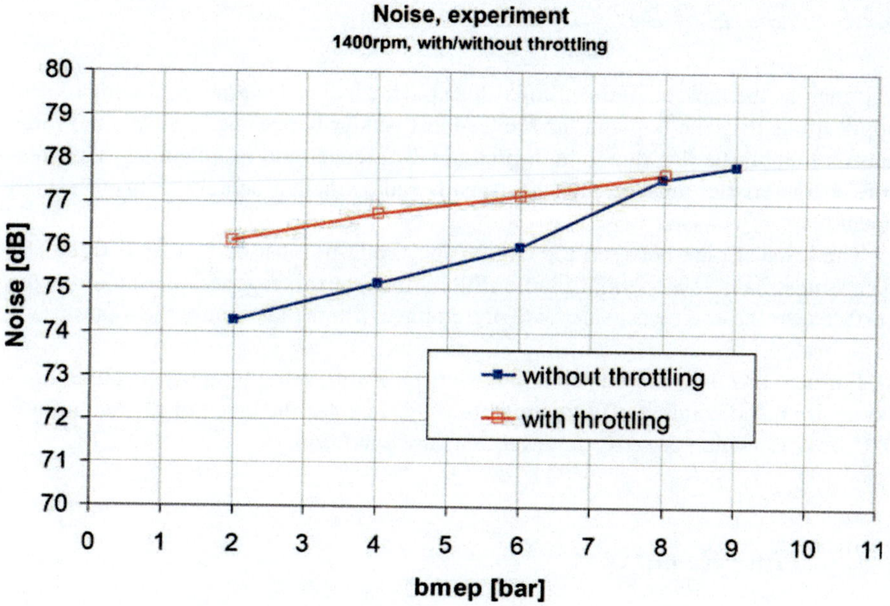

Fig. 6.15 The noise of combustion with and without throttling

- Also, closed-loop control is possible, based, for example on the exhaust oxygen detected using a λ-sensor [3, 4]. The control strategy must take care of the response time of the sensor which is about 50 ms.
- A simple mechanical solution would be to control the flap by the turbocharger's waste-gate position which may be suitable for retrofits.
- If the full load is demanded during the throttling operation, the flap must be moved to the neutral position with no delay, so that the full load can be achieved without reducing the excess air.
- The flap may be controlled such that the filter is heated up to the highest allowable exhaust temperature. After the ignition of soot is imitated, the throttle is reduced to increase oxygen for a speedy chemical reaction. If the reaction becomes very fast, it can be slowed by throttling again.
- The additional parameters for the control of the operation are the back-pressure of the filter and the temperatures before and after the filter.
- The flap may be simultaneously used to apply EGR for NOx control [5]. If an additional valve is used to control the gases from the exhaust to intake, it must be closed during the regeneration of the DPF.
- The flap may be used, for reducing the duration of white smoke during cold start by reducing the airflow to a minimum.

6.1.8 Aspects of Design

In principle, the method of throttling is feasible in all diesel engines including retrofit applications [6]. The flap and the hinges must be sturdy because a substantial force acts on the throttle flap due to high-pressure differences and fast changes. The pressure characteristics must be such that sensitive adjustment is achievable at even small openings.

The exhaust pipe between the engine and the DPF must be insulated well. The filters must have low thermal inertia. Filters that start to regenerate as soon as the surface near the engine is heated, are preferable to filters that require the total surface to be heated.

Further, throttling must be combined with catalytic methods, applying either fuel-borne or coated catalysts. The temperature increase for the onset of regeneration is 100–150 °C, which is easily achievable in any condition.

6.1.9 Final Remarks

In the case of mere passive regeneration, major failures cannot be avoided in real-world applications, as there can be situations when regeneration conditions are not reached. A backup should be available to trigger regeneration.

Sometimes the original equipment manufacturers use computer-controlled injection in the exhaust to increase exhaust temperature when required, which is not feasible for retrofit.

Throttling of the intake, preferably downstream of the turbocharger compressor is interesting and can be applied for retrofits with catalysts.

6.2 Part 2: Retrofitting TRU-Diesel Engines with DPF-Systems Using FBC and Intake Throttling for Active Regeneration

6.2.1 Introduction

In 1998, the California Air Resources Board (ARB) identified Particulate Matter (PM) emissions from diesel-fueled engines as a Toxic Air Contaminant (TAC). The diesel exhaust is a mixture of many gases and fine particles that contain more than 40 TACs, such as benzene, arsenic, and formaldehyde that cause cancer. Studies have linked high particle levels in the air to increased respiratory problems, hospital admissions, emergency room visits, asthma attacks, and premature deaths.

Due to this, the ARB approved a diesel Risk Reduction Plan [7] in 2000, outlining the steps to reduce PM emissions from both diesel-fueled engines. The plan is to reduce cancer risk by diesel PM, by 75 and 85% by 2010 and 2020.

The ARB adopted a regulation In February 2004 to reduce diesel PM from existing Transport Refrigeration Units (TRU) and generator sets used with these refrigeration systems [8]. TRUs are powered by diesel engines rated 5 ~ 27 kW. The most common rating is, however, 26 kW. TRUs are designed to refrigerate temperature-sensitive products that are transported in insulated semi-trailer vans, truck vans, shipping containers, and rail cars. About 40,000 TRUs operate in California and contributed about two tons (t/d) per day of PM, and about 19 t/d of NOx in 2000.

The regulation requires existing TRU engines to meet an in-use standard that varies by model year and power and is phased in over 12 years, beginning 2009, which can be met by using an engine that meets a certification value that is aligned with the U.S. EPA's Tier 4 non-road standards or by a Verified Diesel Emissions Controls Strategy (VDECS) involving a retrofit. Tables 6.2 and 6.3 show the performance standards and the compliance schedule. An option of "alternative technologies" is identified in the regulation, e.g. electric standby engines, cryogenic systems, alternative fuels, and alternative diesel fuels that have been verified by ARB as a VDECS.

Unfortunately, exhaust temperatures are relatively low in TRU applications, affecting the effectiveness of regeneration. A fuel-borne catalyst with a particle

Table 6.2 Low emission in-use performance standards and ultra-low emission in-use perform standards

Horse power	Engine certification	Verified diesel emission control strategy
a Low emission		
<25	0.30 g/bhp-h	Level 2 or better (at least 50% PM reduction)
≥25	0.22 g/bhp-h	Level 2 or better (at least 50% PM reduction)
b Ultra-low emission		
<25	N/A	Level 3 (at least 85% PM reduction)
≥25	0.02 g/bhp-h	Level 3 (at least 85% PM reduction)

Table 6.3 In-use TRU compliance schedule

Engine model year	Compliance date for low emission standard	Compliance date for ultra-low emission
2001 or older	December 31, 2008	December 31, 2015[a]
2002	December 31, 2009	December 31, 2016[a]
2003	N/A	December 31, 2010
Future years	N/A	December 31st of the model year + 7 years

[a]Early compliance (from 2005 to 2007) with the low emission in-use standard for the model year 2002 or older may extend the compliance date for ultra-low emission standard by up to three years

filter is a promising approach to treat the PM burden in the filter and enhancing regeneration on the filter surfaces.

In this part of the chapter, a retrofit technology for small TRU diesel engines using a new highly efficient SiC wall-flow filter, a fuel-borne catalyst, and an intake air throttle valve to promote filter regeneration is discussed. This system meets the stringent PM standard of 0.02 g/bhp-h or an 85% reduction in PM emissions (Table 6.2b).

6.2.2 TRU Design and Operation Conditions

TRU refrigeration systems are powered by diesel engines, to refrigerate perishable products that are transported by semi-trailer vans, truck vans, shipping containers, and rail cars, Figs. 6.16 and 6.17.

The small diesel engines in the TRU application run unattended for about 2400 h per year on average, without any operator assistance or interface. The TRU equipment maintains temperature control within the truck or trailer refrigerated body, and diagnoses and protects itself if there is some malfunction in refrigeration or the diesel engine. It only needs maintenance once per 2000–3000 h of operation, including oil and filter changes. It runs at a relatively heavy load of up to 34 hp when it is cooling a warm trailer, though 80% of the time the load is only at 12 hp, to maintain the set temperature in the refrigerated trailer body. The engine runs for about 25,000 h before requiring a rebuild.

This unique application places severe constraints on the design of a DPF. The exhaust temperature can be up to 350 °C (Fig. 6.18) but is mostly 180 °C (Fig. 6.19). An active control system detects exhaust backpressure and then initiates the regeneration of the trap when backpressure is high.

The FBC tank and dosing system hold the fuel additive for up to 3000 h. Also, the DPF can accommodate ash buildup for 3000 h before cleaning. The system diagnoses itself and makes corrections in operation so that the regeneration is not noticed by

Fig. 6.16 A typical TRU installation on the trailer front side

Fig. 6.17 An engine sits at the bottom of the TRU with a compressor on the left, intake throttle in the center of the picture, and the CPU, as it is arranged during the tests

Fig. 6.18 Exhaust gas temperature distribution during operation in summer throughout the US. *Source* ThermoKing

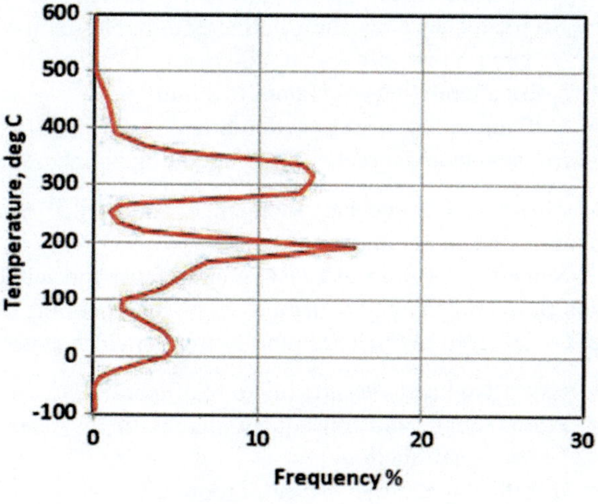

the customer. The life of DPF should be 25,000 h, the same as the engine. Therefore, the DPF element must be carefully designed for maximum durability.

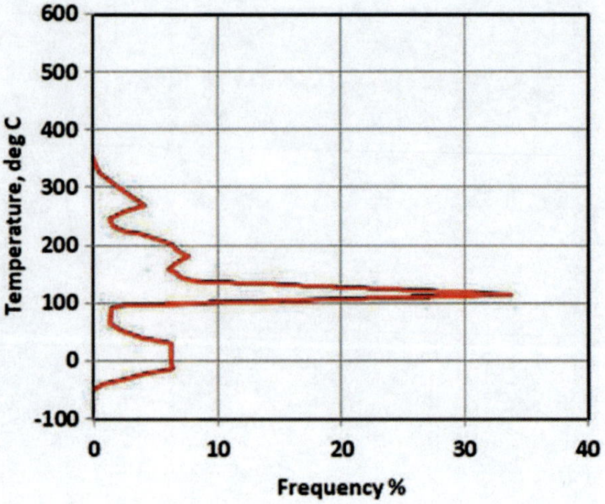

Fig. 6.19 Exhaust gas temperature distribution during operation in Minnesota winter. *Source* ThermoKing

6.2.3 *Emission Reduction Objectives*

The objectives for this engine are derived from the criteria of the VERT [9, 10]

- Filtration efficiency > 95%, for solid particles in the size range 20–300 nm
- Not exceed the legislated limits for CO, HC, and NOx
- No secondary emissions, no increase in NO_2 emissions
- Maximum pressure loss = 200 mbar
- Noise attenuation equivalent to a muffler
- DPF durability = Engine durability
- Maintenance interval = 3000 operating hours.

This research and development of the major parameters of the new emission control system are discussed here.

Since the DPF on the TRU engine must run without any external intervention and automatically adjust itself to changing operating conditions, any fault should be communicated to the TRU monitoring unit. The alarm is necessary when:

- Excessive backpressure due to malfunction
- Filter damage and consequent increase in emissions
- General malfunction
- Insufficient additive, i.e. refill required.

The targeted cost of the completely automatic retrofitted trap should not exceed US $2000 even for small production numbers of retrofits.

Table 6.4 Typical emissions of the engine in the test cycle ISO 8178-4

	NOx	HC	CO	PM
g/kWh	6.52	1.48	2.51	0.45

Source ThermoKing

6.2.3.1 Tested TRU and Engine

ThermoKing supplied the refrigeration unit for the trailer deployment with emissions as in Table 6.4.

Type	SB TC1-30 TRU
Engine	Manufacturer Yanmar
Type	4TNE86-TK, DI-diesel without TC
Engine capacity	2.1 L
Rated rpm	2200
Rated power	25.3 kW
Operating rpm	2200 and 1450
Average fuel consumption	2.5–3 L/h
Lubricant	API Cl-4 10W30
Fuel	Any US-available diesel fuel (S < 500 ppm)
The unit	Completely autonomous and can operate on a parked trailer

6.2.3.2 Preliminary Concept Evaluation

The challenges in developing a dependable particle filter system for TRU:

- The average exhaust temperatures are low. Under unusual conditions, long periods of part-load operation or idling occur at low ambient.
- The DPF on the roof of the trailer is connected to the engine, by two meters long and thin pipe exposed to cool airflow. Inevitably, much exhaust heat is lost here.
- Nobody can intervene between maintenance periods.

The following concepts are rejected in the preliminary evaluation after careful examination:

Concept	Reason for rejection
Electrical regeneration	Expensive and requires a large battery capacity
NO_2 regenerating systems like CRT	NOx/PM ratio is too low and the S-content of the market fuel high
Catalytic post-combustion [11]	Insufficient experience in similar applications
FBC with electrical ignition [12, 13]	Large battery capacity

6.2.3.3 Selected Concept and Its Component, Fig. 6.20

The DPF is located on the trailer roof at a distance of about two meters from the engine and the thin exhaust pipe is cooled by ventilation air. Therefore, heat shielding against the fan flow and effective insulation to prevent heat loss to the ambient are needed. More than this, since most of the energy is lost during the heating periods in just reheating the exhaust pipe and raising the temperature of the filter intake sufficiently. A thin-walled inner pipe characterized by lower thermal inertia and conduction losses is used to limit the temperature dip during transients, conserving the heat within the exhaust gases.

The FBC, from a five-liter canister, is either dosed into the fuel backflow line or the fuel inflow line. The dosage is regulated by the main ECU. Normally, the dosage is continuous and independent of engine operation to ensure sufficient time-averaged concentration of catalyst on the soot deposit [14]. The dosage can also be controlled in response to changing operating conditions automatically.

A butterfly valve is upstream of the engine, usually in the neutral position. To initiate regeneration, the flap is turned to any selected angle within 100 ms.

The ECU monitors the engine speed, the gauge pressure at the DPF inlet, temperatures after the engine, and in front of DPF as well as oxygen concentration. To start regeneration after reaching a certain time or backpressure, the ECU closes the throttle till the required temperature in the filter element is attained. Next, the flap is opened as a strategy to permit a higher concentration of oxygen in the heated

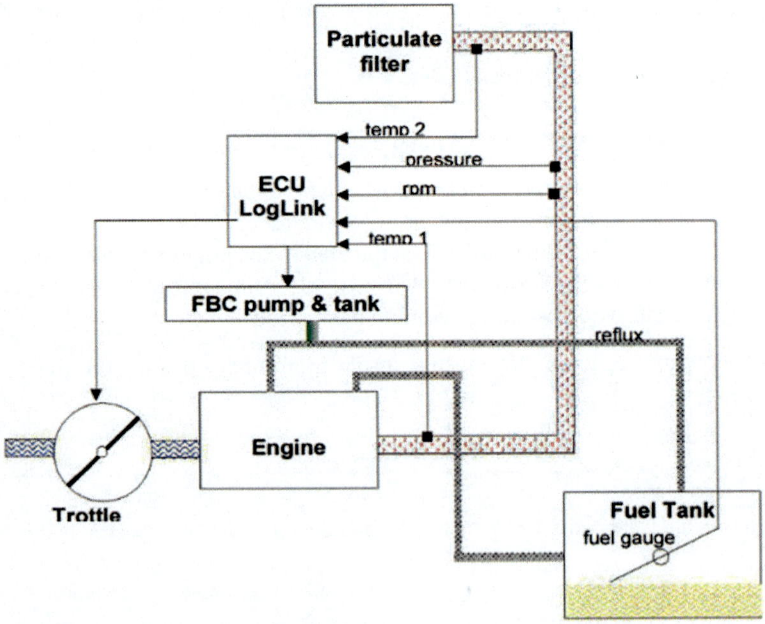

Fig. 6.20 Scheme of the TRU retrofit DPF system

Fig. 6.21 Elliptic SiC cell filter from LiqTech (2 segments)

filter to accelerate the regeneration process and completely clear the soot in the filter element. Thus, the process uses the thermal inertia of the ceramic filter element to decouple temperature and oxygen availability; within certain limits, the parameters namely, the temperature, oxygen, and the timing can be selected to either accelerate or decelerate the combustion of the soot deposit. If the back pressure does not reach the required low level after regeneration, it is automatically repeated after modifying the parameters.

If the operating conditions change (ambient temperature, altitude, ash accumulation, and raw emission of the engine) then the ECU can modify all parameters, including the FBC dosage rate.

1. **Filter**

A compact elliptic filter, Fig. 6.21, was fitted as shown in Fig. 6.22. The specifications are given in Table 6.5. It was cost-effective as well.

This filter type is used in several DPF systems, which comply with the VERT criteria and are approved for Swiss deployment [13]. The traps attain (at all ISO 8178-operating points) a filtration efficiency of 99.8%, for solid particle numbers in the range 20–300 nm, reduction of EC-mass by 97%, and reduction of PM by 87%.

2. **Regeneration Additive**

To lower the ignition temperature of the soot and promote the combustion reaction, an effective FBC based on the metal-organic substance Satacene® is selected, where iron-oxide is the catalytic active component. FBC is particularly attractive for the given application:

- No aging
- Adaptable to the mass of soot deposit
- Independent of soot cake thickness

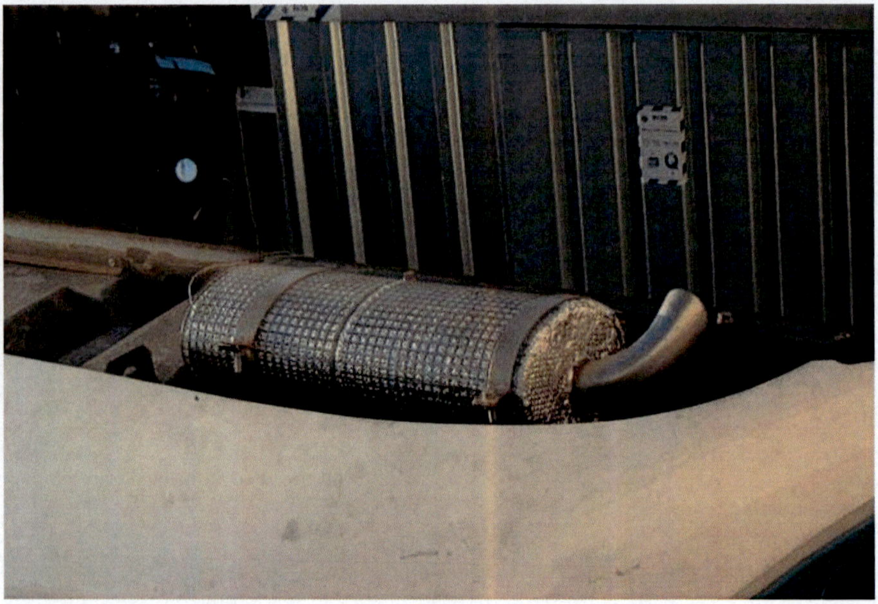

Fig. 6.22 Insulated DPF installed within the truck silhouette

Table 6.5 Properties of the filter element

Manufacturer	LiqTech	
Material	SiC-bound SiC	
Cell density	90	cpi
Cell dimension	2 × 2	mm × mm
Wall thickness	0.8	mm
Pore size	23	μm
Porosity	45%	
Thermal conductivity	40	W/mK at 25 °C
Thermal expansion radial	4.2×10^{-6}	per °C at 25 °C
Elasticity modulus, E	50	GPa
Bending strength	>25	MPa
Density	1.8	g/cm^3
Dimensions	115 × 185	mm × mm
Length	254	mm
Volume	4.2	L

- Combustion 5–10 times faster than coatings [15]
- Thorough regeneration of the filter [15]
- Regeneration conditions are required daily once [16]
- Well suited for dynamic control
- Independent of the composition of fuel and lubricant
- No secondary emissions, no increase of NO_2 [13].

The disadvantages are:

- Dosage equipment and its control unit
- Refilling additive at maintenance intervals
- Ashes of additive raising DPF back-pressure
- Soot ignition temperature 350–400 °C.

Iron-based additives [14], in three different formulations, are chosen for the evaluation. The dosage is in the range of 15–25 mg metal per liter of fuel.

3. **Throttling**

Throttling the gas at a low load is proven to raise the exhaust temperature. VOLVO utilized it in Scandinavia, in the 1970s, to shorten the white-smoke phase of truck diesel engines. Today most car diesel engines practice intake throttling for EGR control and also supporting the regeneration of DPF. Figure 6.23 illustrates the principle [17] for a turbocharged engine.

Throttling either ahead of the engine or in the exhaust stream after the engine is possible. The physics: lower air density when throttled before the engine, but higher exhaust gas retention when throttled after the engine. Throttling before the engine is preferable because of lower component temperatures and is chosen here. The throttle valve is with integrated electronic position control (Fig. 6.24) and is used in high volume car production for EGR control. The opening time is 70 ms.

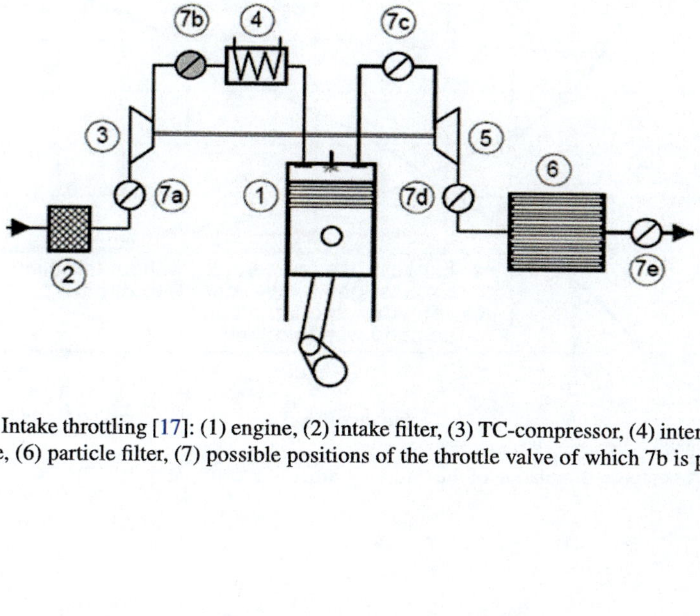

Fig. 6.23 Intake throttling [17]: (1) engine, (2) intake filter, (3) TC-compressor, (4) intercooler, (5) TC-turbine, (6) particle filter, (7) possible positions of the throttle valve of which 7b is preferred

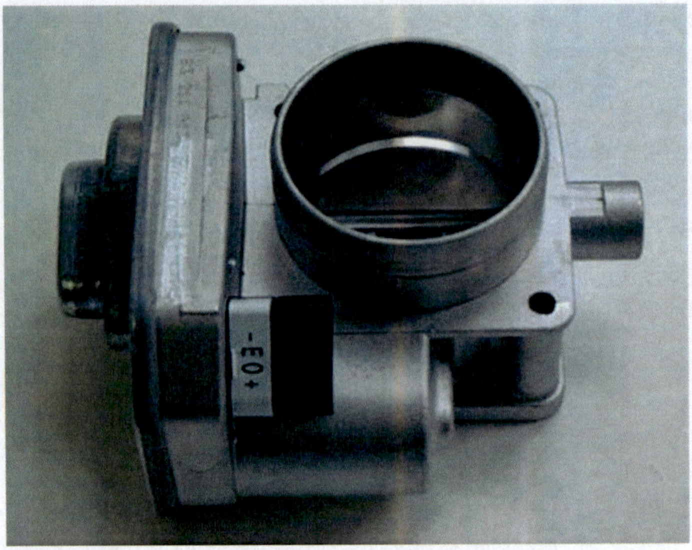

Fig. 6.24 Throttle valve for automobile deployment

The fresh airflow is restricted to be just sufficient for engine combustion at all loads. Figure 6.25 shows the results of the computer simulation of the throttling process [17]. Throttling can nearly achieve the full-load temperature at part-load

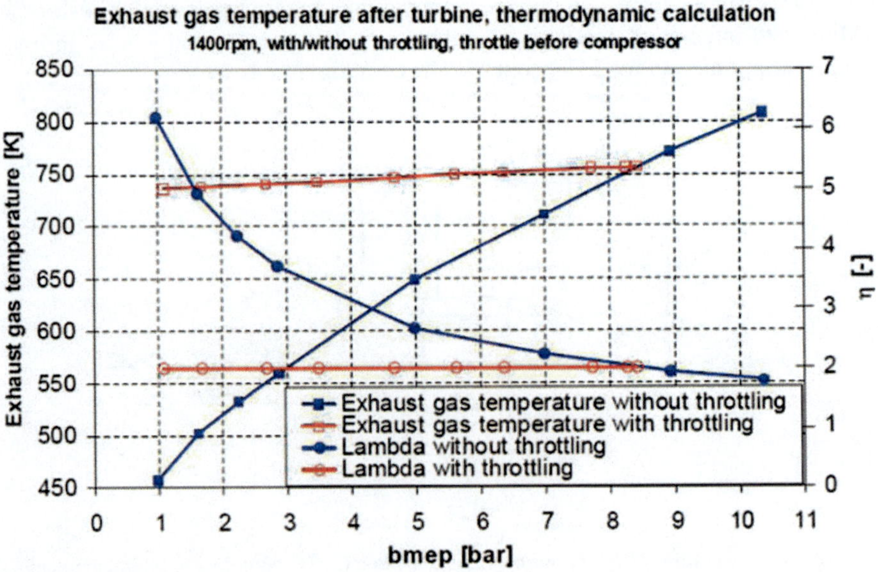

Fig. 6.25 A computer simulation of the throttling process at constant rpm; [17]

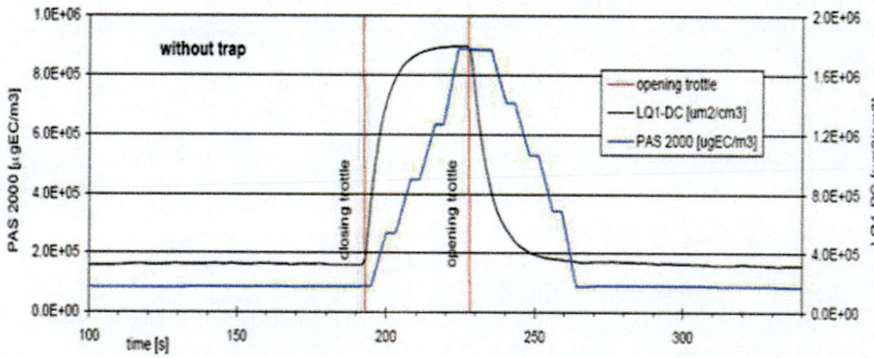

Fig. 6.26 Transient particle emissions during intake throttling, measured by NanoMet [18]

also. Between the two curves with or without throttling, a spontaneous transition is possible. Closing the flap increases the exhaust gas temperature from 450 to 750 K at this operating point without any change in the total enthalpy. The mass flow is correspondingly lower, which also results in a drop in temperature by heat transfer.

Throttle valves for EGR control in passenger car engines are mostly electrically actuated and the flap can be positioned to any desired angle. The selected throttle closes within 70 ms from a fully open position to a fully closed position.

The penalty for throttling is a 10–15% increase in fuel consumption. This is however irrelevant since the flap acts once or twice a day, for about 2 min. The engine particle emissions increase when the air surplus is below the critical limit of 1.3 ~ 1.4, depending on the type of the engine. In Fig. 6.26, the minimum level λ is set to 1.3. The resulting particle emissions are measured using the NanoMet instrumentation, under transient conditions when closing and reopening the valve [19].

The increase in particle emissions is detected with two sensors: the total surface by the DC sensor, and the elementary carbon (EC) in the fine particle range by the photoelectric aerosol sensor PAS. Both sensors show similar results. Intense throttling increases particle emissions fourfold, which reverts spontaneously to the base when the flap is reopened. The delay seen in Fig. 6.26 is the delay of the measuring instrument and not the process.

Thus, the throttling dynamically raises the exhaust gas temperature sufficient for the catalytic soot combustion process assisted by the FBC.

4. **Servo Control**

The closed-loop control begins as soon as high backpressure detected is by the sensor, corresponding to excessive deposits in the DPF. The controller then closes the throttle to a preset position and awaits the rise in exhaust gas temperature ahead of the DPF. If the desired temperature is not attained within the set time, due to low ambient temperature or other reasons, then the throttling is intensified. After the required

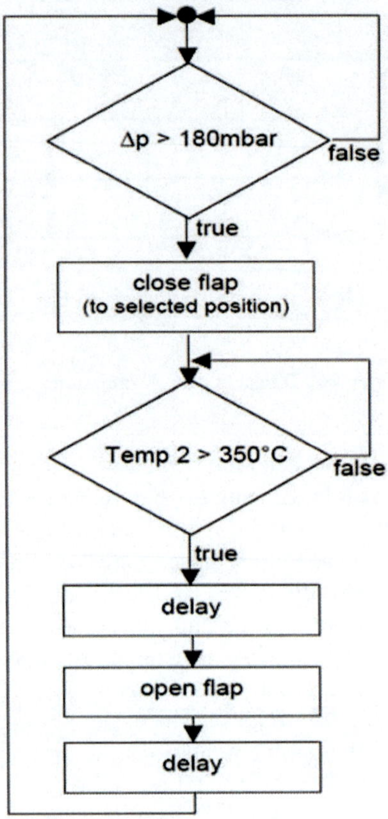

Fig. 6.27 Throttle control

temperature is reached, the controller waits to ensure sufficient DPF heating, before reopening the flap.

The control scheme, Fig. 6.27 illustrates the special feature of the servo control of the valve. Usually, in diesel engines, high temperatures are always associated with low oxygen and low temperatures with high oxygen. The scheme cleverly decouples temperature and oxygen. After the flap opens, the DPF remains hot for several minutes and simultaneously, the oxygen content is high. This favorable combination helps rapid soot combustion. This could even result in extreme thermal conditions damaging the DPF. To prevent such failures, with a temperature signal and a λ-signal from the DPF exit, the flap can be partially closed to regulate oxygen and enable optimum regeneration [3]. Uncontrolled run-away regenerations can be extinguished by a sudden reduction of oxygen. Supplementary responsibilities of the ECU are: data logging, calculating the trends, alarms, and monitoring the throttle, additive pump, and additive tank level.

6.2.4 Test Cycle

The emission testing is carried out at the two usual working points: high rpm (2200)/full load and low rpm (1600)/load. Since the tests are done on the TRU itself the load is established by setting the desired refrigeration. Also, following the VERT protocol [9], opacity and PM emission during free acceleration [20] are explored during regeneration.

6.2.5 Reducing Emissions

1. ***Particle number***

The method and instrumentation [18, 19]:

- Hot sampling directly from the exhaust.
- Conditioning the sample by dilution with rotating diluter followed by heating to separate volatile constituents (ME-Instruments).
- DMA for size classification by their mobility diameter, in 60 logarithmic equidistant sizes (TSI-instrument).
- CNC to determine the particle number, #/cm^3 (TSI-instrument).

The PN from the engine is high (~ 10^8 #/cm^3) relative to many modern engines which emit less than 10^7 #/cm^3. The bimodal distribution with a peak in the range 10–20 nm is because of the additive ash, which appears either attached to the soot particles or as suspended iron oxide particles [21]. Downstream of the DPF, the distribution is parallel shifted down with a filtration efficiency of about 99.8%, Figs. 6.28 and 6.29.

Here, filtration is satisfactory for all sizes up to the finest size detected by the instrument, 10 nm. Hence it may be concluded that filtration is equally good beyond the range of sizes that are not measured.

The same measurement set-up is useful to study the composition of the particles: (1) raw sample from the exhaust, (2) heated to 300 °C to evaporate all volatiles, Fig. 6.30. Since there is no higher concentration below 50 nm the nucleation is not dominating, and it can be concluded that all particles are dry: soot and additive ash.

2. ***Emissions of solid particles by surface***

Two sensors of the NanoMet are used [18]

- DC to find the total particulate surface area in the range of 20–1000 nm.
- PAS to determine the EC.

Figure 6.31 shows low penetration 0.002 is related to 99.8% filtration efficiency demonstrating the high quality of the DPF and corroborates the results of Figs. 6.28 and 6.29.

The consistency of results from PAS and DC is observed despite different physical units: (1) DC finds the total surface of all volatile and solid particles, (2) PAS sees

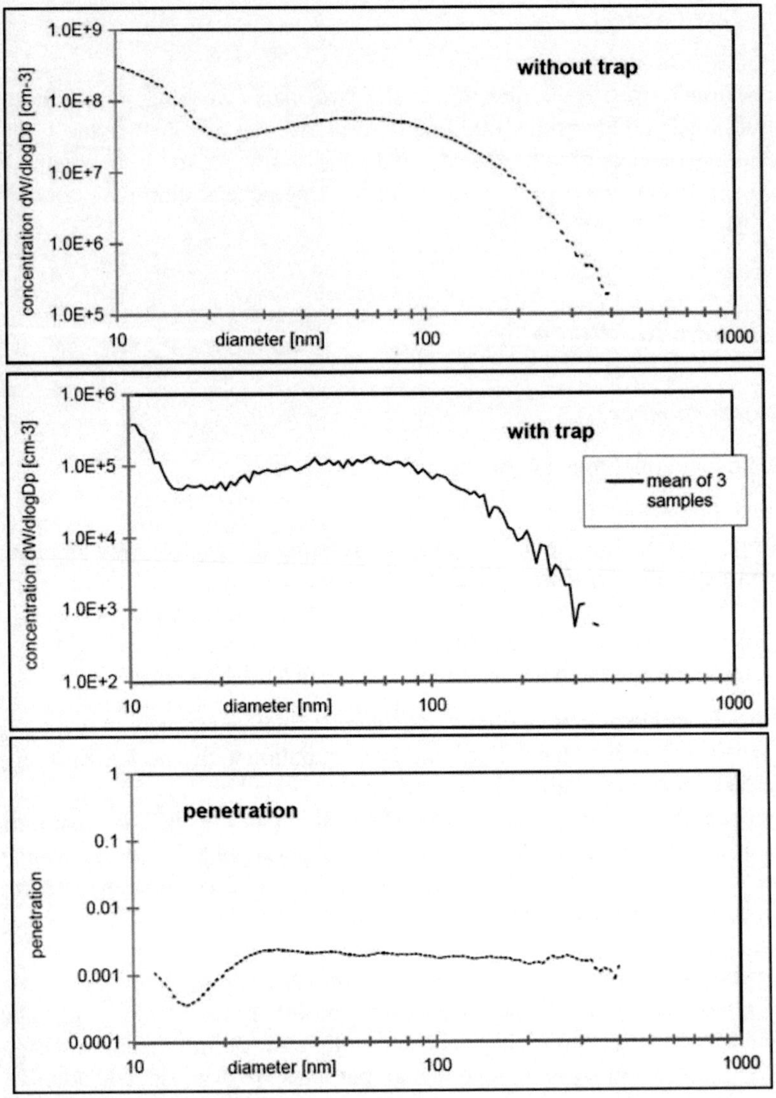

Fig. 6.28 Size distribution at high RPM and maximum load; upper chart without DPF; middle chart with DPF; lower chart: penetration = 1 − filtration rate

the soot particles generated by combustion only. It indicates the lack of presence of condensates. Particle emissions measured by PAS during regeneration as per VERT protocol are shown in Fig. 6.32.

The effect of throttling can be seen by the fourfold increase in particle emission during the short interval. This is acceptable because the filtration never wanes below 99%. When the flap opens, the emissions decrease to immeasurable levels.

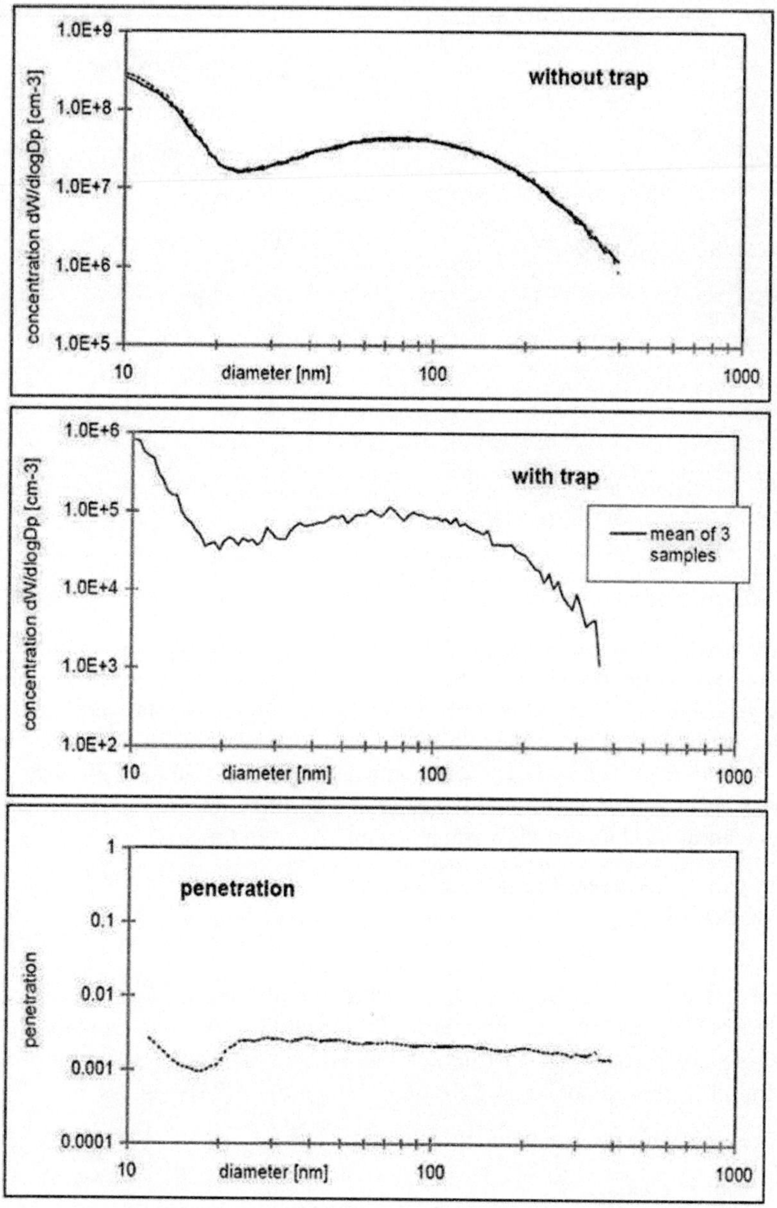

Fig. 6.29 Size distribution at low RPM and low load, without (top) and with trap (middle); the penetration distribution (bottom)

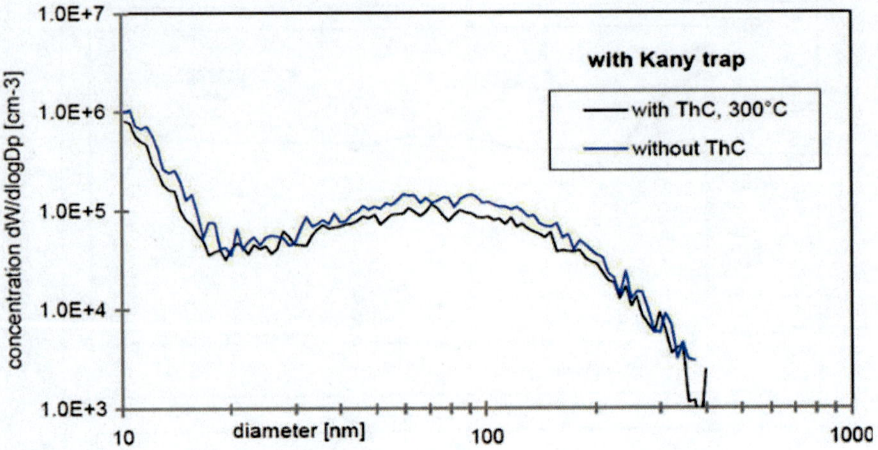

Fig. 6.30 Particle distribution at low load with and without thermo-conditioner showing that the exhaust gas contains little volatile material

3. ***Emissions of particulate mass and gases***

The filter system installed on the TRU was studied by measuring only the filter smoke number (FSN) corresponding to black smoke. No provisions for PM measurement existed in the setup (see Table 6.6). Steady-state and transients were studied by converting FSN to EC according to the MIRA-correlation [22]. With the filter in place, the estimated PM is nearly zero confirming the high filtration efficiency.

However, this filter is thoroughly tested in a certification engine [13] in a fully-fledged laboratory. The filtration efficiency measured in the lab:

- solid particle number: 99.3%
- EC-mass: 96.7%
- PM: 86.8% (CVS-tunnel).

Total PM is less than EC-mass since when the dilution ratios are 6 ~ 8 vapors including water condense. Downstream of an ideal DPF does not contain soot nor any other solid particles but just liquids like HC, sulfuric acid, and a lot of water. Therefore, PM should not be used for describing a filter [23].

4. ***Gaseous emissions and smoke***

The instruments used:

- rbr-ecom-KD for CO, NO. NO_2 and HC
- AVL 407 Smoke-Tester for FSN
- ETAS LAMBDA Meter LA 4.

Iron-based FBC does not affect CO but reduces NO and NO_2 [16]. Also, HC is a third of the raw gas emission, explaining the deNOx-effect, where HC and some soot

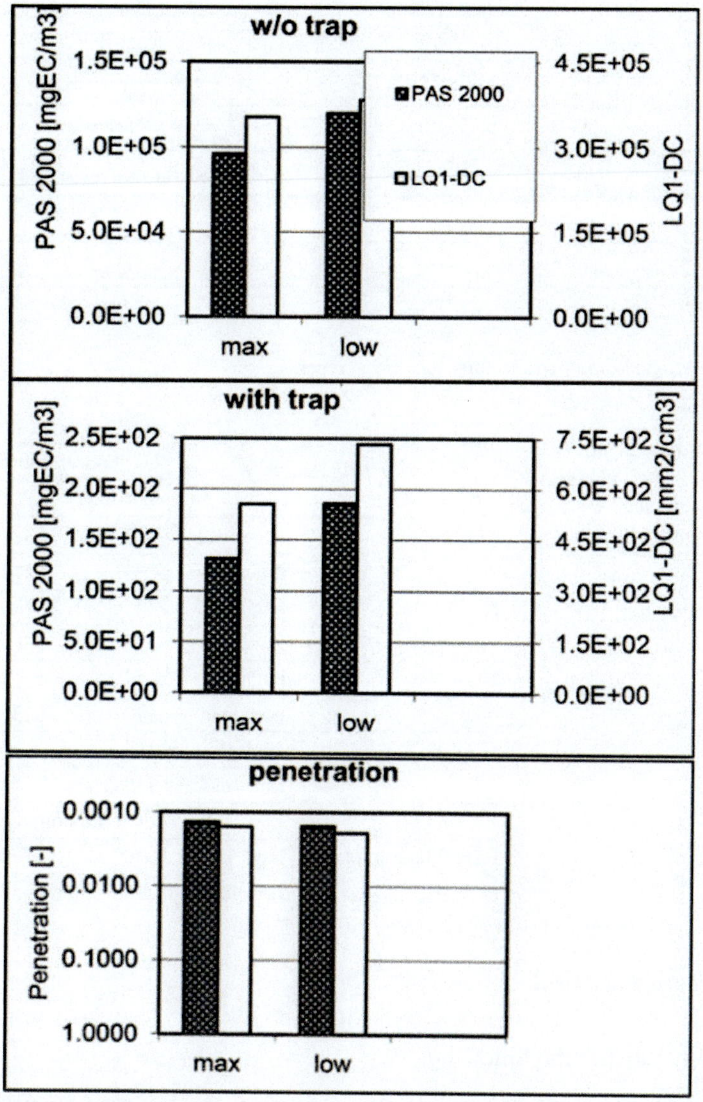

Fig. 6.31 DC and PAS result without DPF (top) and with a (middle) trap; bottom. Penetration = 1 − filtration efficiency

act as reductants. No smoke is detectable after the DPF as measured by the sensitive AVL 407 instrument that correlates well with EC [22].

Besides PN, PAH is accountable for carcinogenicity. Tests [13] proved that DPF, even without a coating or FBC, limits PAH to about the same extent as the particles, because the PAH adsorbed on the particles are trapped in the DPF and converted to harmless products during regeneration [24, 25].

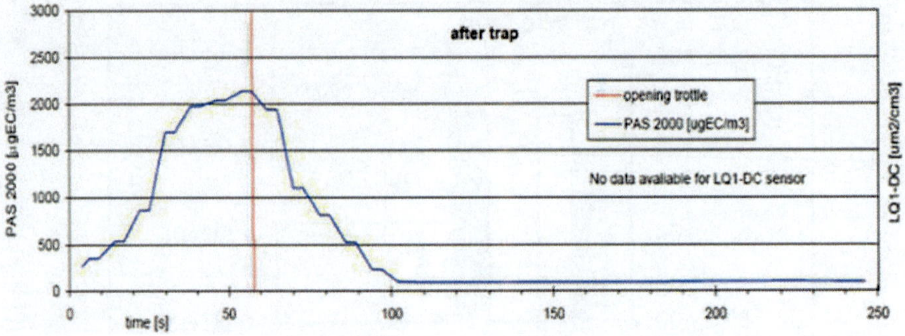

Fig. 6.32 PAS measured particle emission during regeneration

Table 6.6 Emissions of gaseous pollutants and black smoke at two typical operating points with and without DPF

		Gas concentration, ppm	
		W/o DPF	With DPF
CO	High load	320	328
	Low load	295	280
NO	High load	425	400
	Low load	385	260
NO_2	High load	40	7
	Low load	70	30
HC	High load	116	36
	Low load	119	42
		EC-emission, mg/m^3	
FSN (black smoke)	High load	10	0
	Low load	18	0

6.2.6 Regeneration

The strategy satisfies the following:

- Start regeneration at low back pressure and reduce the effect on temperatures in the filter and fuel economy.
- Complete burning of the soot deposit without residues in the filter boundary.
- No emission peaks of gaseous pollutants or metal oxides (from additive) during regeneration.
- Prevent thermo-mechanical stress on the filter monolith, by controlled regeneration.
- Consider other parameters, e.g. ambient temperature, altitude, changes in raw emissions.

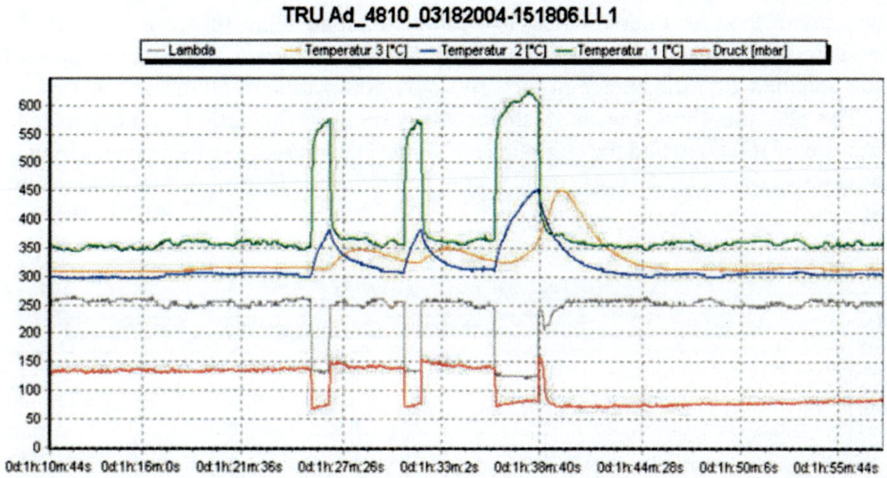

Fig. 6.33 Optimization of exhaust-gas temperature and heating time for the automatic regeneration procedure (Druck—pressure drop across the DPF, Temperatur 1—temperature before the DPF, Lambda—air–fuel ratio relative to stoichiometric ratio)

First, the important process parameters were optimized e.g., threshold pressure, temperatures, and duration of heating, Fig. 6.33. When backpressure increases from 70 to 140 mbar, the flap closes and the heating starts. Throttling lowers the flow and hence backpressure decreases. Air excess is reduced to 1.3. The engine out temperature first jumps up to ~550 °C. There are heat losses in the pipes leading to the DPF and hence the temperature at the DPF inlet rises more slowly. The temperature at the DPF exit is delayed.

The first regeneration experiment is unsuccessful since the duration of heating is very short and the DPF inlet temperature increased to 380 °C only, which is not enough to light off the catalyst. Nevertheless, a weak pressure drop is observed, implying the DPF is not cleaned well. Similarly, when the procedure is repeated after some time, the results are still unsatisfactory.

Next, when heating is doubled for two minutes, the temperature at the inlet to the DPF increases to 450 °C, regeneration begins and completes amazingly fast. The reduced air excess indicates oxygen consumption. The pressure drop across the DPF drops rapidly decreases in less than 30 s indicating a clean filter again. The moderate temperature spike after the DPF indicates mildness of the regeneration process, thanks to the low soot load.

Heat transfer

During the period, between two successive regenerations, the temperatures before and after the DPF are practically identical, indicating the DPF is well insulated. The temperature drop between the engine outlet and the DPF inlet is also only 50 °C despite the long thin pipe. However, during throttling (heating) the temperature drop is markedly larger. As stated before this is more due to the energy needed to heat the mass of the pipe than any heat loss to the ambient. Thus, this drop can only be decreased by lowering the thermal inertia of the pipe. Figure 6.34 charts two days of data obtained after the choice of the FBC and optimization of all the parameters.

The pressure chart, Fig. 6.35 shows the periodic regeneration at the threshold pressure of 160 mbar and the complete cleaning. The hot summer day trend is seen in the discrete rise of exhaust gas temperature and reduced air surplus. Simultaneously, the soot burden increases and results in smaller intervals of regeneration. The first

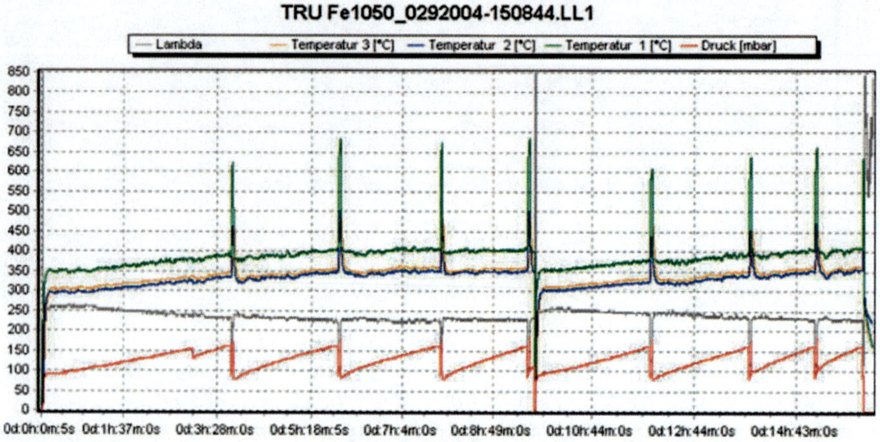

Fig. 6.34 Prolonged test of automatic regenerations (Druck—pressure drop across the DPF, Temperatur 1—temperature before the DPF, Lambda—air–fuel ratio with respect stoichiometric ratio)

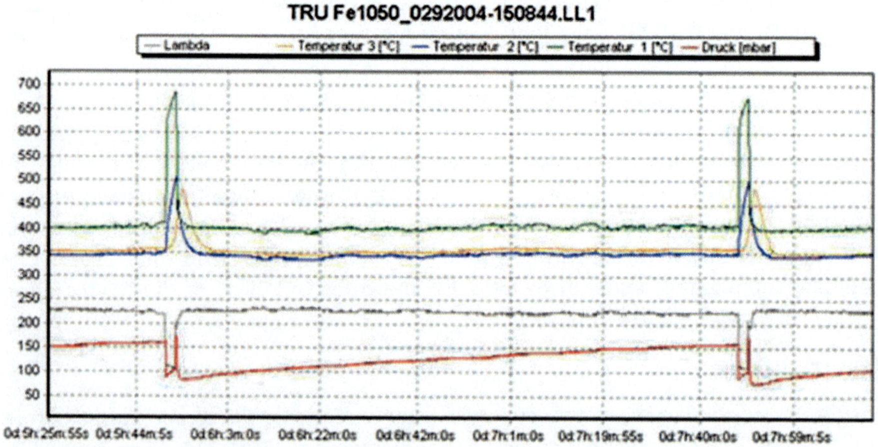

Fig. 6.35 The zoom of the prolonged test shows the regularity of the active regenerations (Druck—pressure drop across the DPF, Temperatur 1—temperature before the DPF, Lambda—air–fuel ratio with respect to stoichiometric ratio)

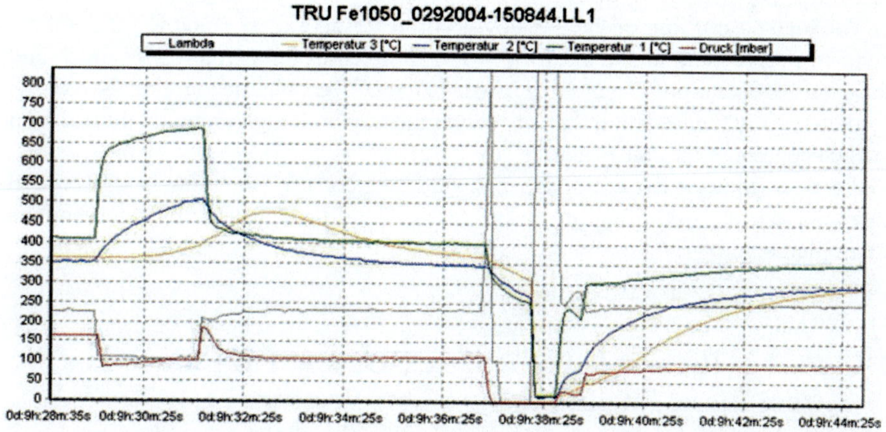

Fig. 6.36 Another zoom from Fig. 6.35; the last regeneration of the first day together with data collected during shut down as well as the engine-restart the next day (Druck—pressure drop across the DPF, Temperatur 1—temperature before the DPF, Lambda—air–fuel ratio with respect stoichiometric ratio)

spontaneous regeneration resulted in partial cleaning only. Later, the regenerations are regular and controlled always by the backpressure.

The CO emission increases temporarily from about 300 to 1600 ppm but quickly drops. The other emissions like HC and NOx are unaffected. The oxygen falls briefly to 7 ~ 8%. During throttling, more smoke is observed is confirmed by the higher particle count (Fig. 6.32). The data from the above test are shown zoomed in Figs. 6.35 and 6.36.

The shut-down at the end of the day is indicated by the gradual cooling and sudden rise of λ to 8.3, Fig. 6.36. At the start of TRU, the specified program runs the engine at a lower rpm.

Shown are three temperatures (°C): at engine exit, DPF inlet, and outlet. Also the pressure difference across the DPF (mbar) (the lowest curve) and the surplus air ratio Lambda (divide by 100).

6.2.7 Conclusions

The segmented SiC DPF with honeycomb structure is reliable and eliminates not only the soot from TRU engines but also all solid particles which affect health [26], like oil-ash and engine wear metals.

FBC encourages quick and total regenerations and is suited to many operating conditions of the high emission engines, and even enables the application of self-adapting actions without any secondary emissions. What is more, a slight decrease in NOx and a substantial reduction of NO_2 and HC are seen.

The intake throttling of the gas flow to limit the air excess suitable for combustion is a simple method and apt for retrofit. It increases temperatures for quick and complete regeneration. The thermal inertia of the DPF enables decoupling the oxygen availability and temperature, thus providing the trap with high temperature and high oxygen content at the same time.

An ECU controls the entire process, monitors correct functioning, logs critical data, and alarms. Additionally, control tasks including self-adapting interventions are implemented.

6.3 Part 3: Retrofit Kit to Reduce NOx and PM Emissions from Diesel Engines Using a Low-Pressure EGR and a DPF System with FBC and Throttling for Active Regeneration Without Production of Secondary Emissions

6.3.1 Introduction

In Swiss cities, the law limits the daily maximum PM to 50 $\mu g/m^3$ to fight the serious problem of air pollution. This limit may not be exceeded more than once a year. However, in the city of Berne alone, PM exceeded the limit 47 times in 2005.

Ground-level ozone is a result of the photochemical reactions between oxides of nitrogen (NOX) and volatile organic compounds (VOCs) in the sunlight. The law requires that the ozone may not exceed 120 $\mu g/m^3$ more than once a year. In May of 2005, 240 $\mu g/m^3$ is measured in the southern part of Switzerland. Of the soot emitted in Switzerland, 17% (3500 tons/year) are contributed by diesel engines [27].

Annually, 19.5 kilotons of nitrogen oxides are emitted in Switzerland by the transport sector (passenger traffic, goods traffic, air traffic, and off-road), which is 58% of the total emissions.

6.3.2 Regulation

Since October 2005 new heavy-duty engines have to meet stringent EURO IV limits of 3.5 g/kWh of NOx and 0.02 g/kWh of PM and the renewal rate of the vehicles is ~15 years. Therefore, to improve the air quality in cities, such old vehicles are retrofitted with a control system to reduce the emission of NOx and PM without any secondary emissions like NO_2 or N_2O.

6.3.2.1 NOx Emission Control Technologies

The NOx control technologies of existing heavy-duty vehicles are selective catalyst reduction (SCR), fuel additives, and EGR. In the case of SCR, an additional fluid (urea) must be carried on board and it requires an exhaust temperature of higher than 225 °C. This can be a problem for garbage trucks or city buses in the cities.

It is possible to emulsify the fuel with 10% water to reduce NOx emissions by about 6% [28] without any modifications to the engine.

The EGR effectively reduces NOx from diesel exhaust independent of the exhaust temperature, which is a big relief for vehicles that run in cities. Here, a portion of the engine exhaust is inducted into the engine intake. EGR introduces the inert products of combustion, such as CO_2 and H_2O, and reduces the combustion temperature. Lower combustion temperature reduces NOx formation [29].

6.3.2.2 EGR

There are two typical configurations of EGR: (1) In a high-pressure system a part of the exhaust gases are cooled in an EGR cooler before being introduced into the intake manifold, (2) A low-pressure system feeds back a portion of the gas after the particulate filter between the air filter and the turbo compressor.

The major advantages of the second system are:

- Since the recycled gases are clean, there is no soot on the intake line.
- The exhaust gases of the low-pressure EGR are cooler than those of the high-pressure EGR due to the longer loop.
- It is easier to retrofit an existing engine without requiring any modifications.

However, the low-pressure EGR suffers from a longer response time in transient operation.

6.3.3 DPF Regeneration

Diesel particulate filter (DPF) significantly reduces PM emissions from diesel exhaust. After a few hours of operation, the DPF is loaded with soot, and increased backpressures lead to a loss of power and an increase in fuel consumption [30]. Therefore, it must be regenerated.

The process of soot combustion with O_2 begins at temperatures of 500–600 °C, which is rarely achieved in a turbocharged diesel engine. Today, with the use of FBCs, the soot can be ignited at 450 °C, which is mostly reached in a diesel engine. However, in the case of applications like community vehicles and garbage trucks that run at low loads and speeds the particulate trap cannot be correctly regenerated without any additional devices. If there is a large concentration of NO_2, the combustion of soot with NO_2 is faster at an exhaust temperature as low as 250 °C e.g., in a Continuous

Regeneration Trap (CRT), where NO_2 is formed by catalytic oxidation of NO. As a result, the amount of NO_2 emitted increases, especially when soot emissions are low, as is the case during city driving [31].

One possibility to increase the exhaust gas temperature at a low load is to reduce the air/fuel ratio by engine throttling.

6.3.4 The Focus of This Investigation

A controlled EGR-DPF system that combines (a) the low-pressure cooled EGR and (b) a DPF system using FBC and throttling for active regeneration. The optimal throttling, the steady-state test results, the control methods for the EGR and the intake throttling, the regeneration strategy, and the emission measurement results are discussed in the following paragraphs.

After the successful development, the system is applied to a heavy-duty engine.

6.3.5 Evaluation of the Optimal Setup

The optimal setup achieves:

- Highest possible NOx reduction
- Highest possible PM reduction
- The lowest increase in BSFC
- Regeneration must work
- Easy to retrofit on a truck.

The original engine is modified and equipped with all actuators, sensors, and DPF necessary for the retrofit kit system and investigated at a test bench in the first phase of the project.

Low-pressure EGR is chosen because it is easier to retrofit than the high-pressure system. The following combinations with low-pressure EGR are evaluated:

1. with throttle 3 for the active regeneration and to increase the EGR rate (Fig. 6.37)
2. with throttle 1 for the active regeneration and to increase the EGR rate (Fig. 6.38)
3. with throttle 2 for the active regeneration and throttle 1 to increase the EGR rate (Fig. 6.39).

The (a) NOx and lambda sensors are placed 15 cm after the turbine, (b) the DPF at 50 cm downstream of the turbocharger and (c) the EGR line is at 20 cm after the DPF, Fig. 6.40. The NOx sensor measures online the nitrogen oxide emissions, while the lambda sensor measures the air/fuel ratio.

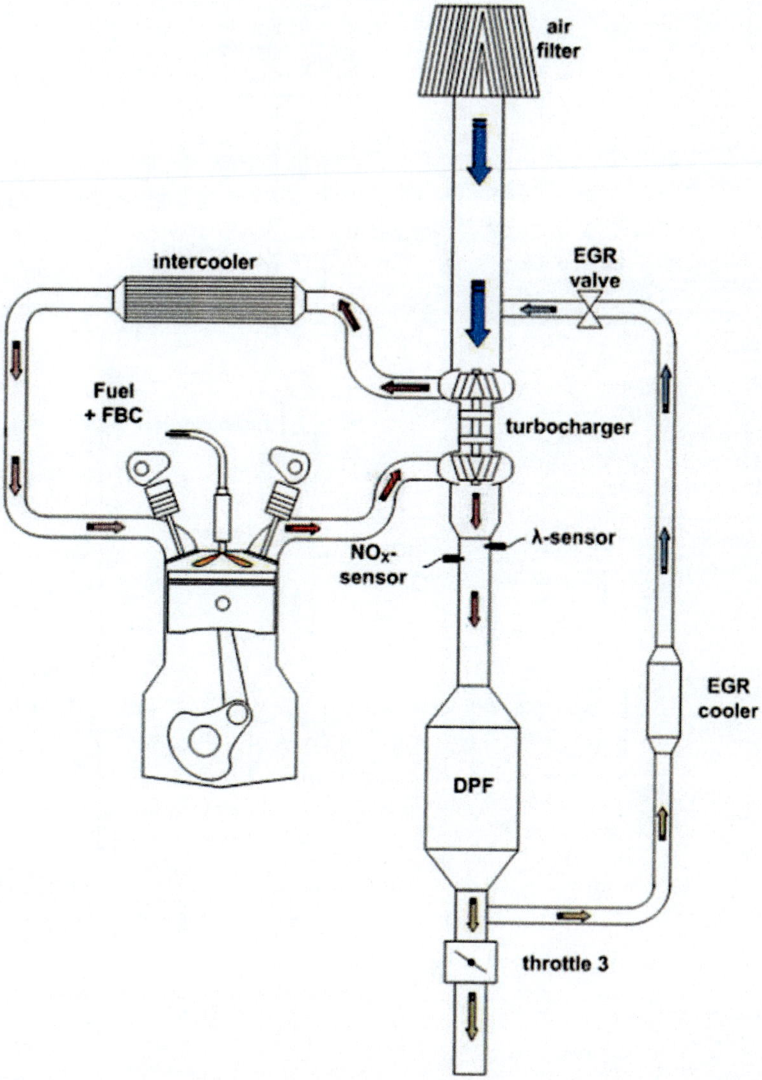

Fig. 6.37 Setup 1: exhaust throttling for the DPF regeneration, and increasing the EGR rate

6.3.5.1 Engine Test Bench

Experimental Facilities

Various pressure and temperature sensors are placed on the engine, the locations of which (Table 6.7) are shown in Fig. 6.40. The heat release rate in the cylinders is calculated from the pressure sensed by a pickup mounted on the cylinder.

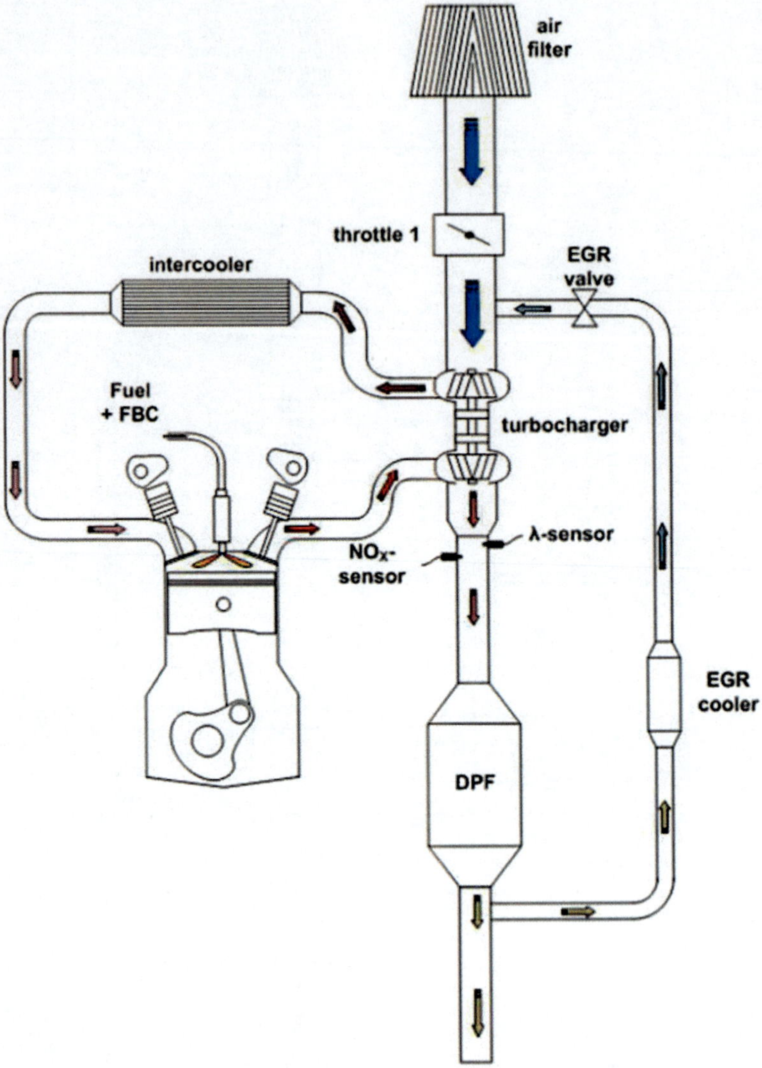

Fig. 6.38 Setup 2: intake throttling before the compressor for the DPF regeneration, and increasing the EGR rate

The exhaust gas measurement devices are installed 30 cm after the DPF. HC is measured by the FID, NO, and NOx by chemiluminescence method, CO, CO_2, and O_2 by NDIR, and the EC with the photo-acoustic soot sensor.

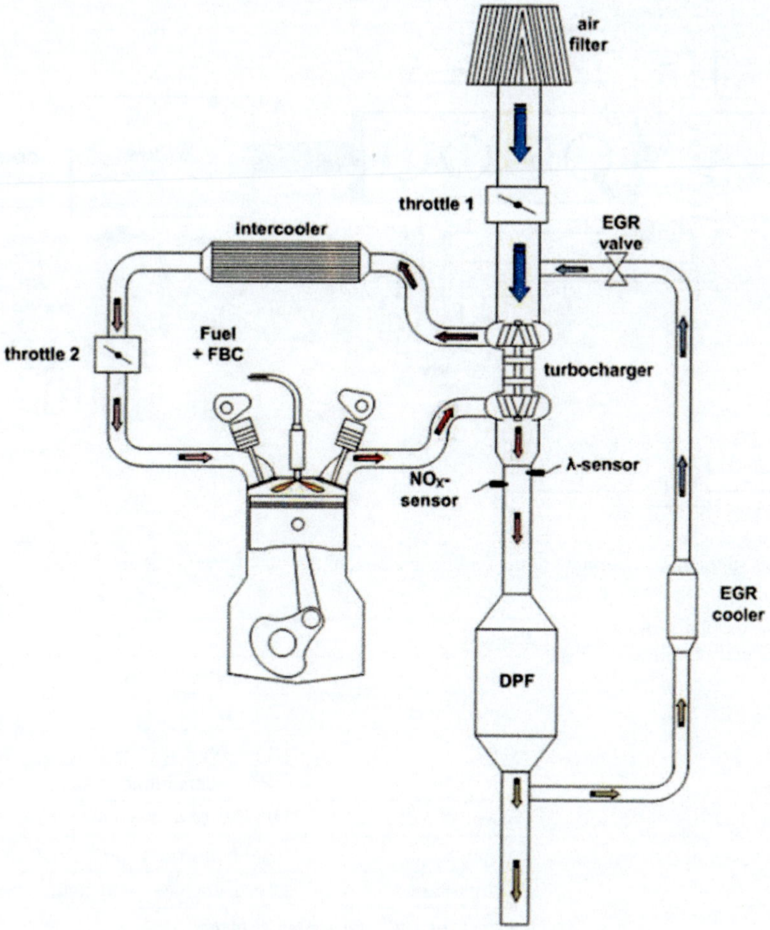

Fig. 6.39 Setup 3: intake throttling before the compressor for increasing the EGR rate, and intake throttling after the compressor for the DPF regeneration

6.3.5.2 Control of the Test Bench

Engine parameters can be tuned with the ES1000-ETK7 units, Fig. 6.41, or by a dSpace online system that handles the entire control algorithm, data acquisition, and monitoring. The communication between the different sensors and actuators on the engine is handled by a CAN-bus system.

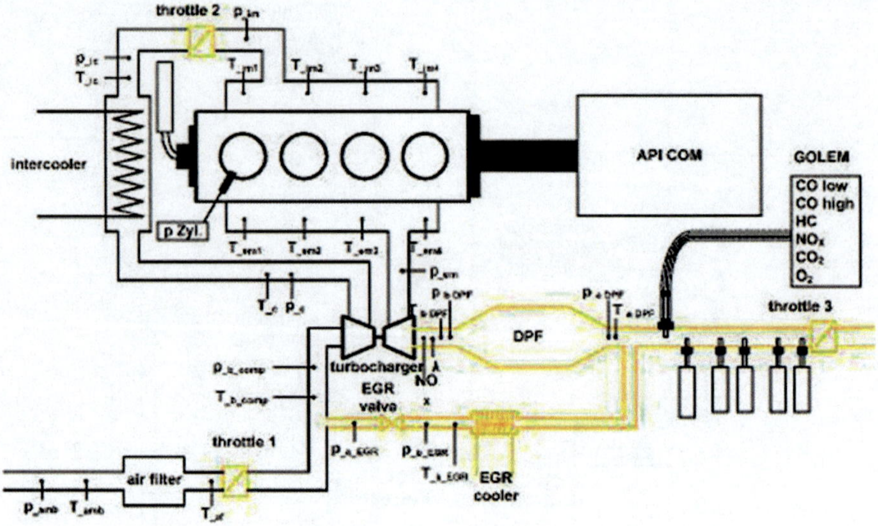

Fig. 6.40 Schematic overview of the engine test bench

Table 6.7 Details of the experimental facilities

Engine	
Model	Mercedes OM 611
Displacement	2.15 L
Type	4-stroke
Cylinder	4 cylinders inline
Power	92 kW @ 4200 rpm
Torque	300 Nm from 1600 to 2600 rpm
Configuration	Turbocharged, with common rail
Devices for stationary measurements	
HC	Flame ionization method (FID)
NO/NOx	Chemiluminescence method
CO, CO_2, O_2	Non-dispersive infrared analysis method (NDIR)
EC (elemental carbon)	Photo-acoustic soot sensor, PASS
Devices for dynamic measurements	
CO, CO_2	Nondispersive infrared analysis method (NDIR)
NO, NOX	Chemiluminescence method
EC (elemental carbon)	Photo-acoustic soot sensor (PASS)

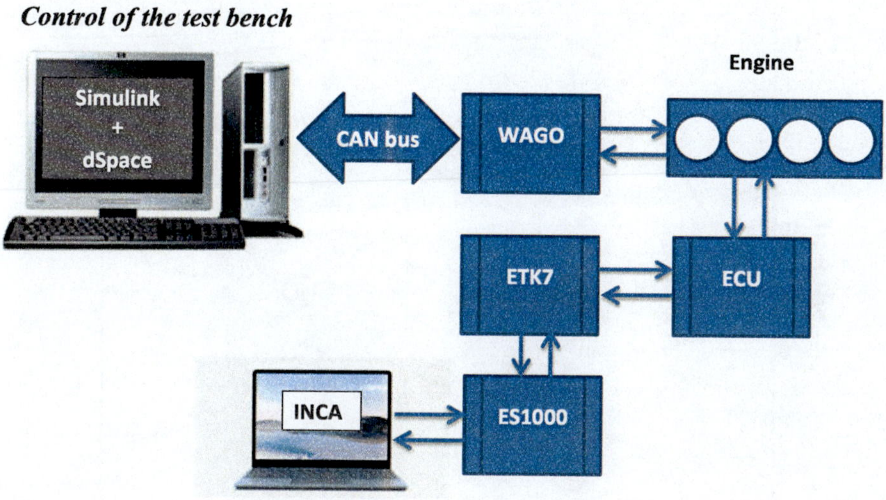

Fig. 6.41 Scheme of the control devices at the test bench

6.3.6 *Result of the Steady State Measurements*

With the use of the fuel-borne catalyst, soot ignition occurs at ~450 °C. This temperature is reached at low load levels, by throttling to reduce the lambda. Intake throttling reduces the air quantity and exhaust throttling increases the residual gases in the cylinder (internal EGR) [30]. Different arrangements are tested at steady-state to evaluate and compare the effects on the exhaust gas temperature and the fuel consumption at five points (Fig. 6.42) to evaluate the engine throttling.

As shown in Figs. 6.43, 6.44, 6.45, 6.46, 6.47 and 6.48, for the five operating points and the three setups, the throttles are closed for 120 s till lambda 1.2 or the limiting factor is reached (see below).

For each throttle, there is a limiting factor:

- Throttle 1: the pressure before the compressor below 0.8 bar absolute can cause oil consumption through the compressor
- Throttle 2: the air/fuel ratio because of the soot emissions
- Throttle 3: the backpressure on the turbine. Above 3 bar, the ECU turns off the variable turbine geometry (VTG) of the turbocharger of this engine.

It is not possible to reach the exhaust temperature necessary for the DPF regeneration with throttle 1. It is possible to reach lambda 1.2 at all points with throttle 2. The temperature (450 °C) for the DPF regeneration is reached after 120 s, except at point 1 (1700 rpm, 2 bar BMEP). At 1, the intake manifold pressure is 0.5 bar as the volumetric efficiency of the cylinder is low, and to maintain the same torque, it is necessary to inject more fuel to raise the exhaust gas temperature from ~250 to ~450 °C.

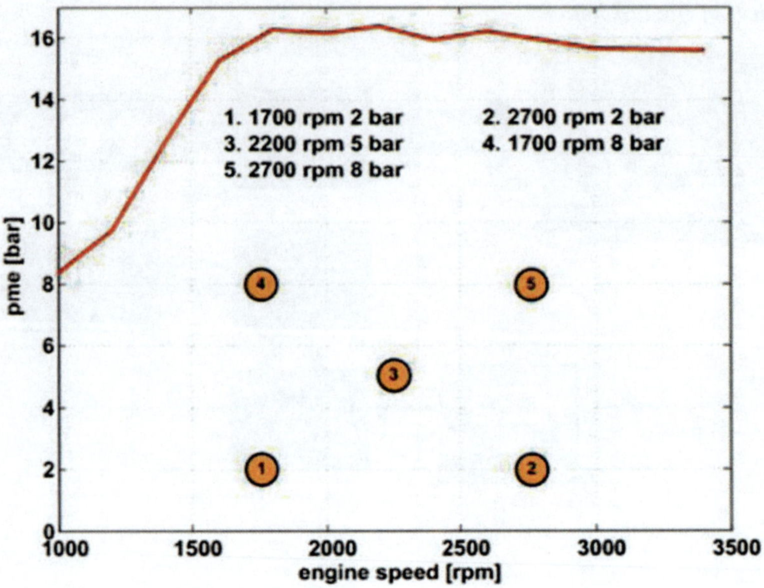

Fig. 6.42 Operating points selected to find the best regeneration setup

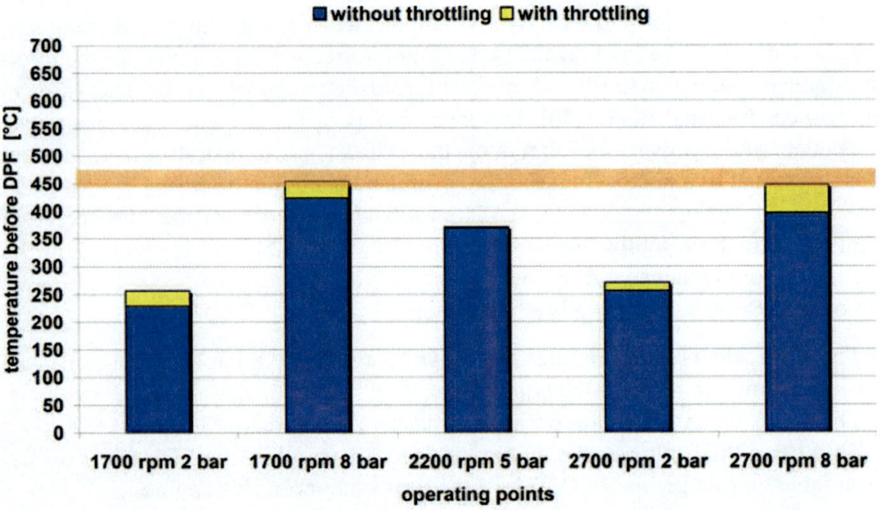

Fig. 6.43 Effects of throttle 1 on the exhaust gas temperature

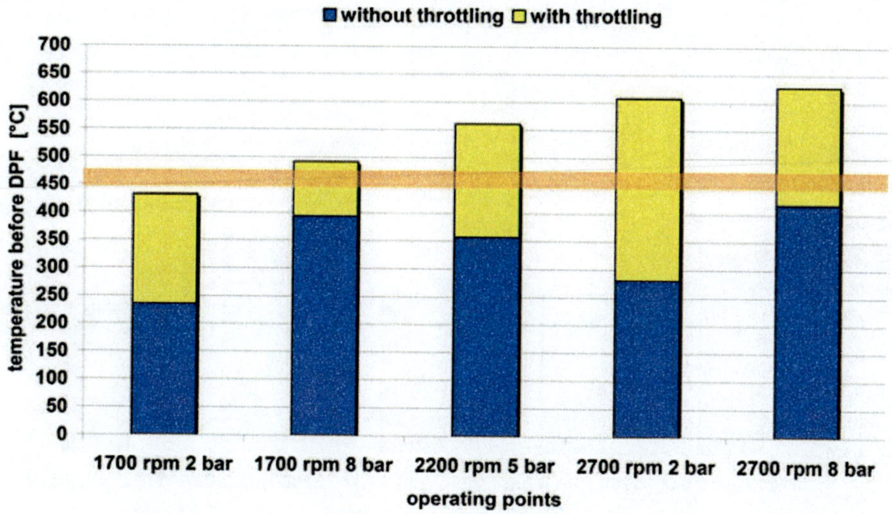

Fig. 6.44 Effects of throttle 2 on the exhaust gas temperature

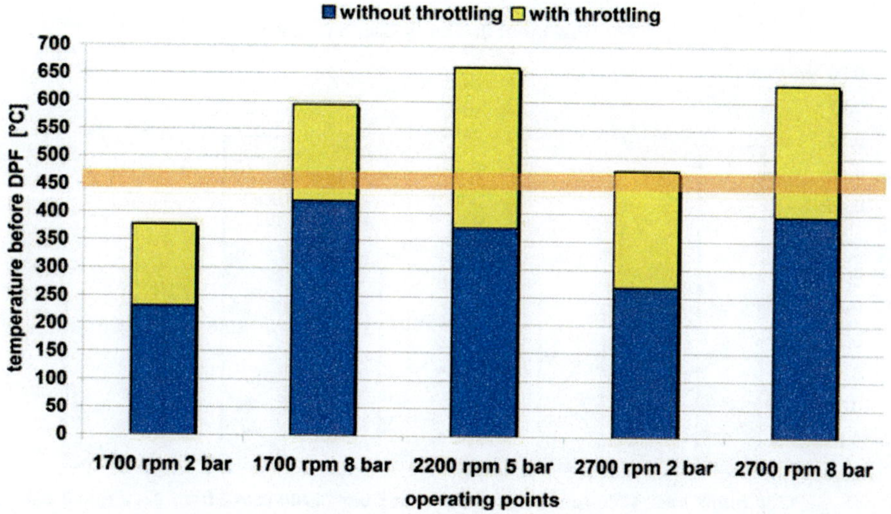

Fig. 6.45 Effects of throttle 3 on the exhaust gas temperature

Throttle 3 is installed after the DPF stays clean but bears a high thermal load. It is not possible to reach lambda 1.2 at low loads, since the throttle has to be closed almost completely. The limiting back pressure on the turbine (3 bar) produces high internal EGR, reduces the internal efficiency of the engine, and thus results in higher fuel consumption.

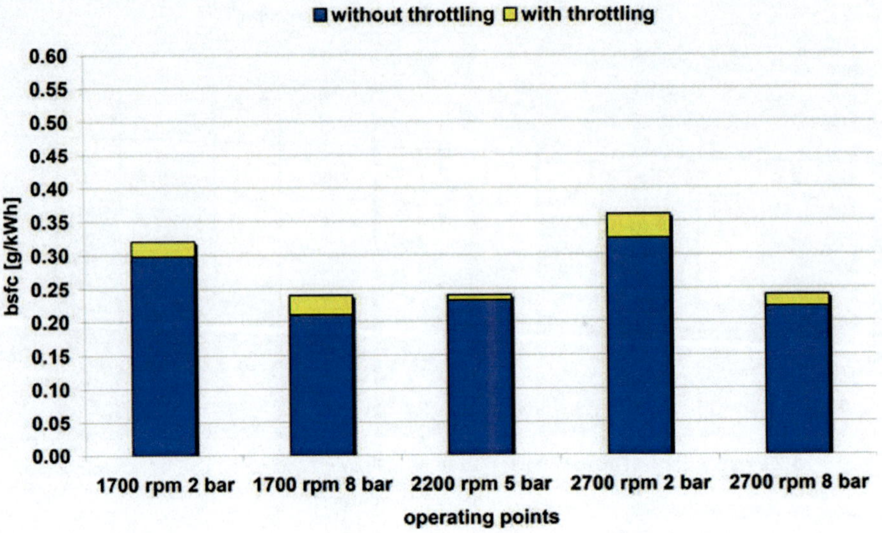

Fig. 6.46 BSFC increase using a throttle

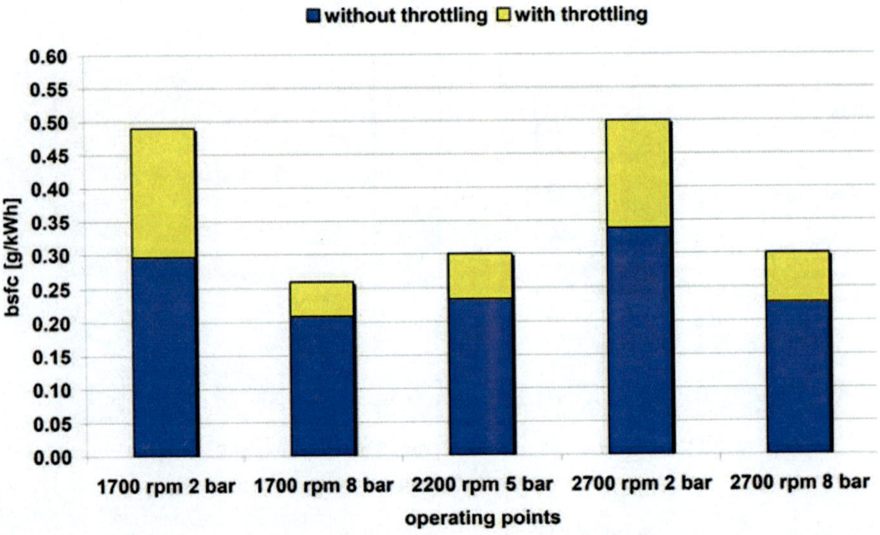

Fig. 6.47 BSFC increase using throttle 2

All these test results lead to the conclusion that throttle 2 after the compressor is the best setup for engine throttling.

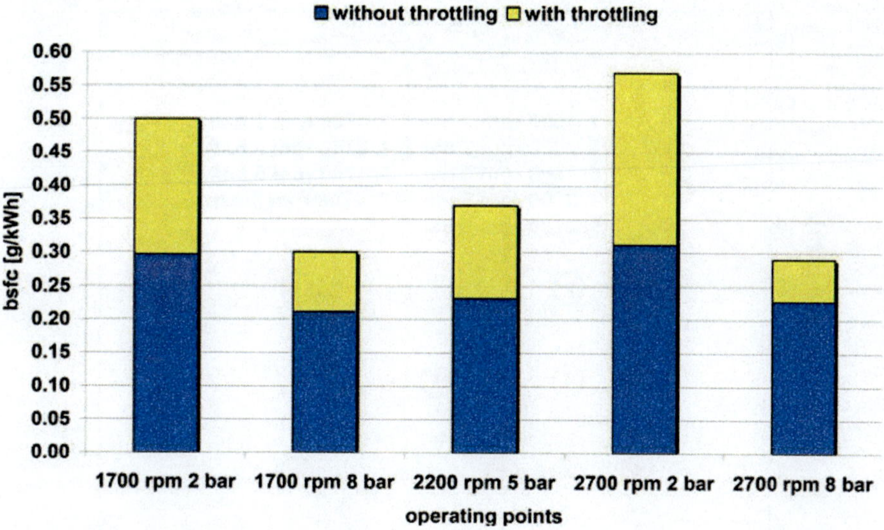

Fig. 6.48 BSFC increase using throttle 3

6.3.7 NOx Reduction Due to EGR

Only throttles 1 and 3 could be used to increase the EGR rate as throttle 2 does not affect the pressure difference between the intake and the exhaust side. The two throttles are closed until the desired lambda value is reached. The efficiency of the low-pressure EGR at the steady-state is evaluated at the points shown in Fig. 6.49.

The maximum NOx reduction (−79%) is measured at 1500 rpm 2 bar BMEP whereas the minimum (−38%) is measured at 2700 rpm 2 bar BMEP, Figs. 6.50 and 6.51. Throttling (1500 rpm 2 bar BMEP, 1700 rpm 2 bar BMEP, and 2000 rpm 2 bar BMEP) is a good solution to increase the EGR at low loads. For the other points, it is not necessary to throttle to reach the desired EGR.

To increase the EGR, the solution with throttle 1 is chosen. Throttle 3 performs like throttle 1, but it has to bear a high thermal load.

6.3.8 Components and Sub-systems of the Retrofit Kit

In this section, the components of the retrofit kit, the extension of steady-state measurements, and the DPF regeneration are described.

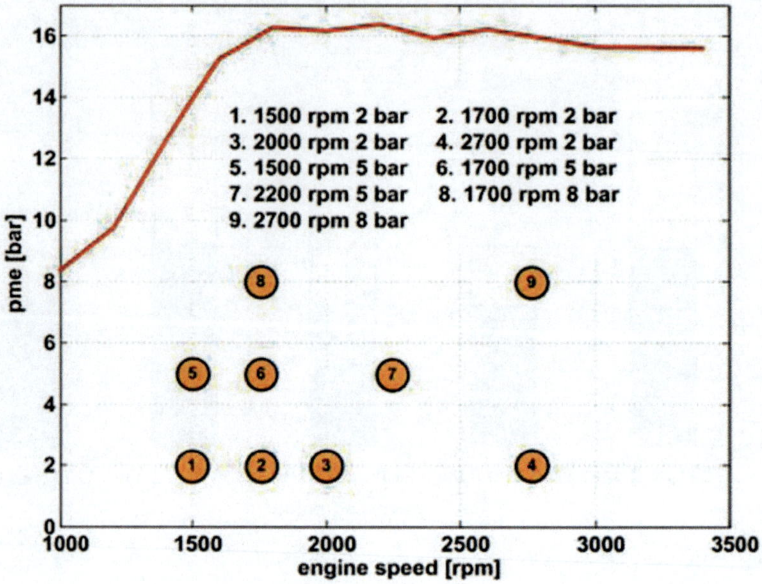

Fig. 6.49 Operating points selected to test the best EGR setup

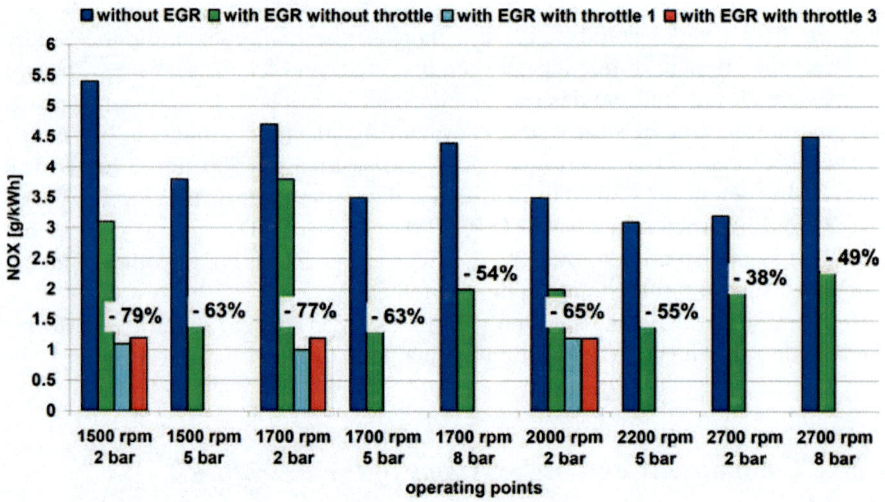

Fig. 6.50 NOx reduction in percent using EGR with throttling

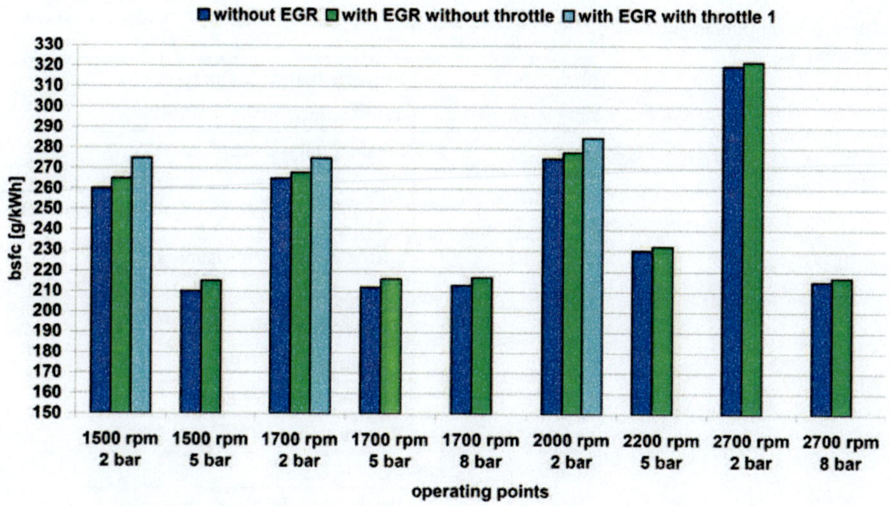

Fig. 6.51 Effect of the EGR on the BSFC

6.3.8.1 Components

The kit consists of four pressure sensors, three temperature sensors, one lambda sensor, and one NOx sensor. The DPF system consists of a SiC particulate trap with FBC and throttle-2 for active regeneration.

The EGR system consists of an EGR valve, an EGR cooler, and throttle-1 to increase the EGR at low loads. The cooler reduces the temperature of the EGR allows the introduction of more inert gases in the cylinder, effectively increasing the EGR.

The location of the retrofit kit components is listed in Table 6.8 is shown in Fig. 6.52 by light lines.

The control algorithm requires the following inputs:

- λ sensor to measure the air/fuel ratio (control variable)
- engine speed
- gas pedal position
- pressure and temperature in the intake manifold (calculation of the air mass flow)
- temperature after the DPF (to stop the regeneration process).

The NOx sensor and the temperature sensor before the DPF are used only as measurement devices and not for influencing the control algorithm. The pressure before the compressor is only used to ensure that it does not fall below 0.8 bar.

Table 6.8 Components of the retrofit kit

Sensors of the kit	
pb comp.	Pressure before compressor
tim	Temperature intake manifold
pim	Pressure intake manifold
rpm	Engine speed
pos gas pedal	Position of the gas pedal
NOx sensor	Nitrogen oxides measurement
λ sensor	Air/fuel ratio measurement
tb DPF	Temperature before DPF
p DPF	Pressure before DPF
ta DPF	Temperature after DPF
pa DPF	Pressure after DPF
Actuators of the kit	
Throttle 1	Increase the EGR rate at low load
EGR valve	Control the exhaust gas recirculation
Throttle 2	Increase the exhaust gas temperature
Hardware	
EGR cooler	Cooled the exhaust gas
DPF	SiC particulate trap
FBC	Fe based FBC (20 ppm)

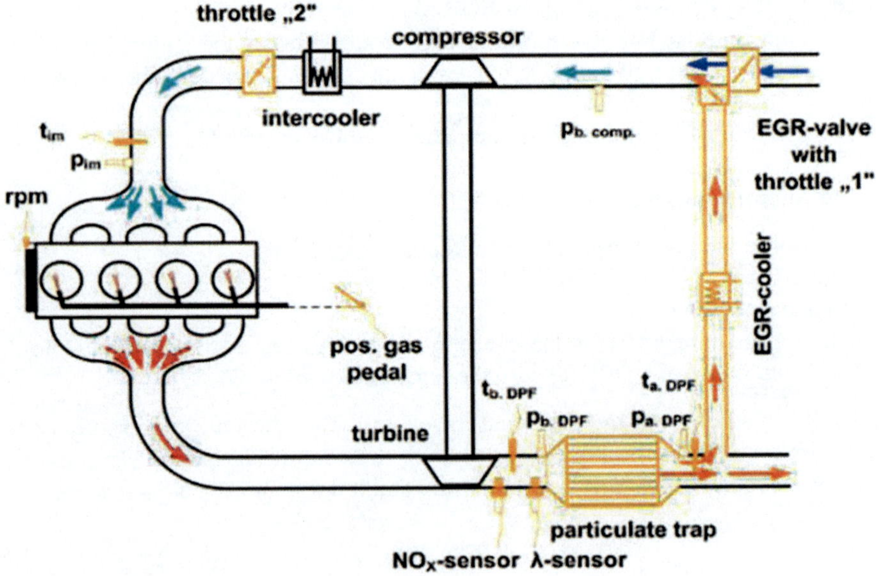

Fig. 6.52 Schematic representation of the retrofit kit

6.3.9 *Steady State Investigation of the Retrofit Kit*

6.3.9.1 Effect of the Intake Throttling with Lambda 1.5

Because of the soot formation, for the next tests, throttle 2 is closed to reach λ 1.5 instead of λ 1.2. Figure 6.53 shows the intake throttling effect on the temperature before the DPF. The temperatures are measured at points between 1500 and 2500 rpm and between 2 and 8 bar BMEP. Without throttling, the maximum measured temperature is 360 °C at 8 bar BMEP. The isolines increase with the load, i.e. with lambda decreasing.

With throttling, the maximum measured temperature is 520 °C at 2500 rpm. The isolines increase with the engine speed; since for a constant air–fuel ratio, a higher quantity of fuel is injected the exhaust temperature increases.

Another reason is that at low engine speeds, the burned gases stay in the cylinder longer resulting in higher heat losses to the cylinder wall, and the exhaust temperature lowers at low speeds.

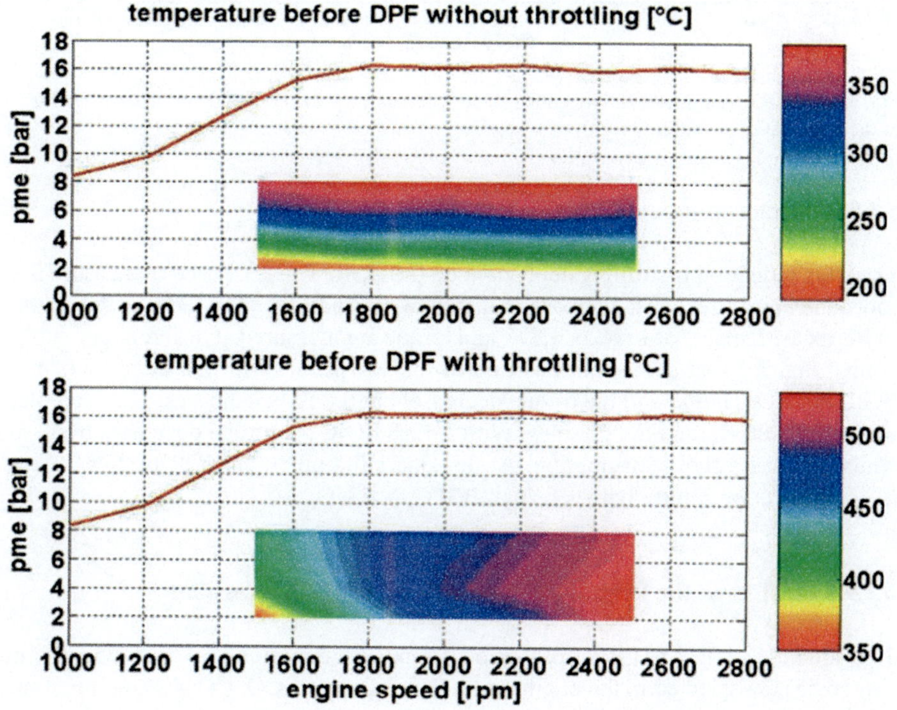

Fig. 6.53 Effect of the throttling on the temperature

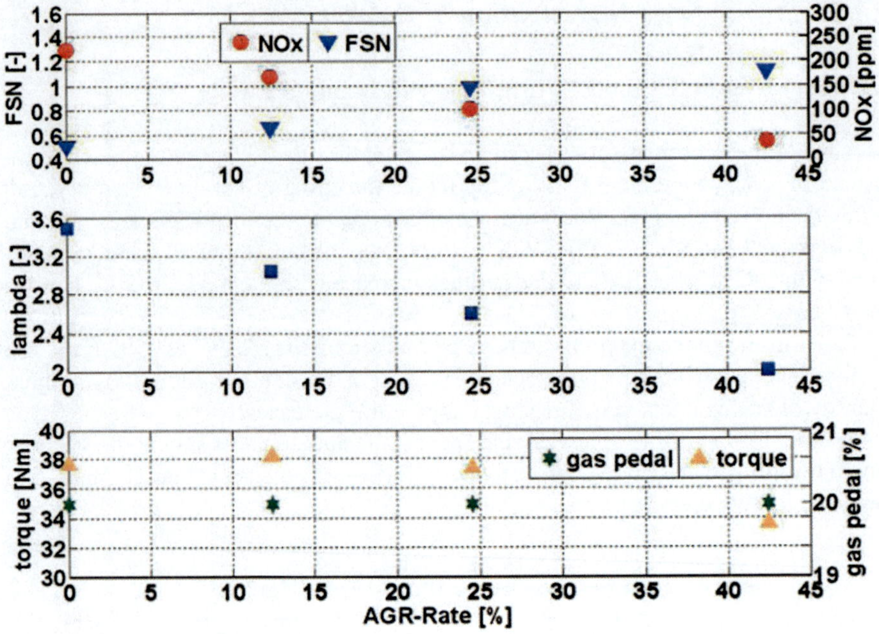

Fig. 6.54 EGR variation at 1700 rpm 4 bar BMEP, to define the better lambda value for the control of the EGR (AGR = EGR)

6.3.9.2 Determination of the EGR Rate

The NOx reduction is strongly dependent on the EGR. The FSN used to measure the soot content in the exhaust, increases with the EGR (trade-off NOx-PM). To define the EGR, measurements of λ, NOx, FSN, and torque are measured at different operating points, Fig. 6.54, at a constant position of the gas pedal. With the increase of the EGR, the NOx, the λ, and the torque decrease, but the FSN increases. For example, with a 25% EGR, the FSN doubles. With a 42% EGR, the torque decreases by 11% which increases fuel consumption. At this operating point, an EGR of 28% offers the optimal compromise between soot, BSFC, and NOx.

6.3.9.3 Effect of the Retrofit Kit on the Exhaust Gas Emissions

To evaluate the effect of the retrofit kit on the exhaust emissions, 30 operating points (Fig. 6.55) are selected in the engine map. The EC, NOx, CO, CO_2, O_2, and HC are measured at these points.

Tables 6.9 and 6.10 show the exhaust emissions on seven selected operating points without and with the kit respectively.

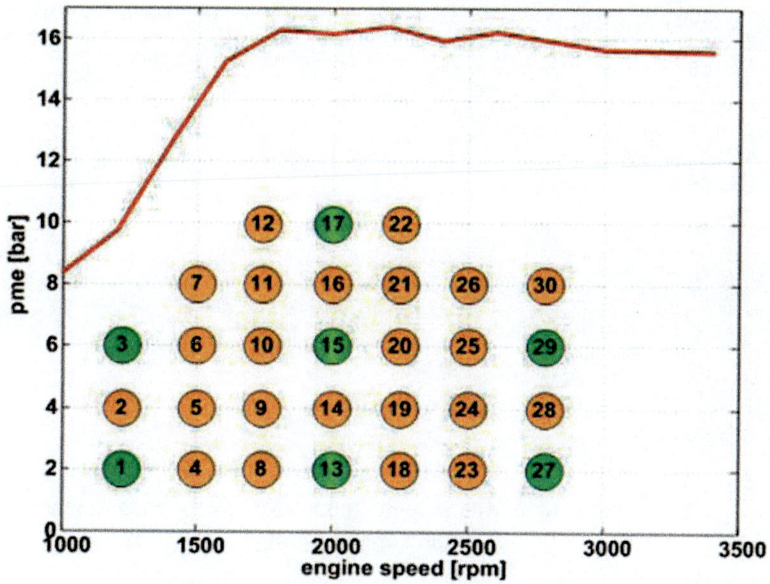

Fig. 6.55 Extended engine map used for the exhaust gas measurement

Table 6.9 Exhaust gas composition without the kit

Nr	CO	HC	NOx	CO_2	EC	BSFC
(–)	(ppm)	(ppm)	(ppm)	(%)	(mg/m^3)	(g/kWh)
1	204	39	**243**	3.09	**0.21**	295
3	192	35	**679**	7.7	**0.39**	224
13	552	49	**165**	3.27	**0.26**	296
15	243	24	**414**	6.8	**0.56**	227
17	103	12	**768**	8.02	**0.43**	211
27	1372	85	**170**	3.36	**0.21**	329
29	343	26	**418**	6.37	**0.62**	229

Table 6.10 Exhaust gas composition with the kit

Nr	CO	HC	NOx	CO_2	EC	BSFC
(–)	(ppm)	(ppm)	(ppm)	(%)	(mg/m^3)	(g/kWh)
1	701	72	**54**	6.21	**0.002**	302+ **2%**
3	274	29	**318**	9.14	**0.004**	228+ **2%**
13	1134	81	**58**	5.63	**0.001**	322+ **9%**
15	290	26	**273**	7.67	**0.002**	230+ **1%**
17	190	8	**612**	8.42	**0.004**	214+ **1%**
27	1617	91	**93**	5.00	**0.001**	343+ **4%**
29	371	24	**365**	6.64	**0.001**	235+ **3%**

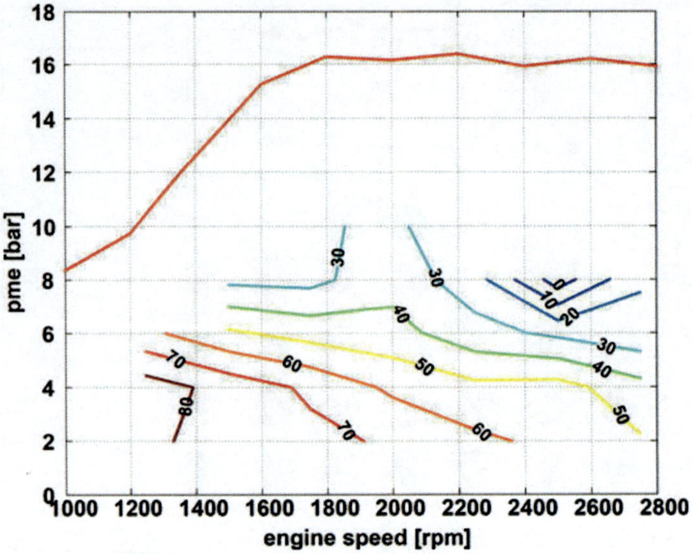

Fig. 6.56 NOx reduction using low-pressure EGR (%)

A comparison of Tables 6.8 and 6.9 shows that when EGR is used CO, CO_2, and HC increase significantly depending on the operating point. The NOX emissions decrease by more than 50%, the filtration efficiency is higher than 99%, and the average BSFC increases by 3%. At some points, the BSFC increases by more than 10%. If the operating point 13 is considered, the EGR rate is about 40%, the NOX reduction is about 70%, and the BSFC increase is about 9%. At these points, it is necessary to decrease the EGR rate to limit the increase in fuel consumption.

Three engine maps obtained by measuring the low-pressure EGR showing the NOX reduction, the EGR rate, and the increase in BSFC as a result of the low-pressure EGR are given in Figs. 6.56, 6.57, 6.58 and 6.59.

As already discussed, the EGR considerably influences fuel consumption. Compared without EGR, the BSFC with EGR on average increases by 3%. However, between 1400 and 2200 rpm and between 2 and 4 bar BMEP the 13%, increase is unacceptably high. For this area of operation, the EGR should be reduced. The filtration efficiency of the DPF is more than 99%, Fig. 6.59, which is the current value for the wall-flow filter.

DPF regeneration occurs at 2300 rpm and 4 bar BMEP, Fig. 6.60. In the beginning, the temperature before and after the DPF is ~340 °C, the air/fuel ratio ~2.8, and the back pressure on DPF ~120 mbar. Throttle 2 is closed after 15 s until lambda 1.5 is reached. Now, the temperature before and after the DPF begins to increase. The temperature after the DPF increases slower than the temperature before DPF because of the thermal inertia of the DPF. After the throttle closes, the backpressure decreases instantly because the airflow through the engine reduces. At the beginning of the regeneration process, the backpressure increases until the "balance point"

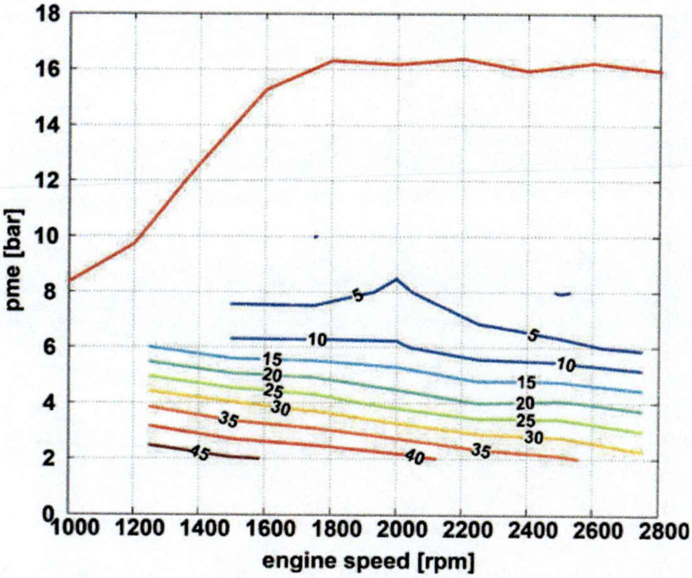

Fig. 6.57 EGR rate (%)

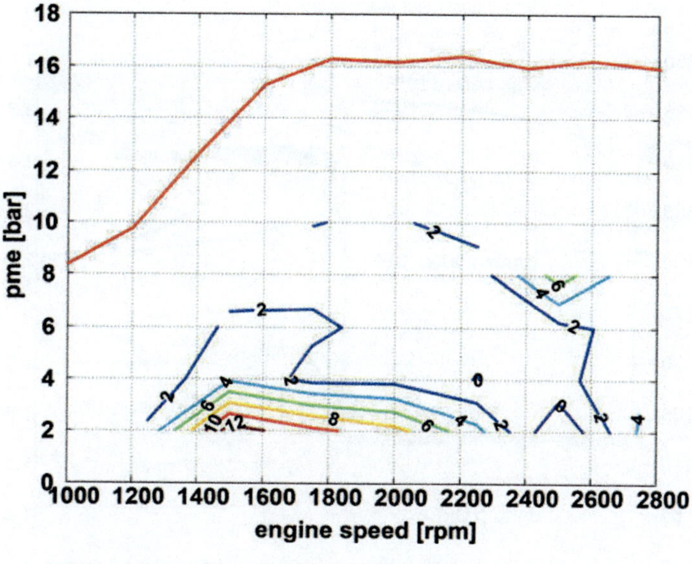

Fig. 6.58 Increase of the BSFC using low-pressure EGR (%)

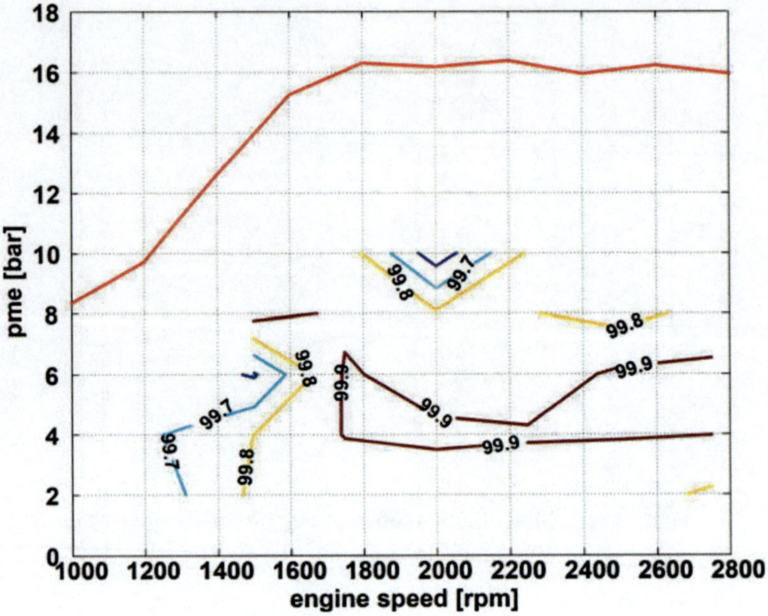

Fig. 6.59 Reduction of the elementary carbon emission, %

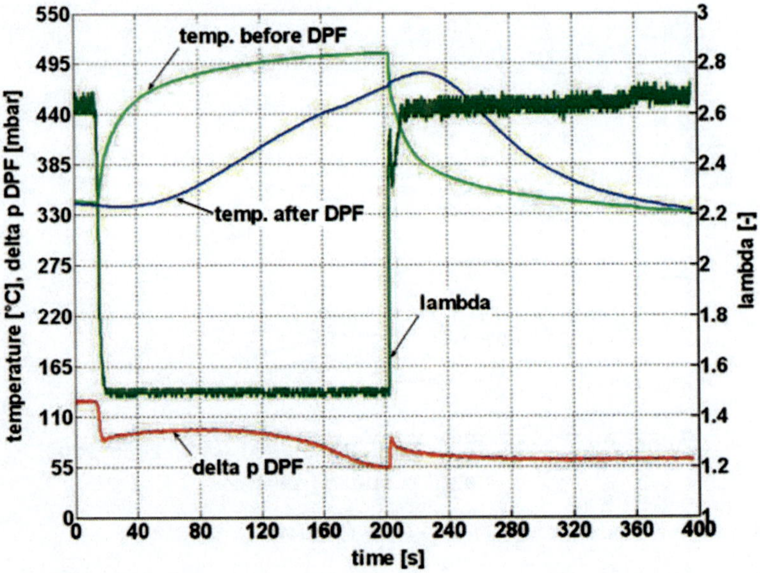

Fig. 6.60 DPF regeneration at steady-state condition

which describes the point at which all the soot from the engine is burned. This point is reached after 80 s (the red curve is flat), after which the backpressure decreases.

At 200 s the throttle is opened. The soot continues to burn for another 100 s during which time the DPF is fully regenerated (backpressure on the DPF = 60 mbar). The duration of the regeneration depends on the initial temperature before and after DPF and the operating point of the engine. For the same load, the duration is shorter with a high engine speed since at high speed more heat energy is admitted into the DPF.

$$H_{ex} = m_{ex} c_p T_{ex} \tag{6.1}$$

The enthalpy depends on the specific heat of the gases, the mass flow, and the temperature of the exhaust gas.

$$m_{ex} = m_{air} + m_{fuel} \tag{6.2}$$

For a diesel engine, the fuel mass flow depends directly on the torque.

$$\dot{m}_{air}(t) = \frac{p_m(t)}{T_m(t) - R} \lambda_l(p_m, \omega_e) \frac{V_d}{N} \cdot \frac{\omega_e(t)}{2\pi} \tag{6.3}$$

Airflow is a function of the pressure and temperature in the intake manifold, the volumetric efficiency, the engine displacement, and the engine speed. The temperature and pressure in the intake manifold are more or less constant, as are the volumetric efficiency and the displacement. Therefore, the airflow and the exhaust enthalpy are nearly proportional to the engine speed [32].

6.3.10 Control Approach

The method to find the control parameters, the EGR control system, the intake throttling control system, and the regeneration strategy are described in the following paragraphs.

6.3.10.1 Description

A control system interconnects components and forms a system that provides the desired response. An open-loop control system is without feedback and utilizes a controller or control actuator to obtain the desired response. In contrast, a closed-loop control utilizes a measure of the actual output called the feedback signal and compares it with the desired output response. Often the difference between the output of the controlled process and the reference input is amplified and used to control the process so that the difference is continually reduced [33]. A closed-loop control system compensates for any disturbance which is not the case for an open-loop control system.

For the EGR system, the loading of the DPF is considered a disturbance. An increase of the backpressure produces an internal EGR. With only a feedforward control, this internal EGR would not be taken into account, but with a feedback system, this internal EGR can be compensated.

6.3.10.2 Controller

A PI controller is used to control every actuator (throttle 2, EGR valve, throttle 1). The control parameters are defined using the Ziegler-Nichols method. It uses a P controller first and increases the gain value until the actuator becomes unstable. If the oscillation period (T_{kr}) and the gain ($K_{\mathrm{p,\,kr}}$) are known, it is possible to define the control parameters [34]:

$$K_P = 0.45 K_{P,k_r} \tag{6.4}$$

$$T_N = 0.85 T_{kr}$$

6.3.10.3 EGR Control

The reaction time of the EGR system is shown in Fig. 6.61. At 1500 rpm 2 bar BMEP, the reaction time for a change of the EGR valve position is 500 ms. The long reaction time is because the long low-pressure EGR delays input to the engine.

Because of this delay, it is necessary to insert a feedforward in the control algorithm and hence reduce the overshoot of λ during a load step.

Figure 6.62 shows the EGR control logic. The values of (1) the desired λ that determine the EGR, the positions of (2) the EGR valve, and (3) throttle-1 depend on the engine speed and load and are saved in maps. The advantage of the feedforward is proved for a step-change in either engine load or engine speed (Fig. 6.63). The actuator is first positioned with the feedforward and then the small residual error is compensated with the feedback. The overshoot of the measured value is less and it is possible to slow down the controller producing fewer problems of instability.

Figure 6.63 shows the three ways of controlling the EGR system. The black curve designates the desired λ which is saved in a lookup table (map). The red curve defines the open-loop control system, the blue a closed-loop control system, and the green a combination of the open-loop and closed-loop systems.

At steady-state, there is a difference between the desired and measured values already at the beginning, when an open-loop control system is used. A disturbance cannot be compensated with an open-loop control system. At the steps, the system

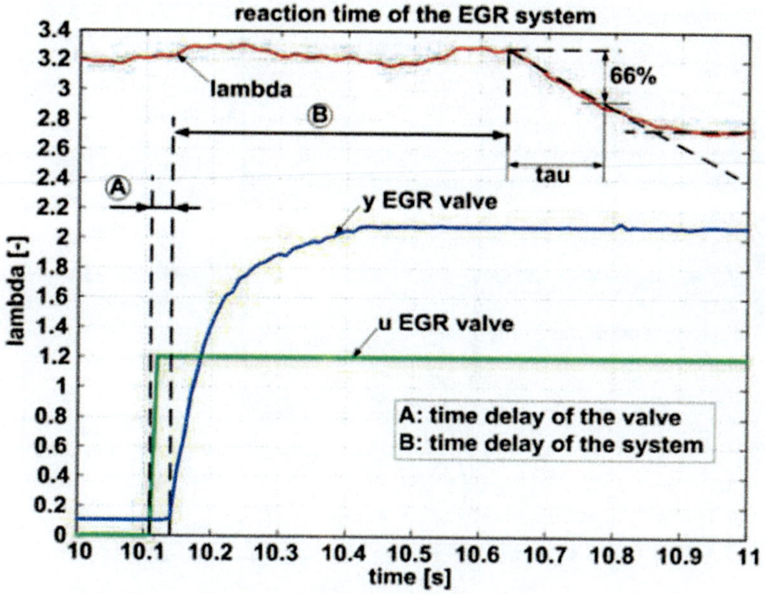

Fig. 6.61 The reaction time of the EGR system at 1500 rpm 2 bar BMEP

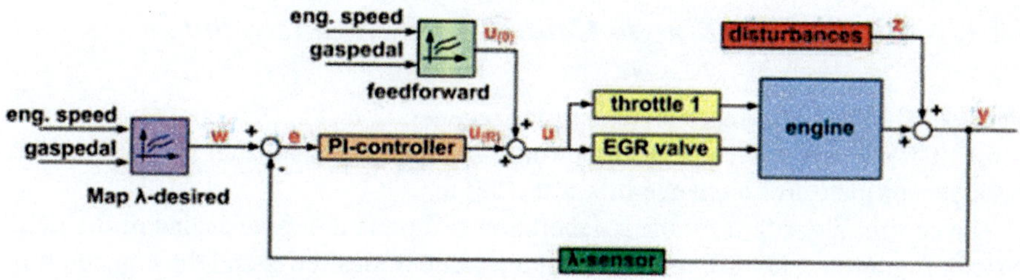

Fig. 6.62 Schematic diagram of the EGR control algorithm

reacts fast, because both the feedforward and the desired λ depend on the engine speed and load.

With a closed-loop control system, the error is zero at steady-state. At the load step, there is a large overshoot, consequent to the large delay by the long pipe of the EGR system. If the control gain is increased to reduce this overshoot it makes the controller faster but makes the system unstable.

The combination of the feedback and feedforward controllers (green curve) shows the best results. At steady-state, the error is zero. During a load step, it causes a smaller overshoot than the feedback controller alone.

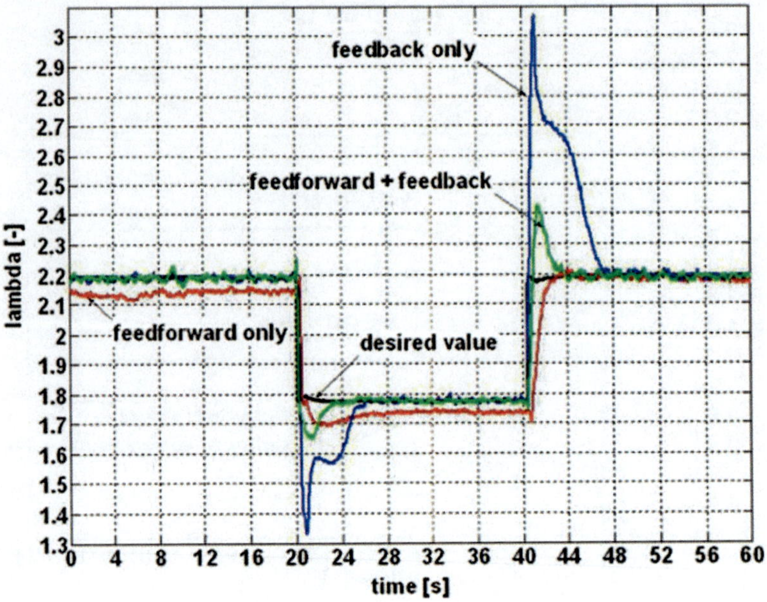

Fig. 6.63 Load steps from 2 to 6 to 2 bar BMEP at 2000 rpm

6.3.11 Intake Throttling to Control DPF Regeneration

Figure 6.64 defines the reaction time of the regeneration. This system is twice faster as the EGR system. After a step with throttle 2, which is placed after the compressor, the λ-sensor measures a change in λ after 200 ms.

The control algorithm for the regeneration system is the same as that of the EGR system, Fig. 6.65. The differences are in the value of desired λ and the regeneration strategy. Although the reaction of the system is twice as fast as that of the EGR system, a controller with a feedforward is necessary, Fig. 6.66. The combination of the feedback and feedforward controllers (green) shows the best results.

6.3.12 Regeneration Strategy

Normally, heavy-duty engines are not stationary engines. The backpressure created by the DPF hinges on the mass flow through the DPF, which in turn depends on the engine speed (Eq. 6.3). To know the soot loading of the DPF at any time, it is necessary to develop a regeneration strategy.

The fuller the DPF is, the smaller the flow area becomes, similar to when the throttle is closed.

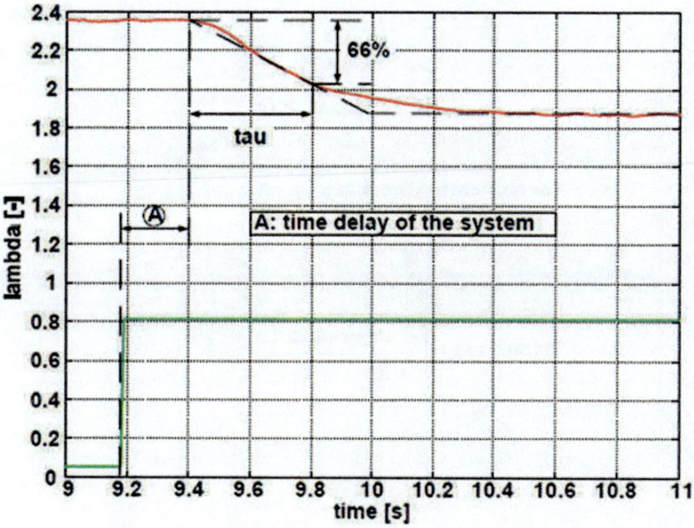

Fig. 6.64 The reaction time of the regeneration system at 1500 rpm 4 bar BMEP

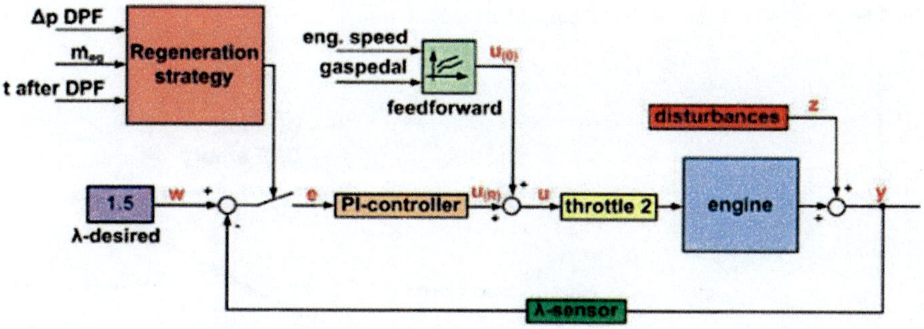

Fig. 6.65 Schematic diagram of the intake throttling control algorithm

$$\dot{m}(t) = c_d A(t)\sqrt{2\rho}\sqrt{p_{in}(t) - p_{out}(t)} \tag{6.5}$$

Equation (6.5) is based on the Bernoulli law for an incompressible flow applied to the throttle. This equation is solved by substituting $p_{in} - p_{out} = \Delta p_{DPF}$:

$$\Delta p_{DPF} = \left\{ \frac{\dot{m}_{ex}}{c_d A_{DPF}(t)\sqrt{2\rho_{ex}}} \right\}^2 \tag{6.6}$$

Figure 6.67 shows the relationship between the exhaust mass flow and the back-pressure at the DPF. The empty DPF generates back pressure as well. The blue line marks the lower limit of the operating area. The values necessary to determine this line have to be measured. The back pressure must not exceed a certain value to limit

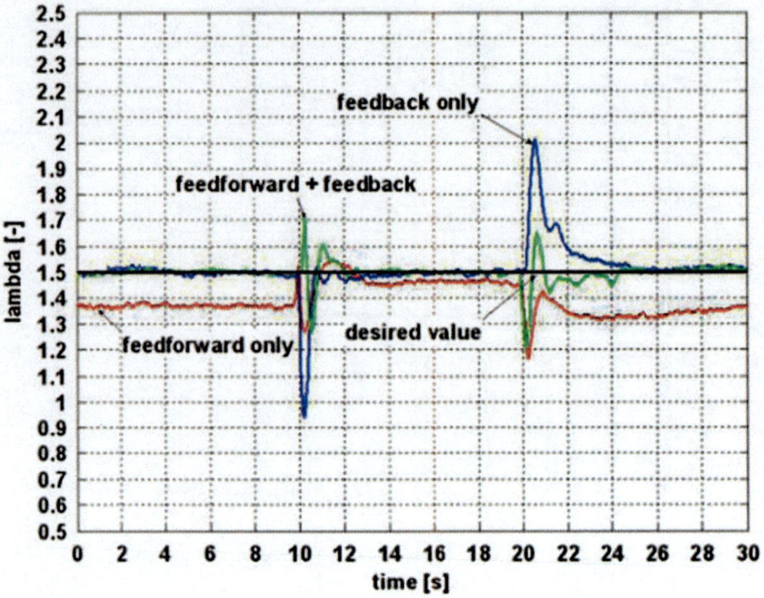

Fig. 6.66 Load step from 2 to 6 to 2 bar BMEP at 2250 rpm

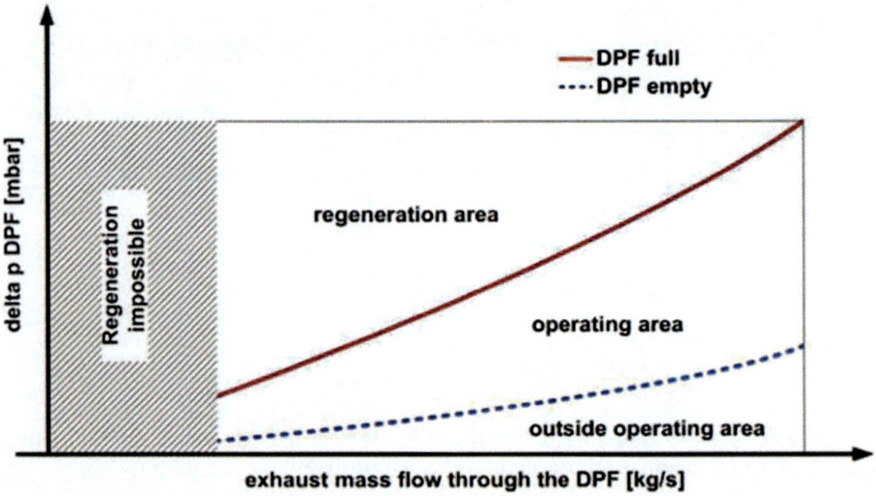

Fig. 6.67 Schematic representation of the regeneration strategy

the loss of power and an increase in fuel consumption. It is possible to determine the line which marks the upper limit of the operating area. If the measured value of the backpressure exceeds the red line, DPF regeneration is necessary.

6.3.13 Dynamic Tests

The measurement results obtained by using the retrofit kit are described here.

6.3.13.1 Test Cycle

During this test cycle, the engine is driven between 1800 and 2700 rpm and between 3.5 bar (21% load) and 7.5 bar (46% load) BMEP. The exhaust gas is measured and regeneration is effected, Fig. 6.68.

This cycle is driven three times: (1) with the original engine without the kit, (2) with the kit and with EGR, and (3) with the kit during a DPF regeneration.

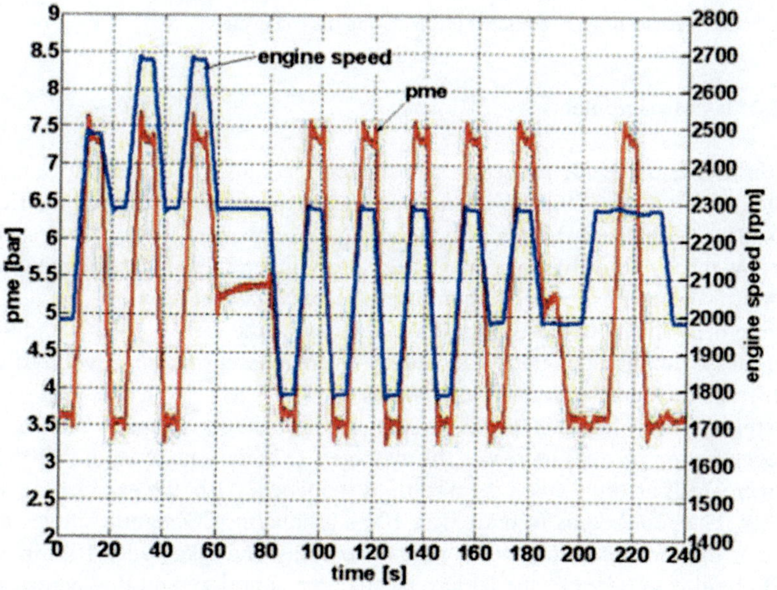

Fig. 6.68 Test cycle used for the exhaust gas measurement

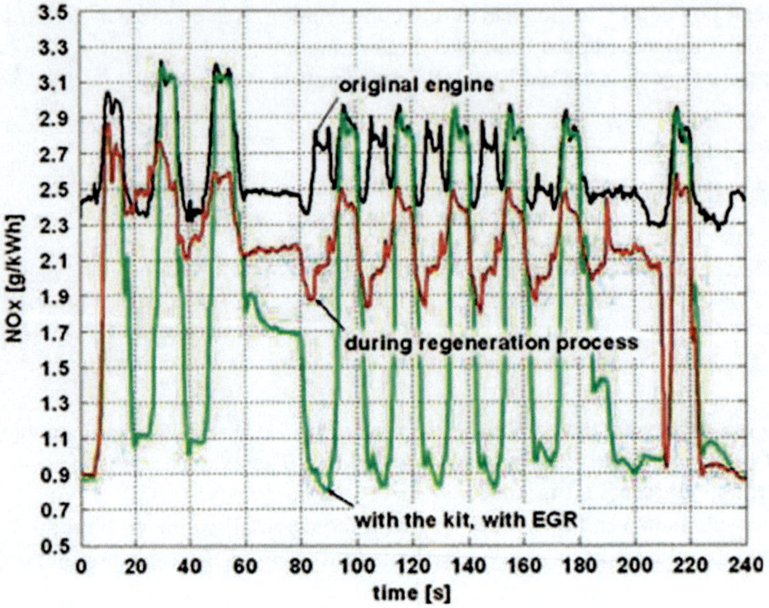

Fig. 6.69 Comparison of NOx emission levels during the test cycle

Exhaust Gas Measurement

With the EGR, the NOX reduction is between 0% and more than 50%, with an average of 30%, Fig. 6.69. This average can be improved by increasing the EGR at high loads. The intake throttling also reduces NOX emissions. If throttle-2 is closed the airflow through the engine decreases. In the cylinder, there is less oxygen, which causes a lower combustion temperature, i.e. lower NOX emissions.

Figure 6.70 shows that the filtration efficiency of the DPF is 99.9% and 98.9% for the cases without regeneration and during the regeneration respectively. After the DPF there is no increase in EC after a load step.

For this test cycle, the EGR produces an increase of CO emissions by 2%, Fig. 6.71. The regeneration process increases the average CO emissions by about 120%. The large increase is at 190 s when the throttle is reopened. Also, the exhaust mass flow is higher. The soot begins to burn after 100 s producing CO significantly. Considering that the regeneration process runs once a day and takes only 10 min out of 8 h-30 min/day (510 min), the additional fuel consumption, and the increased CO emissions are acceptably low.

For this test cycle, the fuel consumption increases by 3% with the EGR system; during the regeneration process, the fuel consumption increases by 9%, Fig. 6.72.

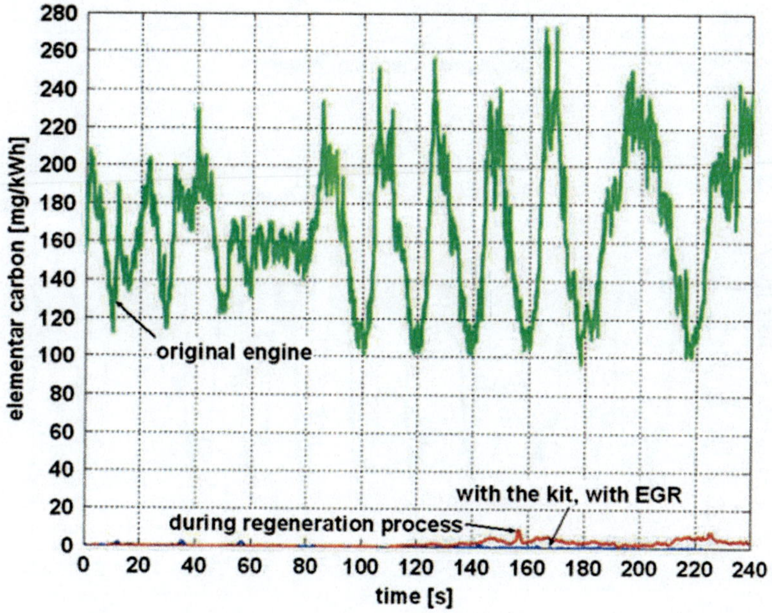

Fig. 6.70 EC emissions during the test cycle

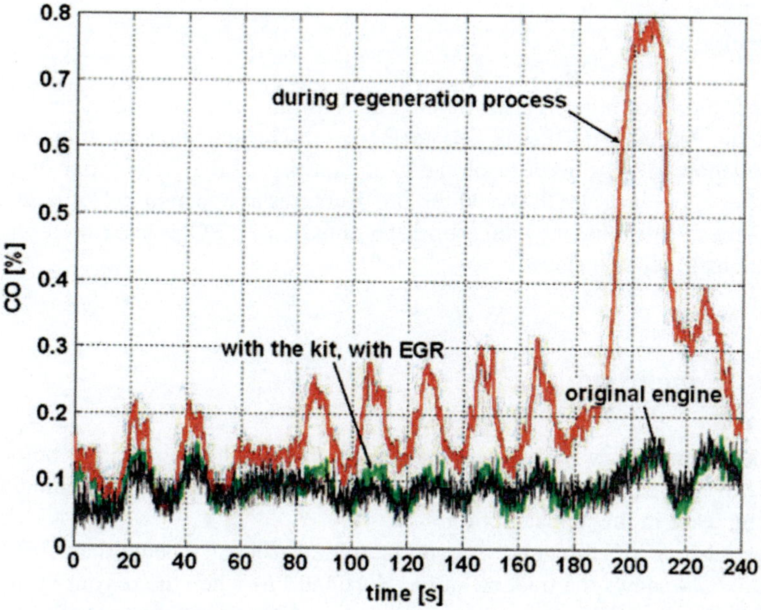

Fig. 6.71 CO emissions during the test cycle

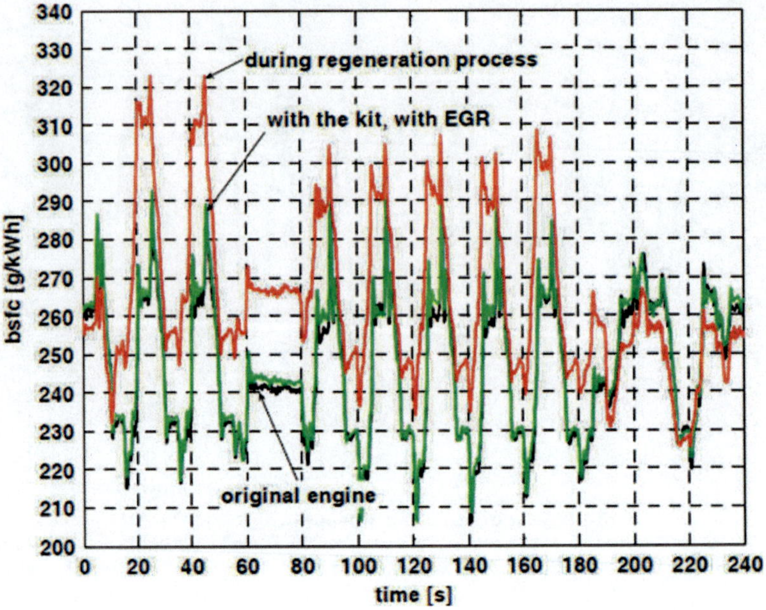

Fig. 6.72 Brake-specific fuel consumption during the test cycle

Regeneration

The upper graph of Fig. 6.73 shows the effect of throttle-2 closing after 10 s. The temperature before the DPF begins to increase. The temperature after the DPF increases more slowly because of the heat capacity of the DPF. After 100 s the soot on begins to burn and the backpressure decreases as apparent in Fig. 6.73. After 100 s there is a significant increase in the emission of EC as measured with the photo-acoustic soot sensor.

6.3.14 Conclusion

The experiments show that depending on the EGR and the operating point, it is possible to control the NOx by more than 50% with low-pressure EGR and without a large increase in fuel consumption.

The wall-flow SiC filter shows a good filtration efficiency better than 99%.

Figure 6.74 shows the tradeoff of the NOx and PM when the retrofit kit is used. The most suitable EGR is found by considering the NOx reduction, the fuel consumption, and the soot emissions. With a higher EGR, the regeneration is more frequent, affecting the BSFC and the exhaust gas emissions.

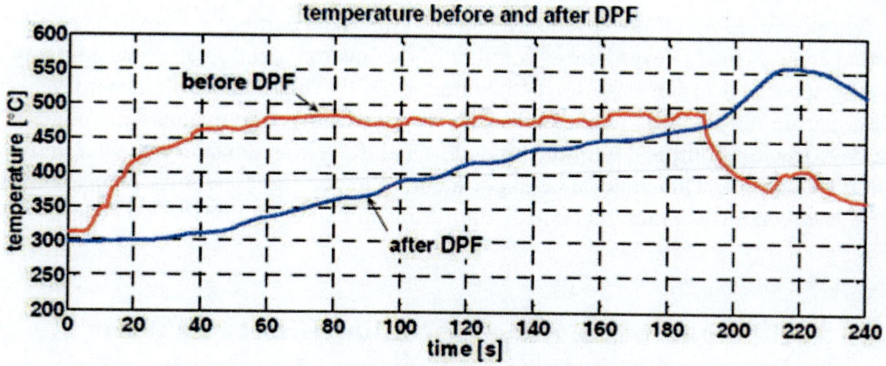

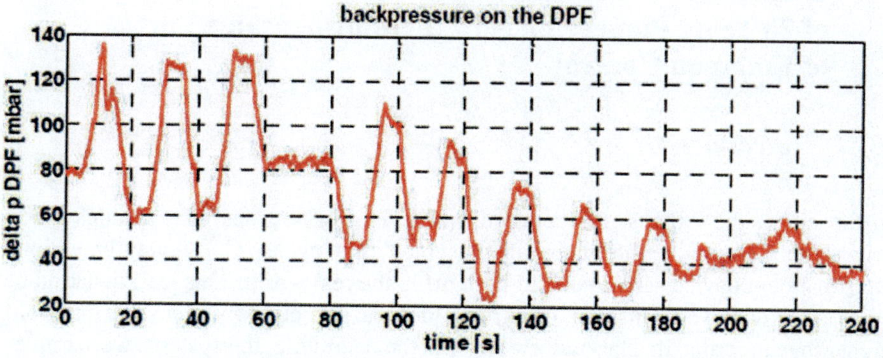

Fig. 6.73 DPF regeneration during the test cycle

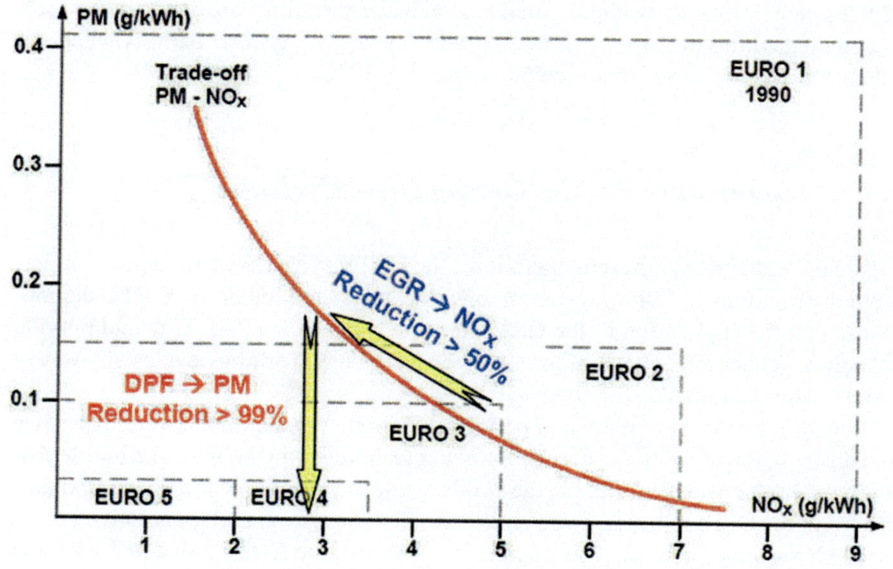

Fig. 6.74 Trade-off PM-NOx and advantages using a retrofit kit

The active regeneration by intake throttling allows the particulate trap to be regenerated at a low load. Nevertheless, because of the low oxygen flow and low enthalpy at low speeds, it is not easy to regenerate the DPF. This is a problem with vehicles that run only in urban areas with frequent stop-and-go. To evaluate the point, a garbage truck is equipped with the kit and successfully was run over one year under the ETH-control. This endurance experiment validated the feasibility of installing the system on a truck and the control algorithms in real conditions.

6.4 Part 4: State of the Art of Throttling Diesel Intake Air or Exhaust Gas in Euro VI Vehicles—The Result of 20 Years Pioneering Now the Indispensable Part of Emission Control

6.4.1 Introduction

In the cities, the exhaust temperature profile is nowhere near the profiles when driven according to the internationally accepted drive cycles. Consequently, various aftertreatment devices are unable to light off in the real world. This lead to a kind of crisis as the observed emissions of nitric oxides and particulate matter were manifold higher than recorded in a laboratory. Also, in the beginning, the systems were implemented either as retrofits or as add-ons in new vehicles by the OEMs. The learnings in baby steps in a short period of ten years were highly rewarding for the developers of later generations of vehicles, Intake or exhaust throttling proved highly useful in complementing the aftertreatment systems during the critical periods when the exhaust temperatures are insufficiently low.

6.4.2 Temperature Profiles During Drive Cycles

Figure 6.75 illustrates different operations: the NRTC cycle shows favorable temperature distributions for DPF regeneration with catalyst and the on-road ETC demonstrates that the application of the CRT requiring greater than 240 °C would provide reliable regeneration. The Braunschweig and the New York bus cycles show very problematic operations.

Because the use of burners and periodic overnight regenerations were expensive and difficult, the retrofitters yearned for better technology in the 90s. The problem of low temperature was alleviated in the 2000s by the introduction of intake and exhaust throttling for DPFs in buses, waste collectors, TRU engines, and locomotives.

MAN was the first OEM to operate bus fleets in the Swiss cities of Basel and Zürich. Figure 6.76 shows a typical application with two throttles. To avoid high EGR in the cylinder, the EGR throttle is closed when the exhaust throttle is operated.

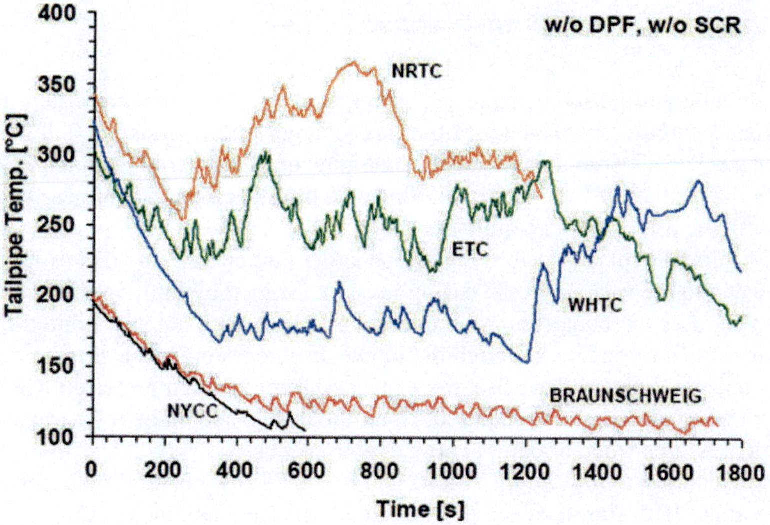

Fig. 6.75 Exhaust temperature in different typical operation profiles; "NYCC" and "Braunschweig" for city buses

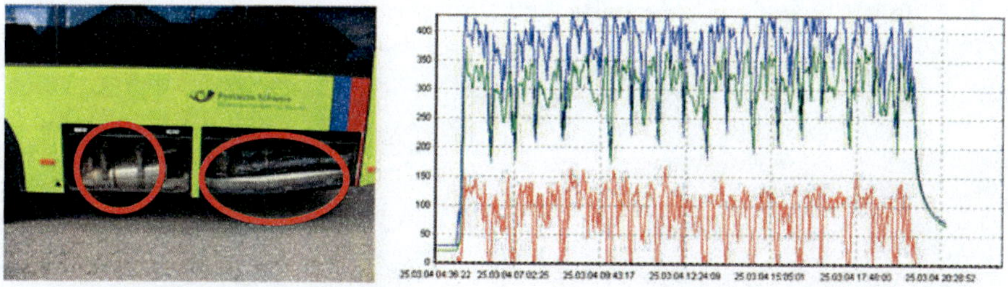

Fig. 6.76 Exhaust throttle upstream DPF lifts exhaust temperature to 400 °C—a good level for CRT regeneration

The upstream (blue) and downstream (green) temperatures of the DPF are shown in Fig. 6.76. However, the pressure difference (red) across the DPF remains low.

Since regeneration by the NO_2 in the CRT system requires long periods, the impact on fuel consumption was relatively high. Therefore, oxygen regeneration became attractive first with the use of FBC and later with catalytic combustion of the fuel injected upstream of the DOC.

6.4.3 Design of the Throttle Valves

The aluminum throttle at the intake can be operated fast using an electric step-motor. The exhaust throttle downstream of the turbocharger turbine runs hot and must be made of steel or cast iron. It is run by the hydraulic or electric drive, through linkages keeping the heat away from the drive. Since the throttle effect is nonlinear with the angle both the drives must be quite accurate.

If the throttle is placed before the turbocharger compressor, the risk of lubricant leakage due to the vacuum in the casing must be avoided by limiting the operation. If installed after the compressor the operation is less risky but still limited by the vacuum in the air system. Although the intake throttle seems much simpler design, easy to operate, light, and precise, the exhaust throttle is often preferred where the exhaust brake is legitimately required. Here, the design should be robust to survive the high pressures when braking at high engine speeds.

The intake throttle in Fig. 6.77 (left) and Fig. 6.78 (left) is used by Liebherr for engines stage IIIB/Tier 4i of the non-road emission regulations in Europe and the USA.

- Intake throttling increases the exhaust gas temperature to light off the DOC to start HC dosing, at low loads.
- The throttle is limited by the maximum allowable vacuum in the intake manifold to avoid excessive oil loss.

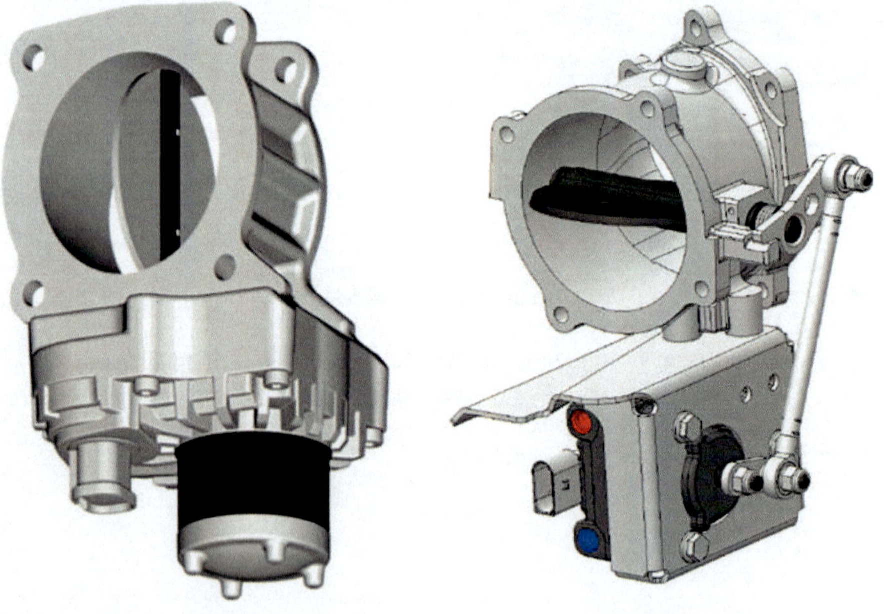

Fig. 6.77 Intake throttle (left) and exhaust throttle (right); pictures by courtesy of Liebherr/Bulle Switzerland [35]

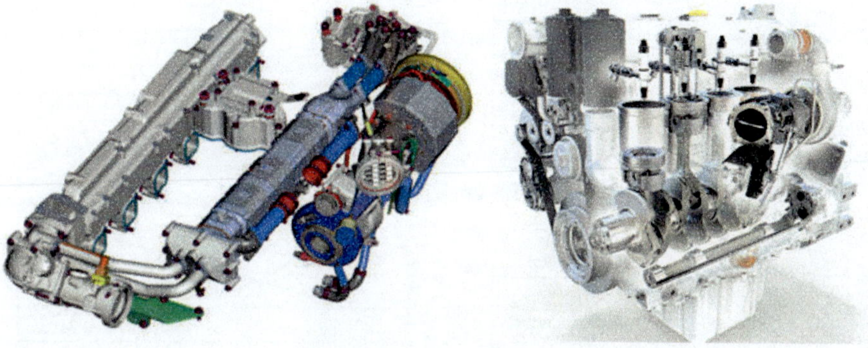

Fig. 6.78 Installation of the intake throttle after charge air cooler (left) and the exhaust throttle downstream of the turbocharger casing (right); pictures by courtesy of Liebherr/Bulle Switzerland [35]

 - This may not allow a sufficient increase in the exhaust temperature to regenerate the DPF. Therefore, HC dosing is employed at low loads to manage the exhaust temperature.

- The throttle control is by feedforward calibration in a closed-loop control based on the DOC temperature.

The exhaust throttle as shown in Fig. 6.77 (right) and Fig. 6.78 (right) is used by Liebherr for engines Stage IV/Stage V and Tier 4F of the non-road emission regulations in Europe and the USA.

- An exhaust flap is used as an engine brake on the road, e.g., Liebherr mobile cranes.
- The flap is also used
  - to increase the temperature for regenerating the DPF.
  - to increase the temperature of the DOC to reach the light-off to burn the post-injected fuel and help in the soot oxidation.
- The flap can be closed almost completely for a fast rise to a high temperature.
- The control of the flap is based on feedforward calibration corrected by a closed-loop control based on the DOC temperature.

6.4.4 Catalytic Combustion of Injected Fuel to Further Increase the Gas Temperature

With fast electronic common rail injection systems, providing heat even at light loads is possible. Peugeot pioneered this in a flexible regeneration strategy for LDV as well as HDV [36].

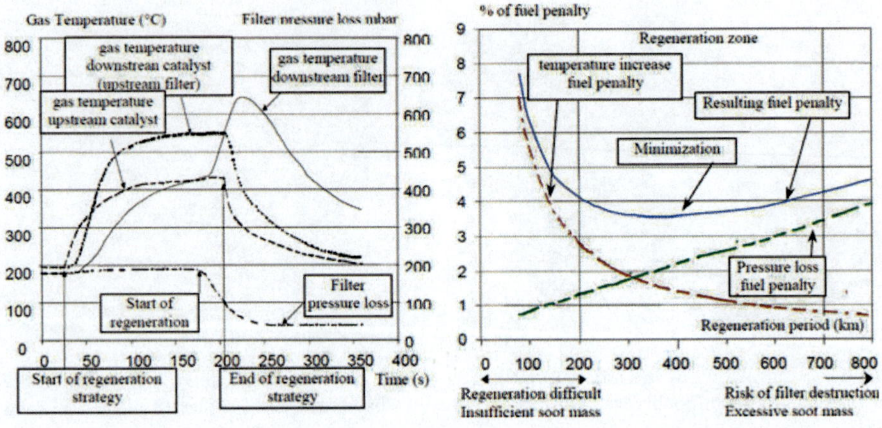

Fig. 6.79 The classic case of temperature management demonstrated by Peugeot in the year 2000 [36] became possible with electronic common rail injection combining late HC injection in the engine, catalytic combustion, and FBC regeneration

The fuel injected late in the expansion stroke (Fig. 6.79 left) burnt partly in the cylinder and raised the exhaust temperature by burning the remaining HC in the DOC for regeneration of the DPF by the FBC. This method proved to be the fastest to clean the DPF and the particulate emissions were ten times cleaner than the environment! The total fuel penalty is limited to 3–4%, by optimizing this strategy. This became a historic milestone after use by PSA and FORD to be followed by many diesel cars.

In some engines, this technology results in a risk of lubricant dilution, by the fuel-injected very late. The fuel reaches the liner wall with a potential risk of higher wear of liner and piston ring. Alternatively, it is possible to inject the fuel downstream of the engine, relying on the DOC for complete combustion. Bosch developed an HC-dosing system that produced a very fine and uniform distribution of droplets. It is reliable for both metal and ceramic DOC and applicable to different engine environments. Figure 6.80 shows the challenging off-road application by Liebherr.

Since the external catalytic combustion is possible only after the DOC has reached 250–300 °C, a two-step strategy was chosen where the first is throttling, to which the engine responds fast even at very low loads as has already been shown in the chapter, Fig. 6.81. This strategy is used by Liebherr since Euro III.

Combining these tools allows safe regeneration under any conditions and could also be upgraded by FBC to

1. lower the mandatory temperature,
2. minimalize thereby the overall fuel penalty,
3. hasten the regeneration time, and
4. improve the combustion of graphitized deposits which otherwise may be difficult to burn.

The next challenge was to combine particle filtration and NOx reduction and hence satisfy the Euro V and Stage IIIB standards for on-road and non-road engines. The

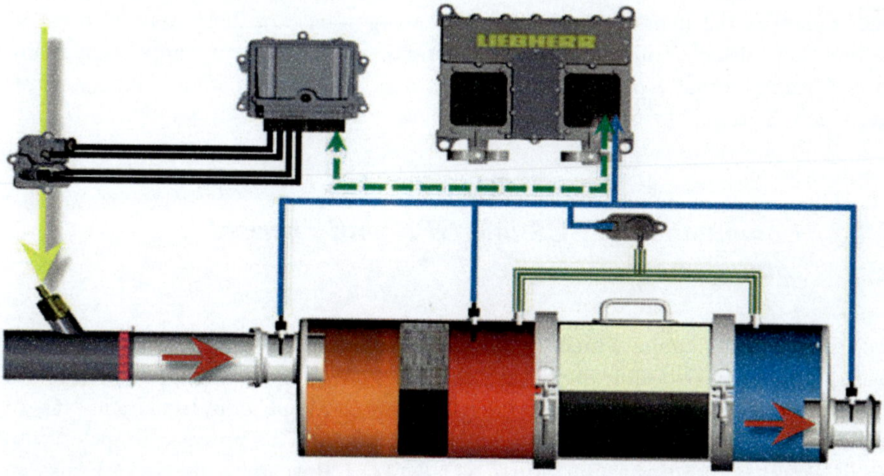

Fig. 6.80 Principle of HC-dosing followed by catalytic combustion on DOC to support DPF regeneration [35]

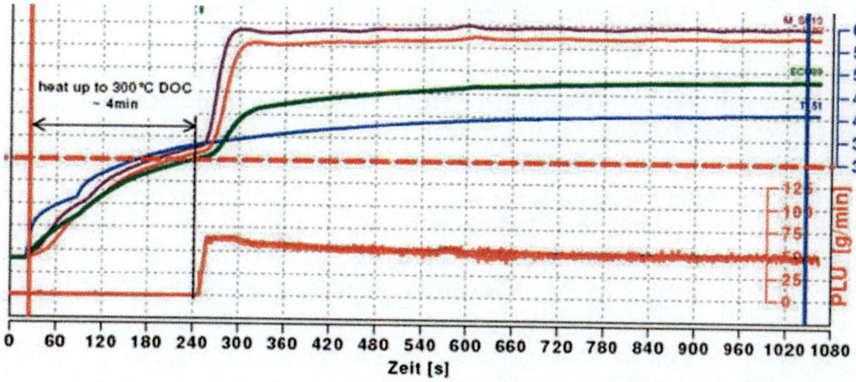

Fig. 6.81 Two-stage temperature management: starting with throttling followed by HC dosing and catalytic combustion, for a fast reaction of the soot with oxygen and proper cleaning of the DPF [35]

usual EGR solutions were not sufficient and the selective catalytic reduction (SCR) with ammonia or urea—used in large engines since the late eighties [37]—became compulsory. The complex and expensive DOC-DPF-SCR-DOC has been installed in many engines in LEV and HDSV as well as non-road vehicles.

The low load problem however continued to trouble apart from many problems like DPF blocking and urea deposits. Also, the public became aware of the real-life NOx emissions of SCR systems much higher than test cycle levels because of typical applications like city buses. In these cases, there are long periods of low load operation resulting in temperatures lower than the light-off and hence the injection of urea was interrupted by the manufacturers. The Volkswagen case of 2015 may be recalled here; indeed, double emission strategies using defeat devices were applied by many manufacturers. Finally, public awareness enforced proper temperature management, for which the tools like throttling were applied.

6.4.5 *Combination of SCR and DPF with Thermal Management*

An impressive example, which also made use of throttling, is the combination of SCR and DPF in one unit, the SCR-coated DPF, or SDPF. This application is not trivial if the CRT process is used where the upstream DOC converts NO to NO_2 to accelerate the combustion of soot at low temperatures. This process competes with the SCR process needing a high ratio of $NO_2/NO = 1$. Research at the Swiss Emission Laboratory AFHB, part of the University of Applied Sciences at Biel started a VERT research program in 2011 and provided a valid solution [38].

The system was tested for transient behavior starting from the idle in two load steps, Fig. 6.82. The interesting competition of the two processes is demonstrated, first with an empty DPF, and all NO_2 produced in the DOC available for a very fast de-NOx process.

Ammonia slip is rather high at the stoichiometric injection of urea, and it requires a small DOC downstream to clean up. Alternatively, a sub-stoichiometric injection could be applied. Also, the PN emission measured by a CNC is a little higher as with empty filters the filtration by the soot cake is missing. While the particle emission is close to zero with a fully loaded filter, the NOx reduction needs more time since the $NO_2/NO = 1$ condition needed for the fast process, cannot be reached.

The industrial partner of the University namely, Liebherr realized an elegant industrial solution in 2014 [35], Fig. 6.83.

Here again, the exhaust throttle is used for all temperature management requirements at low loads for:

- DPF regeneration, and
- SCR function.

This may be combined all possibilities of the temperature management toolbox:

- HC dosing either in the engine or downstream,
- FBC and
- Energy recuperation. See Chap. 8.

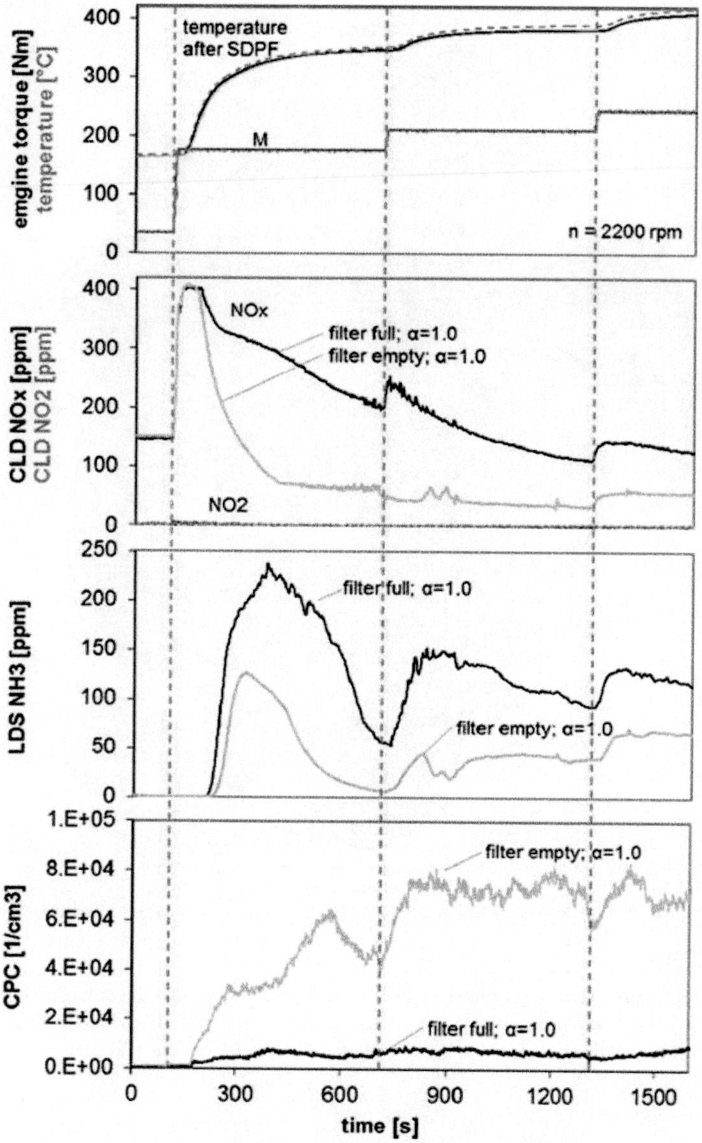

Fig. 6.82 SDPF: the combination of filtration and catalytic NOx reduction

Fig. 6.83 SDPF (red) compared to SCR only by Liebherr for non-road applications [35]

Acknowledgements The copyrights from the SAE for publishing many excerpts from the contents of the papers (copyright license in brackets) of the authors: SAE 2003-01-0381 (License 1136850-1), SAE 2005-01-0662 (License 1136868-1) and SAE 2008-01-0330 (License 1136880-1) are gratefully acknowledged.

References

1. Pattas K et al (1990) On-road experience with trap oxidiser systems installed on urban buses. Department of Mechanical Engineering, Aristotle University of Thessaloniki, SAE 900109
2. Mayer A, Pauli E (1988) Emissions concept for vehicle diesel engine supercharged with COMPREX. BBC, SAE 880008
3. Amstutz A (1991) Geregelte Abgasführung zur Senkung der Stickoxid-und Partikelemissionen beim Dieselmotor mit Comprex-Aufladung. Dissertation No. 9421, ETH

4. Pauli E, Amstutz A (1988) Regelstrategie des COMPREX-aufgeladenen Dieselmotors zur Emissionsminimierung, ABB. In: 3. Aufladetechnische-Konferenz, Zürich
5. Gleisberg T (2001) Regelklappe in Kunststofftechnik zur Dosierung der Abgasrückführmenge in Dieselmotoren. MTZ 62:4
6. Legerer FJ, Mayer A. Intake throttling for trap regeneration, a cursory engineering estimate. AKPF-home-page, www.akpf.org
7. ARB (2000) Risk reduction plan to reduce particulate matter emissions from diesel-fueled engines and vehicles. California Air Resources Board, Sacramento, CA
8. ARB (2003) Staff report: initial statement of reasons for proposed rulemaking. In: Proposed airborne toxic control measure for in-use diesel-fueled transport refrigeration units (TRU) and TRU generator sets, and facilities where TRUs operate. California Air Resources Board, Sacramento, CA
9. Mayer A et al (2002) VERT particulate trap verification, SAE 2002-01-0435
10. Mayer A et al (2004) Reliability of DPF-systems: experience with 6000 applications of the Swiss retrofit fleet, SAE 2004-01-0076
11. Heimlich F et al (2004) Externe Nacheinspritzung zur Regeneration von Dieselpartikelfiltern. MTZ 65:5
12. Zikoridse et al (2000) Particulate trap technology for light-duty vehicles with a new regeneration strategy, SAE paper 2000-01-1924
13. VERT filter list. www.umwelt-schweiz.ch/buwal/de/fachgebiete/fg_luft/vorschriften/industrie_gewerbe/filter
14. Vincent MW, Richards PJ, Catterson DJ (2003) A novel fuel borne catalyst dosing system for use with a diesel particulate filter, SAE 2003-01-0382
15. Bélot G (2003) More than half a million ultra-clean PSA diesel vehicles with FAP in the field. In: 7th ETH conference on combustion generated particles, Zurich, 18–20 Aug 2003
16. Richards R, Vincent M, Kalischewski W (2003) Additional NOx-benefits as a result of using FBC to aid DPF regeneration. In: 7th ETH-conference on combustion generated nanoparticles, Zürich
17. Mayer A et al (2003) Engine intake throttling for active regeneration of diesel particle filters, SAE 2003-01-0381
18. Kasper M (2004) The number concentration of non-volatile particles—design study for an instrument according to the PMP recommendations, SAE 2004-01-0960
19. Kasper M, Matter U, Burtscher H (2000) NanoMet: on-line characterization of nanoparticle size and composition, SAE 2000-01-1998
20. SAE (1996) SNAP-acceleration smoke test procedure for heavy-duty diesel-powered vehicles, J1667
21. Skillas G, Qian Z, Baltensperger U, Matter U, Burtscher H (2000) The influence of additives on the size distribution and composition of particles produced by diesel engines. Combust Sci Technol 154:259–273
22. Hardenberg H, Albrecht H (1987) Limits of soot mass determination from optical measurements. MTZ Motortech Z 48:2
23. Burtscher H (2004) Physical characterization of particulate emissions from diesel engines: a review. J Aerosol Sci (in press). Received 29 June 2004. Available at www.sciencedirect.com
24. Mayer A, Heeb N, Czerwinski J, Wyser M (2003) Secondary emissions from catalytic active particle filter systems, SAE 2003-01-0291
25. Mayer A, Ulrich A, Czerwinski J, Matter U, Wyser M (2003) Retention of fuel borne catalyst particles by diesel particle filter systems, SAE 2003-01-0287
26. Costantini M (2000) Relation between particle metal content (with focus on iron) and biological responses. 4. In: Internat. ETH-conference on nanoparticle measurement, Zürich
27. Fact sheet (2005) Stickstoffhaltige Luftschadstoffe in der Schweiz. Schriftenreihe Umwelt Nr. 384. Eidgenössische Kommission für Lufthygiene (EKL), 16 Jan 2005. Swiss Federal Office for the Environment FOEN, Bern

28. Hohl Y, Amstutz A, Onder C, Guzzella L, Mayer A (2008) Retrofit kit to reduce NOx and PM emissions from diesel engines using a low-pressure EGR and a DPF-system with FBC and throttling for active regeneration without production of secondary emissions, No. 2008-01-0330. SAE technical paper
29. Chatterjee S, Conway R, Viswanathan S, Blomquist M, Klüsener B, Andersson S (2003) NOx and PM control from heavy-duty diesel engines using a combination of low-pressure EGR and continuously regenerating diesel particulate filter. SAE technical paper 2003-01-0048
30. Mayer A, Lutz Th, Lämmle Ch, Legerer F (2003) Drosselung des Gasstroms von Dieselmotoren, ein Werkzeug zur aktiven Regeneration von Partikelfiltern, SAE 2003-01-0281
31. Lemaire J (2010) How to select efficient diesel exhaust emissions control strategies for meeting air quality targets in 2010. In: ETH nanoparticle conference 2010. www.nanoparticles.ethz.ch
32. Guzzella L, Onder CH (2004) Introduction to modeling and control of internal combustion engine systems, Springer edn
33. Dorf RC, Bishop RH (2004) Modern control system, 10th edn, international edn. Geering HP (2005) Regelungstechnik, 6. Auflage. Springer
34. Das S, Chakraborty A, Ray JK, Bhattacharjee S, Neogi B (2012) Study on different tuning approach with the incorporation of simulation aspect for ZN (Ziegler-Nichols) rules. Int J Sci Res Publ 2(8):1–5
35. Ellensohn R, Pfeifer A, Jaussi J (2014) New four-cylinder in–line diesel engine for mobile machines diesels. ATZ
36. Salvat O, Marez P, Belot G (2000) Passenger car serial application of a particulate filter system on a common rail direct injection diesel engine. PSA Peugeot Citroen, SAE-paper 2000-01-0473
37. Hug Th, Mayer A, Hartenstein A (1993) Off-highway exhaust gas after-treatment: combining urea-SCR, oxidation catalysis, and particulate traps, SAE 930363
38. Czerwinski J, Zimmerli Y, Mayer A (2015) Emission reduction with diesel particle filter with SCR-coating (SDPF). Emiss Control Sci Technol

Chapter 7
Decoupling Temperature and Oxygen for DPF Regeneration

A. Mayer and Chr. Lämmle

7.1 Important Parameters for the Regeneration of Soot-Laden DPFs

The regeneration of a loaded diesel particulate filter (DPF), i.e. the burn-up of the stored soot to CO_2, depends on numerous factors, but the most important ones are temperature and oxygen content of the exhaust gas, as can be seen from a simplified approach [1] for the reaction rate, RR, which thus says something about the burn-up rate of the stored soot quantity $\frac{1}{M}\frac{dM}{dt}$.

$$RR = \frac{1}{M}\frac{dM}{dt} = A \cdot [O_2] \cdot m \cdot e^{-\frac{E}{RT}}$$

where

RR	reaction rate, particle mass oxidised per time unit
M	total mass of combustible soot stored in the filter
A	constant
$[O_2]$	oxygen concentration
m	order of the reaction
E	activation energy
R	gas constant
T	temperature

A. Mayer (✉)
TTM, Fohrhölzlistrasse 14 b, CH 5443 Niederrohrdorf, Switzerland

Chr. Lämmle
Combustion and Flow Solutions GmbH, Technoparkstrasse 1, CH 8005 Zürich, Switzerland
e-mail: laemmle@combustion-flow-solutions.com

P. A. Lakshminarayanan and A. K. Agarwal (eds.), *Handbook of Thermal Management of Engines*, Energy, Environment, and Sustainability,
https://doi.org/10.1007/978-981-16-8570-5_7

This shows first of all the strong influence of the oxygen content, which must be assumed to have a reaction order of unity when the concentration drops below 9%, i.e. the reaction becomes very slow with decreasing oxygen content. The temperature influence becomes exponentially effective, as usual. Since carbon black is not chemically uniform, a staged combustion results, whereby the first fraction (attached hydrocarbons) burns quickly with an activation energy of approx. 50 kJ/mol, while following up the "dry" elemental carbon requires activation energy of approx. 150 kJ/mol, i.e. it is much more difficult to ignite and will burn slowly.

In addition to the composition of the soot, many other factors play a role, such as the compression of the soot cake, the composition of the exhaust gas, the temperature distribution, the potential use of the exothermically released combustion heat (see Chap. 5) and the use of catalytic coatings or catalytically active fuel additives, with which the activation energy can be reduced by half [2–5].

Given the continuous change of parameters due to the dynamic driving behavior of numerous vehicle applications, situations with sufficiently high temperatures are often rare and short-lived. Rapid regeneration is therefore advantageous. In this context, it is of utmost importance that the regeneration leads to a complete burn-off to avoid that, as a result of aborted regenerations, the soot layer is compacted, thus hardly permeable to gas anymore and subsequent regeneration attempts are therefore strongly impeded.

For the present investigation, it remains to be noted that with a given system, rapid and complete regeneration is only possible if both the exhaust gas temperature and the oxygen concentration are high and remain on a high level during the entire process sequence.

> The fundamental fact: in internal combustion engines exhaust temperature is always inversely related to oxygen availability.

Unfortunately, this basic requirement to have both, temperature and oxygen at a high level is never fulfilled in an internal combustion engine. It is rather the case that temperature and oxygen content develop in opposite directions, i.e. the temperature is generally coupled with the lack of oxygen; after all, the temperature can only reach high values if the available oxygen was used for combustion in the engine. For a typical commercial vehicle engine [6], this is shown in Fig. 7.1 map section by measurement and computational modelling.

This does not change if the intake airflow is throttled to trigger regeneration or if more fuel is used in the engine through post-injection or after the engine through so-called HC-dosing [7]: in all cases, these interventions for temperature management result in raising the temperature, but only by accepting the loss of oxygen concentration.

The postulated ideal case, in which high temperature and high oxygen content can be achieved at the same time, thus seems fundamentally unattainable, and this situation corresponds to the current state of the art.

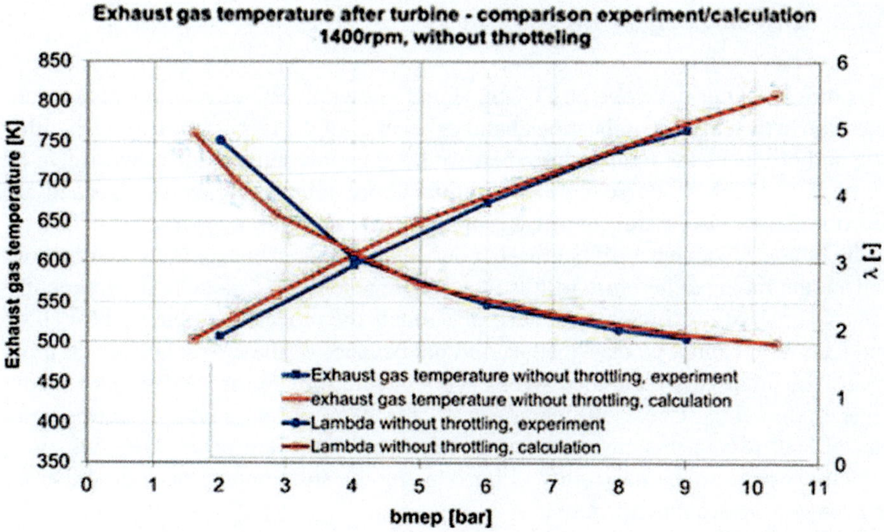

Fig. 7.1 The map section at constant rpm of a commercial vehicle engine (computational modeling and measurement) shows the opposing development of temperature and oxygen content—shown here as excess air λ. At $\lambda = 1$, all the oxygen is consumed

7.2 Shortcomings of Existing Systems

7.2.1 Engine-Internal Post-injection

Engine-internal post-injection is described in detail in Chap. 5. In this process, first implemented by PEUGEOT [5], an additional amount of fuel is injected well after the top dead center using the possibilities of electronic injection technology. However, the thermal energy thus released can no longer be converted into mechanical energy by expansion, but heats the exhaust gas. Further heating often takes place in the downstream oxidation catalyst, which catalytically converts unburnt hydrocarbons employing so-called flameless combustion, so that even at partial engine load, temperatures are reached which are sufficient to ignite the soot stored in the filter. However, the rate of combustion is slow due to the lack of oxygen, so the heating state must be maintained for a very long time, which leads to painful energy losses. Slow regenerations promote the drying out of the soot layer and it was clearly shown in [5] that no satisfactory burnout was possible with this method. For this reason, Peugeot additionally used a catalytically active fuel additive with which the process could be significantly accelerated. This process is still in use after 20 successful years of this technology, although the additive naturally results in increased costs.

7.2.2 *Injection After the Engine*

This process, later developed by BOSCH and known as HC-dosing [7], avoids late injection in the engine combustion chamber because of the risk of lubricating oil dilution and produces the temperature increase by catalytic combustion downstream of the engine, which of course requires an initial temperature sufficient for the catalytic reaction. Thanks to careful control, it is possible to reach entry temperatures of 650 °C in the DPF, i.e. soot burn-off begins, but due to low oxygen concentration with a low conversion rate and therefore results in a long regeneration time with the frequently observed risk of insufficient soot burn-off. Today, the process is often used for HDV and LDV and cannot be classified as optimal because of the energetic loss. In most cases, the process is combined with CRT regeneration, so that a large part of the regeneration already takes place—albeit very slowly—through NO_2 regeneration at partial load [8]; in this case, a DOC is connected upstream of the DPF, which in principle opens up the possibility of introducing the solution approach described in the chapter on dynamic throttling.

7.2.3 *Static Throttling*

The effect of static throttling has already been described in Chap. 1 and is shown again in Fig. 7.2. Here, the throttling is regulated to a minimum value of $\lambda = 2$

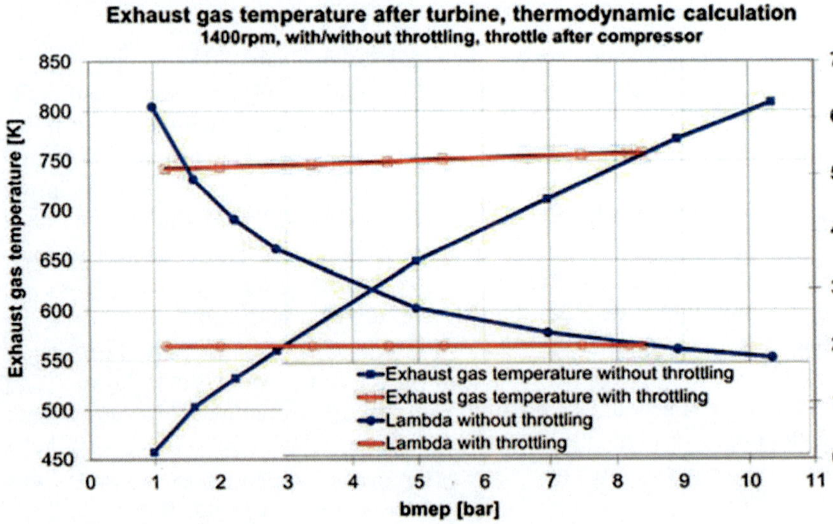

Fig. 7.2 Exhaust gas temperature and Lambda without throttling before compressor (blue) and with throttling (red) for Lambda = constant = 2

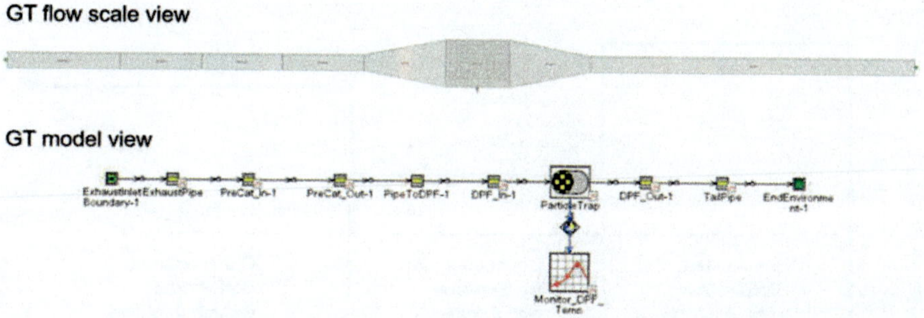

Fig. 7.3 Process-model

(corresponds to an oxygen concentration of 10.5%) to counteract the problem of too slow regeneration.

The temperature that can be achieved with this method is almost constant over the entire map range up to the low partial load, but at 750 K = 477 °C it is not high enough to trigger regeneration without catalytic assistance. For this reason, the throttling method was often combined with the application of catalytically effective fuel additives (see Chap. 2). Where this was not possible, throttling was increased to higher temperatures, although the problem of oxygen deficiency inevitably arises. To counter this problem, dynamic throttling was introduced.

7.2.4 Dynamic Throttling

The idea of dynamic throttling is based on the idea that the process is carried out in two stages, namely first maximum throttling to heat the DPF to the highest possible temperature and then opening the throttle to feed oxygen into the hot DPF at the highest possible concentration. This process was expected to lead to rapid regeneration. The process was modelled in detail and the results are shown in Figs. 7.3, 7.4, 7.5 and 7.6.

7.3 Software to Investigate the Transient Temperature Behavior

The simulation model was build up with GT-POWER [9] which is the leading software for engine performance simulations. The code solves the 1-dimensional unsteady Navier–Stokes equations. Furthermore, the software captures the effects of combustion, heat transfer, evaporation, in-cylinder motion and turbulence, and engine and tailpipe out emissions.

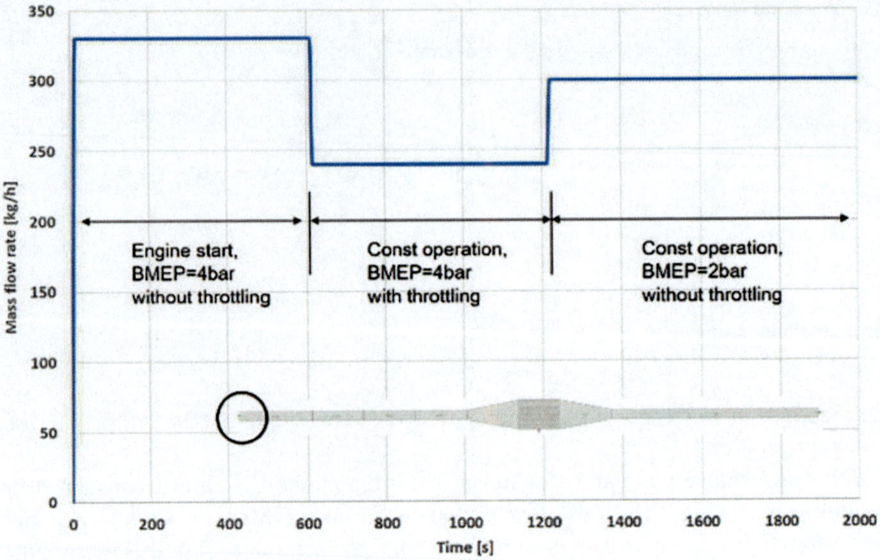

Fig. 7.4 Mass flow with closing and opening the throttle

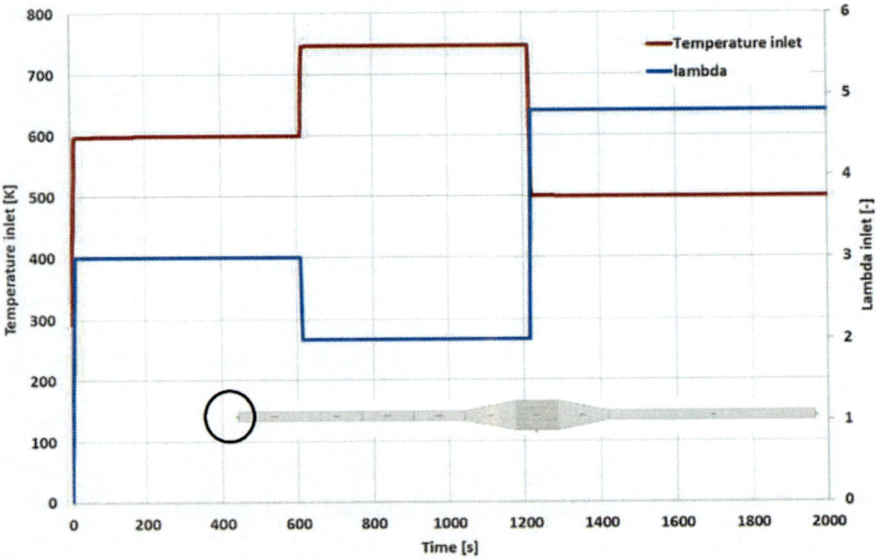

Fig. 7.5 Temperature and oxygen (Lambda) at engine out

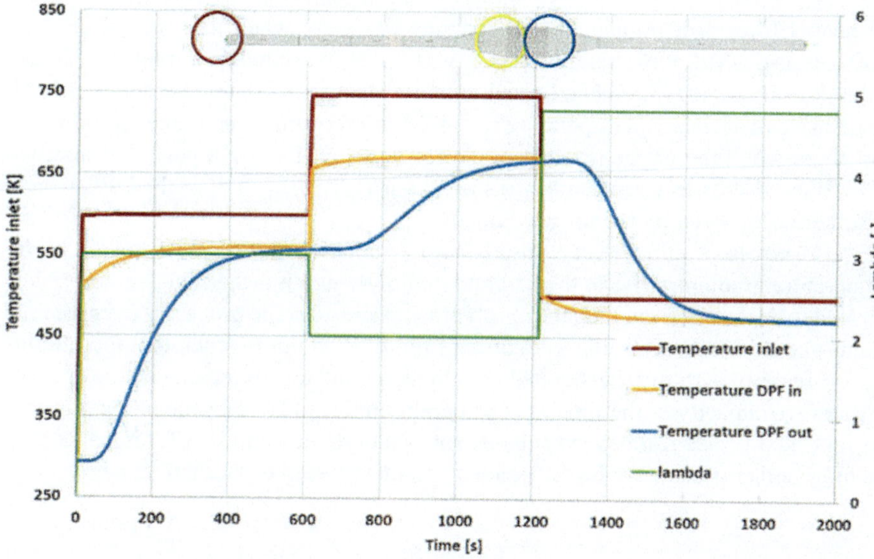

Fig. 7.6 Temperatures at DPF entry and exit

The model here was set up as a transient time-based simulation model that covers all relevant physical processes to analyze thermal decoupling.

Figure 7.3 shows the process model with the DPF and its pipes upstream and downstream, whose thermal behavior was taken into account in individual model steps. This arrangement is preceded by the engine with its (intake or exhaust) throttle, which, when actuated, can increase the exhaust temperature almost instantaneously. The diameter of the exhaust pipe is 100 mm and the pipes are made of steel with a wall thickness of 2 mm. The pipe length up to the entry into the DPF is 2700 mm. The properties of the DPF were taken from an actual VERT DPF certification test [10]. It was also assumed that convective heat transfer and heat radiation to the environment take place. The heat transfer coefficient due to convection is 15 W/mK, the emission factor for the calculation of the radiation component = 0.8. These values are good estimates in many comparable simulations.

Figure 7.4 shows the change in mass flow caused by the throttling process, which is responsible for the temperature change because in pure throttle regeneration no additional energy is added, but the fuel quantity remains the same and with reduced air mass the temperature changes accordingly.

Figure 7.5 shows the course of temperature and oxygen content (Lambda) in this two-stage process. The times are chosen to be reasonably realistic but can change due to control interventions.

Figure 7.6 shows what changes in the DPF over time as a result of this intervention; plotted are the conditions at the inlet and outlet of the DPF. It is obvious that when the throttle is opened, the oxygen content rises abruptly as desired, but the temperature at the inlet level immediately drops to the value of the cold gas, i.e. the hope that the heated filter will immediately regenerate perfectly when the oxygen content increases, is dashed. Presumably, the soot will not even ignite in the inlet area. In the outlet area, however, which had heated up more slowly due to the thermal inertia of the ceramic material, the temperature remains high for longer, and regeneration will start there when the oxygen-rich exhaust gas flows in. During a period of approx. 200 s, there is thus an opportunity for soot to burn off in this part of the DPF, which is further accelerated by the released heat.

The result is of course very unsatisfactory; because during this process high-temperature gradients arise in the ceramic monolith and thus thermal stresses which can lead to destruction (so-called ring-off cracks) and after the process is over we find a heterogeneous situation: the inlet area of the filter is not regenerated, the middle part is probably partially regenerated and dried out and the outlet area is free of soot. The storage capacity of the filter is considerably reduced for subsequent operations, the soot quality deteriorates for subsequent regenerations and it is foreseeable that the filter will subsequently be overloaded, which can lead to thermal damage.

The solution comes with transient heat storage upstream of the DPF [11].

How can we create a situation in which the temperature in the entire filter reaches a maximum level with simultaneous availability of maximum oxygen concentration without additional energy input? The solution is to add a heat-inertia element upstream of the DPF, and this process has also been modelled and shown in Figs. 7.7, 7.8, 7.9 and 7.10.

Figure 7.7 again shows the structure of the system and the process model. This time, the DPF is preceded by a DOC whose physical parameters (mass, material, heat storage) are taken from a typical exhaust gas aftertreatment system of a commercial vehicle with a CRT system [10].

Figure 7.8 shows the temperature curve as well as the oxygen availability due to throttling in the same way as for the previously considered system. The course of the mass flow is not shown, because it is the same as before in Fig. 7.5.

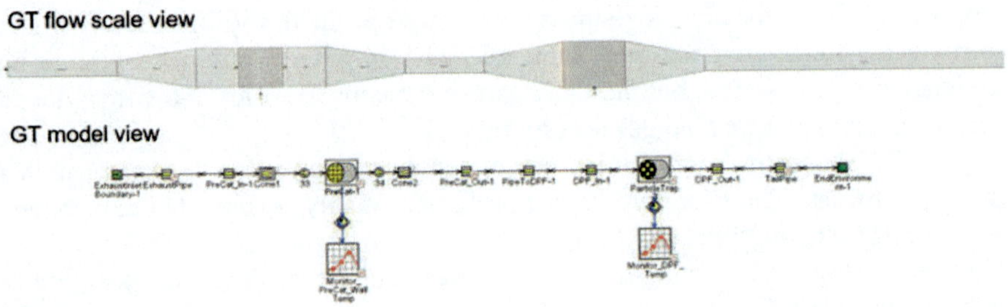

Fig. 7.7 A process model of DPF with a heat storage element in front (a DOC with high thermal inertia)

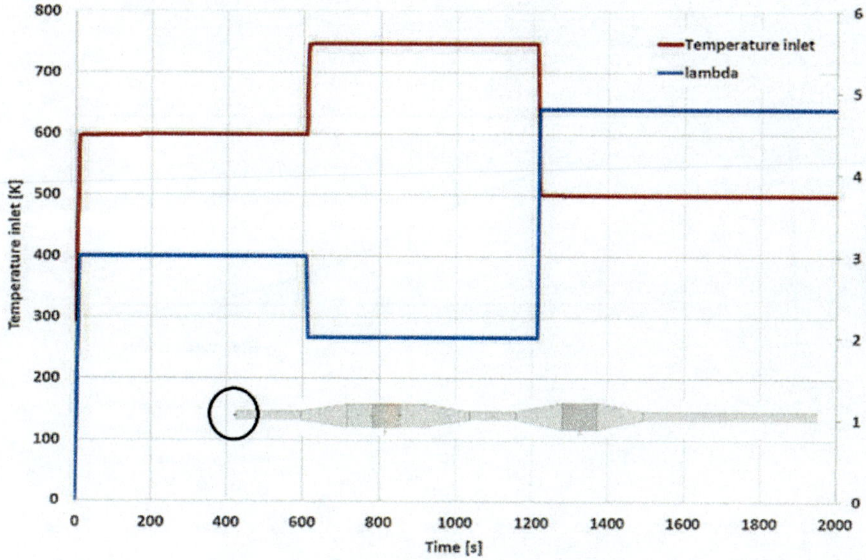

Fig. 7.8 Temperature and oxygen (Lambda) at engine outlet when closing and opening the throttle

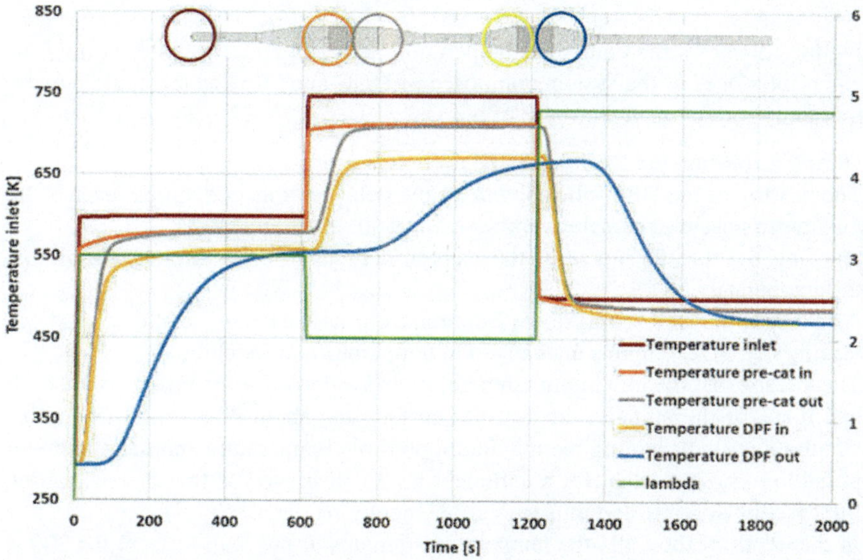

Fig. 7.9 Temperature and oxygen (Lambda) at the four important process points: DOC in, DOC out, DPF in, DPF out as functions of time during regeneration

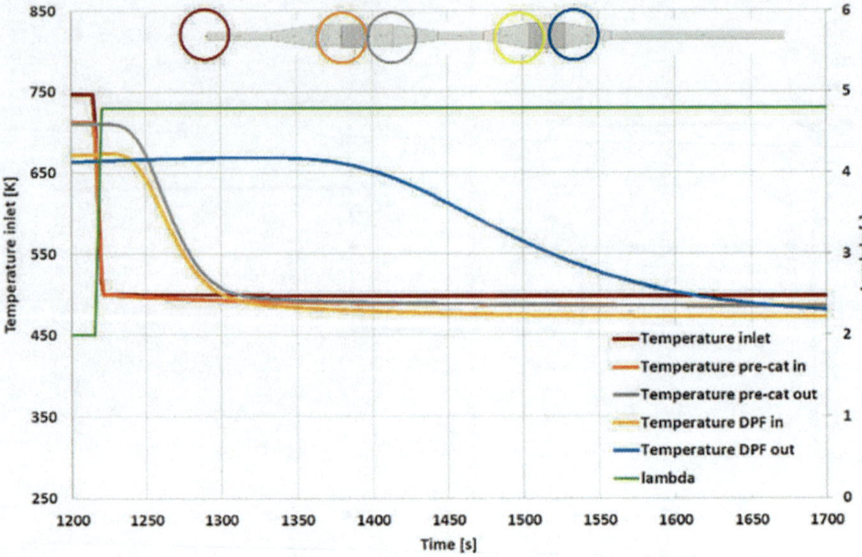

Fig. 7.10 Zoom during regeneration

Figure 7.9 shows the course of temperature and oxygen availability at the four essential interfaces of the system composed of DOC and DPF, which could now be substantially improved in the sense of our task:

- During throttling, the DOC is heated up first.
- The heating of the DPF follows with a time delay, and its outlet area reaches the maximum possible temperature after about 500 s.
- Opening the throttle increases the oxygen content in the entire system almost instantaneously.
- As a result, the DOC cools down from the front with a delay, i.e. the exhaust gas leaving the DOC remains unchanged in temperature for a while.
- This means that the maximum temperature is available for the inflow to the DPF and thus for the situation at its front surface for about 30 s, and the goal of simultaneously providing the maximum possible temperature and the maximum possible oxygen content for a sufficient time is achieved so that the soot at the filter front can ignite and thus trigger regeneration.
- In the depth of the DPF, the temperature remains at the highest level for 200 s, i.e. the burnout can start almost in the entire filter at the same time. This achieves a homogeneous temperature distribution in the DPF, which alleviates the main problem of thermal stress damage to ceramic DPF.
- This process is supported by the exothermic reaction, which is not yet included in this model.

Figure 7.10 is a zoomed view of Fig. 7.9, which further illustrates the conditions during regeneration.

It can thus be shown that with transient heat storage the paradoxical sounding task can be solved, namely to simultaneously provide the highest possible temperature of the throttled system and the highest possible oxygen content of the given operating point for a sufficiently long time to ensure complete regeneration of the filter, i.e. burning off of all stored soot.

It is important to note that this system is controllable in many ways:

- Depending on the temperature requirement (with/without catalysis), the degree of throttling can be selected.
- The duration of heating can be selected as well as the duration of regeneration.
- In case of insufficient regeneration (backpressure control), the process is repeated, whereby learning effects are stored.
- Altitude effects and aging effects can thus be largely compensated for.
- Additional measures such as additional fuel injection can be integrated.
- Escalating regenerations with very high-temperature outbreaks, which are dangerous for the ceramic monolith, can be detected by the temperature measurement after DPF and reduced to a safe turnover practically without delay by a rapid shut-off of the oxygen supply.

7.4 Technical Solutions That Already Use This Approach

The DPF system in a Transport Refrigeration Unit TRU described in Chap. 2 was already designed in this way. The staged intake throttling in this system would not have been sufficient to carry out such perfect regeneration if the DPF had not been preceded by a heat accumulator. In this case, it was not a DOC but the inner pipe of the air gap insulator.

The design can be done in many different ways. For example, the catalytic converters arranged in series in today's exhaust systems are inherently suitable for periodic storage of the heat generated by throttling, in almost classic form the so-called CRT system with the DOC and DPF closely connected in series. The combination of such a system with staged throttling, as described and claimed in the patent specification [11], can therefore bring a great advantage for today's systems at little additional expense.

References

1. Lepperhoff/TU Aachen (1984) The calculation of regeneration limits of diesel particulate traps for different regeneration methods. SAE 1984-02-01
2. Mayer A et al (2004) Minimierung der Partikelemissionen von Verbrennungsmotoren. Haus der Technik 36. Expert-Verlag. ISBN 3-8169-2430-1
3. Mayer et al (2005) Elimination of engine generated nanoparticles. Haus der Technik 58. Expert-Verlag. ISBN 3-8169-2552-9

4. Mayer et al (2008) Particle filter retrofit for all diesel engines. Haus der Technik 97. Expert-Verlag. ISBN 978-3-8169-2850-8
5. Belot G et al (2000) Passenger car serial application of a particulate filter system on a common rail direct injection diesel engine. SAE 2000-01-0473
6. Mayer et al (2003) Engine intake throttling for active regeneration of diesel particle filters. SAE 2003-01-0381
7. Bosch. HC dosing. http://auto2015.bosch.com.cn/ebrochures2015/energizing_powertrain/ds/cv/departronic2_diesel_dosing_system_en.pdf
8. Walker AP et al (2002) The use of the continuously regeneration trap (CRT) to control particulate emissions. SAE 2002-01-1271
9. https://www.gtisoft.com/gt-suite-applications/propulsion-systems/gt-power-engine-simulation-software/
10. Mayer A et al. VERT filterlist. www.vert.certification.eu
11. Verfahren zum Betreiben eines Verbrennungsmotors inclusive Partikelfilterregeneration. Deutsches Patent Nr. 10 2005 012 525

Chapter 8
Thermal Management of the DPF, DOC, and SCR Processes by Heat Recovery

A. Mayer and H. Stieglbauer

8.1 Low Temperatures at Urban Driving Conditions Create a Fundamental Problem

The combustion process in an internal combustion engine at very high temperatures with ambient air produces inevitably high concentrations of nitrogen oxides namely, NO and NO_2 in the exhaust gas. This concentration can only be lowered internally if the combustion temperature is lowered, e.g. by water injection or exhaust gas recirculation. According to Carnot, however, lowering the combustion temperature leads to deterioration in thermal efficiency. Therefore, there is a serious conflict of goals.

Toxic air pollutants namely, NO and also NO_2 (produced at lower temperatures and with excess air) are subject to increasingly stringent emission limits worldwide. A reduction of NO_x by at least 90% is required today to meet the limits at the tailpipe. In the case of petrol engines, the problem is already solved with the three-way catalytic converter with a precise lambda = 1 control invented by John J. Mooney et al. in the 1970s [1]. Until a few years ago, diesel engines could comply with emission limits with internal engine measures like injection delay and exhaust gas recirculation. However, in the wake of stricter legislation of emission limits and requirement of reduction of CO_2 from Euro V onwards, catalytic after treatment measures are required. For diesel engines operated with excess air (lambda > 1), the selective catalytic reaction (SCR) process using urea or ammonia as a reducing agent [2] is the most suitable method. This method has long been applied to large stationary power plants.

A. Mayer (✉)
TTM, Fohrhölzlistrasse 14 b, CH 5443, Niederrohrdorf, Switzerland

H. Stieglbauer
STARFILTER, Am Katzernbuckel 59, 82377 Penzberg, Germany

P. A. Lakshminarayanan and A. K. Agarwal (eds.), *Handbook of Thermal Management of Engines*, Energy, Environment, and Sustainability,
https://doi.org/10.1007/978-981-16-8570-5_8

As with all catalysts, the conversion rate depends on operating temperature. With SCR, NO_x conversion rates of over 98% can be achieved when the exhaust gas temperature exceeds 300 °C. Below that, the conversion is weak and below 240 °C the dosing of the reactant must be stopped to prevent crystallization and deposition of urea and formation of ammonium sulphate or sulphite. On the other hand, since the exhaust gas temperature of vehicles in urban use—where a particularly large number of people are exposed to pollutants—rarely exceeds 250 °C; the problem arises in the city traffic in the vehicles that demonstrate low emissions in the European test cycle ETC. The conditions are now referred to as the "real world". The real NO_x emissions exceed the cycle emissions many times over, rendering SCR catalysis practically ineffective (Figs. 8.1 and 8.2).

While the exhaust gas temperature in the European test cycle ETC reaches 300 °C and NO_x can thus be reduced by 90% with SCR, in the New York City cycle (NYCC) or the Braunschweig cycle, which are typical for city buses, an average of only 150 °C is reached and the reduction of NO_x with SCR becomes insufficient. This means that the problem is true for public transport buses, taxis, and vehicles used in dense traffic.

	ETC		WHTC		NYCC		Braunschweig	
	1800 s		1800 s		598 s		1740 s	
	OEM	DPF+ SCR	OEM	DPF+ SCR	OEM	DPF+ SCR	OEM	DPF+ SCR
Fuel [l/test]	4.15	4.30	2.58	2.64	0.25	0.26	1.02	1.05
T Engine Out [°C]	282	310	247	261	144	151	149	156
T Tailpipe [°C]	248	291	210	241	139	182	127	150

Fig. 8.1 Urban driving cycles like NYCC and Braunschweig cycle, typical for public transport, provide much lower exhaust temperatures than standard homologation cycles ETC and WHTC

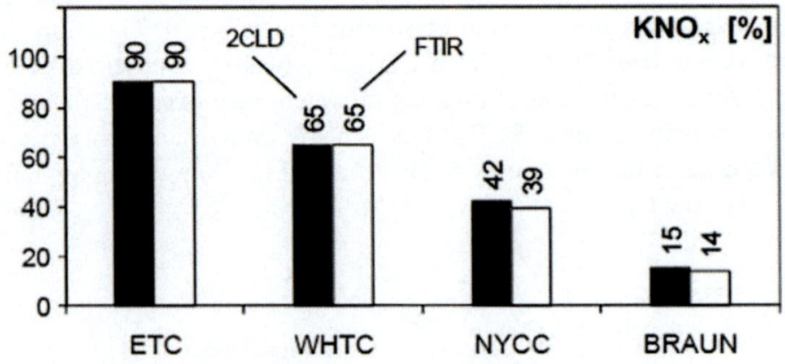

Fig. 8.2 NO_x-Reduction by SCR inefficient at urban driving due to low exhaust temperatures [3]

On the other hand, SCR combined with EGR is successfully used in most other applications like long-distance driving, motorway operation, where the average temperatures are high enough; also, during common chassis-dyno homologation tests, all new vehicles did reach the required legal emission limits. However, to avoid technical risks in the urban traffic operation most manufacturers introduced defeat devices using an illegal two-way strategy, which guaranteed good emission results during the chassis test cycle but stopped urea injection and EGR for real-world operation. This triggered the diesel scandal in 2015 [4]. Once this became obvious, the legislator added real-life on-road testing to chassis dyno testing which required a new technology known as thermo-management.

This problem was already recognized in 1993 [5], and new catalytic solutions were developed for it by ETH Zurich [6], but unfortunately, they were not able to withstand longer aging cycles at that time.

Since modern engines must be further improved to reduce CO_2 and since this generally leads to an increase in NO_x formation, it is becoming increasingly difficult to achieve that goal. Aggravating the problem, the engine-out temperature will further decrease in the future with increasing degrees of supercharging and exhaust heat utilization measures.

In many cases, a temperature increase in the SCR range of 50 °C would already be sufficient to bridge critical phases but for energy reasons. "simple" heating of the exhaust gas and also throttling as successfully used for DPF regeneration is out of the question, as the time during which this heating would be necessary can be very long.

So the question arises on how the temperature of the exhaust gas can be increased without heating the exhaust gas as a whole to this temperature. At first glance, this sounds like a physical impossibility.

8.2 The Heat Recovery Approach, Still not Widely Used Provides an Option

A simple "recuperator" is shown in Fig. 8.3, in which the gases flowing out cross the gases flowing into the heat exchanger. Of course, this makes sense only if exothermic processes take place in the reactor. This could be the catalytic oxidation of CO and HC in a DOC or the combustion of soot in the particulate filter (DPF) during regeneration. In the example shown, the DOC is connected upstream of the DPF as widely used in the CRT circuit by Johnson Matthey [7], where heat is generated in both reactors. The combustion of soot generates a great deal of heat, which is recovered to accelerate the regeneration and complete the combustion of the stored soot. This is a self-regulated process because the speed of combustion is limited by the availability of oxygen, and the process immediately slows down once the soot is combusted. The principle is efficient and proven in practice by the cleverly designed STARFILTER [8].

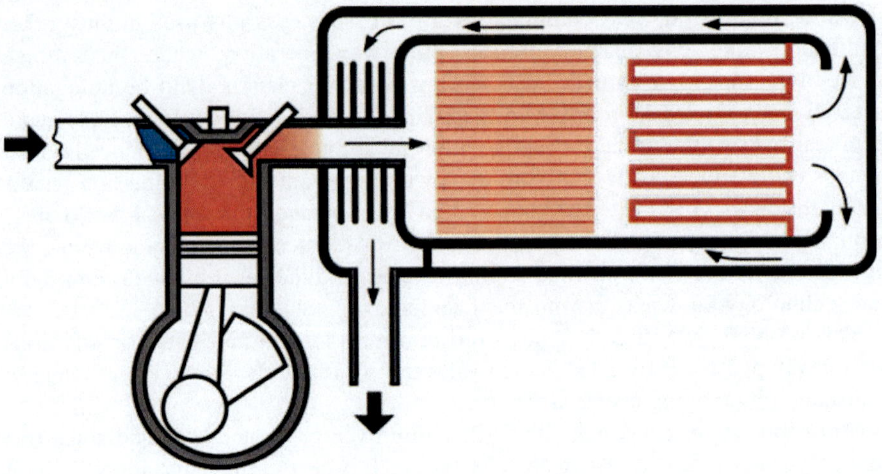

Fig. 8.3 Heat Recovery for exhaust aftertreatment by DOC and DPF

However, the chemical process of SCR catalysis, i.e. the reduction of the pollutants NO and NO_2 to N_2 with the stoichiometric addition of ammonia (or urea) is not exothermic. At first look, a recuperation circuit is therefore useless for the SCR catalytic converter. The additional processes namely, (a) catalytically increasing the NO_2 content to hasten the reaction in the SCR and (b) converting urea to ammonia also generate little process heat. There is hardly any difference between the exhaust gas temperatures before and after the SCR system. If the temperature is below the minimum operating temperature of the SCR system, i.e., less than 230 °C, not only is little NO_x converted, but because the Adblue injection is switched off there is no NO_x reduction at all. Therefore, during light-load operation, the emission corresponds to the raw engine-out NO_x emission, which can be 20 times above the permissible limit.

Raising the temperature within the engine, for example, by delayed fuel injection, as is often used to regenerate the DPF, must be ruled out because this measure is not limited to a short time. The resulting increase in fuel consumption or CO_2 emissions would be unacceptable. The same applies to a temperature increase through a burner or even through electric heating or throttling.

8.3 Solution by a Heat Source Within the Recovery Loop

A heat source EH, for example, an electrical heat source, is used in the recuperation loop. The location can be between the heat exchanger HE and the SCR, i.e. in the direction of flow upstream of the SCR, or between the SCR and the HE, i.e. in the direction of flow downstream of the SCR (Fig. 8.4).

The heat supplied by the EH increases the temperature of the gases circulating in the recuperation loop and this is in turn transferred to the gases entering the heat

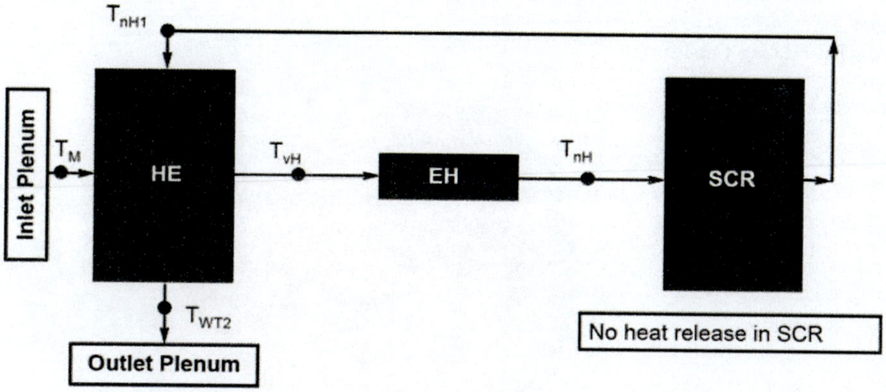

Fig. 8.4 Heat recovery with a small heat source within the recovery loop [9]

exchanger. The second round starts at a higher level and this repeats in the subsequent cycles. After each cycle, the gases become a little warmer. The circulation is repeated at full load very frequently at a space velocity of 20 s^{-1} in the SCR catalytic converter. For example, at a space velocity of 10 s^{-1}, one circulation takes 0.1 s. At partial loads and idling, the frequency of circulation is correspondingly slower.

If there were no heat losses to the surroundings and the heat exchanger ideally transferred 100% of the heat to the incoming gas, an arbitrarily high temperature could be reached in the regeneration loop even at low power of the heat source, while the gases would remain cold at the entry and the exit of the loop.

This sounds a bit like "Münchhausen" or "Perpetuum mobile", but one must distinguish well between energy and temperature: the amount of energy used per unit of time is low, but the temperature can certainly become very high if the energy is supplied over an appropriate time and losses are avoided.

Of course, this ideal case is not achievable, since the heat exchange is imperfect and heat losses to the surroundings are not be completely avoidable. Therefore, this device must be designed to ensure these two sources of loss are kept small. In other words, the energy must remain typically in the core and the recuperation loop.

The so-called heating catalysts, such as those used by EMITEC for many years to accelerate the response of three-way catalytic converters in large petrol engines, can be used as a heat source. These heating elements [10]:

- have a very compact design in metal technology,
- transmit high power at low on-board voltages,
- are insensitive to contamination,
- are proven in robust vehicle use, and
- integrate into the recuperation loop of an SCR catalytic converter easily without any problems (Fig. 8.5).

A simple mathematical model of the process from the moment the heat source was switched on was presented by El-Nashar [9], with realistic assumptions for heat

Fig. 8.5 EMITEC-Heating-Catalyst, an option for the heat recovery loop [10]

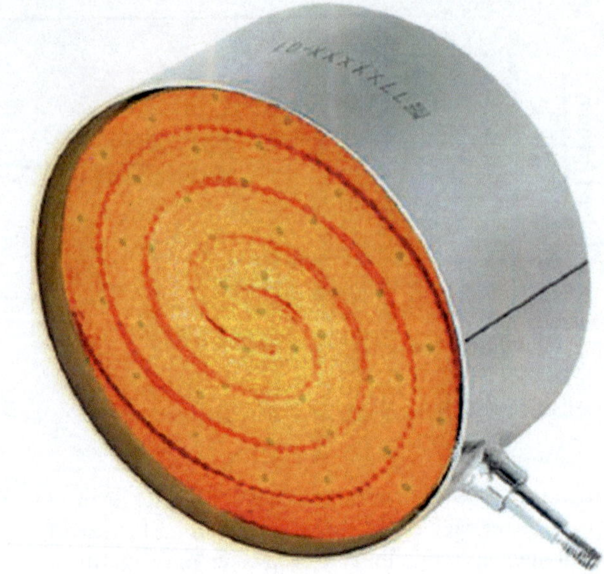

transfer and heat losses. Further, it was assumed that the heat source increases the gas temperature in each loop by 10 °C (Fig. 8.6).

The model showed a temperature increase of 40 °C at the entry to the SCR catalyst in ten cycles, i.e. a fourfold temperature rise compared to continuous heating of the gas flow. After ten, the temperature saturates, i.e. the heat that was added only covered

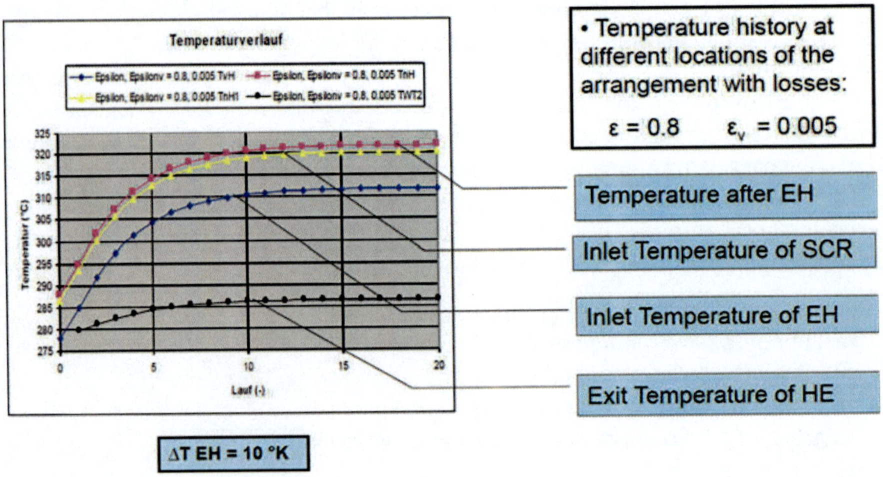

Fig. 8.6 Computer-Model of the temperature rise by heat recovery with a very small heat source within the recuperation loop [9]

the losses. This calculation does not, however, take into account the thermal inertia of the SCR substrate, and hence, in reality, the process runs somewhat slower.

Thus, it is possible to minimize the heat losses to the ambient and optimize the recuperative heat transfer, by this simple and clever design of the circuit with a low energy input to always keep the temperature in the SCR reactor sufficiently high so that even at low partial load the NO_x emissions of the engine can be sufficiently reduced. Many different solutions are conceivable for the design; electrical heating is not necessarily the only possible one.

8.4 Conceivable Heat Recovery Options

8.4.1 Predecessors of Heat Recovery Applications

In chemical engineering, the use of process heat by recuperation to accelerate chemical reactions is common.

8.4.2 Vehicle Gas Turbine

Recuperation is the technical state of the art for gas turbines, especially for gas turbines for vehicle use, as they cannot achieve sufficient efficiency at partial load without recuperating the exhaust gas heat into the process and would therefore not be competitive with the internal combustion engine.

8.4.3 QuadCAT

In-vehicle emissions-technology, recuperation was applied by the US manufacturer CERYX and published in 1999 [11]. At that time, the system went by the trade name QuadCAT, because CERYX claimed to reduce all four legally limited pollutants simultaneously, using the usual catalytically active components. A new feature was the use of process heat through recuperation to make the chemical reactions faster and more complete.

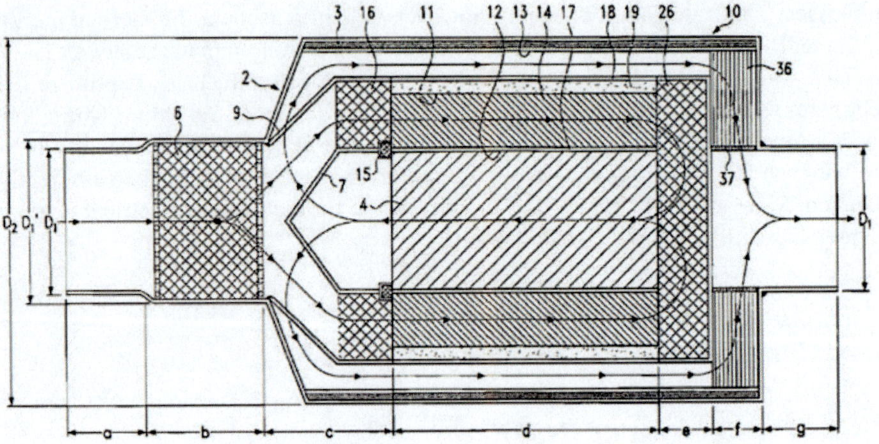

Fig. 8.7 Flow pattern in the catalyzed particle filter "STARFILTER" according to a Stieglbauer patent with direct heat exchange via the annular gaps [8]

8.4.4 Particle Filter Regeneration

The German patent by H. Stieglbauer DE 10 2005 062 050.7 [8] for a particle filter with integrated catalytic converters and heat recovery loop, from which the following sketch is taken, should be mentioned first (Fig. 8.7).

The STARFILTER contains several special features regarding flow guidance and catalysis, which are described in the patent. Of special interest in the present context is the highly efficient effect of the recuperation of process heat at the onset of soot combustion, which aggressively accelerates this process and thus ensures perfect cleaning of the ceramic filter within a short time. Additional energy is, therefore, not required for the STARFILTER.

8.4.5 EMITEC-DOC-Recuperator

Brück et al. presented a simple design proposal for the recuperation of process heat in an oxidation catalytic converter at the Vienna Motor Symposium 2006, which was backed by convincing data [12] (Fig. 8.8).

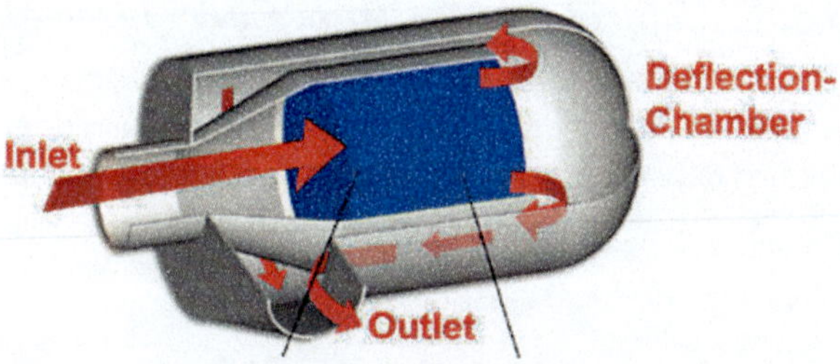

Fig. 8.8 Oxidation-Catalyst embedded in a heat recovery structure acc. to EMITEC [12]

8.5 Further Elements to be Applied with Heat Recovery for Exhaust Gas After Treatment

8.5.1 Urea Mixing

Perfect mixing of the exhaust gas with the reactant is of utmost importance. If there is no uniform distribution, little NO_x is converted where the concentration is low, and ammonia slip is formed where the concentration is high—two processes that cannot be reversed. Usually, therefore, a long mixing section with mixers is provided, which is difficult to accommodate structurally in the vehicle. In the present concept, the long path through the jacket gap and the heating catalyst is used for this distribution and designed accordingly.

8.5.2 Urea Deposition and Crystallization

Adblue hitting cool surfaces can hardly be avoided in the usual arrangement in an exhaust pipe cooled by ambient air. This allows the formation of urea deposits which can grow rapidly. In the recuperation system, the mist of urea solution reaches warm internal surfaces and evaporates rapidly; thus, both mixing and conversion to ammonia are accelerated.

8.5.3 Downstream Ammonia-Slip Catalyst ASC at a High-Temperature Level

The effectiveness of the slip catalyst depends on the temperature. Therefore it is placed in the recuperation loop directly after the reactor i.e., before the heat exchanger.

8.5.4 Control for Minimum Energy Demand

As the method of internal heating in the recuperation loop is efficient, it should be applied only when it is necessary. As soon as the exhaust gas temperature rises above 250 °C, the heating can be switched off.

8.5.5 Staged DPF Regeneration

Filter regeneration can indeed become a big problem in part-load operation if the exhaust gas temperature is not sufficient to initiate combustion of the soot by NO_2 or by O_2. In the staged DPF regeneration, the NO_2 regeneration starts first, followed by the increase in temperature in the recuperation loop to a level for initiating the oxidation of soot supported by an appropriate catalytic coating. This process accelerates regeneration and enables complete burnout.

8.5.6 SDPF

SDPF is an extremely attractive proposition where DPF and SCR are combined in a single ceramic substrate [13] where the pores are optimally large enough for an effective and at the same time, small enough for proper SCR coating. However, SDPF has a problem with the sufficiency of NO_2 regeneration because NO_2 is needed by both processes namely, CRT regeneration and denitrification in the SCR. This competitive situation leads to problems especially when the temperature level is low, i.e. when both processes are slow. In the recuperation loop, however, the DPF regeneration takes place quickly by substituting the NO_2 process by the O_2 combustion process, so that the NO_2 remains available for the fast reactions in the SCR.

References

1. Mooney JJ et al (1985) Polyfunctional catalysts and method of use; US-Patent 4552733 A. Priority date: 1997-08-27
2. Katalytische Verfahren der Abgasreinigung; VDI Richtlinien 3476
3. Czerwinski J et al, Diesel Emission with DPF+SCR in VERTdePN—Testing and Potential SAE 2011-01-1139
4. Thompson G, Besch et al (2014) In-use emission testing of light duty diesel vehicles in the United States. Report West Virginia University, 15 May 2014
5. Mayer A et al, Off-Highway Exhaust Gas After-Treatment: Combining Urea-SCR, Oxidation Catalysis, and Traps; SAE 930363
6. Baiker et al (1990) Amorphous chromia for low-temperature selective catalytic reduction of nitric oxic. I+EC Research 29

7. Walker AP et al (2002) The use of the continuously regeneration trap (CRT) to control particulate emissions. SAE 2002-01-1271
8. Stieglbauer H (2005) STARFILTER. Deutsches Patent DE 10 2005 062 050.7
9. Mayer A, Nashar I et al (2008) Particle filter retrofit for all diesel engines. Haus der Technik 97; Expert-Verlag. ISBN 978-3-8169-2850-8
10. Emitec Heizkat, https://www.emitec.com
11. QuatCAT/CERYX: Moore R, Gottesfeld S, Zelenay P (1999) A comparison between direct-methanol and direct-hydrogen fuel cell vehicles. SAE Technical Paper 1999-01-2914. https://doi.org/10.4271/1999-01-2914
12. Brück R et al (2006) Emitec DOC recuperation. Wiener Motorsymposium
13. Czerwinski J, Zimmerli Y, Mayer A, D'Urbano G, Zürcher D (2015) Emission reduction with diesel particle filters with SCR coating (SDPF). Emission Control Sci Technol 1(2):152–166

Chapter 9
Evaluation of Next-Generation SCR Concepts for Heavy-Duty Applications by Using Catalytic Simulation

Tae Joong Wang

Nomenclature

•

a_c	Surface area per unit reactor volume (m^2/m^3)
C_i	Molar concentration of species i (mol/m^3)
E_i	Activation energy of species i (J/mol)
$E_d{}^o$	Activation energy of NH_3 desorption rate at zero coverage (J/mol)
k_i	Reaction rate constant of species i (mol $K/m^2/s$)
$k_i{}^o$	Pre-exponential factor (mol $K/m^2/s$)
R_i	Reaction rate of species i ($mol/m^2/s$)
R_u	Universal gas constant (J/mol/K)
S	Active surface site
t	Time (s)
u_D	Superficial velocity (m/s)
x	Axial position (m)

Greek symbols

α	Parameter for NH_3 surface coverage dependence
ε	Porosity of the monolithic substrate
η	SCR efficiency

T. J. Wang (✉)
Aftertreatment System Development Part, Engine Business Group, HYUNDAI DOOSAN INFRACORE (Co., Ltd.), 489 Injung-ro, Dong-gu, Incheon 22502, Republic of Korea
e-mail: taejoong.wang@hyundai-di.com

P. A. Lakshminarayanan and A. K. Agarwal (eds.), *Handbook of Thermal Management of Engines*, Energy, Environment, and Sustainability,
https://doi.org/10.1007/978-981-16-8570-5_9

θ Surface coverage (non-dimensional)

Subscripts

a Adsorption
d Desorption
g Gas phase
p Pore or void fraction
s Solid phase

9.1 Introduction

The progress of powertrain electrification is extremely fast in the passenger-car automobile industry and this trend is expanding into heavy-duty vehicle sectors. However, it is expected that a much longer time would be required for electricity to be a mainstream power source in heavy-duty sectors, especially in off-road applications such as construction, agricultural and industrial equipment. Therefore, in near future, diesel engines would still play an important role as a primary power source in various on- and off-road applications thanks to their superior features in terms of efficiency and durability among internal combustion engines [1].

Major challenges for diesel engines are the tailpipe emissions of criteria pollutants as well as greenhouse gases. Emissions legislation has been increasingly stringent over the world. Especially, a strong focus has been on reducing nitrogen oxides (NO_x) as well as particulate matter (PM) emissions [2–5]. In recent years, carbon dioxide (CO_2) emission is also being tightly regulated [6, 7]. These regulation drivers pose a significant technical challenge due to the trade-off nature of NO_x and CO_2 emissions.

9.1.1 Heavy-Duty NO_xEmission Regulations

Although the current NO_x emission standards are already quite tough, it is expected that these NO_x limits would be further tightened in near future. In the United States (US), the current NO_x limit for on-road heavy-duty diesel engines is 0.2 g/bhp-h (i.e., 0.27 g/kWh) and this NO_x limit is possibly expected to be much lower to 0.05 g/bhp-h from 2024 and to 0.02 g/bhp-h from 2027 [2]. This corresponds to a 75% reduction in 2024 and a 90% reduction in 2027 compared to the current NO_x limit. In the European Union (EU), the current NO_x emission limit of Euro VI for on-road heavy-duty diesel engines is 0.4 g/kWh in steady-state mode and 0.46 g/kWh in transient mode [3]. Similar to the US, the post Euro VI NO_x standard is also predicted to be much lower than the current level.

Also, next-step emissions legislations are expected to call for a tough real-driving NO_x limit. It has been well known that the level of NO_x emission in real-driving conditions could be much higher than those measured in controlled operations or laboratory conditions [8–11].

Also, the same trend is observed for off-road applications. In the US, Tier 4 standard claims a substantial NO_x reduction, which needs a 90% NO_x reduction compared to its previous Tier 3 [4]. In the EU, Stage IV is a similar regulation to the US Tier 4 [5].

9.1.2 NO_xControl Technologies (EGR Versus SCR)

As a useful method to reduce NO_x emissions from heavy-duty diesel engines, exhaust gas recirculation (EGR) technology has been widely employed. However, an EGR system has a negative impact on an engine system in various aspects such as durability, maintenance, cost, etc. since it requires additional components including a dedicated circuit, cooler, valve, and controller. Also, an EGR system generally gives a negative impact on PM emission due to the recirculation of exhaust gas into engine cylinders.

On the other hand, as a post-combustion technology to comply with NO_x emission regulations in various heavy-duty applications, ammonia/urea selective catalytic reduction (SCR) removes NO_x through catalytic reactions using ammonia as a reducing agent has been also broadly used [12–20]. To meet increasingly stringent NO_x limits, the NO_x removal efficiency of SCR systems has been continuously increased with the improvement of related technologies regarding catalysts, sensors, controls, etc. The cycle-averaged NO_x conversion efficiency of heavy-duty SCR systems has been raised from 94% in 2012 to 96% in 2014, and it has reached around 98% in 2016 [21, 22].

9.1.3 Current Trends in Heavy-Duty Diesel Engine and SCR System

In recent years, the improvement of SCR technology has been impressive. Thanks to these highly efficient SCR systems, modern heavy-duty diesel engines can eliminate an EGR system while meeting rigorous NO_x limits. Several heavy-duty diesel engine manufacturers such as IVECO [23] and DOOSAN INFRACORE [24] have successfully complied with on-road Euro VI emissions regulation by adopting a high-efficiency SCR technology without an EGR system. However, it should be noted that a relatively large number of manufacturers have employed both EGR and SCR technologies to meet Euro VI standards.

More recently, in the off-road sector, the EU Stage V emissions regulation came into effect in 2019. With the advancement of high-efficiency SCR technologies, an increased number of manufacturers such as CUMMINS [25] and SCANIA [26] as well as IVECO and DOOSAN INFRACORE have met off-road Stage V regulation by employing 'SCR only' architecture without an EGR system.

9.1.4 The Perspective of Next-Generation SCR Technology

There is a consensus that current SCR technology should be further improved to meet next-step NO_x standards, in which a strong focus will be on reducing NO_x in low-temperature operations. Low catalytic activity and DEF evaporation are the major challenges for low-temperature NO_x control so that temperature rise would be a crucial remedy to further improve current SCR technology. However, temperature rise normally causes the deterioration of fuel economy. Moreover, engine-side techniques for temperature rise could give a negative impact on the dynamic performance of vehicles or equipment. Therefore, an appropriate thermal management strategy would be highly important in next-generation SCR systems.

9.1.5 Use of Catalytic Simulation in the Design of SCR Systems

Over the decades, numerical simulations have been extensively employed in the research and development of SCR technologies [27–37]. They are mainly divided into one- and three-dimensional simulations according to their primary usage. Evaluating the performance and potential of an SCR system as early as possible is a very important element in the new product development process. One-dimensional catalytic simulation is a very useful tool in the concept design of a new SCR system since various system configurations and their proper control strategies are required to be agilely evaluated at this early development stage. On the other hand, three-dimensional simulation is proper for the detailed design of the hardware.

9.1.6 Next-Generation SCR Concepts Studied in This Article

This article is intended to provide a quantitative insight into the answers to the questions: "What is the best next-generation SCR concept to meet future heavy-duty NO_x limits?" and "What is a proper thermal management strategy for next-generation SCR systems?"

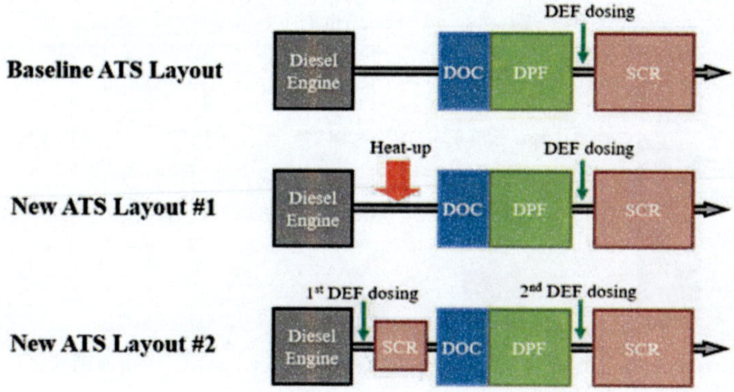

Fig. 9.1 Next-generation ATS layouts

In this study, the baseline diesel engine and its exhaust aftertreatment system (ATS) meet heavy-duty Euro VI-d emission standards and they have been manufactured by DOOSAN INFRACORE Co. Ltd. This system features 'No-EGR' and 'SCR-only' architecture. As shown in Fig. 9.1, the baseline ATS layout is composed of a diesel oxidation catalyst (DOC) and a diesel particulate filter (DPF) followed by an SCR catalyst. Also, a diesel exhaust fluid (DEF) injection system and a DEF mixer were located between the DPF and SCR catalyst.

Figure 9.1 illustrates two different next-generation ATS layouts studied in this article. The first kind of next-generation ATS denoted as 'New ATS Layout #1' is an addition of a heat-up function to the baseline ATS. This heat-up can be obtained by using various methods. Dedicated heating devices such as an electrical heating system or a burner system can be used. Also, engine-side techniques can be applied, which include post fuel injection, timing shift of fuel injection, intake or exhaust throttling, exhaust cam phasing, and cylinder de-activation (Chap. 2).

The second kind of next-generation ATS denoted as 'New ATS Layout #2' is an addition of a close-coupled SCR catalyst with its exclusive DEF dosing system. There is a clear trend that this dual-dosing SCR system is becoming a mainstream technology pathway to prepare next-step NO_x limits. This is mainly because a close-coupled SCR can utilize the thermal energy of exhaust gas before it cools down through DOC and DPF, especially in low-temperature operating conditions.

9.1.7 Next-Generation NO_xReduction Target

Before evaluating the potential of 'New ATS Layout #1' and 'New ATS Layout #2' shown in Fig. 9.1, a fundamental question is raised on "What is the NO_x conversion efficiency target of each ATS layout?

Figure 9.2 shows three different NO_x limit scenarios which are now being considered as next-step emissions regulations. Based on the heavy-duty on-road Euro VI

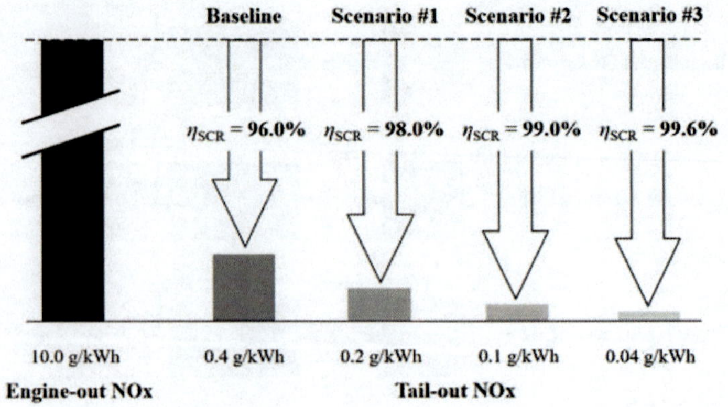

Fig. 9.2 The target of tail-out NO_x and corresponding SCR efficiency

NO_x limit of 0.4 g/kWh, three different NO_x limits of 0.2 g/kWh, 0.1 g/kWh, and 0.04 g/kWh have been targeted in this article. The reduction rate of each NO_x limit respectively indicates 50%, 75%, and 90% compared to the Euro VI NO_x limit. It should be noted that, in actual product development situations, an engineering margin for tail-out NO_x emission is considered. This means that those SCR efficiency targets would be higher than the levels presented in Fig. 9.2.

This article provides a quantitative estimation of how much the baseline SCR efficiency can be improved by adopting 'New ATS Layout #1' and 'New ATS Layout #2'. Based on the result, the readers of this article would have an opportunity to deliberate on the best ATS concept and its proper thermal management strategy to meet the requirement of their next-generation SCR systems.

9.2 Catalytic Simulation of SCR System

One of the important purposes of this article is to show that the catalytic simulation technique is a quite useful tool in the concept design of SCR systems. One-dimensional SCR catalytic simulation combined with thermal flow solver is highly applicable to the concept design of SCR products. For this, transient SCR simulations were carried out using the modified version of a FORTRAN in-house code which was originally developed by Wang et al. [38]. This one-dimensional computer code was well-validated under various steady and transient conditions as reported in the previous works [38–40]. This in-house simulation code has been used in the design and development of various SCR system products manufactured by DOOSAN INFRACORE Co. Ltd. [41, 42].

To deepen theoretical understanding of the numerical simulation of an SCR system, this section describes its mathematical modeling procedure along with some comments on kinetic parameter calibration for its proper use.

9.2.1 Modeling of Ammonia Adsorption/desorption

Due to inherent toxicity and handling problems with pure ammonia, urea-water solution (UWS) is practically used for mobile SCR applications. Once UWS spray is injected into a hot exhaust gas stream before an SCR catalyst, water content first evaporates from UWS. Then, ammonia (NH_3) is generated in-situ through thermal-decomposition of urea and hydrolysis of isocyanic acid as follows [43–45].

$$(NH_2)_2CO \rightarrow NH_3 + HNCO \tag{9.1}$$

$$HNCO + H_2O \rightarrow NH_3 + CO_2 \tag{9.2}$$

Modeling of ammonia adsorption/desorption over an SCR catalyst is an essential part of the overall SCR catalytic reaction modeling because ammonia plays a primary role in deNO$_x$ processes [46–48]. It is well known that the deNO$_x$ performance and the dynamics of an SCR catalyst are governed by the reactivity of adsorbed ammonia [49]. Hence, realistic modeling of ammonia adsorption and desorption processes is needed for an accurate prediction of SCR performance.

In a urea/ammonia SCR system, ammonia molecules are chemically adsorbed both on active metal sites and on Brönsted acid sites [50]. It is known that several ammonia molecules can be bound to one active metal site. Komatsu et al. [51] reported that each copper ion can coordinate up to four ammonia molecules. Based on that, Olsson et al. [52] presented an SCR model with detailed surface descriptions. However, for model simplicity, the assumption that there exists only one kind of surface site (denoted by S in Eq. 9.3) on which gaseous ammonia molecules are adsorbed or desorbed with 1:1 adsorption stoichiometry, has been widely adopted and reported in the literature [53–56].

Adsorption and desorption of ammonia on active sites are described as the following forward and backward reactions.

$$NH_3 + S \rightleftarrows NH_3 \cdot S \tag{9.3}$$

Here, the symbol S denotes active surface site, and ammonia adsorption and desorption rates are respectively described as

$$R_a = k_a C_{NH_3}\left(1 - \theta_{NH_3}\right) \tag{9.4}$$

$$R_d = k_d \theta_{NH_3} \tag{9.5}$$

where each rate constant is formulated by an Arrhenius form such that

$$k_i = k_i^o \exp\left(-\frac{E_i}{R_u T_s}\right), i = a, d \tag{9.6}$$

Regarding the determination of the adsorption/desorption activation energies, E_i in Eq. 9.6, it is reported from Ref. [50, 51] that ammonia adsorption is assumed to occur through a non-activated process so that the adsorption activation energy, E_a is set to zero. On the other hand, the ammonia desorption process is assumed to be strongly dependent on the surface condition so that the desorption activation energy, E_d is represented by Tempkin type expression in which E_d decreases linearly with increasing ammonia surface coverage, θ_{NH3} as follows.

$$E_d = E_d^0(1 - \alpha\theta_{NH_3}) \tag{9.7}$$

Mass balance of ammonia in the gas phase is described as

$$\varepsilon_p \frac{\partial C_{g,NH_3}}{\partial t} = -u_{D,p}\frac{\partial C_{g,NH_3}}{\partial x} + \alpha_{c,p}(R_d - R_a) \tag{9.8}$$

where the left-hand-side term indicates unsteady accumulation of gas-phase ammonia within monolith reactor-averaged control volume which consists of both pores and solid parts. On the right-hand side, the first term physically means convective flux of gas-phase ammonia. Note that the superficial velocity (or often referred to as Darcian velocity), $u_{D,p}$ is equal to pore velocity (i.e., actual velocity through channels) multiplied by the void fraction of the monolith reactor. Also, the second term accounts for the source or the sink of gas-phase ammonia through its adsorption or desorption on active sites. There must be a careful identification in defining catalytic active site because it is diversely appeared in the literature [48, 51].

Mass balance of ammonia on a solid surface is written as

$$\frac{\partial\theta_{NH_3}}{\partial t} = R_a - R_d \tag{9.9}$$

Equation 9.9 physically describes that the difference between ammonia adsorption and desorption rates makes a temporal change of ammonia accumulation on active sites.

9.2.2 *Modeling of SCR Reactions*

As a next step, several important reactions occurring in SCR catalysts are modeled and solved numerically. Main deNO$_x$ reactions under a typical diesel exhaust environment can be summarized as follows [52, 53].

$$4NH_3 + 4NO + O_2 \rightarrow 4N_2 + 6H_2O \tag{9.10}$$

$$4NH_3 + 2NO + 2NO_2 \rightarrow 4N_2 + 6H_2O \tag{9.11}$$

$$4NH_3 + 3NO_2 \rightarrow 7/2N_2 + 6H_2O \tag{9.12}$$

9.2.2.1 Standard SCR Reaction

Over a wide range of operations with typical engine-out exhaust gas, nitrogen monoxide (NO) is a major NO_x component. It is well known that the SCR reaction of NO by NH_3 (i.e., Eq. 9.10) which consumes NH_3 and NO with 1:1 stoichiometry is a dominant deNO_x reaction pathway in ammonia/urea SCR catalysts. Therefore, the modeling of this standard SCR reaction is essential to establish an SCR model set.

9.2.2.2 Fast SCR Reaction

It is also known that the deNO_x reaction consuming equal amounts of NO and nitrogen dioxide (NO_2) with NH_3 (i.e., Eq. 9.11) is much faster than Eqs. 9.10 and 9.12 [57, 58]. Therefore, this fast SCR reaction is considered a practical means to increase the performance of the ammonia/urea SCR system. Therefore, the fast SCR reaction consuming equimolar NO and NO_2 plays an important role in constructing an SCR model set.

9.2.2.3 Slow SCR Reaction

In general, the deNO_x activity of zeolite- or vanadium-based SCR catalysts which are widely employed for on- and off-road mobile applications is enhanced with an increase in the ratio of NO_2 over NO_x. This is because the reaction pathway consuming NO and NO_2 simultaneously (i.e., Eq. 9.11) is very fast. However, as NO_2/NO_x ratio becomes too high, the deNO_x efficiency becomes lower again since the reaction pathway removing NO_2 only (i.e., Eq. 9.12) is quite slow.

Koebel et al. [59] reported that, for a vanadium-based catalyst, the ratio of NO_2/NO_x should not exceed 0.5 to maximize deNO_x performance, while Baik et al. [60] presented that the optimum NO_2/NO_x feed ratio for the best deNO_x activity is 0.75 for both Fe-ZSM5 and Cu-ZSM5 catalysts and 0.5–0.75 for V_2O_5/TiO_2 catalyst. Also, Baik et al. [60] concluded that the optimum NO_2/NO_x feed ratio depends on the kind of catalyst and reaction temperature. Therefore, for an accurate prediction of SCR performance under various NO and NO_2 compositions and operating conditions, this slow SCR reaction of NO_2 by NH_3 (i.e., Eq. 9.12) should be included in an SCR model set.

9.2.2.4 NH_3 Oxidation

Accurate prediction of NH_3 concentration in an SCR catalyst is essential for a reliable SCR model since NH_3 relates to various deNO$_x$ reactions as well as NH_3 slip. Therefore, the modeling of NH_3 oxidation with O_2 as described in Eq. 9.13 is included in an SCR model set. An NH_3 oxidation capability of general SCR catalysts is quite weak up to quite high temperatures. Wurzenberger and Wanker [61] reported that NH_3 oxidation over SCR catalysts is not pronounced until temperature reaches as high as 500 °C.

$$4NH_3 \cdot S + 3O_2 \rightarrow 2N_2 + 6H_2O \tag{9.13}$$

9.2.3 *Kinetic Parameter Calibration*

In the mathematical modeling and numerical simulation of a catalytic reactor system, reaction chemistry is one of the most important factors determining soundness as well as accuracy. But, since catalytic reactions are affected by numerous factors, their chemical kinetics has a case-by-case nature depending on individual system configuration. Thus, the direct re-use of kinetic parameters borrowed from other sources is required to be carefully reviewed. In many cases, it cannot be justified. Therefore, calibration of reaction kinetic parameters with emissions measurement data is often required, which takes an important part to obtain an accurate SCR simulation result.

9.3 The Emissions Test Result of the Baseline SCR System

In this article, the baseline diesel engine and exhaust aftertreatment system meet heavy-duty on-road Euro VI-d emissions regulation. This system features 'No-EGR' and 'SCR-only' architecture, which is manufactured by DOOSAN INFRACORE Co. Ltd. The specifications of the baseline diesel engine and SCR catalyst are summarized in Tables 9.1 and 9.2 respectively.

Engine bench tests were conducted over both WHTC hot and cold modes with measuring NO$_x$ emissions at both inlet and outlet of the baseline SCR catalyst. In this study, the first run of the WHTC cold test was followed by the second run of the WHTC hot test.

The WHTC test is a transient engine dynamometer schedule defined by the global technical regulation (GTR). Two representative test cycles of WHTC with both hot and cold start and world harmonized stationary cycle (WHSC) with hot start have been created to cover typical driving conditions in the EU, USA, Japan, and Australia. These WHTC test requirements were adopted for the first time by the Euro VI

Table 9.1 Specification of the baseline diesel engine

Parameter	Value
Bore × Stroke	100 mm × 125 mm
Number of cylinders	6 (in-line)
Displacement	5.89 dm^3
Rated power	141 kW @ 1900 rpm
Maximum torque	804 Nm @ 1400 rpm
Fuel injection	Common-rail direct injection
Aspiration	Turbocharged/aftercooled
EGR	None
Turbocharger	Waste-gate type
Application	Heavy-duty trucks and buses

Table 9.2 Specification of the baseline SCR catalyst

Parameter	Value
Diameter	266.7 mm
Length	355.6 mm
Volume	19.87 dm^3
Cell density	62 cm^{-2} (400 cpsi)
Substrate wall thickness	101.6 μm
Catalyst type	Cu-zeolite

emission regulation for heavy-duty engines. The WHTC is a transient test of 1800s duration, with several motoring segments [62]. The Euro VI emission certification requires a weighted sum of hot and cold mode tests with a 10 (±1) minute soak period between them, where the hot mode takes a weighting factor of 86% and the cold mode occupies the remaining 14%.

The diesel fuel used in the current emissions test is an ultra-low sulphur diesel containing 6.4 wt ppm sulphur and no oxygen, which was supplied from GS Caltex Co. Ltd. The Cetane number of this fuel is 55.4, and its density is 0.8232 g/cm^3 at 15 °C.

Figures 9.3, 9.4 and 9.5 display the measurement results of engine performance over WHTC cold and hot modes, which include engine speeds, brake torques, and exhaust mass flow rates.

Figures 9.6, 9.7 and 9.8 present several input conditions for the deNO$_x$ performance of the baseline SCR system, which includes SCR-inlet NO$_x$ concentrations, exhaust gas temperatures at SCR catalyst, and DEF dosing rates. Note that the measured exhaust gas temperatures at an SCR catalyst shown in Fig. 9.7 are the average gas temperature of the inlet and outlet of the SCR catalyst. Also, it is worthwhile to note the injection profiles of the reducing agent. The DEF is supplied during the entire test period of WHTC hot mode. On the other hand, the DEF supply begins

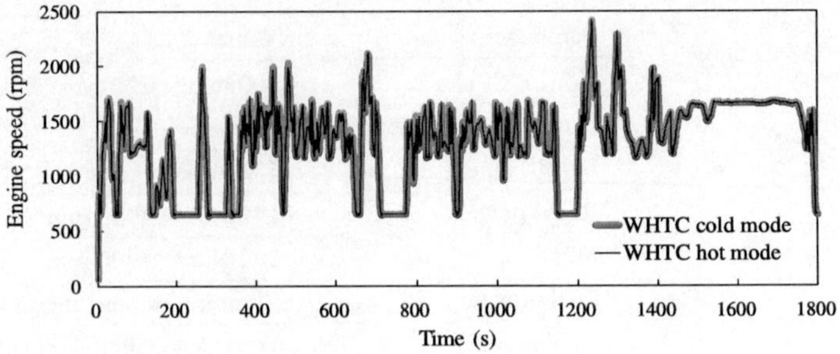

Fig. 9.3 Measured engine speeds over WHTC cold and hot modes

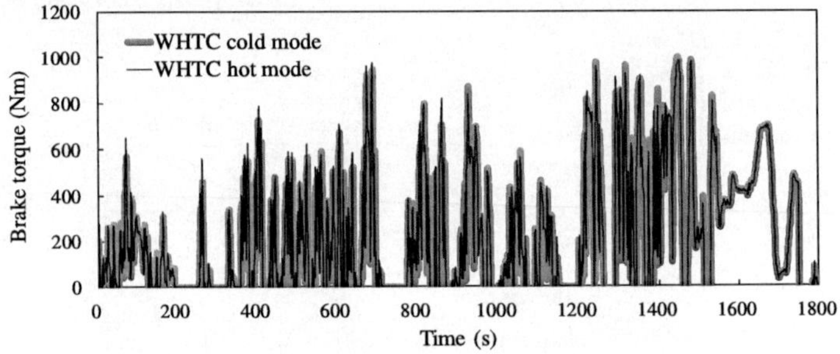

Fig. 9.4 Measured brake torques over WHTC cold and hot modes

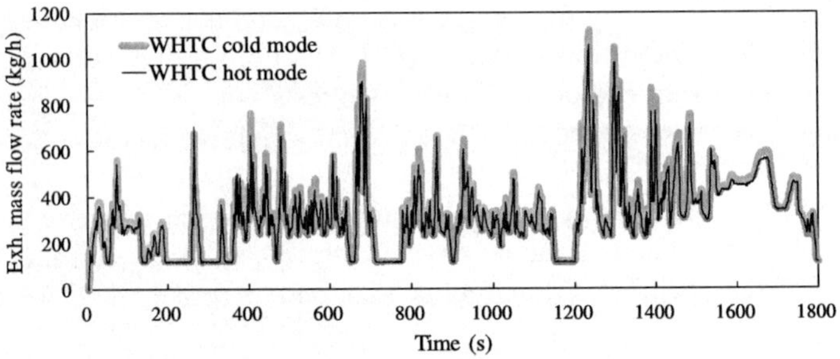

Fig. 9.5 Measured exhaust mass flow rates over WHTC cold and hot modes

from 473 s for WHTC cold mode since the exhaust gas temperature at SCR inlet becomes higher than 180 °C which is the initiation temperature of DEF dosing.

Figure 9.9 shows the measured NO_x concentrations at the SCR outlet. For WHTC cold mode, NO_x is emitted from 0 s since the DEF is not supplied until 473 s. On the other hand, for WHTC hot mode, NO_x conversion is initiated from 0 s since the DEF is supplied from the beginning of the mode operation. The measure of NO_x conversion performance of the baseline SCR system is summarized in Table 9.3. The WHTC-averaged SCR efficiency of this baseline SCR system is 96.54%. Note that

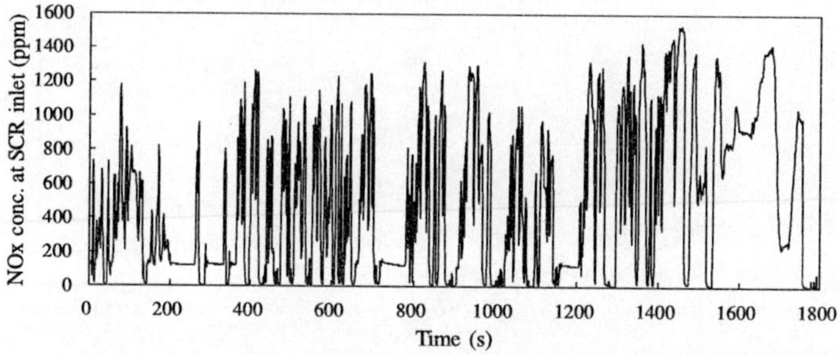

Fig. 9.6 Measured NO_x concentrations at SCR inlet over WHTC cold and hot modes

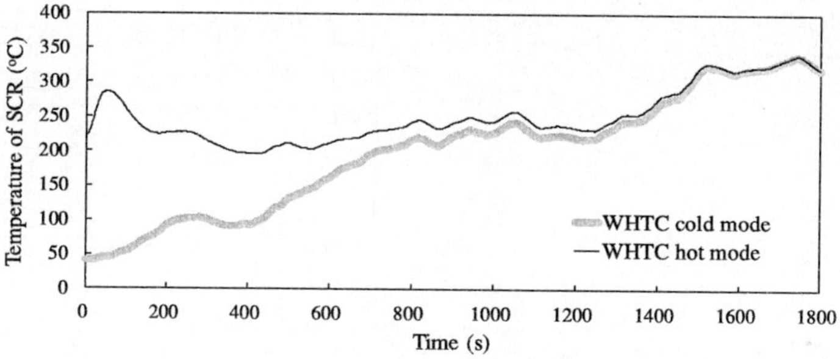

Fig. 9.7 Measured exhaust gas temperatures at SCR catalyst over WHTC cold and hot modes

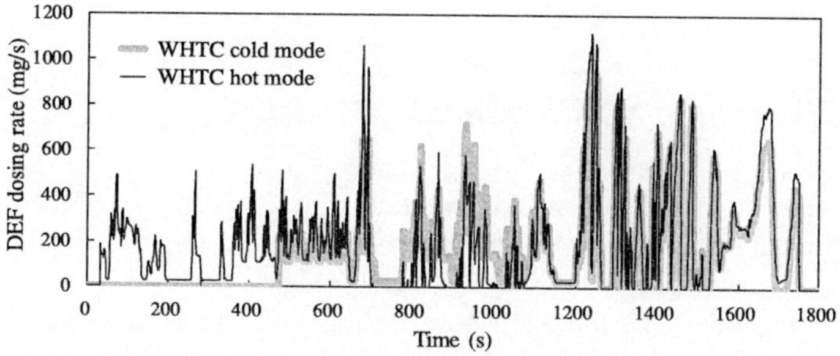

Fig. 9.8 Measured DEF dosing rates over WHTC cold and hot modes

this WHTC-averaged SCR efficiency was weight-averaged with a weighting factor of 86% for hot mode and 14% for cold mode.

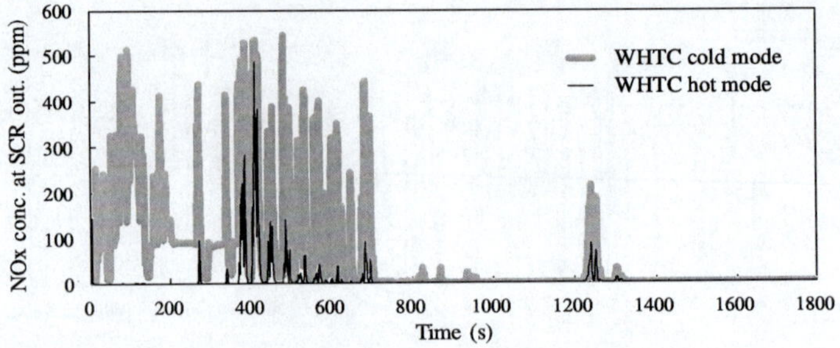

Fig. 9.9 Measured NO_x concentrations at SCR outlet over WHTC cold and hot modes

Table 9.3 NO_x conversion performance of the baseline SCR system

Parameter	Unit	WHTC cold mode	WHTC hot mode
Cycle work	kWh	17.29	17.42
Engine-out NO_x	g	160.3	153.1
Tail-out NO_x	g	24.86	2.294
Engine-out NO_x	g/kWh	9.269	8.787
Tail-out NO_x	g/kWh	1.438	0.1317
SCR efficiency	%	84.48	98.50
WHTC-avg. tail-out NO_x	g/kWh	0.3146	
WHTC-avg. SCR efficiency	%	96.54	

9.3.1 *Validation of SCR Simulation Model*

Confirmation of the validity of an SCR simulation by comparing simulation data with measurement data is an important task as it is essential to obtain a reliable concept design result in developing a new SCR system product.

Figures 9.10, 9.11, 9.12 and 9.13 show the comparison result of SCR simulation data with measurement data. Figures 9.10 and 9.11 present WHTC hot mode results, while Figs. 9.12 and 9.13 exhibit WHTC cold mode results. Instantaneous SCR-out NO_x concentration trends and peak values are in good agreement for hot mode, while there is some discrepancy for cold mode.

Note that those simulation results of Figs. 9.10, 9.11, 9.12 and 9.13 were obtained by applying an appropriate SCR kinetic parameter set. Figure 9.13 shows there is scope for improving the simulation by introducing more accurate parameters. An exemplary work of the estimation of a new SCR kinetic parameter set can be referred from Wang et al. [38].

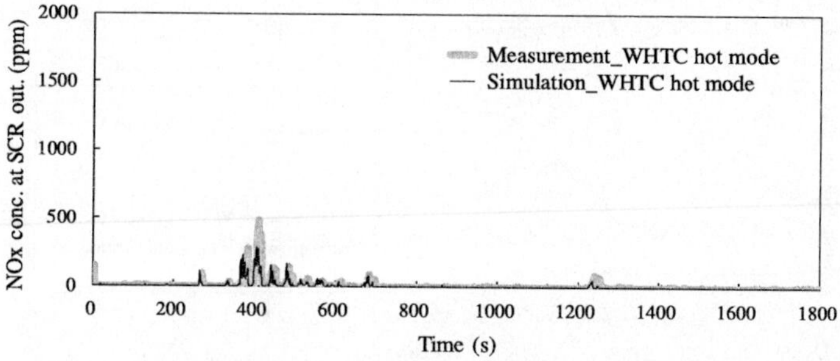

Fig. 9.10 Comparison of simulated and measured instantaneous SCR-out NO_x concentration over WHTC hot mode

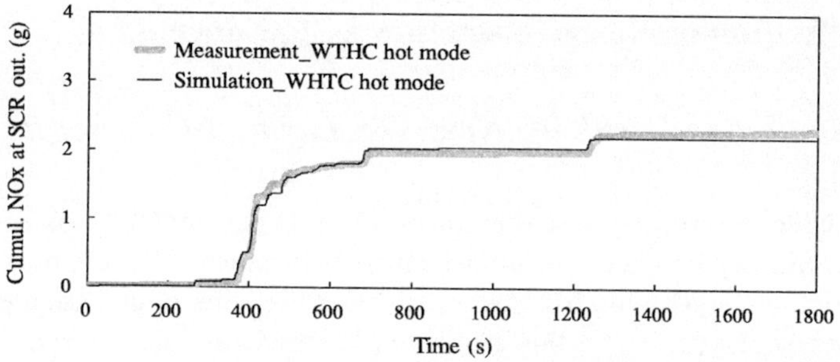

Fig. 9.11 Comparison of simulated and measured cumulative SCR-out NO_x concentration over WHTC hot mode

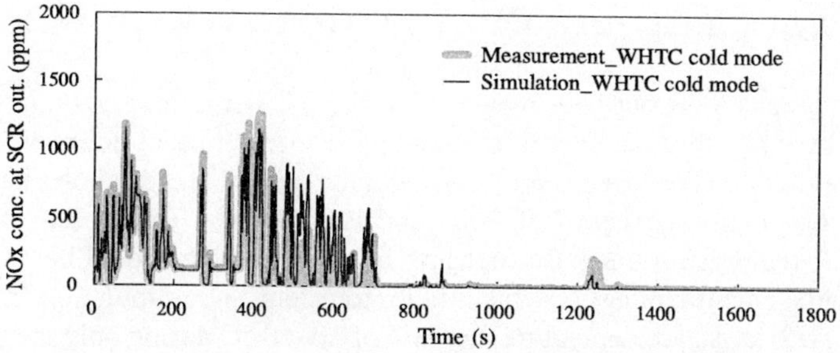

Fig. 9.12 Comparison of simulated and measured instantaneous SCR-out NO_x concentration over WHTC cold mode

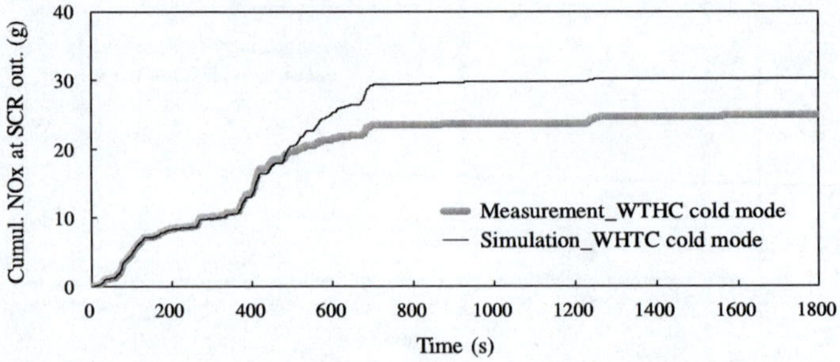

Fig. 9.13 Comparison of simulated and measured cumulative SCR-out NO_x concentration over WHTC cold mode

9.4 Evaluation of Next-Generation SCR Concepts

9.4.1 Simulation Result of 'New ATS Layout #1' Concept

After reviewing the profiles of SCR temperature and SCR-out NO_x concentration of the baseline SCR system, it can be inferred that the increase of SCR temperature and DEF injection amount would become an effective measure to increase the baseline SCR efficiency. Therefore, in this article, SCR simulations have been conducted for an evaluation of how much the baseline SCR efficiency would be improved by increasing SCR temperature and DEF injection amount.

9.4.1.1 Effect of Temperature Increase at WHTC Hot Mode

The instantaneous SCR-out NO_x measured at WHTC hot mode shown in Fig. 9.10 reveals that temperature increase throughout the entire period of hot mode is not an efficient way. This is because most of the hot mode period already shows nearly zero SCR-out NO_x emission except 200–800 s and 1200–1400 s. Therefore, it is thought that an SCR temperature increase focusing on these periods would be an efficient thermal management strategy. In this article, transient SCR simulations have been performed with an SCR temperature increase of 10–60 °C during only these periods (i.e., 200–800 s and 1200–1400 s).

Table 9.4 summarizes the simulation results, which indicate how much the baseline SCR efficiency at WHTC hot mode would be improved by increasing the SCR temperature. Figure 9.14 shows the input temperature profiles, while Figs. 9.15 and 9.16 present the results of SCR simulations. As shown in Fig. 9.16, only a 10 °C increase gives a substantial reduction in SCR-out NO_x. Around half of cumulative NO_x has been decreased (i.e., from 2.20 to 1.18 g). Furthermore, a 60 °C increase leads to almost zero NO_x emission, which shows 0.05 g of cumulative SCR-out NO_x (i.e., 99.97% of SCR efficiency) at WHTC hot mode.

Table 9.4 Transient SCR simulation results: Improvement of SCR efficiency at WHTC hot mode with SCR temperature rise of 10–60 °C during 200 –800 s and 1200–1400 s (Cumulative SCR-in NO_x obtained from baseline simulation = 153.1 g)

Temperature increment (°C)	Cumulative SCR-out NO_x (g)	SCR efficiency (%)
Baseline	2.20	98.56
10	1.18	99.23
20	0.57	99.63
30	0.25	99.84
60	0.05	99.97

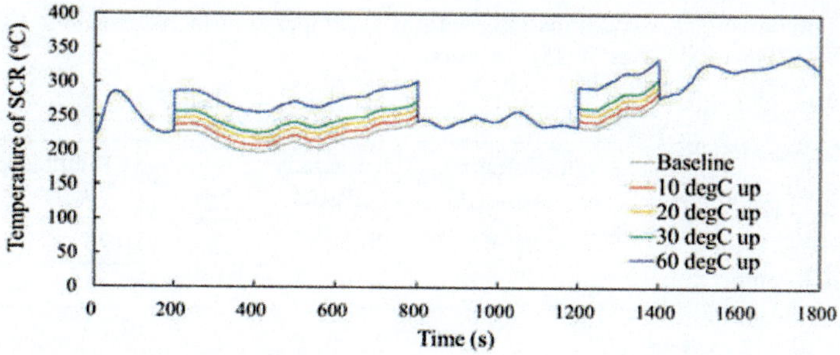

Fig. 9.14 Temperature profiles with SCR temperature increase with a 10–60 °C at 200–800 s and 1200–1400 s over WHTC hot mode

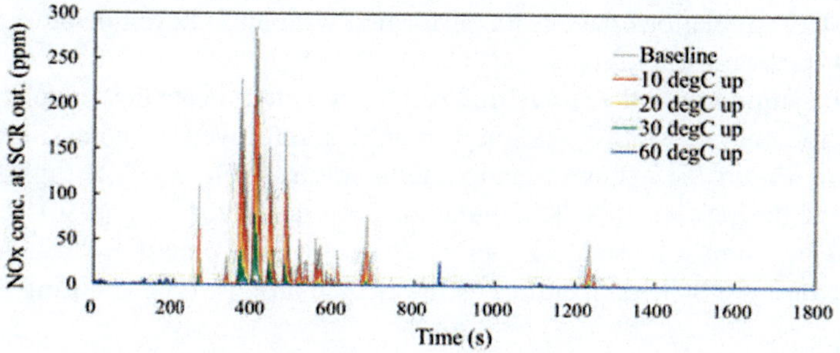

Fig. 9.15 NO_x concentrations at SCR outlet with SCR temperature increase with a 10–60 °C at 200–800 s and 1200–1400 s over WHTC hot mode

9.4.1.2 Effect of Temperature Increase at WHTC Cold Mode

Similar to the previous discussion for WHTC hot mode, the instantaneous SCR-out NO_x measured at WTHC cold mode shown in Fig. 9.12 also indicates that temperature increase throughout the entire period of cold mode is not an efficient way. Most of the hot mode period already exhibits nearly zero SCR-out NO_x emission except 0–710 s. Therefore, it is also thought that an SCR temperature increase focusing on this period would be an efficient thermal management strategy. In this article,

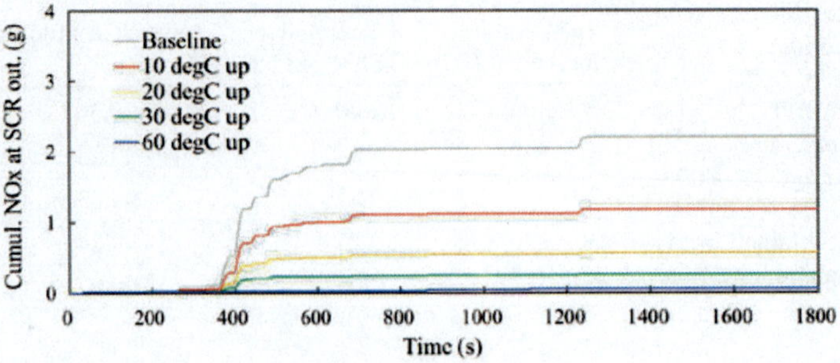

Fig. 9.16 Cumulative NO_x mass emissions with SCR temperature increase with a 10–60 °C at 200–800 s and 1200–1400 s over WHTC hot mode

Table 9.5 Transient SCR simulation results: Improvement of SCR efficiency at WHTC cold mode with SCR temperature rise of 30–120 °C during 0–710 s (Cumulative SCR-in NO_x obtained from baseline simulation = 160.3 g)

Temperature increment (°C)	Cumulative SCR-out NO_x (g)	SCR efficiency (%)
Baseline	30.22	80.26
30	18.81	87.71
60	13.42	91.23
90	8.44	94.49
120	4.78	96.88

transient SCR simulations have been performed with an SCR temperature increase of 30–120 °C during only this period (i.e., 0–710 s).

Table 9.5 summarizes the simulation results, which indicate how much the baseline SCR efficiency of WHTC cold mode would be improved by increasing the SCR temperature. Figure 9.17 shows the input temperature profiles, while Figs. 9.18 and 9.19 present the results of SCR simulations. As displayed in Fig. 9.18, a 30 °C increase shows almost no reduction in the SCR-out NO_x emission during 0–400 s. This is because the SCR temperature is lower than around 100 °C during this early

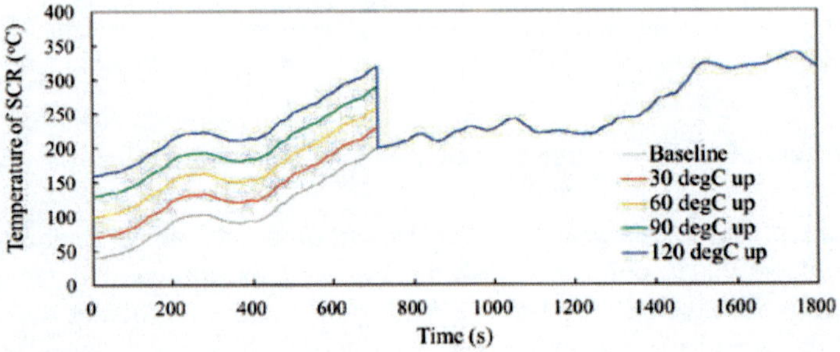

Fig. 9.17 Temperature profiles with SCR temperature increase with a 30–120 °C at 0–710 s over WHTC cold mode

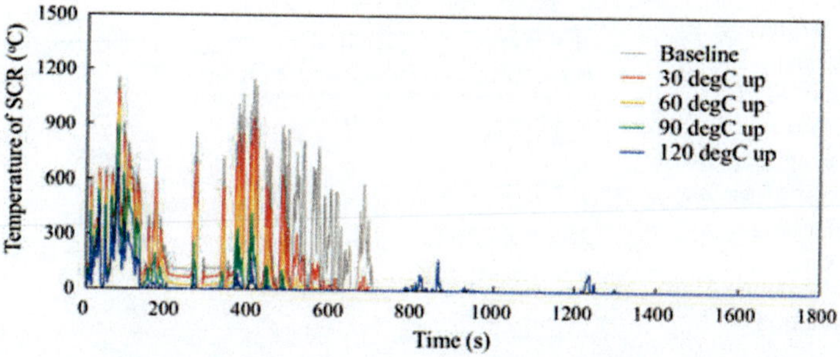

Fig. 9.18 NO_x concentrations at SCR outlet with SCR temperature increase with a 30–120 °C at 0–710 s over WHTC cold mode

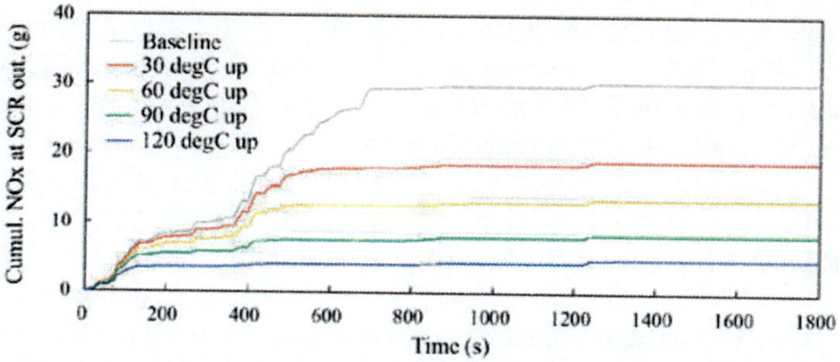

Fig. 9.19 Cumulative NO_x mass emissions with SCR temperature increase with a 30–120 °C at 0–710 s over WHTC cold mode

period of WHTC cold mode. Moreover, even a 120 °C increase leads to almost no reduction in the SCR-out NO_x emission during 0–50 s due to a very low SCR temperature. During this very early period of WHTC cold mode (i.e., 0–50 s), the SCR temperature was measured to be lower than around 50 °C.

Table 9.6 combines the SCR efficiency result of both WHTC hot and cold modes given in Tables 9.4 and 9.5. The SCR efficiency shown in Table 9.6 was weight-averaged with a weighting factor of 86% for hot mode and 14% for cold mode.

Table 9.6 Improvement of WHTC mode-averaged SCR efficiency which was weight-averaged with a weighting factor of 86% for hot mode and 14% for cold mode by applying a thermal management strategy with selective temperature rise (unit: %)

η_{cold} η_{hot}	98.56	99.23	99.63	99.84	99.97
80.26	96.00	96.57	96.92	97.10	97.21
87.71	97.04	97.62	97.96	98.14	98.25
91.23	97.54	98.11	98.45	98.63	98.74
94.49	97.99	98.57	98.91	99.09	99.20
96.88	98.33	98.90	99.24	99.42	99.53

Table 9.7 Transient SCR simulation results: Improvement of SCR efficiency at WHTC hot mode with DEF dosing increase of 10–50% from the baseline during 200–800 s and 1200–1400 s (Cumulative SCR-in NO_x obtained from baseline simulation = 153.1 g)

DEF dosing increment (%)	Cumulative SCR-out NO_x (g)	SCR efficiency (%)
Baseline	2.20	98.56
10	1.88	98.77
30	1.45	99.05
50	1.17	99.24

The hot-mode SCR efficiency could be improved up to 99.97 from the baseline of 98.56% by increasing SCR temperature with 60 °C, while the cold-mode SCR efficiency could be raised to 96.88 from the baseline of 80.26% by increasing SCR temperature with 120 °C. As a consequence, the WHTC-averaged SCR efficiency could be increased up to 99.53% from the baseline of 96.00% by adopting the thermal management strategy that was described above.

9.4.1.3 Effect of Increase in DEF Dosing Amount

In this article, the impact of an increase in DEF dosing amount has been also evaluated through SCR simulations. For this, the DEF dosing rate of the WHTC hot mode shown in Fig. 9.8 was increased by 10% to 50% from the baseline. Also, similar to the temperature increase strategy that was previously discussed, this additional DEF supply was applied only during 200–800 s and 1200–1400 s over WHTC hot mode.

Table 9.7 summarizes the simulation results, which indicate how much the baseline SCR efficiency of WHTC hot mode would be improved by increasing the DEF dosing amount. Figure 9.20 shows the input NH_3 concentrations obtained from the additional DEF supply, while Figs. 9.21 and 9.22 present the results of SCR simulations over WHTC hot mode. As summarized in Table 9.7, even a 50% increase in

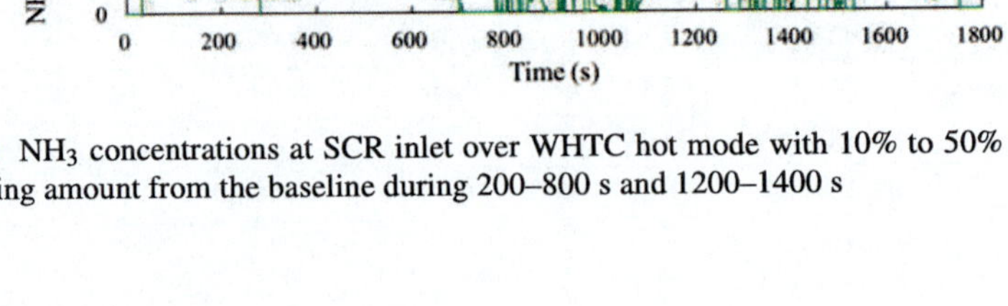

Fig. 9.20 NH_3 concentrations at SCR inlet over WHTC hot mode with 10% to 50% increase in DEF dosing amount from the baseline during 200–800 s and 1200–1400 s

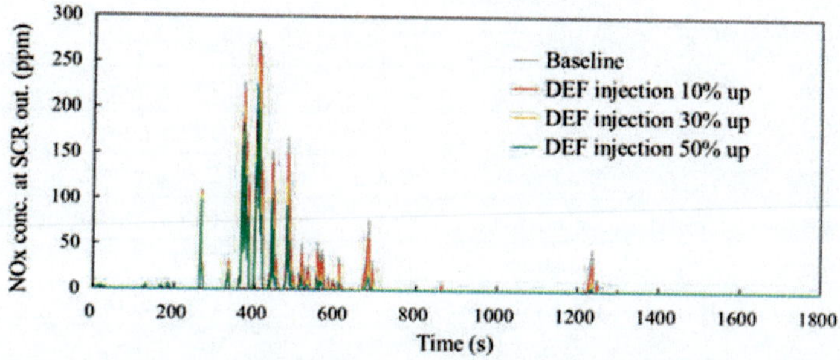

Fig. 9.21 NO_x concentrations at SCR outlet over WHTC hot mode with 10% to 50% increase in DEF dosing amount from the baseline during 200–800 s and 1200–1400 s

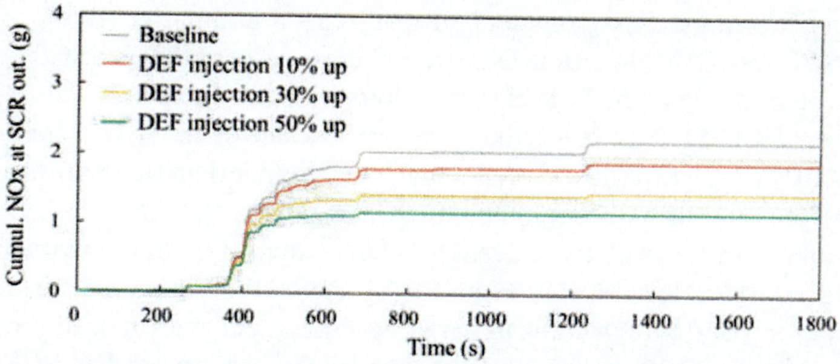

Fig. 9.22 Cumulative NO_x mass emission over WHTC hot mode with 10% to 50% increase in DEF dosing amount from the baseline during 200–800 s and 1200–1400 s

DEF dosing amount would reduce around half of the baseline cumulative SCR-out NO_x (i.e., from 2.20 to 1.17 g). Note that almost the same result could be obtained by increasing only 10 °C of SCR temperature.

9.4.2 Simulation Result of 'New ATS Layout #2' Concept

The second kind of next-generation ATS layout evaluated in this article is a dual-dosing SCR concept, which is denoted as 'New ATS Layout #2'. This ATS concept is an addition of a close-coupled SCR (CC-SCR) catalyst with its exclusive DEF dosing system to the baseline ATS. A close-coupled position gives a large benefit in terms of temperature. In low-temperature operating conditions such as cold start for the baseline ATS, the temperature of exhaust gas decreases as it passes through DOC and DPF located upstream of SCR due to the heat loss of exhaust gas to the substrates of DOC and DPF. This temperature drop makes NO_x emission control in low-temperature conditions difficult. Therefore, a dual-dosing SCR system that includes a CC-SCR with an exclusive DEF dosing system, as well as the main SCR

Table 9.8 Specification of CC-SCR

Parameter	Value
Diameter	143.8 mm
Length	177.8 mm
Volume	2.89 dm^3
Cell density	62 cm^{-2} (400 cpsi)
Substrate wall thickness	76.2 μm
Catalyst type	Cu-zeolite

system, would dramatically improve NO_x conversion efficiency at low-temperature conditions.

To evaluate the potential of a CC-SCR system, its steady-state emission test was carried out on an engine dynamometer bench. Table 9.8 summarizes the specification of the CC-SCR tested in this article. Note that the diameter of the CC-SCR catalyst is smaller than that of the main SCR catalyst summarized in Table 9.2. This is because the package of CC-SCR is restricted compared to the main SCR. Therefore, the substrate wall thickness of the CC-SCR catalyst is thinner than that of the main SCR to lower the pressure drop through the CC-SCR catalyst.

The simulation results of the current CC-SCR catalyst were compared with the steady-state emission test results. Figure 9.23 shows the validation result of CC-SCR simulation data, which indicates a good agreement with measurement data. As displayed in Fig. 9.23, the measurement results of the current CC-SCR catalyst reveal the NO_x conversion efficiency of 46.4% at 274 °C, 10.3% at 182 °C, and 6.4% at 172 °C.

Figure 9.24 depicts the exhaust temperature profiles measured upstream of both CC-SCR and main SCR catalysts in a transient operation with a cold start. Figure 9.24 indicates that the temperature of exhaust gas considerably drops through DOC and DPF during 0–700 s. The maximum temperature drop exceeds 100 °C during 100–200 s.

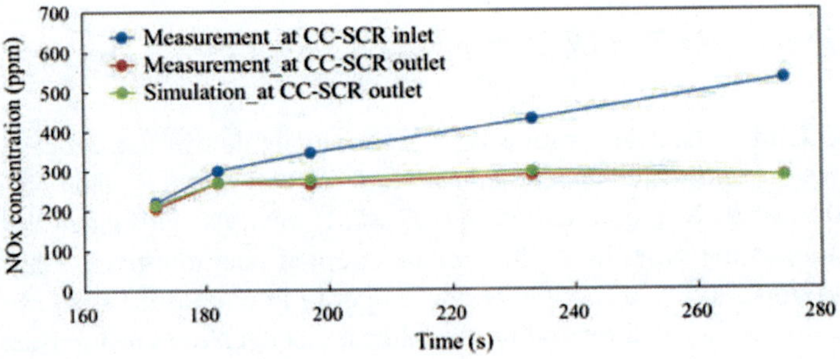

Fig. 9.23 Validation of CC-SCR simulation results with measurement data

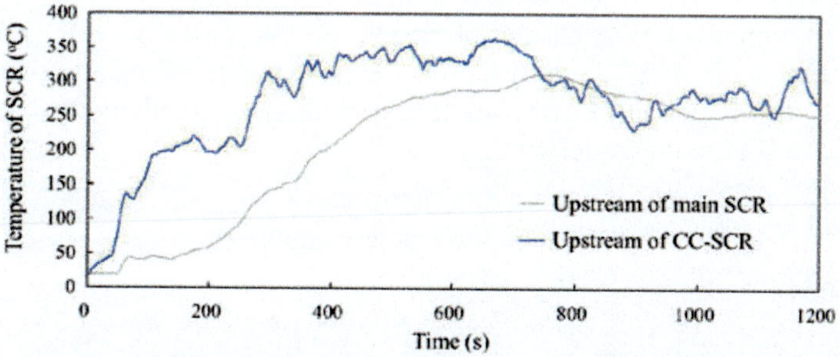

Fig. 9.24 Comparison of the measured exhaust temperatures at the upstream of both CC-SCR and main SCR catalysts in a transient operation with cold start

Based on this temperature drop result, transient CC-SCR simulations were performed over WHTC cold mode. Figures 9.25 and 9.26 display the transient simulation results of the CC-SCR catalyst. Figure 9.25 shows the instantaneous NO_x and Fig. 9.26 presents the cumulative NO_x mass emission at the CC-SCR outlet. For comparison, Figs. 9.25 and 9.26 also include the NO_x emission results of the main

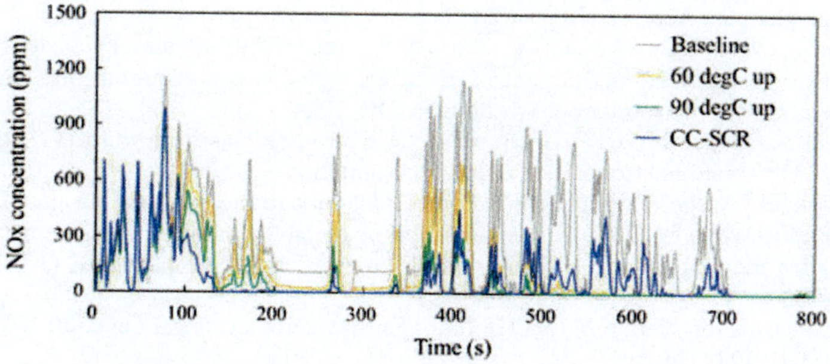

Fig. 9.25 NO_x concentrations at the outlet of CC-SCR and main SCR with a temperature increase of 60 and 90 °C at 0–710 s over WHTC cold mode

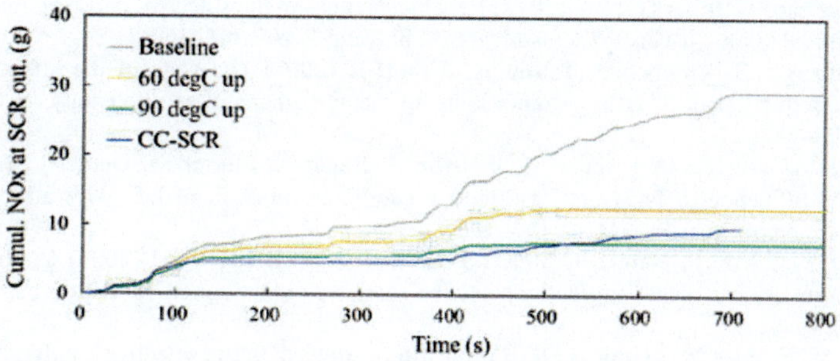

Fig. 9.26 Cumulative NO_x mass emissions of CC-SCR and main SCR with a temperature increase of 60 and 90 °C at 0–710 s over WHTC cold mode

SCR with its temperature increase of 60 and 90 °C. As presented in Figs. 9.25 and 9.26, the addition of CC-SCR catalyst gives the same impact of exhaust temperature increase between 60 and 90 °C in terms of the potential of cumulative NO_x emission reduction at WHTC cold mode.

References

1. Jääskeläinen H, Khair MK (2020) DieselNet technology guide, engine fundamentals, https://www.dieselnet.com/tech/diesel_engines.php. Accessed 31-Oct-2020
2. Emission Standards, United States: Heavy-Duty On-road Engines, https://www.dieselnet.com/standards/us/hd.php. Accessed 30-Jan-2021
3. Emission Standards, EU: Heavy-Duty Truck and Bus Engines, https://www.dieselnet.com/standards/eu/hd.php. Accessed 31-Oct-2020
4. Emission Standards, United States: Non-road Diesel Engines, https://www.dieselnet.com/standards/us/nonroad.php. Accessed 31-Oct-2020
5. Emission Standards, EU: Non-road Engines, https://www.dieselnet.com/standards/eu/nonroad.php. Accessed 31-Oct-2020
6. Emission Standards, United States: Heavy-Duty Vehicles: GHG Emissions & Fuel Economy, https://www.dieselnet.com/standards/us/fe_hd.php. Accessed 31-Oct-2020
7. Emission Standards, EU: Heavy-Duty Vehicles: GHG Emissions, https://dieselnet.com/standards/eu/hd_ghg.php. Accessed 31-Oct-2020
8. Muncrief R (2016) NO_x emissions from heavy-duty and light-duty diesel vehicles in the EU: comparison of real-world performance and current type-approval requirements. International Council on Clean Transportation: Washington, DC, USA
9. Lowell D, Kamakaté F (2012) Urban off-cycle NOx emissions from Euro IV/V trucks and buses. The International Council on Clean Transportation
10. Giechaskiel B, Vlachos T, Riccobono F, Forni F, Colombo R, Montigny F, Le-Lijour P, Carriero M, Bonnel P, Weiss M (2016) Implementation of portable emissions measurement systems (PEMS) for the real-driving emissions (RDE) regulation in Europe. JoVE (J Visual Exp) 118:e54753
11. EU Commission Regulation 2016/1718 (http://eur-lex.europa.eu/legal-content/EN/TXT/?uri=CELEX:32016R1718)
12. https://www.dieselnet.com/tech/cat_scr.html
13. Barreau M, Tarot M-L, Duprez D, Courtois X, Can F (2018) Remarkable enhancement of the selective catalytic reduction of NO at low temperature by the collaborative effect of ethanol and NH3 over silver supported catalyst. Appl Catal B: Environ 220:19–30
14. Wang J, Zhao H, Haller G, Li Y (2017) Recent advances in the selective catalytic reduction of NO_x with NH3 on Cu-Chabazite catalysts. Appl Catal B 202:346–354
15. Brandenberger S, Kröcher O, Tissler A, Althoff R (2008) The state of the art in selective catalytic reduction of NO_x by Ammonia using metal-exchanged zeolite catalysts. Catal Rev 50(4):492–531
16. Devadas M, Kröcher O, Elsener M, Wokaun A, Söger N, Pfeifer M, Demel Y, Mussmann L (2006) Influence of NO_2 on the selective catalytic reduction of NO with ammonia over Fe-ZSM5. Appl Catal B 67(3–4):187–196
17. Baik JH, Dae Yim S, Nam I-S, Mok YS, Lee J-H, Cho BK, Oh SH (2004) Control of NO x emissions from a diesel engine by selective catalytic reduction (SCR) with urea. Top Catal 30(1–4):37–41
18. Koebel M, Elsener M, Madia G (2001) Reaction pathways in the selective catalytic reduction process with NO and NO_2 at low temperatures. Ind Eng Chem Res 40(1):52–59
19. Koebel M, Elsener M, Kleemann M (2000) Urea-SCR: a promising technique to reduce NO_x emissions from automotive diesel engines. Catal Today 59(3–4):335–345

20. Busca G, Lietti L, Ramis G, Berti F (1998) Chemical and mechanistic aspects of the selective catalytic reduction of NO_x by ammonia over oxide catalysts: a review. Appl Catal B 18(1–2):1–36
21. Johnson T, Joshi A (2018) Review of vehicle engine efficiency and emissions. SAE Int J Engines 11(6):1307–1330
22. Johnson T (2014) Vehicular emissions in review. SAE Int J Engines 7(3):1207–1227
23. https://www.iveco.com/en-us/press-room/release/pages/techdayiveco.aspx
24. https://www.doosanengine.com/en/innovation/euro6-stage5
25. https://www.cummins.co.kr/en/service/training/stage-v-emissions
26. https://www.scania.com/uk/en/home/experience-scania/features/scania-eu-stage-v.html
27. Berg A, Widd A, Abildskov J et al (2017) Parameter estimation and analysis of an automotive heavy-duty SCR catalyst model. Chem Eng Sci 161:167–177
28. Metkar PS, Harold MP, Balakotaiah V (2013) Experimental and kinetic modeling study of NH3-SCR of NO_x on Fe-ZSM-5, Cu-chabazite, and combined Fe-and Cu-zeolite monolithic catalysts. Chem Eng Sci 87:51–66
29. Song X, Naber J, Johnson JH et al (2013) An experimental and modeling study of reaction kinetics for a Cu-Zeolite SCR catalyst based on engine experiments. SAE technical paper 2013-01-1054
30. Chen P, Wang J (2014) Control-oriented model for integrated diesel engine and aftertreatment systems thermal management. Control Eng Pract 22:81–93
31. Colombo M, Nova I, Tronconi E (2012) Detailed kinetic modeling of the NH_3-NO/NO_2 SCR reactions over a commercial Cu-zeolite catalyst for diesel exhausts aftertreatment. Catal Today 197:243–255
32. Tang W, Wahiduzzaman S, Wenzel S et al (2008) Development of a quasi-steady approach-based simulation tool for system-level exhaust aftertreatment modeling. SAE technical paper 2008-01-0866
33. Katare SR, Patterson JE, Laing PM (2007) Diesel aftertreatment modeling: a systems approach to NO_x control. Ind Eng Chem Res 46:2445–2454
34. He Y (2007) Development of an integrated diesel exhaust aftertreatment simulation tool with applications in aftertreatment system architecture design. SAE technical paper 2007-01-1138
35. Gong J, Narayanaswamy K, Rutland CJ (2016) Heterogeneous ammonia storage model for NH3–SCR modeling. Ind Eng Chem Res 55(20):5874–5884
36. Colombo M, Koltsakis G, Nova I et al (2012) Modelling the ammonia adsorption-desorption process over a Fe-zeolite catalyst for SCR automotive applications. Catal Today 188:42–52
37. Skarlis SA, Berthout D, Nicolle A et al (2012) Modeling NH3 storage over Fe-and Cu-zeolite based, urea-SCR catalysts for mobile diesel engines. Proc Soc Behav Sci 48:1672–1682
38. Wang TJ, Baek SW, Kwon HJ, Kim YJ, Nam IS, Cha MS, Yeo GK (2011) Kinetic parameter estimation of a commercial Fe-zeolite SCR. Ind Eng Chem Res 50(5):2850–2864
39. Wang TJ, Kim DS, Ahn TS (2017) Simulation study on improving the selective catalytic reduction efficiency by using the temperature rise in a non-road transient cycle. Proc Inst Mech Eng Part D: J Automob Eng 231(6):810–827
40. Wang TJ, Baek SW, Jung MG, Yeo GK (2006) A study of NH3 adsorption/desorption characteristics in the monolithic NH3-SCR reactor. Trans Korean Soc Autom Eng 14(3):125–132
41. Wang TJ (2019) Effects of insulation on exhaust temperature and subsequent SCR efficiency of a heavy-duty diesel engine. J Mech Sci Technol 33(2):923–929
42. Wang TJ, Kim JH (2018) Simulation study on thermal management strategy to achieve 99% SCR efficiency of a heavy-duty diesel engine over a transient cycle. Int J Automot Technol 19(4):597–603
43. Koebel M, Elsener M (1995) Determination of urea and its thermal decomposition products by high-performance liquid chromatography. J Chromatogr A 689(1):164–169
44. Yim SD, Kim SJ, Baik JH, Nam I-S, Mok YS, Lee J-H, Cho BK, Oh SH (2004) Decomposition of Urea into NH3 for the SCR process. Ind Eng Chem Res 43(16):4856–4863
45. Kleemann M, Elsener M, Koebel M, Wokaun A (2000) Hydrolysis of isocyanic acid on SCR catalysts. Ind Eng Chem Res 39(11):4120–4126

46. Skarlis SA, Berthout D, Nicolle A, Dujardin C, Granger P (2012) Modeling NH3 storage over Fe-and Cu-zeolite based, Urea-SCR catalysts for mobile diesel engines. Procedia Soc Behav Sci 48:1672–1682
47. Sjövall H, Blint RJ, Olsson L (2009) Detailed kinetic modeling of NH3 and H2O adsorption, and NH3 oxidation over Cu-ZSM-5. J Phys Chem C 113(4):1393–1405
48. Lietti L, Nova I, Camurri S, Tronconi E, Forzatti P (1997) Dynamics of the SCR-DeNO$_x$ reaction by the transient-response method. AIChE J 43(10):2559–2570
49. Nova I, Tronconi E (eds) (2014) Urea-SCR technology for deNOx after treatment of diesel exhausts, vol 5. Springer, New York
50. Felix SP, Savill-Jowitt C, Brown DR (2005) Base adsorption calorimetry for characterising surface acidity: a comparison between pulse flow and conventional "static" techniques. Thermochimica Acta 433(1–2):59–65
51. Komatsu T, Nunokawa M, Moon S Il, Takahara T, Namba S, Yashima T (1994) Kinetic studies of reduction of nitric oxide with ammonia on Cu^{2+}-exchanged zeolites. J Catal 148(2):427–437
52. Olsson L, Sjövall H, Blint RJ (2009) Detailed kinetic modeling of NO$_x$ adsorption and NO oxidation over Cu-ZSM-5. Appl Catal B 87(3–4):200–210
53. Wurzenberger JC, Wanker R (2005) Multi-scale SCR modeling, 1D kinetic analysis, and 3D system simulation. SAE Trans 344–362
54. Chatterjee D, Burkhardt T, Weibel M, Nova I, Grossale A, Tronconi E (2007) Numerical simulation of zeolite- and V-based SCR catalytic converters. No. 2007-01-1136. SAE Technical Paper
55. Olsson L, Sjövall H, Blint RJ (2008) A kinetic model for ammonia selective catalytic reduction over Cu-ZSM-5. Appl Catal B 81(3–4):203–217
56. Schuler A, Votsmeier M, Malmberg S, Gieshoff J, Drochner A, Vogel H, Kiwic P (2008) Dynamic model for the selective catalytic reduction of NO with NH3 on Fe-zeolite catalysts. No. 2008-01-1323. SAE Technical Paper
57. Kato A, Matsuda S, Kamo T, Nakajima F, Kuroda H, Narita T (1981) Reaction between NO$_x$ and NH3 on iron oxide-titanium oxide catalyst. J Phys Chem 85:4099–4102
58. Tuenter G, Leeuwen WF, Snepvangers LJM (1986) Kinetics and mechanism of the NO$_x$ reduction with NH3 on V2O5-WO3-TiO2 catalyst. Ind Eng Chem Prod Res Dev 25(4):633–636
59. Koebel M, Elsener, M.; Madia, G (2001) Recent advances in the development of Urea-SCR for automotive applications; SAE Paper 2001-01-3625; SAE International: Warrendale, PA
60. Baik JH, Roh JH, Yim SD, Nam I-S, Lee J-H, Cho BK, Oh SH (2006) Synergistic roles of NO and NO_2 in selective catalytic reduction of NOx by NH3. Stud Surf Sci Catal 159:441–444
61. Wurzenberger JC, Wanker R (2005) Multi-scale SCR modeling, 1D kinetic analysis and 3D system simulation. SAE Paper 2005-01-0948. SAE International: Warrendale, PA
62. https://www.dieselnet.com/standards/cycles/whtc.php

Part IV
The Methane Conundrum

Chapter 10
Cold Phase Methane Emissions, a Challenge to Overcome in Spark Ignited Natural Gas Engines

Parthiban Rajamani

10.1 Introduction

Carbon dioxide, by definition, has a global warming potential (GWP) of unity regardless of the period used, because it is used as the reference. The gas is left over in the environment for thousands of years. Methane (CH_4) lasts a decade on average, much less than CO_2. However, CH_4 also absorbs more energy than CO_2. The net effect of the shorter lifetime and higher energy absorption is reflected in the GWP. The GWP also accounts for some indirect effects, such as the fact that CH_4 is a forerunner to ozone which itself is a GHG. Hence, CH_4 has a GWP of 28–36 over 100 years and 80 over 20 years [1, 2]. Hence the newer standards regulate CH_4 emission by combining it with other hydrocarbons [3].

Continuous research and technological improvements such as multi-spark, control of in-cylinder flow through tumble port or tumble and swirl port, optimized squish, and combustion chamber shape improved combustion efficiency of spark-ignited engines and the engine-out emissions. However, reducing engine-out emissions below a certain level is highly challenging, especially in the cold cycle phase. It needs a holistic approach by considering all sources of emission and aftertreatment for meeting the current and future emission standards, by focusing on calibration and aftertreatment strategy. During calibration, lambda is tightly controlled with reduced deviation with the help of improved models to predict the behaviour of the emission-related components, and adaptation for the deterioration and environmental conditions. The after-treatment strategy includes coating technology like zone coating (for example, one catalyst with two axial zones can replace two separate substrates coated with the respective catalysts or through radial zones, one can put more precious

P. Rajamani (✉)
Mahindra & Mahindra Ltd., Mahindra Research Valley, Mahindra World City, Chengalpattu, Tamil Nadu 603004, India
e-mail: rajamani.parthiban@mahindra.com

P. A. Lakshminarayanan and A. K. Agarwal (eds.), *Handbook of Thermal Management of Engines*, Energy, Environment, and Sustainability,
https://doi.org/10.1007/978-981-16-8570-5_10

metal in the center channels, where the gas flow rate is usually higher), the use of core structures for improved flow uniformity index, and optimizing precious metal ratios.

10.2 Emission Cycles and Bharat Stage 6 and 4 Standards

Emission cycles and standards for Bharat stages 4 and 6 are closely aligned with Euro 4 and 6 respectively. The emission limits of unburned total and non-methane hydrocarbons are given in Tables 10.1 and 10.2 [4].

10.2.1 BS4

In the case of BS4, methane-powered engines have to meet only non-methane HC (NMHC) which is calculated as 0.3 times the total hydrocarbon measured if a methane cutter is not available.

10.2.1.1 Light Vehicles

A vehicle with a GVW of less than 3.5 tons is tested at a chassis dynamometer. The new European Drive Cycle (NEDC) modified to limit the maximum speed to 90 km/h with a grace period for warm-up is adopted, Fig. 10.1.

Table 10.1 Comparison of emission standards BS6 versus BS4 light vehicles of M and N category

Category	Class	Reference mass (kg)	BS6		BS4
			Total HC (mg/km)	Non-methane (mg/km)	HC (mg/km)
M1 and M2		All	100	68	100
N1	I	RM < 1305	100	68	100
	II	1305 < RM < 1760	130	90	130
	III	1760 < RM	160	108	160

Table 10.2 Emission norms comparison BS6 versus BS4 for heavy vehicles of GVW more than 3.5 ton

Norms	Cycle	NMHC	CH_4
		mg/km	mg/km
BS6	WHTC	160	500
BS4	ETC	550	1100

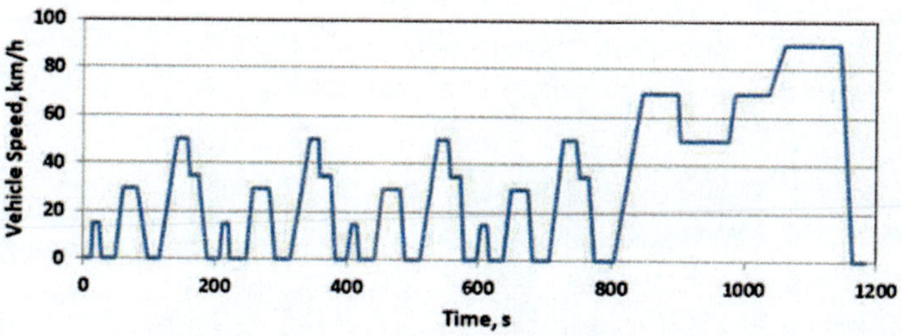

Fig. 10.1 Test cycle for BS 4 and 6—vehicle speed (km/h) versus time (s)

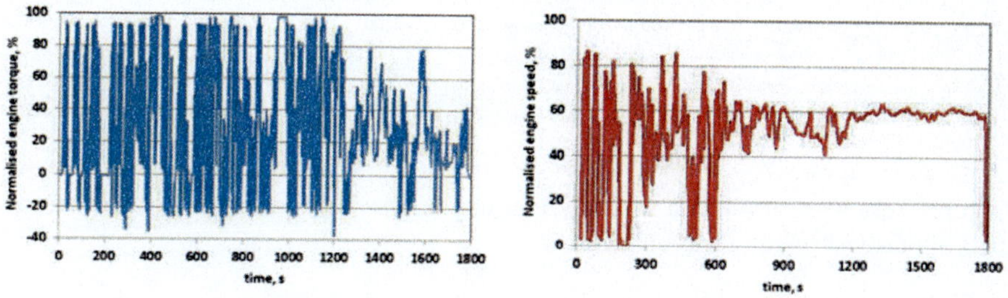

Fig. 10.2 BS 4-ETC—engine transient cycle, urban: 0–600 s, Rural: 600–1200 s, motorway: 1200–1800s

10.2.1.2 Steady-State and Transient Testing of Engines in Heavy Vehicles

For GVW greater than 3.5 tons emissions are measured at the transient engine dynamometer as per the load/speed cycle given in Fig. 10.2. For spark-ignited engines running either on gasoline or methane gas testing for emissions as per World Harmonized Stationary Cycle (WHSC) is not obligatory; for compression ignition engines it is, however, mandatory.

10.2.2 BS6

All engines have to meet total HC (THC) and non-methane (NMHC) emission limits.

10.2.2.1 Light Vehicles

The test cycle has been the NEDC for low-powered vehicles (max. speed limited to 90 km/h) lighter than 3.5 tons, Fig. 10.1. In the EU and other countries, the vehicles are tested at a chassis dynamometer according to the Worldwide Harmonized Light

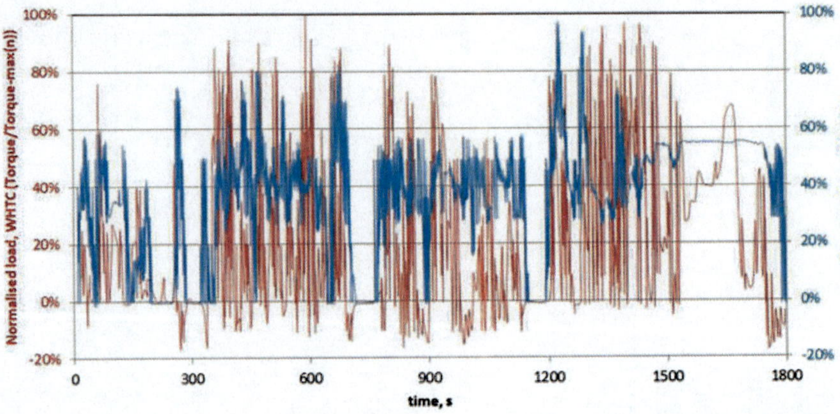

Fig. 10.3 BS6-WHTC—world harmonized transient cycle for heavy-duty engines

Vehicles Test Cycle (WLTC). India is looking forward to introducing WLTC from 2023 in place of the modified NEDC.

10.2.2.2 Steady-State and Transient Testing of Engines in Heavy Vehicles

For gasoline or diesel vehicles heavier than 3.5 tons, the engine is tested at a transient engine dynamometer according to the World Harmonized Transient Cycle (WHTC), Fig. 10.3. The test cycle is very aggressive compared to the NEDC cycle recommended for BS4. Two tests, one cold and one hot, using the same cycle are conducted. The "soak time" between the cold and hot cycles is 10 ± 1 min. The weighting factors are 14% cold test and 86% hot test.

10.2.2.3 Comparison of BS4 and BS6 Standards for Total Hydrocarbons and Methane Emissions

For BS4 SI engines running on natural gas with GVW less than 3.5 tons have to meet only NMHC, which is calculated as 0.3 times the measured THC. For BS6, all spark ignition (SI) engines have to meet the limits of total hydrocarbon (THC) and non-methane hydrocarbon (NMHC) by actual measurement. Treating the exhaust of a SI engine running on natural gas for THC needs very high effort as 90% of the unburned hydrocarbons are plain methane, as explained later. Emission norms for GVW greater than 3.5 tons, methane limits are reduced as shown in Table 10.2.

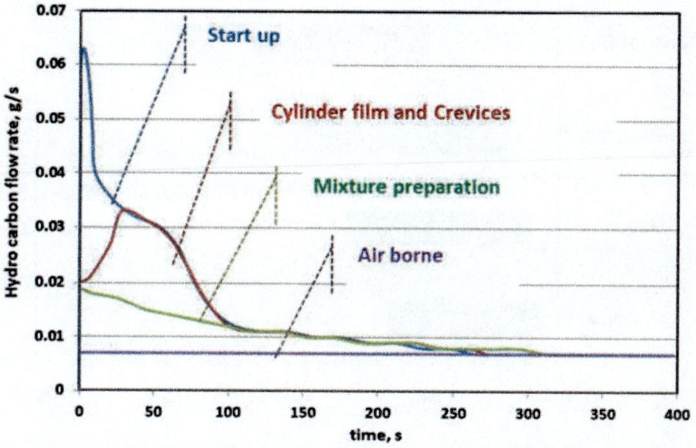

Fig. 10.4 Exhaust hydrocarbon flow rate

10.3 Cold Start Emissions

10.3.1 Sources of Unburned Hydrocarbons

To study the unburned hydrocarbons (UHC) in the exhaust of a spark-ignition engine, a clever idea is proposed in Ref. [5]. The oxygen in the exhaust is measured using a wideband oxygen sensor containing a pump cell operated by the powertrain control module and the corresponding air–fuel ratio utilised in burned products is calculated. The measurement of the instantaneous airflow and the fuel flow into the engine gives the air–fuel ratio at the inlet. A typical plot of the UHC estimated from the difference in the two air–fuel ratios is plotted against time from a cold start, in Fig. 10.4, with annotations of different sources of UHC.

10.3.2 Factors Affecting UHC in the Cylinder

10.3.2.1 Non-uniform Mixing

The lower flammability limit of methane in the air is 5%, Fig. 10.5. Relatively, gasoline has a better limit of 1%. During starting an engine, the fuel injected in the ports is not well mixed with air flowing relatively at a lower velocity corresponding to low engine speeds. The non-uniformity results in zones where lambda is beyond flammability limits and it results in UHC.

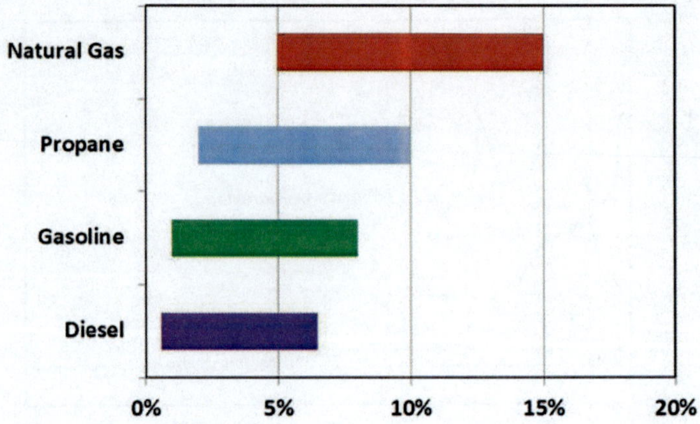

Fig. 10.5 Flammability range of methane (natural gas), propane, gasoline, and diesel

10.3.2.2 Lambda Control

In port fuel injection (PFI) engines, the control of the air–fuel ratio is poorer than in gasoline direct injection (GDI) engines because of the fuelling delay and the liquid film in the intake port and can thereby add to slower initial firing [6] and it increases the HC emissions during start-up. However, fortunately for engines running on methane gas, this problem is absent.

The contribution to HC emissions due to other factors is common for both liquid and gaseous fuel engines. The manifold pressure drops very quickly from the initial atmospheric pressure with the increase of engine speed, which leads to more air induction in the cylinder during a short time. As the fuel injected is not regulated properly, which often is the case during start-up, this can cause a lean charge and thereby poor combustion and risk of misfire.

10.3.2.3 Coefficient of Variation of IMEP

In a well-optimized SI engine coefficient of variation (COV) of IMEP at full load conditions operating at stoichiometry is has a 5–7% and it cannot be reduced gainfully below this level [7, 8]. High COV results in higher engine-out emissions, particularly the hydrocarbons [9].

10.3.2.4 Quenching at the Cold Walls and in Crevices

The flame quenches at cold walls, where the rate of loss of heat to the walls is greater than the rate of production by combustion at the flame front. Since the walls of the piston, cylinder liner, and cylinder head are extremely cold during the start, the

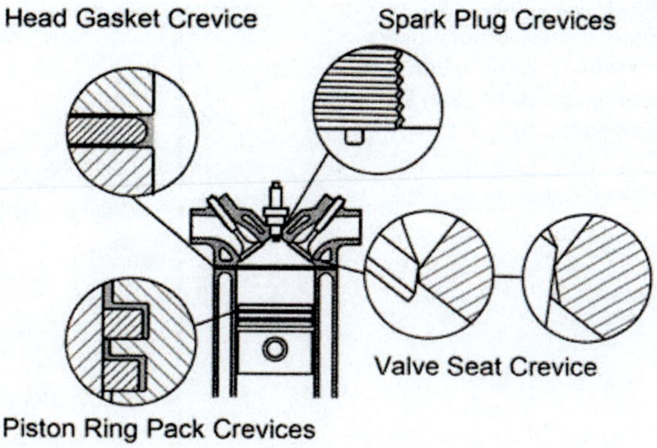

Fig. 10.6 Quench zones in the combustion chamber

quench distances are larger than when the engine is warmed up, resulting in relatively higher UHC during the cold start.

The crevices near the valve seat, piston ring grooves, and the piston lands are narrower than the local quench distances and hold substantial UHC, Fig. 10.6. This is so when the engine is not sufficiently warmed up i.e., when cold. However, when the engine is cold, the crevice volumes are high since the aluminium piston has not fully expanded to provide lesser clearances which are designed for the piston to operate safely at the rated power. Hence, the UHC contributed by crevices is higher at cold than when warm.

Further, when the parts are cold the density of gases in the crevices is high which leads to an increased mass of hydrocarbons left unburnt in the crevices and quench distances.

10.3.2.5 Contribution of Oil Film

The oil film on the liner walls adsorbs hydrocarbons when the gases in the cylinder are at high pressure during compression, combustion, and early expansion strokes. The hydrocarbons are desorbed in low-pressure gases during late expansion and exhaust strokes and hence escape combustion. This phenomenon contributes to UHC emissions as much as by the crevices.

The Henry constant characterising the phenomenon is dependent on the type of hydrocarbon. Higher molecules are adsorbed less preferentially. For example, the constant at 300 K is 10,000 times higher for methane than for the octane molecule which is closely related to gasoline, Fig. 10.7. In other words, unburned hydrocarbons due to the phenomenon at the oil film are severe in natural gas engines [10].

The involvement of oil film is considerable when the walls are cold. An example is given in Ref. [11], Fig. 10.8. The effect of temperature is partly due to the increase in minimum oil film thickness (MOFT) with a decrease in temperature. An instance is given in Ref. [12] Fig. 10.9.

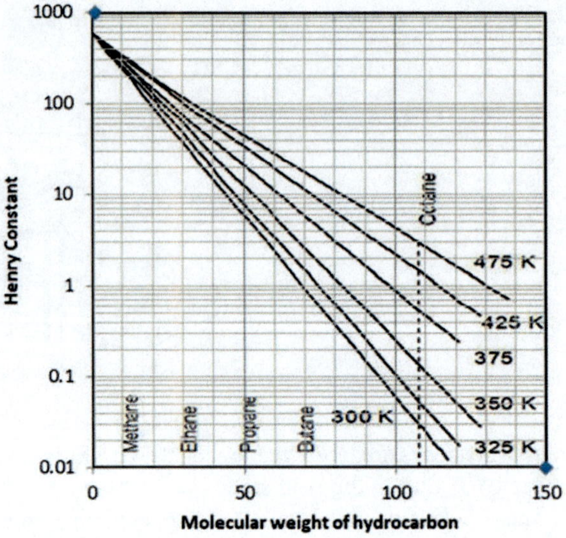

Fig. 10.7 Variation of Henry's constant for a range of kinds of paraffin dissolved in Squalane (C30 H62) Data from Ref. [10]

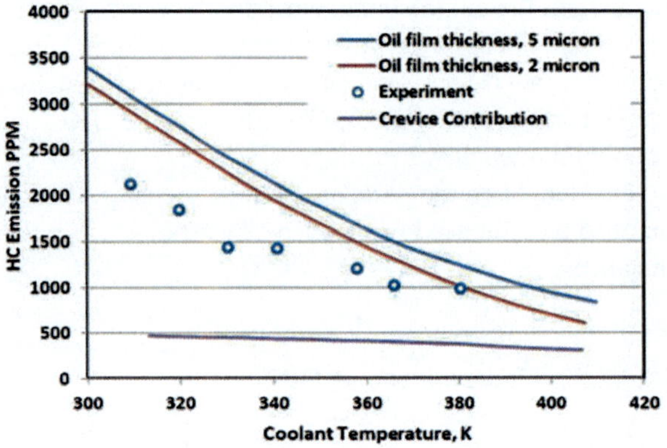

Fig. 10.8 Effect of coolant temperature on emissions due to adsorption/desorption phenomenon; experimental condition: Test conditions: 1250 rpm, 300 kPa IMEP, 20 deg BTDC, equ. ratio 0.9, shrouded valve, Isooctane fuel [11]

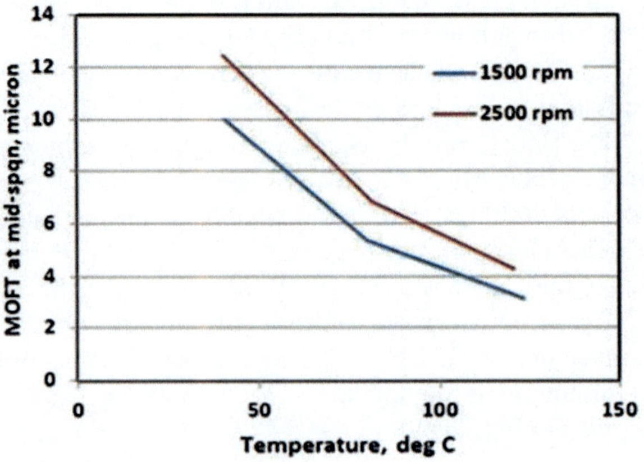

Fig. 10.9 Reduction in minimum oil film thickness at mid-speed with the rising operating temperature at different speeds [12]

10.3.2.6 Operating Parameters

Though methane has a higher flame speed compared to gasoline, the ignition delay is higher for methane which increases the combustion duration [13 L]. Therefore, advancing spark timing overcomes this problem to a large extent in hot conditions, with an attendant increase in UHC.

The problem is compounded when the intake air is cold because the ignition delay is significantly enhanced. To compensate for the delay, the ignition timing is further advanced by sensing the coolant temperature. Any such advance leads to higher cycle-cycle variation, higher UHC, and lower exhaust temperatures. A case is given in Fig. 10.10 [13]. Lower exhaust temperature delays the warm-up of the catalyst. Further, it also reduces the rate of oxidation of the hydrocarbons in the exhaust, Fig. 10.11 [13]. However, a small drop in the exhaust mass flow rate saves grace, Fig. 10.10.

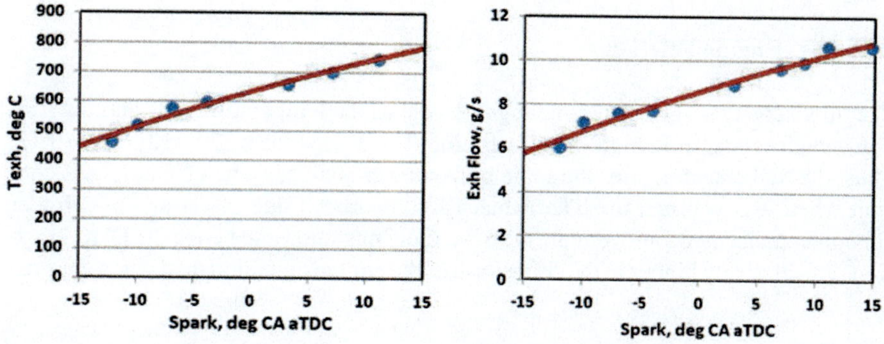

Fig. 10.10 Effect of spark timing on exhaust gas temperature and mass flow rate [13]

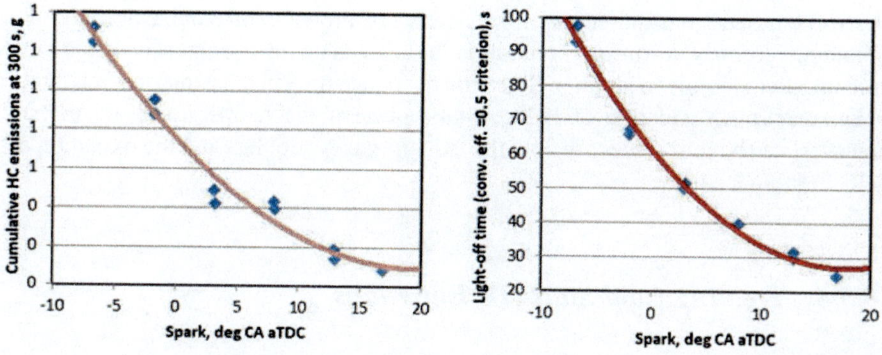

Fig. 10.11 Late ignition decreases cumulative hydrocarbon emissions through faster warm-up and reduced emissions due to post-flame oxidation [14 M]

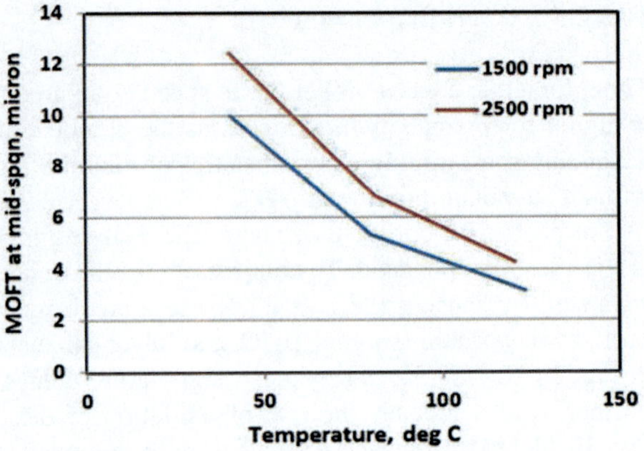

Fig. 10.12 Variation of total friction with temperature [12]

10.3.3 *Factors After the Cylinder*

10.3.3.1 Engine Friction

The oil viscosity surges as per the log–log rule as the temperature falls and the friction enhances manifold high. An illustration is extracted from Ref. [12], Fig. 10.12. Thus, the fuel consumed for the same brake power is higher when the engine is cold than when it is warmed up. The exhaust hydrocarbons due to various in-cylinder phenomena discussed above scale with the fuel consumed and hence the UHC emissions are relatively higher in the cold phase of the engine operation than when warmed up.

10.3.3.2 Oxidation of UHC in the Exhaust

The HC emitted from the engine mixes with hot products of combustion and the remaining air fairly uniformly. It oxidises in the exhaust flow progressively slowing from turbulent regime to laminar. Since the residence time in the exhaust is relatively higher reactions, oxidation of HC can take place if the temperatures are higher. However, when the surfaces are cold the exhaust gases cool fast and the oxidation of UHC is dampened.

10.3.4 *Treating Cold Start HC Emissions*

A detailed look into a typical CH_4 emission trace of an engine in an NEDC cycle throws light on the different phases, Fig. 10.13. The HC emissions are peaking during the initial 80 s when the engine is cold, particularly during the acceleration. Also,

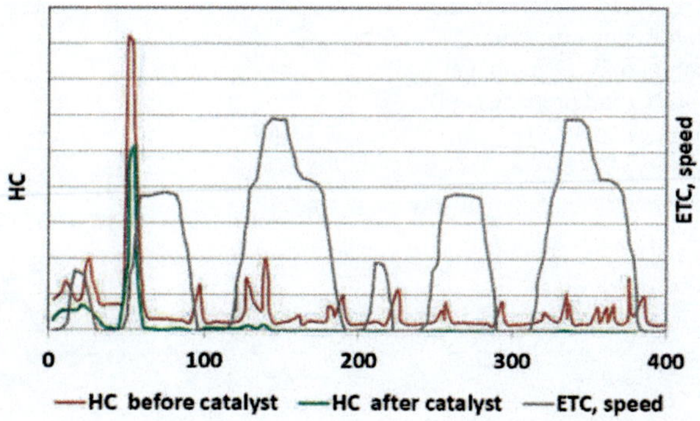

Fig. 10.13 Typical engine-out and catalyst-out HC emissions in the NEDC cycle

the catalytic converter conversion efficiency is poor during the initial period as the exhaust gas temperature is low.

When the engine coolant temperature is less than 25 °C, the engine-out emissions are far higher than when the coolant temperature is above 85 °C, which is optimum for engine combustion. Emissions can be reduced in two ways.

1. Optimization of in-cylinder combustion
2. Conversion of pollutants using an after-treatment device.

Two types of devices for engines with port fuel injection are:

1. An oxidation catalyst, coated with Pt and Pd to convert HC and CO
2. A 3-way catalyst, coated with Pt, Pd, and Rd to convert HC, CO, NO_x, with proper calibration to maintain lambda close to unity always.

10.4 After-Treatment of Methane

10.4.1 Catalytic Light-Off Temperature

The oxidation of methane with higher light-off temperatures in the after-treatment system is difficult compared to the other higher alkanes and alkenes. This is mainly because of the high bond energy of C-H in methane.

Figure 10.14 shows the comparison of the efficiency of conversion of methane and propane. The light-off temperature of methane oxidation with 10% conversion efficiency is typically 300 °C. It is high compared to the temperature of 200 °C at which 90% of propane oxidises. The exhaust temperature takes some time to reach the light-off temperature of hydrocarbons and during this time the HC emissions at the tailpipe are very high.

Fig. 10.14 Light-off and conversion efficiency comparison of methane (black) and propane (red)

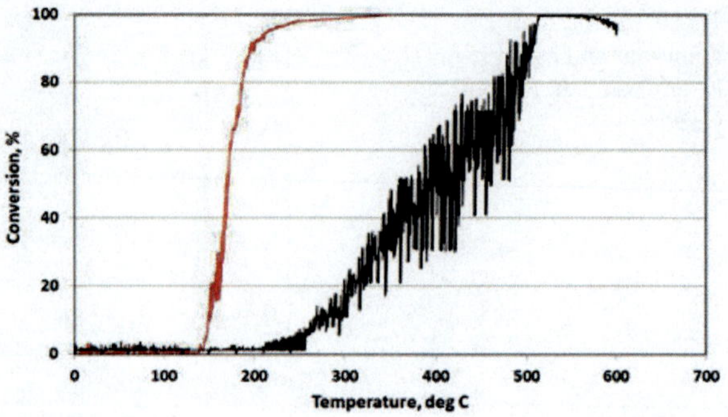

10.4.1.1 Light off Temperature of the Catalyst

From a natural gas engine, almost all the UHC comprises methane, unlike a gasoline engine. The activation energy for the oxidation of methane is high even with a catalyst. For a case in point, gasoline and methane are converted at 50% efficiency at temperatures are 175 and 400 °C by a catalyst loaded with 120 g of PGM per cubic foot. See Fig. 10.15 [14]. In other words, in the cold phase of the engine cycle till the catalyst is sufficiently warmed up hydrocarbons are not treated by the catalyst, especially methane.

Typically, methane-powered engines have a higher compression ratio when compared to gasoline-powered engines and it leads to higher work done during expansion. This lowers the engine-out exhaust temperature and challenges the after-treatment of methane. The time taken for exhaust temperature to rise during the cold start in the NEDC cycle is compared for engines powered by methane and gasoline in Fig. 10.16.

Fig. 10.15 Effects of PGM (Pd-Rh) loadings (g ft^{-3}) on HC conversion of catalyst: gasoline (TWC) versus CNG [14]

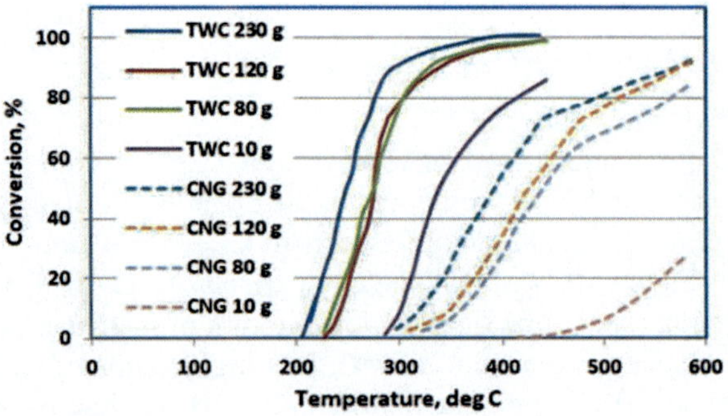

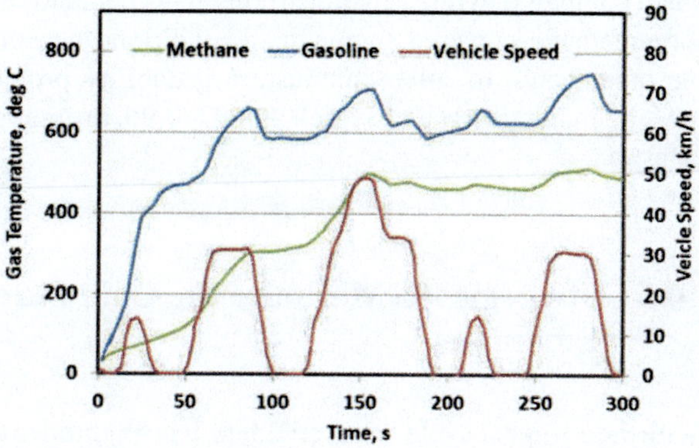

Fig. 10.16 A higher compression ratio of CNG engines result in a decrease in exhaust temperature

10.4.1.2 Lambda Window for HC Conversion

At lambda = 1, there exists an equilibrium between the oxidation and reduction reactions. The window of lambda available for the simultaneous control of the oxidation of HC and CO and reduction of NO_x is very narrow. Therefore, the formation of the air–fuel mixture must be controlled in a closed-loop with feedback from the lambda (oxygen) sensor placed in the engine exhaust. The accuracy of the feedback is typically 5% in the dynamic range, i.e. the fluctuations (about lambda = 1) are unavoidable. The catalyst substrate is coated with cerium oxide which stores some excess oxygen available in the lean phase of the lambda fluctuation to release it in subsequent rich phase and thus, can compensate for small fluctuations. The oxygen stored is in the range of 0.1–1.0 g. This oxygen storage capacity reduces with aging. Thus, the task of the engine management is clearly to maintain the time-averaged lambda upstream of the catalyst precisely at one. In other words, the deviations in lambda should not overtax the oxygen available in the catalyst [15].

Figure 10.17 shows the dependence of catalytic convertor efficiency on lambda for methane and propane. Methane has a lambda window highly tuned about unity

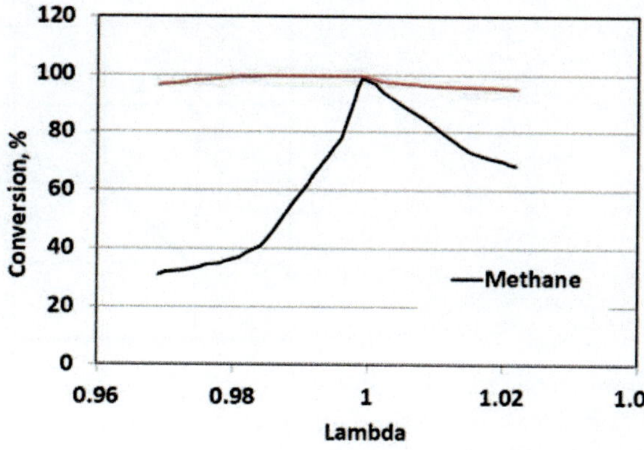

Fig. 10.17 Effect of lambda on the catalytic converter conversion efficiency (red line—higher alkanes than methane, black line—methane)

when compared to other hydrocarbons. This demands a more precise lambda control for a methane-powered engine at all conditions, especially during the cold start. On the other hand, to solve combustion instability problems during starting, lambda must be maintained slightly rich at 0.97–0.95, dropping conversion efficiency of the catalyst.

10.5 Strategies to Minimise the Cold Start Methane Emissions

With the stringent cycle, reduced limits, it is required to have a multipronged approach with improved fuel control, management of ignition timing, and an optimum catalyst.

Current engine control is based on torque management. The driver gives the desired input through the accelerator pedal and engine control receives the input and delivers the torque based on the load and state of the vehicle. The torque delivery is based on the below inputs and calculations

1. Engine friction, load, and vehicle accessories
2. Engine load to meet the driver demand
3. Moderating the actual torque delivered by filtering the high-frequency components of the torque demand by delivery, Fig. 10.18: this is required for meeting the good driveability as well as emissions.

Based on the above inputs and calculations, the controller decides the amount of air, fuel, and spark time to deliver the torque. To respond to the driver torque demand, two types of torque control strategies are coordinated properly to have acceptable vehicle reactions and emissions.

Fig. 10.18 Torque filtering

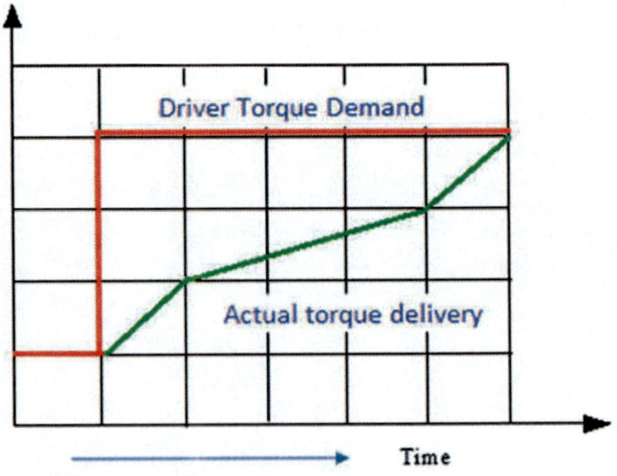

1. Fast path—Ignition time and fuel injection
2. Slow path—Air control.

10.5.1 Catalytic Converter

Stoichiometric combustion is preferred to lean-burn even though the latter was touted some years back for its marginally better efficiency because treating lean-NO_x emissions invites an oxidation catalyst and a selective catalytic reduction after the engine. The system is more complicated than an elegant solution of the 3-way catalyst system for stoichiometric combustion.

10.5.1.1 Catalyst Selection

Parameters like precious metals loading and its ratio, cells per square inch, flow characteristics inside the catalytic converter, wash coat technology, material, wall thickness across the cells, need to be optimized for the methane-powered engine. For treating the exhaust from engines powered by methane the Pa-Rd (10:1) catalyst is found to be ideal. Palladium helps in better oxidization of methane compared to platinum; hence palladium is in higher proportion for methane-powered engines. Overall loading (g/ft^{-3}) is increased by 2–4 times for BS6 compared to BS4, Fig. 10.15.

10.5.1.2 Core Structure

Engines of Light Vehicles Less Than 3.5 Tons

A metal-based substrate was chosen for it warms up quickly unlike a ceramic substrate. The behaviour is highly useful for treating stubborn methane emissions during cold start. Several structures developed by *Emitec* are available for trial. They increase residence time, turbulence as well as thorough mixing, Fig. 10.19 [16]. Structure A (LS) was chosen as it was found optimum for the BS6 engines. The close-up view of the foil is given in Fig. 10.20, showing schematically the intense turbulence, mixing, and interaction of gases with the substrate. The structure is divided into short circular discs with axial clearances. Tests clearly showed the potential of "turbulent" flow control.

Theory

In the structure of the substrate, the conventional smooth channel walls were replaced with walls containing periodic openings designed along the flow direction, thus effectively repeating the flow at the entrance and reducing diffusion paths. New production technologies, process technologies, and materials, as well as coating processes, enable mass production of such substrates.

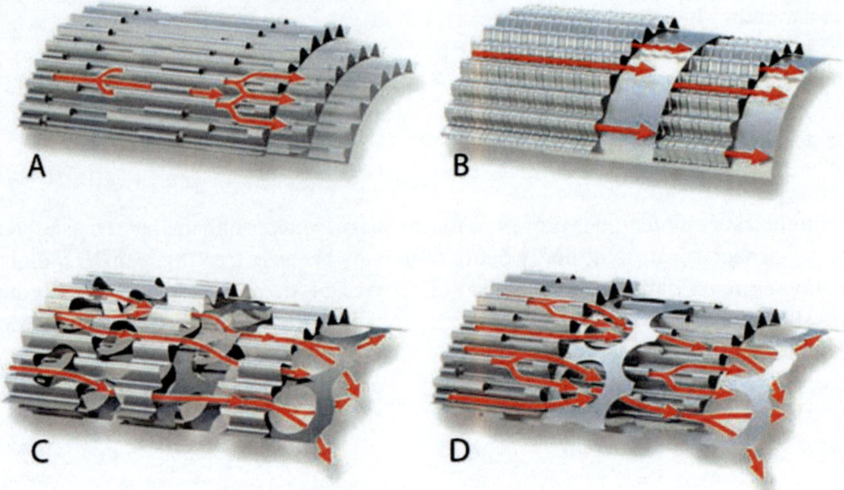

Fig. 10.19 Structured foils: **a** A reduction of diffusion paths and the hydraulic diameter and improved inlet flow (LS® design), **b** Radial flow near the channel wall (TS design®), **c** Radially open, perforated structures (PE™ design), **d** A combination of PE™ and LS® structures. *Courtesy* Emitec

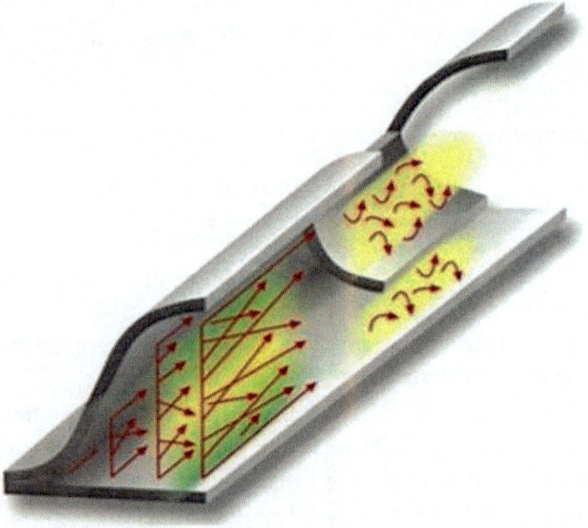

Fig. 10.20 Close-up view of LS® foil. *Courtesy* Emitec

Main Features

- The particles do not have to rely solely on diffusion to travel from the core flow in the channel to the wall.
- A new wall projecting into the channel in the centre of the core flow (at the point of the highest pollutant concentration) reduces the diffusion length nearly to zero.
- Such systems are already successful in the chemical industry.

Advantages

- Creates the same positive turbulent mass transfer conditions of the inlet of the channel over the entire length of the channel.
- Thus,
  - It increases the catalyst efficiency even when the engine-out emissions are very low.
  - It reduces the required substrate volume (without increasing the backpressure), usage of material, weight, and cost.

Large Engines for Vehicles Heavier Than 3.5 Tons

Whereas fast warming up metallic substrate is advantageous for small CNG engines ceramic substrate is preferred for bigger engines. Ceramic though has poorer warm-up characteristics than the metallic substrate, has the advantage of better thermal stability in cycles after warmup. The structure is a simple honeycomb type but with proprietary wash coat technology specially devised for CNG engines.

10.5.2 Mixture Control

Precise air–fuel mixture control in terms of the desired lambda is very important to meet the emissions of THC and NO_x.

$$Fuel\,quantity = f(desired\,torque\,fuel, lambda\,correction, adaptation)$$

10.5.2.1 Fuel for Desired Torque

Fuel for desired torque is calculated based on the airflow required to meet the desired torque. The calibration strategy is derived from a lookup table of speed, absolute manifold pressure, and air quantity for the standard conditions. Then the calculation is carried out taking care of the corrections/compensations for the delay, as well as cold start, warming up, transient conditions, and protection of various components.

10.5.2.2 Lambda Correction

Lambda needs greater attention than any other parameter for achieving a precise mixture. It is corrected by sensing it in the exhaust. Figure 10.21 shows the typical engine coolant temperature measured at the thermostat during the emission cycle for the cold and hot starts of a typical vehicle with GVW greater than 3.5 tons. It can be seen that the optimum temperature of 80 °C is reached only at the end of the cycle when started from cold. As explained before, estimating the airflow is a challenge during this period. A correction enables maintaining the desired lambda and thereby limiting the engine-out emissions.

In BS4 engines the lambda sensor used is of a binary type wherein the output shows the fuel–air mixture is either rich or lean and it does not quantify the absolute value of lambda. From Fig. 10.22 it can be seen that the CH_4 emissions are much higher during the cold start than during the hot start. Typical methane emissions measured in the emission cycle for hot and cold starts are:

- Hot start ~0.4 g/kWh;
- Cold start ~2.8 g/kWh.

In the case of BS6 engines, a linear or broad-band lambda sensor is used to quantify the degree of richness or leanness. Such a sensor enables control of lambda within a tight band, using the more sophisticated PID control algorithm.

The top half of Fig. 10.23 shows the result of fuel corrections (short-term trim) in terms of the percentage of lambda for BS4 and BS6 to the same scale. There is a substantial improvement in lambda control during the cold start phase. A reduction in fluctuation by 100 times is achieved. For brevity, the result of the new lambda control is plotted to an enlarged scale in the bottom half of the figure. The precise control of fuel enables the catalytic converter to perform at its best efficiency and

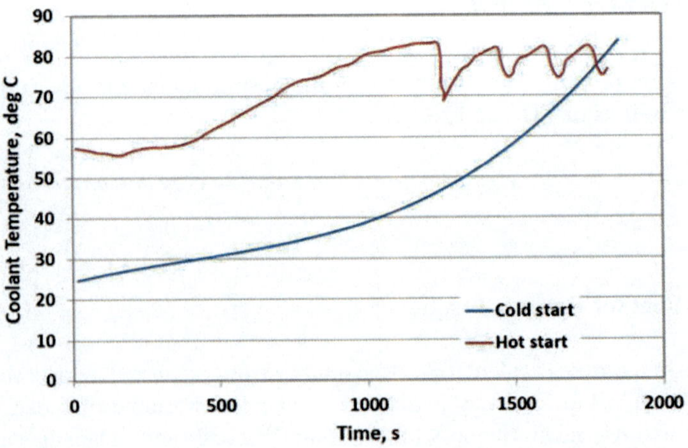

Fig. 10.21 Coolant temperature of hot (red) and cold (blue) start during the emission cycle

UHC, especially methane is converted completely at all loads, speeds, and operating conditions.

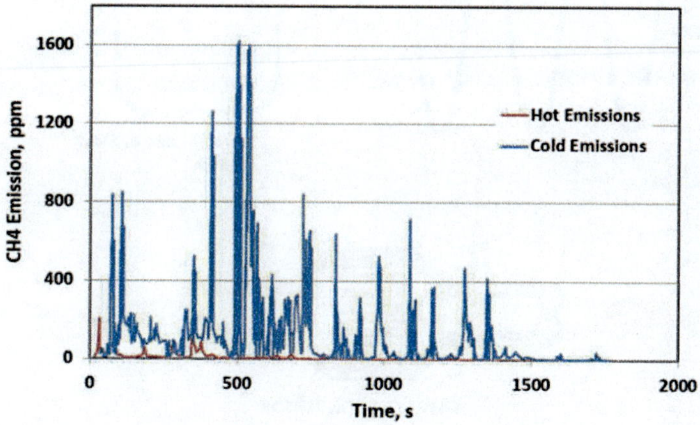

Fig. 10.22 CH4 emissions of cold and hot start during the emission cycle for the BS4 layout (*Note* The cold start emissions are not included in the standard)

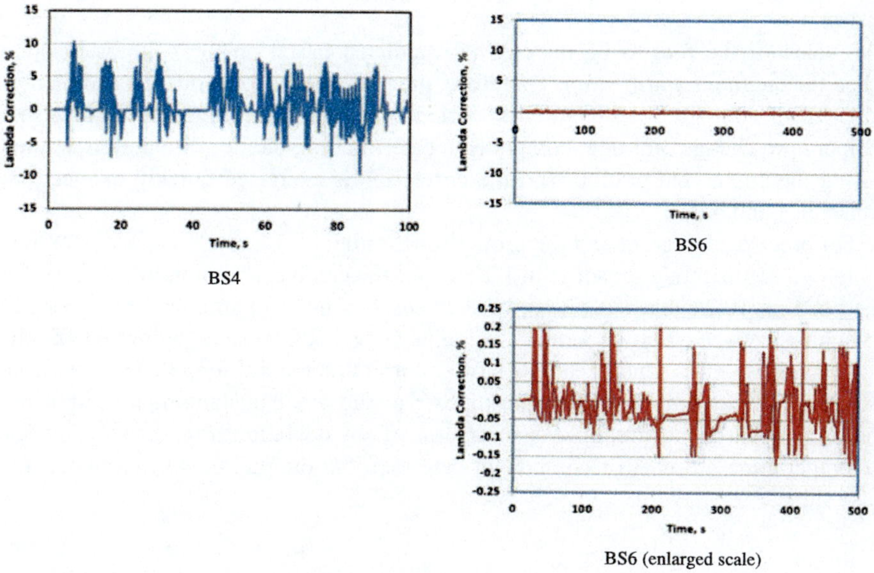

Fig. 10.23 Fuel correction comparison for the first 500 s during cold cycle

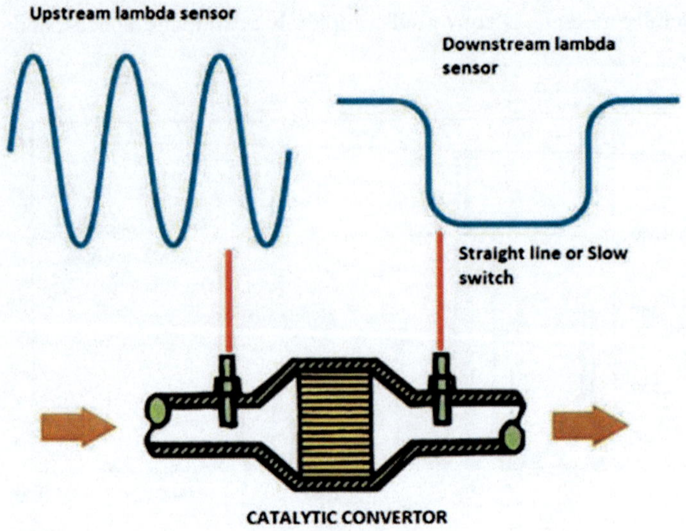

Fig. 10.24 Signals from lambda sensors upstream and downstream of the catalyst

10.5.2.3 Adaptation

The value of the fuel to be injected is calculated based on the calibration done when the engine is warmed up. Therefore, the fuel calculations must be trimmed or adapted for cold-start. Adaptation is carried out by monitoring the conditions like temperature change, altitude, components deterioration, etc. for or a period defined during the engine calibration, and the correction is exercised quickly to meet the emission standards.

For precise adaptation and the proper functioning of the catalytic convertor, one more two-step lambda sensor is installed after the catalyst and monitored. It can be seen in Fig. 10.24 that the output of the sensor is either a straight line or slowly switching between rich or lean. A setpoint (e. g., 600 mV) is calibrated for the downstream lambda sensor. Here, a slower correction control loop (using the signal from the downstream sensor) is superimposed on the wideband lambda control (using the signal from the upstream sensor). In case of any deviation from the setpoint due to temperature, altitude, or component deterioration, the fuel trim is applied to the base signal.

10.5.2.4 Ignition Timing

Ignition timing plays a major role in controlling the in-cylinder combustion parameters namely, the crank angle for 50% burned-mass fraction (MFB 50), combustion duration and efficiency, exhaust gas temperature, peak firing pressure, etc.

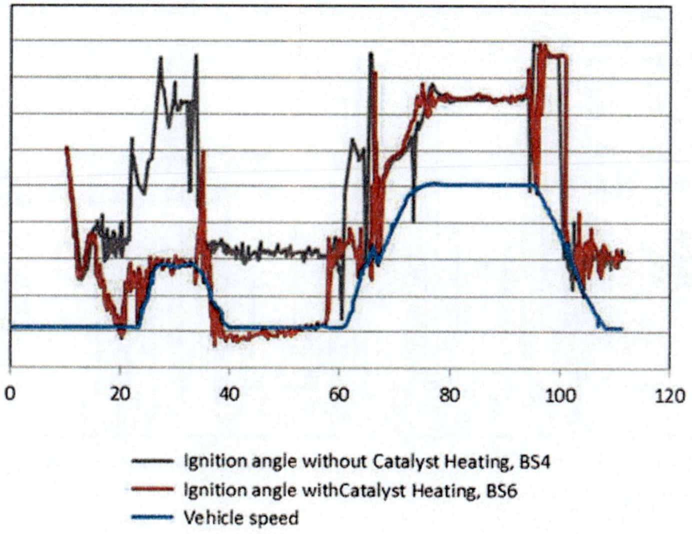

Fig. 10.25 Ignition angle comparison with and without catalyst heating

During a cold start, as explained before, the catalytic converter must be heated quickly to reach its light-off temperature. For this, the ignition timing is retarded during the cold start (Fig. 10.25) and the mixture is made richer. Though enriching increases the engine-out HC emissions slightly for a few seconds, the net result is lesser emissions after the catalyst where. A substantial part of the UHC including methane burns and warms up the catalyst quickly. Figure 10.26 shows the improved tailpipe emissions from a well-developed BS6 engine compared to its predecessor BS4 engine. It may be noted that the emission measuring equipment saturates at 1600 ppm and all the overshoots are truncated as 1600! The BS4 layout floods more often this territory while the BS6 design remains subdued close to zero.

10.6 Summary

Stoichiometric combustion finds a cost-effective and neat solution in a three-way catalyst to resolve all the regulated emissions. Delay in light-off of the catalyst at cold start is solved by enriching and delaying spark. Methane is the most obstinate of all hydrocarbons with an extremely narrow lambda window for treatment and high light-off temperature. While BS4 standard pardoned methane and asked for measurement of emissions only in the warm-up phase, BS6 is very strict. The close control of lambda using a wide-band sensor upstream and a two-step sensor downstream enables the reduction of methane emissions by two orders of magnitude during the cold phase. A substrate structure enabled high turbulent diffusion of pollutants towards catalyst

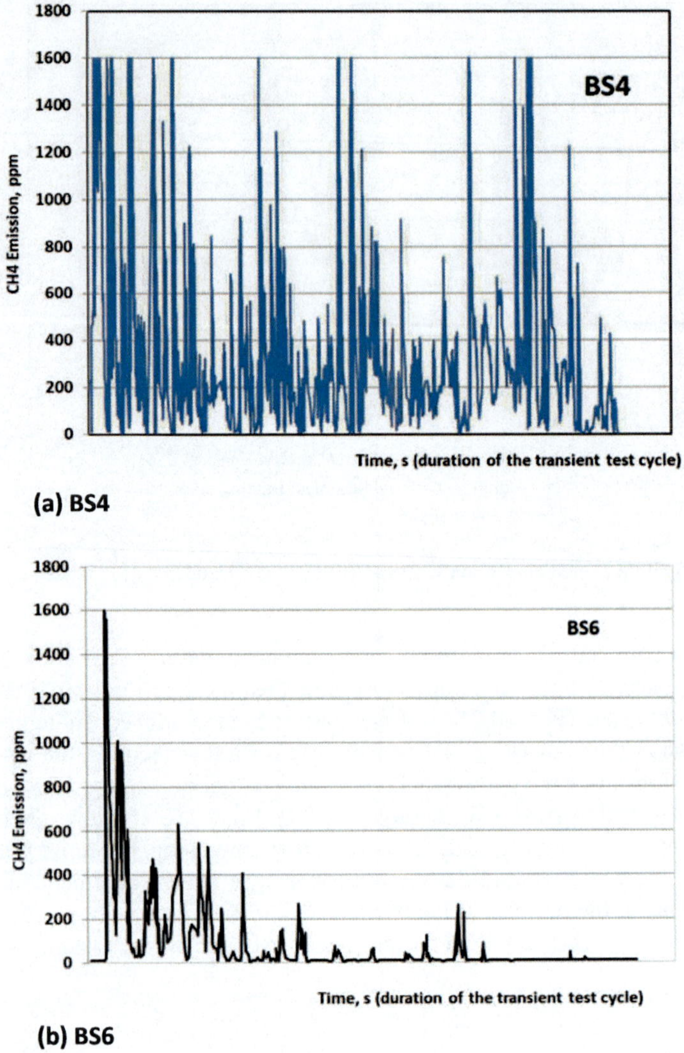

Fig. 10.26 Engine-out emission of BS4 and BS6

and reduced the size and weight of the converter. Also, a metallic substrate improved the warm-up time.

References

1. EPA US (2017) Understanding global warming potentials. Recuperado el 8
2. Heywood JB (1988) Internal combustion engine fundamentals. McGraw-Hill, New York
3. Central Motor Vehicles Rules, India, 1989 with 11th amendment dated September 2016
4. https://dieselnet.com/standards/in/hd.php, India: Heavy-Duty Truck and Bus Engines
5. Shayler PJ, Belton C, Scarisbrick A (1999) Emissions and fuel utilisation after cold starting spark-ignition engines. SAE Trans 140–156
6. Tong K, Quay BD, Zello JV, Santavicca DA (2001) Fuel volatility effects on mixture preparation and performance in a GDI engine during cold start. SAE Trans 2301–2318
7. Karvountzis-Kontakiotis A, Ntziachristos L, Samaras Z, Dimaratos A, Peckham M (2015) Experimental investigation of cyclic variability on combustion and emissions of a high-speed SI engine. No. 2015-01-0742. SAE Technical Paper
8. Cai X, Hu YH (2019) Advances in catalytic conversion of methane and carbon dioxide to highly valuable products. Energy Sci Eng 7(1):4–29
9. Jensen TK, Schramm J, Narusawa K, Hori S (2001) Analysis of UHC emission from a natural gas SI engine using fast response FID and a heat release model. No. 2001-01-3533. SAE Technical Paper
10. Chappelow III CC, Prausnitz JM (1974) Solubilities of gases in high-boiling hydrocarbon solvents. AIChE J 20(6):1097–1104
11. Dent JC, Lakshminarayanan PA (1983) A model for absorption and desorption of fuel vapour by cylinder lubricating oil films and its contribution to hydrocarbon emissions. No. 830652. SAE Technical Paper
12. Rahmani R, Rahnejat H, Fitzsimons B, Dowson D (2017) The effect of cylinder liner operating temperature on frictional loss and engine emissions in piston ring conjunction. Appl Energy 191:568–581
13. Cedrone KD (2013) Control strategy for hydrocarbon emissions in turbocharged direct-injection spark-ignition engines during cold-start. Ph.D. diss., Massachusetts Institute of Technology
14. Raj BA (2016) Methane emission control. Johnson Matthey Technol Rev 60(4):228–235
15. Reif K, Dietsche K-H et al (2014) Bosch automotive handbook, 9th edn. Translated from the German. Karlsruhe: Robert Bosch GmbH
16. Structured foils, https://www.emitec.com/en/technology/catalyst-substrates/structured-foils/. Accessed on 5 Jan 2020

Part V
Simulation of Heat Under Body and Hood

Pan Engine Front: Thermal Simulation *Courtesy* Dassault

Chapter 11
Cooling System Study and Simulation

C. Vijay Ram, R. Sneha Priya, U. G. Remesh, P. T. Haridas, M. Sathya Prasad, and Patro Sambit Kumar

11.1 Introduction

The peak efficiency of an automobile engine is less than 50% for typical vehicle applications. This is because the energy generated by the fuel is not only used to power the engine, but also absorbed as frictional and heat losses, and in cooling. The cooling system is essential in removing the excessive heat generated to maintain the engine system and its surroundings at an optimum temperature for efficiency and durability.

Excessive heat for a prolonged duration on any component is detrimental to its life and performance. On the other hand, too low a temperature leads to non-favourable operating conditions, resulting in problems such as failure to reach combustion temperatures, and thereby leading to poor fuel burning. This leads to further lowering of efficiency and increased pollutants in the form of unburnt fuel in the exhaust. Hence, the role of the cooling system is to maintain an optimal temperature, rather than a low temperature. The cooling system does this by transferring heat from the engine and its related components to the ambient using a coolant. The coolant is driven by a pump through the engine from where it absorbs the heat and transfers it to the ambient in a heat exchanger namely, the radiator. To effectively cool the engine, these components must be sized in a manner that ensures that the engine is maintained at the optimal temperature under all working conditions of the vehicle.

The primary consideration for sizing the engines is the amount of heat rejected by the engine to the coolant. A sufficient flow rate of coolant needs to be maintained to extract this heat at the desired temperature of the engine. The engine has a complex

C. Vijay Ram (✉) · R. Sneha Priya · U. G. Remesh · P. T. Haridas · M. Sathya Prasad
Ashok Leyland Technical Center, Ashok Leyland Ltd, Chennai, Tamil Nadu, India
e-mail: Vijayram.C@ashokleyland.com

P. Sambit Kumar
SRM Institute of Science and Technology, Chennai, Tamil Nadu, India

P. A. Lakshminarayanan and A. K. Agarwal (eds.), *Handbook of Thermal Management of Engines*, Energy, Environment, and Sustainability,
https://doi.org/10.1007/978-981-16-8570-5_11

flow path of the coolant around the cylinders to remove this heat. It is important to design the flow path in a manner such that no part of the engine attains temperatures beyond desired levels. This temperature should be such that the coolant itself does not go beyond its boiling point. The radiator should be large enough with enough density of fins to allow the required area for heat exchange between the coolant and ambient air. Radiators are typically constructed such that the coolant passes through several thin tubes with fins to increase the area of heat transfer. The tubes create various restrictions to flow. The pump should be sized to maintain an adequate flow rate of the coolant through these restrictions and the restriction in the flow path of the engine and other pipes and valves.

Conventional methods of carrying out these sizing calculations are analytical using basic laws of physics, and empirical equations. While the first law of thermodynamics helps calculate the heat balance in an engine and thereby the rate of heat release, empirical relationship based on other engines provides the portion of this dissipated to the coolant. Similarly, using relationships for flow and heat transfer and known characteristics of components, each component of the cooling system is designed. Such a design is however carried out without considering the spatial variation of any of the flow or heat transfer parameters i.e. the system is modelled using lumped parameters. As the sizing in this method is done based on approximations and assumptions derived from past designs, it is often possible that those are not valid for the new design, making them insufficient to meet the service demands. This leads to design modification after prototyping and testing the system.

Computation Fluid Dynamics (CFD) helps reduce the assumptions and approximations used in conventional design calculations by involving several physical phenomena such as flow, heat transfer by conduction, convection, and radiation, boiling, etc., making the calculations more robust and reducing the requirement for design modification after prototyping. CFD analysis is carried out for a three-dimensional space and does not aggregate all properties of the system to a single value. The calculations are carried out for several small finite regions in the space of interest such that the conditions of physics applicable for those regions are satisfied in each of them. This helps predict complex flow and heat transfer phenomena that cannot be solved by analytical calculations. CFD solvers have several models to simulate different physical phenomena and the analyst should select the right set of models corresponding to the actual phenomenon occurring in the problem under consideration. CFD helps make predictions for components that have designs that are significantly different from the past since it does not require empirical data related to the component. However empirical information about components is often used to reduce the computational effort, particularly when simulating a system involving several components.

Some of the common applications of CFD in the design of automotive cooling systems are those of the flow of coolant through the engine to estimate the temperature on the engine components and identifying the pressure restriction in the engine itself, to size the pump. At a component level, the analysis can be carried out for the heat exchangers as well to identify the effectiveness and the pressure restriction. One of the most important vehicle level applications of CFD in the design of the cooling system is known as the underhood analysis, where it is estimated whether the fan can draw sufficient air through the heat exchangers of the cooling system against the restriction provided by the vehicle. It is also determined if the heat exchangers can dissipate the heat with the available airflow distribution over the radiator. Based on these analyses, design solutions are identified to improve the flow, or the components are resized or redesigned to meet the cooling requirements. Multiple operating conditions of the vehicle can be simulated using CFD to ensure adequacy in all those situations.

While CFD analysis provides a lot of insight into the design, it is also very demanding on computational infrastructure. Parallel processing is practically a necessity to solve industry-level problems. Modern-day CFD solvers, as well as computing technology, allow CFD problems to be solved across hundreds and thousands of cores. The advantage of the speed that comes from parallelization allows CFD to be used for design optimization by carrying out several design iterations even during the concept stage.

A cooling system designer needs to understand the working of the cooling system and the methods for its thermal analysis. Understanding the function and objective of each component is important to develop new design concepts. Knowledge of the working of analysis tools helps use the right models to derive meaningful results that would help arrive at designs that meet the requirements or optimize performance over its operational range.

11.2 Automotive Cooling System

The automotive cooling system regulates the heat in the automotive engine system by sending a coolant through the engine block and head. As the coolant passes through the passages, it picks up excess heat and reaches the radiator. Once inside the radiator, the coolant flows through narrow tubes that are exposed to the atmosphere and get cooled by the air entering the engine compartment through the grill in the front of the vehicle. The coolant then returns to the engine block to continue the cycle. A water pump is used to maintain the flow of the coolant through the cooling system.

11.2.1 Cooling System Layout and Components

Figure 11.1 shows the layout of the cooling circuit of a vehicle with the important components. The thick arrows show the main path of the coolant. There are vent lines that allow the escape of air and vapours to the DAT. The makeup line allows for loss of coolant through leaks or the flow of coolant through the vent line. The coolant and thereby the engine must attain a minimum required temperature. Therefore, a thermostat is used which regulates the flow of the coolant such that if it has not attained the required temperature it flows back into the engine through the bypass line without getting cooled in the radiator.

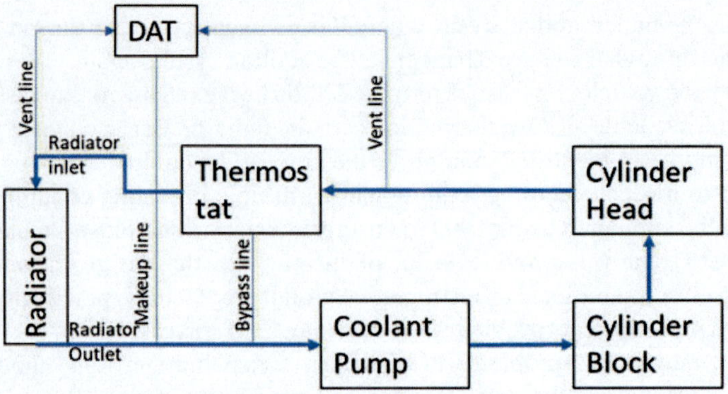

Fig. 11.1 A generic layout of an automotive cooling system

11.2.2 DAT

Internal combustion engines, and particularly diesel engines, are susceptible to the entry of combustion gases into the cooling systems causing the water or coolant pumps to become air locked. The entrapment of air or combustion gases in the coolant reduces the efficiency of the engine cooling system. A mechanism for deaeration should be provided to ensure that this air is not drawn back into the radiator and engine [1].

A DAT, as shown in Fig. 11.2, is a reservoir with a minimal level of coolant located at a sufficient height above the radiator and the pump. Vent lines connect the engine block and the radiator top tank to the DAT so that trapped gases might pass through them into the DAT. Being at the topmost point the deaeration tank has the minimum hydrostatic pressure. Therefore, buoyancy carries any trapped air to the deaeration tank. However, the pressure difference in the line also causes the flow of coolant

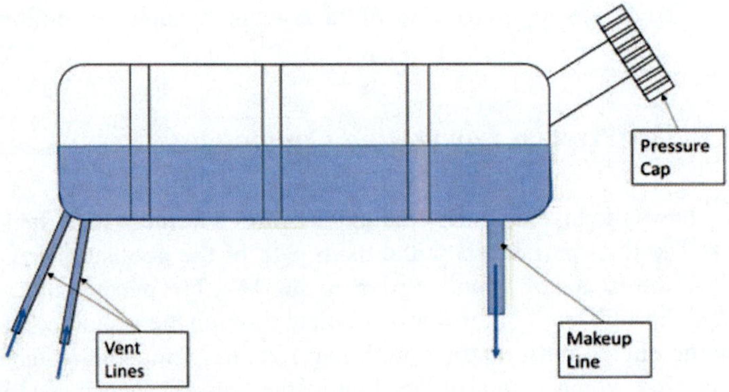

Fig. 11.2 Deaeration Tank (DAT)

into the deaeration tank from the vent lines. These are replenished by flow through the makeup line. The pressure cap in the de-aeration tank is a spring-loaded valve and ensures the pressure in the system due to coolant evaporation does not raise the system pressure to very high levels.

11.2.3 Coolant Pump

Initially, radiators used downward vertical flow, driven solely by a thermosiphon effect which is a natural convection cycle. Coolant heated in the engine becomes less dense and rises. As the radiator cools the fluid, the coolant becomes denser and falls. This effect is sufficient to cool low-power stationary engines but cannot be used for today's automobiles. All automobiles currently use centrifugal pumps to circulate the engine coolant because natural circulation has very low flow rates and the coolant is unable to pass through the water jacket of the engine with ease. The pump is usually driven by a belt taking power from a pulley connected to the crankshaft of the engine.

11.2.4 Heat Exchangers

Heat exchangers aid cooling by dissipating the heat extracted from the engine by the coolant to the environment, either to water or atmospheric air. The functions fulfilled by heat exchangers are classified as below.

- Engine cooling (Radiator): to maintain appropriate operating temperature range in components such as liners, valves, pistons, etc.
- Charge Air Cooler (CAC): to cool the compressed intake air in pressure-charged air; that has been compressed to increase the density and the mass of charge air entering the combustion chamber. While the CAC does not heat the coolant directly as the coolant does not flow through it, it is usually located in front of the radiator heating the air entering the radiator. This reduces the cooling capacity of the radiator compared to ambient air entering directly into the radiator.
- Heating, ventilation, and air conditioning (HVAC) condenser: The refrigerant of the HVAC system condenses by releasing heat to the ambient in the condenser. As in the case of CAC, it causes heating up of the air entering the radiator and CAC as it is usually placed ahead of the CAC.
- EGR Cooler: This heat exchanger cools the exhaust gas which is being recirculated into the engine. The heat is transferred from the exhaust gas to the coolant.
- Oil cooling: to control temperature, and thus maintain the required viscosity of lubricating oil in the cylinder.

Fig. 11.3 Radiator

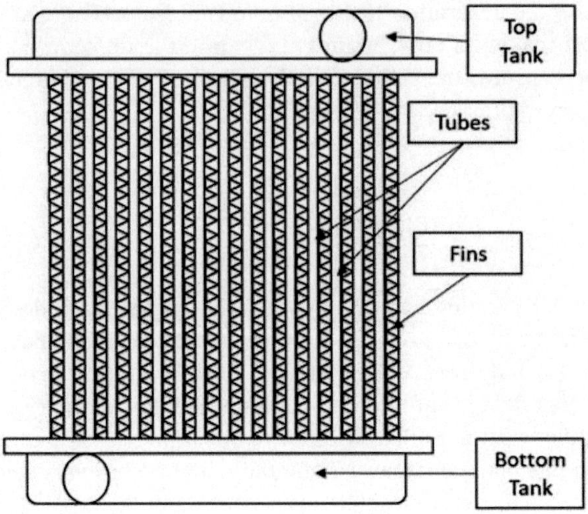

Radiator

The engine coolant, after cooling the engines through the engine block where it is heated, passes through a radiator where it loses the heat to the atmosphere and then returns to the engine to repeat the cycle.

The radiator comprises top and bottom tanks, with a bank of thin tubes that make up the core transporting coolant from the former to the latter, Fig. 11.3. These tubes (pipes) are designed to pass through a series of holes in a stack of thin sheet metal fins to increase the area of contact between the air and the core material. This enables the coolant to lose heat rapidly to the cooler air passing through it [2].

An axial fan is employed to force air through the radiator and improve heat transfer through the means of forced convection. The fan increases the flow of atmospheric air coming into contact with the radiator tubes.

EGR Cooler

The Exhaust Gas Recirculation (EGR) system recirculates exhaust gases back into the combustion chamber of the engine to decrease cylinder temperatures and thus, NOx emissions. The EGR cooler is used to lower the temperature of the exhaust gases recirculated by the EGR system. It is usually a shell and tube heat exchanger where the coolant passes through the outer shell and the exhaust gas passes through the tubes, Fig. 11.4.

Oil Cooler

In addition to providing lubrication, engine oil also acts as a coolant to remove surplus heat from an internal combustion engine. The hot engine transfers heat to the oil which then passes through a heat exchanger, typically an oil cooler. It allows cooling of the oil using the coolant.

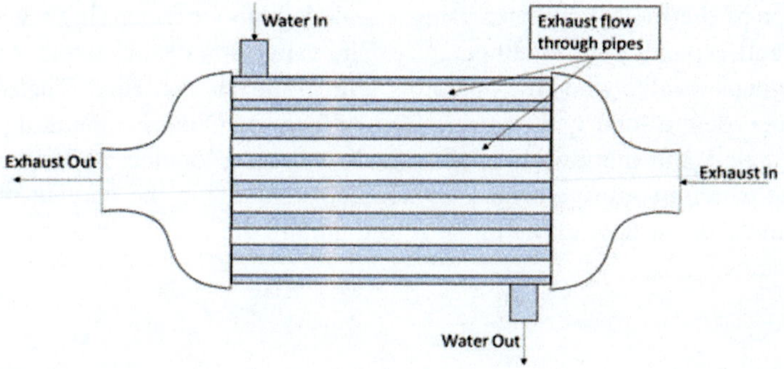

Fig. 11.4 EGR cooler

11.2.5 Engine Water Jacket

The engine water jacket is a pathway either cast or bored through the main engine block and connected by inlet and outlet valves to a radiator. It is used to cool the cylinder liners, cylinder covers, and head with the exhaust valves of the main engine.

11.2.6 Thermostat

The thermostat in a cooling system measures the coolant temperature and controls the bypass valve to regulate the flow. If the coolant is above a predetermined temperature, the valve opens and lets the coolant into the radiator to be cooled and if it lower, the valve is blocked the coolant is recirculated into the engine.

The type of thermostat predominantly used in the automotive cooling system is the wax thermostat, Fig. 11.5.

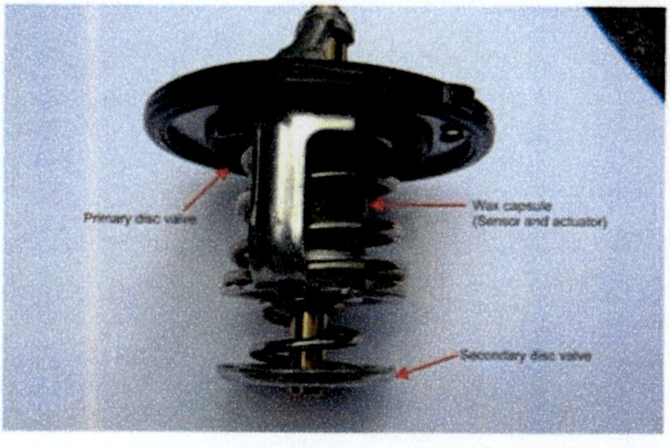

Fig. 11.5 Double valve automotive engine cooling thermostat [3]

This type of thermostat operates using a sealed chamber containing a wax pellet that melts and expands at a set temperature. The expansion of the chamber operates a rod that opens a valve when the operating temperature is exceeded. The operating temperature is determined by the composition of the wax. Once the operating temperature is reached, the thermostat progressively increases or decreases its opening in response to temperature changes, dynamically balancing the coolant recirculation flow and coolant flow to the radiator to maintain the engine temperature in the optimum range.

11.2.7 Valves, Pipes, and Hoses

In addition to the thermostat valve, a pressure cap provided on top of the radiator limits the pressure build-up inside the radiator due to the hot fluids. This cap is operated by a valve, which opens the radiator cap when the pressure inside the radiator goes above a certain predetermined value.

The cooling system consists of multiple hoses that connect the different components within the system. The radiator is linked to the engine by rubber hoses that carry the coolant from and to the radiator, between the radiator and engine, and the radiator and the DAT.

11.2.8 Fan

The higher the flow rate of the air, the higher the rate of heat transfer from the radiator to the atmospheric air. Hence a radiator fan is used to increase the volume of air coming into contact with the radiator. When this air passes through the radiator fins, it absorbs heat from the coolant flowing through the radiator. Automobile cooling systems use one of two types of radiator fans: electrical or mechanical.

Mechanical Fan

This type of fan is driven by the engine using a pulleys drive mechanism. The fan is controlled by a thermal fan clutch [4]. It senses radiator heat and turns it on when the engine is running at a higher temperature than desired.

Electrical Fan

Unlike a mechanical fan, the electric fan, as the name suggests, is driven by electric power. It is controlled by a computer, based on inputs from a coolant with the sensor monitoring the temperature of the engine coolant [5]. The electric drive provides flexibility in the location of the fan and heat exchanger pack, unlike a mechanical drive where it has to be aligned to the engine crankshaft.

Fan Clutch

A fan clutch is designed to let the fan freewheel at low temperatures when cooling is not needed, allowing the engine to warm up faster. Also, this removes unnecessary load on the engine. As temperatures increase, the clutch engages so that the fan is driven by engine power and moves air to cool the engine.

Types of fan clutch include the viscous type coupled with a bi-metallic strip, and the electronically controlled.

The fans are further classified into two types based on the operation: pull-type and push type. The pull-type draws air towards the radiator and is placed such that it follows the radiator in the path of the cooling air. The push-type fan, mounted between the radiator and the outside of the car pushes air from the outside through the radiator.

11.3 Operating Conditions

To determine the size and configurations of a heat exchanger to be used, one needs to know the rate of heat dissipation required from the engine.

The rate of heat rejection depends on factors such as the speed of rotation of the engine, the power output of the engine, the combustion characteristics of the engine including whether or not it uses pre-combustion, turbulence chambers, direct injection, etc.

The maintenance of the temperature of each hot fluid at or near design value contributes not only to the safe and efficient operation of the engine but to the extension of its life. Very high temperatures can lead to the breakdown of lubrication, whereas low-temperature operation usually results in high wear rates of rings and liners. The charge air, however, can be cooled to a temperature substantially below the maximum recommended operating value without affecting the engine, although very low temperatures can lead to starting difficulty in engines, and may lead to condensation of water vapour in conditions of high humidity.

11.4 Heat Transfer in Engines and Influencing Factors

The rate of heat rejection in an engine is proportional to the flow of air and fuel and hence is higher as the speed of the engine increases. Heat rejection is higher at the rated power or a higher speed. For engines using turbulence chambers, combustion efficiency improves, hence increasing the heat rejected.

Conversely, at higher loads, the percentage of fuel power lost to heat rejection is lowered due to an increase in the proportion of engine brake power in fuel energy, and the situation in lower loads is opposite, with a higher fuel power spent on coolant heat rejection and exhaust gases.

The study of heat transfer in engines is an attempt to learn about the combustion process from measured cylinder pressure data. Heat release calculations can provide useful information even though these are based on models that contain many assumptions, some of which could be invalid. Several approaches to heat release analysis can be found in the technical literature but the Krieger and Borman [6] have given the one that is most widely used [7].

The apparent heat release rate is the rate that fuel and air would need to react to create equilibrium combustion products that result in the observed cylinder pressure curve. This rate is calculated by applying the First Law of Thermodynamics to a control volume consisting of the engine cylinder. No mass flow of air is assumed to occur into or out of the cylinder. With these assumptions, the energy equation becomes:

$$\frac{dU}{dt} = \dot{Q} - P\frac{dV}{dt} + \dot{m}_f h_f \tag{11.1}$$

where

U	*internal energy of the cylinder contents*
t	*time*
$\dot{Q}$	heat transfer rate
P	cylinder pressure
V	cylinder volume
$\dot{m}_f$	fuel addition rate (rate fuel burning)
hf	fuel enthalpy

The heat transfer rate from the cylinder can be estimated using empirical correlations that are based on heat transfer measurements from other engines. The internal energy and gas constant values are functions of the temperature, pressure, and composition. The pressure, volume, and temperature themselves are related to each other using the ideal gas equation. The cylinder volume is calculated from geometry as a function can be calculated as a function of the crank position by geometry. Applying these relationships, Eq. (11.1) can be rearranged to predict the rate of fuel burning. Multiplying it with the calorific value gives the apparent heat release rate.

Figure 11.6 shows the heat release curve for an engine. Since the compression is adiabatic, the heat release is zero till shortly after the start of fuel injection. As the fuel evaporates and endothermic pre-flame reactions occur, the temperature dips slightly. Subsequently, auto-ignition occurs, and the heat release rate rises rapidly as the premixed fuel burns. Once the premixed fuel has burnt, the heat release rate is decided by the rate at which the fuel and air mix due to which it slows down.

The heat release rate thus gives several insights into the combustion process such as the start of combustion and the effect of design changes on the combustion process.

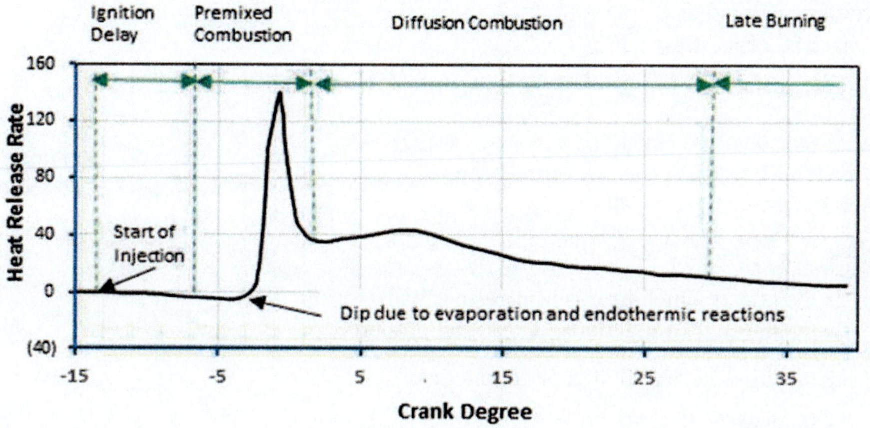

Fig. 11.6 Rate of heat release [8]

11.5 Modes of Heat Transfer in Engine Cooling System

Once the heat rejected by the engine to the coolant is estimated, we need to calculate the temperatures attained by the components as heat is transferred through the cooling circuit to the ambient. Heat transfer can occur by one of the three methods—conduction, convection, or radiation. In an engine, all three modes of heat transfer are present. It is important to understand the mechanisms by which the heat is transferred at different locations on its path from the combustion chamber to the ambient.

11.5.1 Conduction

In an automotive cooling system, heat transfer by conduction takes place between.

- Cylinder liner and the engine block
- Cylinder and the valves and valve stems
- Engine block and the coolant
- Coolant and the radiator fins
- Oil and the pipes in the oil cooler
- Exhaust gas and the tubes carrying it in the EGR.

Conduction is the transfer of thermal energy through direct contact. At a microscopic level, it is the transfer of the internal energy of particles by collisions and the movement of electrons within a body. Factors affecting the rate of heat transfer by conduction.

- temperature difference
- length

- cross-sectional area
- material properties

$$Q = \frac{kA\Delta T}{l} \quad (11.2)$$

where

Q is the rate at which heat is transferred, W
k is the thermal conductivity; a material property
A is the cross-sectional area of conduction
l is the length.

For effective conduction, the area of contact between the surfaces needs to be maximized while keeping the distance of heat transfer minimal.

Material plays a prominent role in the conduction of heat. Hence aluminum, a metal with thermal conductivity of 240 W/m K (at 127 °C) is used for the radiator tubes over steel, which has a thermal conductivity of only 33 W/m K (at 122 °C). The resultant reduced weight is an added benefit.

11.5.2 Convection

The following are some examples of instances of heat transfer taking place by convection.

- The combustion chamber and the cylinder walls
- Cylinder liner and coolant
- Engine lubricant oil and the cylinder wall
- Engine block and coolant at the water jacket
- Engine block and the atmospheric air
- Radiator tubes and the coolant
- Atmospheric air and the radiator tubes
- CAC tubes and the atmospheric air
- EGR tubes and the coolant at the EGR cooler
- Exhaust gas and atmospheric air.

Convective heat transfer is caused by the mass motion of a fluid. Heat convection occurs to the surface of an object where the surrounding fluid is heated and energy is moved away from the source of heat. It only occurs if there is a temperature difference between the surface and that of the surrounding fluid. Although often discussed as a distinct method of heat transfer, convective heat transfer involves a component of conduction. The governing equation of heat convection is given by:

$$q_h = hA(T_s - T_f) \tag{11.3}$$

where

q_h convective heat on the surface
A cross section of the boundary surface
T_s surface temperature
T_f temperature of the surrounding fluid

11.5.2.1 Types of Convection

Natural Convection

Also known as free convection, it occurs when fluid motion is caused by buoyancy forces that result from the density variations due to differences in fluid temperatures.

Natural convection, however, is much less efficient than the forced convection method due to the lesser volume of fluid that comes in contact with the heat transfer surface, effectively reducing the area of contact.

Natural Convection is buoyancy-driven, while viscous forces oppose the motion is determined by Grashof number, a ratio of the two opposing forces

$$Gr = \frac{g\beta \Delta T L^3}{\nu^2}$$

g *acceleration due to gravity*
β *coefficient of thermal expansion*
ΔT *temperature difference between heating surface and bulk of fluid*
ν *kinematic viscosity*
L *length scale*

Forced Convection

Forced convection occurs when a fluid is forced to flow over the surface by an internal source such as fans, stirrers, and pumps, creating an artificially induced convection current. In many real-life applications, mixed convection of natural and forced convections occurs.

The ratio of convective heat transfer to conductive heat transfer in the fluid is the Nusselt number:

$$Nu = \frac{hl}{k}$$

h *convective heat transfer coefficient*
l *length scale*
k *conductivity*

Empirical relationships have been established between Nusselt number, Reynolds number, and Prandtl number for typical flow cases.

$$Nu = Re^{x} Pr^{y} \tag{11.4}$$

x and y take different values based on the nature of the flow.

11.5.3 Radiation

Thermal radiation is the transfer of heat by electromagnetic waves. It is different from conduction and convection as it requires no matter or medium to be present. The radiative energy passes perfectly through a vacuum as well as clean air.

Although radiation heat transfer plays a minimal part in the cooling system, the contribution to the cooling load is higher. When parked in the sun, the interior of the vehicle is heated predominantly through radiation. The power output through radiation is given by:

$$Q = \epsilon \sigma A T^4 \tag{11.5}$$

where

Q *heat*
ϵ *emissivity*
σ *Stefan Boltzmann constant*
A *surface area*
T *Absolute temperature, K.*

Radiosity from a surface is defined as the total radiant heat leaving a surface. For an opaque body, this includes emitted radiation and reflected radiation, Fig. 11.7.

$$J = (1 - \varepsilon)I + \varepsilon \sigma T^4$$
$$Q = 1 - J$$
$$Q = \frac{\varepsilon}{1 - \varepsilon}\left(\sigma T^4 - J\right)$$

When there are several radiating surfaces the heat transfer between any two surfaces i and j is given by

Fig. 11.7 Radiosity

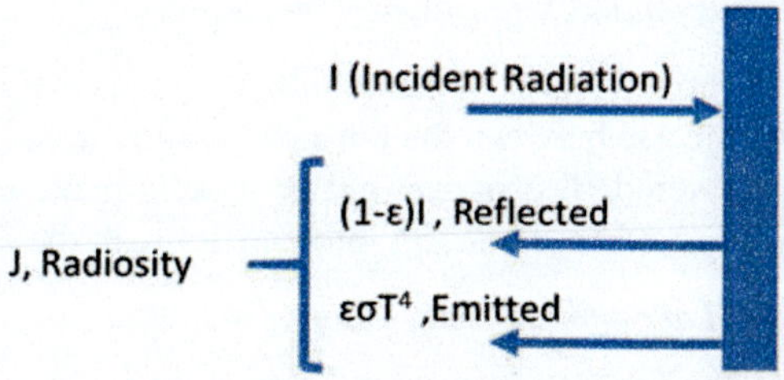

Fig. 11.8 Network method

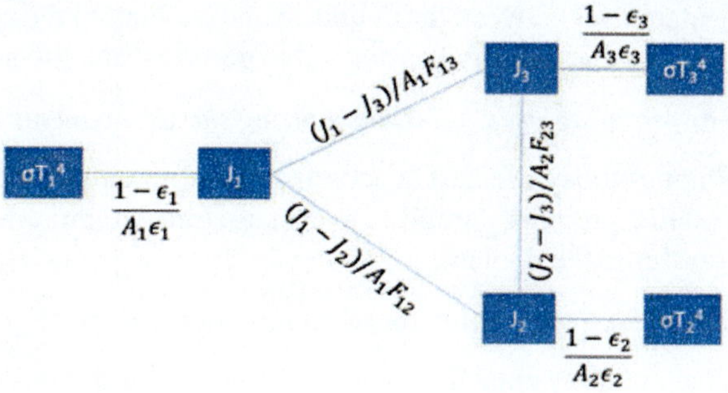

$$Q_{ij} = A_i F_{ij} \sigma (J_i - J_j) \tag{11.6}$$

F_{ij} is the view factor that indicates the proportion of radiation from a surface *i* that is received by surface j. It is a geometric parameter and the mathematical formulation is presented in the section on Surface to Surface heat transfer. The above equations are analogous to the electrical resistance equation. Therefore, when radiation transfer occurs between multiple surfaces, it is modelled similar to an electrical network. Figure 11.8 shows an example of such a network between 3 surfaces.

11.6 Design Considerations in the Cooling System

Based on the heat energy to be dissipated and the mechanism by which it is transferred in various components, the design of the cooling system is modified to ensure appropriate temperatures in the engine components and the cooling system. Maximum possible efficiency of the cooling system is attained by controlling several design characteristics such as:

Cooling Capacity

Appropriate heat exchangers need to be chosen to handle the heat load generated by a vehicle. The need for multiple types of heat exchangers (radiator, CAC, EGR cooler, etc.) makes it possible to compartmentalize the cooling loads.

Location and Size of Cooling Channels

The effectiveness of the cooling system is mainly due to the proximity of the coolant. As the gap between the hot surface and the cooling jacket increases, the heat transfer is lowered. Too narrow water jackets make for poor heat transfer by limiting the volume of fluid coming into contact with the surface of the engine block.

The Layout of Cooling Channels

The water jacket goes around the cylinders in the engine block to draw the excess heat produced within. Care should be taken to surround the cylinders to ensure uniform heat transfer between the cylinders and coolant. In the cylinder head, the water jacket needs to reach the valves and valve ports, especially on the exhaust side.

The Material Used in Manufacturing the Components

Engines today are cast predominantly from either iron or aluminum alloys. While cast iron provides durability to the structure, aluminium castings allow for better heat transfer to the coolant.

Chemical Composition and Additives of Coolant

Water is an acceptable coolant if the only feature to be considered is the heat-carrying capability. But the automotive coolants used are glycol-based, with ethylene glycol–water mixtures being the most common. The reason for this is that water has a lower boiling point and higher freezing point than glycol-based coolants [9, 10]. This means that glycol in the coolant prevents it from freezing in winters and boiling in presence of extreme temperatures of the engine.

Another glycol-based coolant used is propylene glycol and water, with the major difference being that the propylene glycol is less toxic.

The coolant has other additives that prevent corrosion and the formation of rust in the cooling channels that may happen when only water is used as a coolant.

Pressure in the System

The pressure created by the coolant pump decides the flow rate of the coolant in the cooling system. Higher pressure means faster flow through the cooling channel and faster heat transfer between the hot surface and the coolant. Too high pressure is also not a desired factor

Flow Rate and Heat Transfer of Coolant

The flow rate needs to be high enough to keep a steady stream of coolant in contact with the hot surface.

11.7 Computational Fluid Dynamics

Computational Fluid Dynamics (CFD) is the field of simulating flow and heat transfer phenomena in three-dimensional space and time using computers. This is done by solving conservation equations of mass, momentum, and energy, and thermodynamic equations at a set of points in the space, and over a desired interval of time to obtain flow and thermal properties at those points, and at specified instances of time within the interval. Flow properties include pressure, density, and velocity. In some specific problems, species or phase fractions are also predicted. The thermal properties of the flow are temperature and enthalpy. The **solution** of a CFD problem is, therefore, a set of these flow and thermal properties, also called **variables**, corresponding to each point in the domain of interest. The value of the solver variables in between any of the two points is obtained through interpolation functions. Once these variables have been solved, other derived values such as fluxes, forces and moments, shear stress, etc. can be obtained through arithmetic operations or integration over space or time.

The first step towards a CFD analysis is to define the regions/volumes of interest, or **domains**, in which it is intended to predict flow and thermal parameters. The set of points for analysis are identified within this domain by a **discretization** or **meshing** process (Fig. 11.9). It should be kept in mind that domains are regions in space and not physical objects or material. Domains should have well-defined **boundaries** that separate them from the rest of the universe and each other. Boundaries are the locations where the domain interacts with the external region through the exchange of mass, momentum, energy, or species. The manner of this exchange is defined by equations known as **boundary conditions (BC)**. Boundary conditions are either provided by specifying the value of the variable (Dirichlet BC) or the gradient of the variable (Neumann boundary condition). Specifying appropriate boundary conditions is a requirement for formulating a mathematically closed problem. An accurate definition of boundary conditions is the key to obtaining a correct solution. In case the boundary conditions vary with time, or we require capturing time-varying effects in the flow, it is necessary to know the initial conditions—the values of the flow variables at every point in the domain at the instance of time from which we are interested in predicting the flow.

The design of a cooling system involves ensuring that the engine and the corresponding cooling system meet the requirements. There are thus two primary flow phenomena happening in a cooling system.

- The flow of coolant in the coolant circuit.
- The flow of air through the heat exchangers.

The study of coolant flow can be done in the entire circuit or parts of the circuit that are of primary interest, provided the boundary conditions are known. Prediction of engine temperatures considering the flow of coolant through it is an important engineering problem in cooling system design based on which the engine coolant flow path or water jacket can be designed. The analysis also gives the required coolant flow rate and restriction offered by the engine at that flow rate which is an important

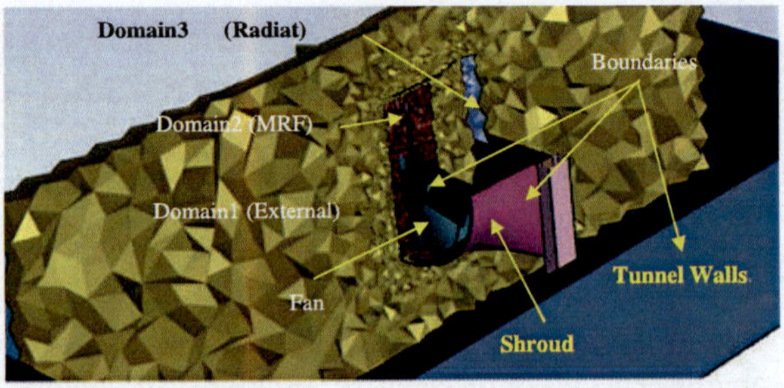

Fig. 11.9 Mesh around a cooling system

input for sizing the pump. The flow of air through the heat exchanger is done in Under-Hood Thermal Analysis. The cooling system is below the hood of the vehicle and the analysis studies the airflow in this region. The entire vehicle along with the heat exchangers and the fan is considered in this analysis to predict the amount of air that would flow through the radiator. Based on the heat released by the engine, the heating up of the air is calculated and the coolant temperatures are estimated. Usually, the flow of the coolant is not solved in an underhood analysis. The effect of the coolant is introduced by incorporating models based on known information such as characteristic curves from physical experiments or component level analysis.

A CFD **solver** is a computer program that can solve the governing equations of flow to satisfy the boundary conditions over the domain. Most CFD solvers solve some form of the Navier Stokes Equation (NSE). Most general-purpose CFD solvers incorporate multiple forms of the Navier Stokes equation to predict different flow phenomena. The Navier Stokes equation describes the transient motion of a compressible, viscous fluid using equations of conservation of mass momentum and energy. The processes of heat transfer and flow are closely coupled by the phenomenon of convection. In most engineering systems, and particularly in the automotive cooling system, convection is an important mode of heat transfer. Convective heat transfer itself is a boundary condition for the conduction problem since in most cases the sink of the heat transfer is a fluid, which is ambient air, and heat transfer is done through the coolant liquid. Radiation acts as an additional heat source for conduction and convection problems. The NSE is of the form:

Conservation of Mass

$$\frac{\partial \rho}{\partial t} + \nabla.\left(\rho \vec{u}\right) = S_m \tag{11.7}$$

Conservation of Momentum

$$\frac{\partial (\rho \vec{u})}{\partial t} + \nabla.\left(\rho \vec{u}\,\vec{u}\right) = -\nabla.\left(\bar{\bar{T}}\right) + \rho \vec{b} + \vec{S_M} \tag{11.8}$$

where

$\overline{\overline{T}} = \overline{\overline{\tau}} - pI$

$\overline{\overline{\tau}} = \mu\left(\nabla\vec{u} + \nabla\vec{u}^{T}\right) - \frac{2}{3}\mu\nabla.\vec{u}$

Conservation of Energy

$$\frac{\partial(\rho E)}{\partial t} + \nabla.\left(\vec{u}\,(\rho E + p)\right) = \nabla.((\lambda + \lambda_t)\nabla T) + S_h \tag{11.9}$$

h	$E + \frac{p}{\rho} + \frac{u^2}{2}$
ρ	*density*
p	*pressure*
$\vec{u}$	*velocity*
h	*enthalpy*
μ	*dynamic viscosity*
t	*time*
S_m	*mass source*
S_M	*momentum source*
k	*thermal conductivity*
S_h	*enthalpy source term*
E	*internal energy*
λ	*conductivity*
λ_t	*turbulent conductivity*
$\vec{b}$	*body force.*

In all the equations the first term on the LHS represents the rate of increase or decrease of the conservation property (mass, momentum, or energy) with time. The second term represents the net influx of conservation property or convective terms. The RHS is a sum of the sources and sinks, and the terms that drive or dissipate the property. e. g., momentum is driven by pressure and dissipated by viscous forces. Energy is driven by temperature difference and dissipated viscous forces. The general form of such a transport equation for any quantity ϕ is, therefore:

$$\frac{\partial\phi}{\partial t} + \nabla.\left(\phi\vec{u}\right) = \nabla.(K\nabla\phi) + S \tag{11.10}$$

$$Time\ Derivative + Convective\ terms = Forcing\ Terms + Dissipative\ Terms + Source\ Terms$$

The NSE as seen in Eqs. (11.6–11.9) is a set of PDEs in the form of a transport Eq. (11.9) and there exist different methods to solve those equations, such as the Finite Difference Method, Finite Element Method, and Finite Volume Method. The Finite Volume Method involves discretizing the domain into small volumes or elements.

The equations are formulated as a conservation law i. e., the rate of increase or decrease of the conserved properties (mass, momentum, energy, species, etc.) in each element is equal to that flowing in and out of the elements respectively added to the sources or sink of the property, respectively, in the domain.

11.8 Forms of NSE

Solving the Navier Stokes equation in the form presented in Eqs. (11.6–11.9) is computationally intensive as it requires that the computational mesh resolves all scales of turbulence from the smallest scales (known as the Kolmogorov scale) to the largest scale, which is the size of the domain. So usually a reduced form of the NSEs such as RANS, LES, DES, etc. is used.

11.9 DNS

Direct Numerical Simulation (DNS) is the method of solving the Navier Stokes equation without any simplifications or assumptions in the equation and resolving all scales of the flow. In turbulent flow, the turbulent energy is produced in the larger eddies and dissipated through molecular viscosity in the smallest ones. The transfer of energy from the large scales to the smallest scales happens through a series of intermediate-range eddies. This phenomenon is known as the energy cascade. So, in the smallest scale of turbulence, the Kolmogorov scale, the kinetic energy is of the same order of energy dissipation through viscosity. This gives the smallest length scale as:

$$\eta = \left(\frac{\nu^3}{\varepsilon}\right)^{\frac{1}{4}} \tag{11.11}$$

The corresponding time scale is:

$$\tau_\eta = \left(\frac{\nu}{\varepsilon}\right)^{\frac{1}{2}} \tag{11.12}$$

$\eta = lengthscale$

$\tau_\eta = timescalecorrespondingtolengthscale\eta$

$\upsilon = kinematicviscosity$

$\varepsilon = turbulentkineticenergy.$

To predict turbulence, it is necessary to resolve these smallest scales necessitating the size of the computational cell to be smaller than the Kolmogorov length scale. The time step would also need to be smaller than the time scale to ensure the Courant-Friedrich-Levy (CFL) condition for numerical stability. It can be seen from Eqs. (11.11 and 11.12) that at a very high Reynolds number (Re) the Kolmogorov length and time scales become very small. This makes use of DNS prohibitive for the flow of high Reynolds numbers. The number of nodes required to resolve all scales is given below [11].

$$N = Re^{9/4} \tag{11.13}$$

$Re = Reynoldsnumber$

The computational cost in terms of the number of operations performed scales as Re^3. So, with the current technology of computers; DNS can only be used for very low Reynolds number turbulent flows [12].

The numerical methods used in DNS include Finite difference methods, spectral methods, and spectral element methods. While the finite difference methods approximate the derivatives using a finite-difference formulation, the spectral methods formulate the solution as a Fourier series and solve for the coefficients to satisfy the differential equations. The spectral methods work well with smooth functions but fail to predict discontinuities such as a shock wave [13].

11.10 RANS

The most commonly used form of the Navier Stokes equation in CFD is known as Reynolds Averaged Navier Stokes (RANS) equation. This method eliminates the requirement of resolving the turbulence scales by considering the velocities as a sum of a time-averaged component $\overline{u}$ and fluctuating component u'.

$$\rho \overline{u}_j \frac{\partial \overline{u_i}}{\partial x_j} = \rho \overline{b}_i + \frac{\partial}{\partial x_j}\left[-\overline{p}\delta_{ij} + \mu\left(\frac{\partial \overline{u_i}}{\partial x_j} + \frac{\partial \overline{u_j}}{\partial x_i}\right) - \rho\overline{u_i' u_j'}\right] \tag{11.14}$$

Except for the last term on the RH, the above equation is similar to Eq. (11.8) except that the velocities in Eq. (11.8) are replaced by time-averaged velocities in Eq. (11.14).

The term $\rho\overline{u_i' u_j'}$ is known as Reynolds stress and is unknown due to which the set of equations is not closed. The closure problem is solved by modelling the Reynolds stress term using additional equations called turbulence models. A complex model is to solve transport equations of the form shown in Eq. (11.10) for each of the terms in the Reynolds stress tensor. Most commonly used turbulence models adopt a simpler formulation by following the Boussinesq approach where the Reynolds Stresses are related to the mean velocities

$$-\overline{u_i^{'}u_j^{'}} = \mu_t\left(\frac{\partial u_i}{\partial x_j} + \frac{\partial u_j}{\partial x_i}\right) - \frac{2}{3}k\delta_{ij} \tag{11.15}$$

It can be seen that this formulation for Reynolds stress is similar to the formulation for $\overline{\overline{\tau}}$ in Eq. (11.8) by relating it to the mean strain rate $\left(\frac{\partial u_i}{\partial x_j} + \frac{\partial u_j}{\partial x_i}\right)$. μ_t is the turbulent viscosity while k is the turbulent kinetic energy defined as

$$k = \frac{1}{2}(\overline{u_i^{'}u_i^{'}} + \overline{u_j^{'}u_j^{'}} + \overline{u_k^{'}u_k^{'}}) \tag{11.16}$$

This method is used in Spalart Allmaras (SA), k-ε, and k-ω models. It can be seen in Eq. (11.15). The Boussinesq approach presents a computationally less intensive method to model turbulence due to isotropic turbulent viscosity, unlike the Reynolds Stress Model where the asymmetric nature of turbulence is modelled. The isotropic assumption of turbulent viscosity works well for a wide variety of flows such as boundary layer flows and jet flows. However, for flows where the anisotropy of turbulence dominates such as highly swirling flows, the computational cost involved in the RSM becomes justified.

11.10.1 Spalart Allmaras Model

The SA turbulence model solves a single transport equation of the form shown in Eq. (11.10) for the Reynolds stress. The turbulence kinetic energy k is not solved and the term involving k in Eq. (11.15) is neglected in estimating Reynolds stress and it is seen that this does not lead to inaccuracies in problems the model is typically applied to. Spalart Allmaras model is widely used for **attached flows** such as the aerodynamics of aerofoils under a low angle of attack. Recent modifications to the model [14] have included mechanisms to address inaccuracies that creep in because of coarse grids that do not resolve the flow adequately and infeasible conditions that result in some transient flow problems. These are formulated to reduce numerical issues near the interface between turbulent and irrotational regions.

11.10.2 Two Equation Models

As discussed in the previous section, the k term is neglected in the one-equation S-A model. Two-equation models handle flow in several regimes, such as separated flows, better than the S-A model.

11.10.2.1 K-ε Models

The $k - \varepsilon$ model is a two-equation model formulated as two transport equations –one for turbulent kinetic energy, k and the other for its rate of dissipation, ε, developed by Launder et al. [15].

$$\frac{D(\varepsilon)}{Dt} = \frac{1}{\rho}\frac{\partial}{\partial x_k}\left[\frac{\mu_t}{\sigma_\varepsilon}\frac{\partial \varepsilon}{\partial x_k}\right] + \frac{C_1 \varepsilon}{k}\frac{\mu_t}{\rho}\left(\frac{\partial u_i}{\partial x_k} + \frac{\partial u_k}{\partial x_i}\right)\frac{\partial u_i}{\partial x_k} - C_2\frac{\varepsilon^2}{k} \quad (11.17)$$

$$\frac{Dk}{Dt} = \frac{1}{\rho}\frac{\partial}{\partial x_k}\cdot\left[\frac{\mu_t}{\sigma_k}\frac{\partial k}{\partial x_k}\right] + \frac{\mu_t}{\rho}\left(\frac{\partial u_i}{\partial x_k} + \frac{\partial u_k}{\partial x_i}\right)\frac{\partial u_i}{\partial x_k} - \varepsilon \quad (11.18)$$

$$\mu_t = \rho C_\mu \frac{k^2}{\varepsilon} \quad (11.19)$$

The term $\frac{1}{2}\left(\frac{\partial u_i}{\partial x_j} + \frac{\partial u_j}{\partial x_i}\right)$ is the strain rate and is represented by the notation S_{ij} in tensor form.

C_{1e}, C_{2e}, Cu, σ_k, and σ_e are model constants determined by tuning the model for the most common turbulent flows such as boundary layers, mixing layers, and jets. They have been found to work well for a wide variety of these flows.

The RNG k-ε model is a modification to the k-ε model using a statistical technique called the renormalization group (RNG) theory. This equation can handle low Re flows better, provided the near-wall treatment is appropriate. It also improves the accuracy of swirling flows by introducing an additional term. Rapidly strained flows show lower turbulent viscosity than that predicted by the standard k-ε model, resulting in better predictions [16].

Both the Standard and the RNG k-ε models can give non-realizable or infeasible results. For example, the formulation for incompressible Reynolds stress is

$$\frac{2}{3}k - 2\nu_t\frac{\partial U}{\partial x} \quad (11.20)$$

From Eqs. (11.19) and (11.20) we can see that Reynolds stress can take negative values when

$$\frac{k}{\varepsilon} > \frac{1}{3C_\mu}$$

However, this is infeasible as Reynolds stress is always positive. This problem is solved by making the C_μ variable. The ε equation has also been modified based on the mean square vorticity fluctuation.

Other variations of k-ε include the V2F model which solves four transport equations and works well for separation-dominated flows [17]. The Abe Kondoh Nagano (AKN) model introduces damping functions in the eddy viscosity term given in Eq. (11.19) making it suitable for low Re flows.

11.10.2.2 K-ω Models

The *k*-*ω* model developed by Wilcox [18], similar to the *k*-ε model is an empirical model that implements transport equations for Turbulent Kinetic Energy (*k*) and specific Dissipation rate (*ω*) is the ratio of ε and *k*. The model equations are

$$\frac{\partial(\rho k)}{\partial t}+\frac{\partial \rho k u_i}{\partial x_i}=\rho\tau_{ij}\frac{\partial u_i}{\partial x_j}--\beta^*\rho k\omega+\frac{\partial}{\partial x_j}\left[\left(\mu+\frac{\sigma^*\rho k}{\omega}\right)\frac{\partial k}{\partial x_j}\right] \quad (11.21)$$

and

$$\frac{\partial(\rho\omega)}{\partial t}+\frac{\partial \rho\omega u_i}{\partial x_i}=\frac{\alpha\omega\rho\tau}{k}\frac{\partial u_i}{\partial x_j}-\beta\rho\omega^2+\sigma_d\frac{\rho}{\omega}\frac{\partial k}{\partial x_j}\frac{\partial\omega}{\partial x_j}+\frac{\partial}{\partial x_j}\left[\left(\mu+\frac{\sigma\rho k}{\omega}\right)\frac{\partial\omega}{\partial x_j}\right] \quad (11.22)$$

where α, β, β^*, σ, σ^*, σ_d are constants of the equation.

The formulation of the diffusive terms has a mechanism to damp turbulent viscosity for low-Re flows.

The Baseline (BSL) k-ω by Menter [19] model blends the *k*-*ω* model with the *k*-ε model to take advantage of the accurate near-wall properties of the *k*-*ω* model and the independence to free stream conditions of the *k*-ε model. The SST *k*-*ω* model, in addition to the BSL *k*-*ω* model, incorporates a limiting term to the eddy viscosity formulation which helps better prediction of flow separation from smooth surfaces [16].

11.10.3 Wall Treatment

The no-slip condition near the wall introduces velocity gradients and shear stresses that generate turbulence. Even in turbulent boundary layers, there is a thin laminar sublayer close to the wall, as shown in Fig. 11.10, from where dissipation due to

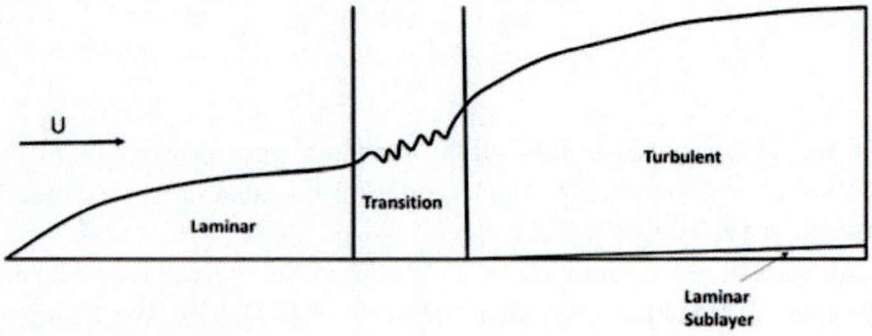

Fig. 11.10 Boundary-layer

molecular viscosity is high. On moving away from the wall the importance of molecular viscosity decreases and turbulence increases till flow becomes fully turbulent and is defined well by the RANS equations. It is necessary to capture the velocity near the wall accurately. In RANS models, where the eddies are not resolved, models are required to determine the variation in wall velocity. Wall functions are the most widely used method of modelling near-wall flow properties. The standard wall function employs a log law to define near-wall velocities outside the laminar sublayer.

In the laminar sublayer, viscosity is defined by the laminar stress–strain relationship. With the standard wall functions, the predictions at very high grid refinements lose accuracy [16]. Scalable wall function incorporates an adjustment to the standard wall functions to improve predictions at all levels of grid refinement. By Reynolds analogy, the temperature also follows the same variation profile as velocity. Therefore, a wall function for temperature with logarithmic variation in the turbulent region and linear variation in the viscous sublayer is also used.

11.11 Large Eddy Simulation

Large Eddy Simulation belongs to a class of simulations known as Scale Resolving Simulations (SRS). These methods play an important role in phenomena that require the resolution of turbulence scales such as aero-acoustics. Additionally, these are more accurate than RANS for problems with large regions of separated flows. The applicability of RANS to separated flows is limited to self-similar flows such as jets.

LES is a method in which the large eddies, being more geometry dependent, are resolved as in the DNS. The smaller eddies which are less problem dependent and for which universal models can be developed are modelled using subgrid-scale (SGS) models. Resolving the larger eddies alone helps reduce the mesh count as the mesh size needs to be just small enough to capture the large eddies, unlike the DNS where it needs to be of the Kolmogorov scale. The smaller scales are filtered out using a low pass filter since they have a higher frequency. Smagorinsky Lilly model is one of the commonly used SGS models in practice. The Germano model [20] also known as the dynamic Smagorinsky model, eliminates certain drawbacks of the Smaogrinsky model, making it applicable to a larger range of flows [21]. The LES equations resolve eddies from the scale of the domain to the size of the filter width. This requires that the grid size, though larger than the Kolmogorov Scale needed for DNS, is smaller than what can be used in RANS. Unlike RANS this method requires grid refinement in all directions—not just normal to the wall. The grid size increases exponentially with the Reynolds number. In wall-bounded flows where the larger scales themselves are small, the application of LES is limited to low-Re flows. Resolving the eddies in wall-bounded flows for high Reynolds numbers would require very small elements. This limits the practical use of LES in industrial flow problems. LES is therefore commonly used in conjunction with RANS models where RANS is used in wall-bounded regions while LES is used in regions of large separation. A few of these models include Detached Eddy Simulation (DES), Wall Modelled LES, and Embedded or Zonal LES models [22].

11.12 Detached Eddy Simulation (DES)

The DES, developed by Spalart [19, 23] and his co-workers [24–26], as explained earlier, uses RANS in wall boundary layers and LES in regions of free sheer flows. The selection between these two models is carried out using simple rules involving the cell length and a parameter, the turbulent length scale. LES is applied in cells with edge lengths smaller than the length scale, else RANS is used. DES is one of the first models that coupled RANS and SRS models allowing SRS models to be used for industrial applications. As the switch from RANS to LES happens based on the element size, it has to be ensured that extremely small sizes which can trigger the LES solver are not used in near-wall regions. The use of LES equations in the wall boundary layer could trigger early separation called Grid Induced Separation (GIS). The delayed DES model addresses this by using a function that is designed to ensure that RANS is used in the wall boundary layer. Similarly, it should be ensured that in regions of separation, the element size is not only taken to limiting size which would enable LES but it should be lower than the filter width as per the requirements of LES.

DES is becoming increasingly used for automotive aerodynamics applications [27, 28] but rarely used for underhood thermal analyses possibly because of the computational cost arising particularly out of the inherently transient nature of SRSs.

11.13 Gridless Methods

A significant portion of the time in a CFD analysis is spent in generating a good grid. A good grid is primary in ensuring convergence and stability of the solver. A poorly generated grid might cause divergence. It can also lead to inaccurate results and long analysis times. Since a mesh is created using CAD models developed for manufacturing, they have small gaps between parts. Retaining these gaps would require very small element sizes to be used in the mesh though it has practically no effect on improving the accuracy of the results. Therefore, these gaps are removed by a process known as CAD clean-up (defeaturing) to create a model suited for CFD analysis. This model has only those small features such as fillets, curves, small surfaces, gaps, etc. which are expected to affect the flow result. The other features are removed. This involves significant manual work even though a high level of automation of this is done with modern mesh generation software through improvements in mesh wrapping algorithms.

Alternate CFD software exists which eliminates the activity of mesh creation. However, the limitations of the formulation of the solver equations allow them to be used only for a few specific applications.

11.13.1 SPH

The Smoothed Particle Hydrodynamics method is a Lagrangian mesh-free method where the fluid is modelled as particles. Apart from being meshless, the computational cost per particle with the meshless method is significantly lower than that per cell for conventional grid-based methods. However, the number of particles used is usually higher than the number of cells used in Mesh-based methods. SPH methods particularly find favor in liquid–gas multiphase simulations as the interface between the liquid and the gas is automatically computed, unlike RANS-based approaches where the interface needs to be tracked. Not having a grid also makes it amenable to massive parallel processing. General-purpose computation GPUs have a large number of cores making parallel processing very efficient. Hence several SPH codes can run on GPU in addition to CPU or GPU alone. The popularity of SPH is more as a qualitative visualization tool than as a tool that accurately predicts flow parameters [29]. This makes it very much suited for rendering fluid motion in video games [30, 31]. While a few studies have established the accuracy of SPH to specific problems [32] and studied sensitivity to smoothening length and domain shape, a more comprehensive sensitivity analysis of all the SPH parameters needs to be carried out.

In SPH the domain is discretized using integration points also called particles that move with their velocity and carry flow and material properties such as density, velocity, and stress. At each instant of time, the new properties of the particle are obtained by interpolating these values between the neighbouring particles. The interpolation is done using a weighting/smoothing function called the kernel which is of key importance in SPH methods. The kernel function selected should fulfill certain basic criteria. The kernel is a function of the distance between two particles and a parameter known as the smoothing length. Some of the kernels used in SPH are quadratic [33], cubic spline [34], quantic [35], and quintic spline [36]. The cubic spline is the most commonly used kernel function seen in literature [29].

SPH can be used to simulate a wide variety of phenomena such as multiphase flow including turbulence [37], evaporation [38] and melting [39], aerodynamics [40], and fluid–structure interactions [41].

11.14 LBM

Lattice Boltzmann Method (LBM) is a CFD technique that has been gaining popularity in several fields particularly the automotive industry due to its being fast for transient computations, eliminating long pre-processing and solving times.

The Lattice Boltzmann method discretizes the domain using groups of fluid particles, with a velocity distribution function, located at the vertices of a lattice structure. These groups of particles move to the neighbouring lattice points in the direction of the linkages. If two particles collide the resultant distribution is calculated using a discretized form of Boltzmann's equation. This equation involves a collision operator. Much of the development of LBM has been in identifying a distribution function and collision function that is suited for fluid flow problems with good accuracy and without numerical noise. The Bhatnagar, Gross, and Krook (BGK) function are one of the most commonly used collision models. Since turbulence is not resolved, appropriate turbulence models have also been introduced in the LBM solver. Teixeira [42] presents the incorporation standard and RNG k-ε model with a wall model in LBM.

The advantages of LBM lie in quick pre-processing, massive parallelization, intrinsic unsteady nature of the formulation, amenability to aero-acoustic simulations.

LBM has been used for a wide variety of automotive analyses such as Aerodynamics, under-hood simulations, aeroacoustics, and vehicle soiling. Kotapati et al. [43] present aerodynamics, under-hood, and aero-acoustic analysis done for a pick-up truck to enable a multidisciplinary design paradigm. Wang et al. [44] present a study using LBM to determine the HVAC performance due to the underhood recirculation flows in the HVAC condenser region.

A wide variety of CFD solvers are currently available to carry out thermal analysis on engines and related components. Each of them has its strengths and applicability corresponding to different aspects such as accuracy, physics phenomena that need to be captured, pre-processing requirements, computational requirements, speed, etc. Depending on the requirement and available computational facilities, the appropriate solver should be used. Modeling turbulence is of importance for accurate simulation of convective heat transfer. Turbulence models have a wide range of accuracy and computational expense. The DNS though highly accurate by resolving all scales of turbulence is computationally prohibitive and practically unusable for full-fledged engine or vehicle analysis applications. While the k-ε and k-ω models provide low-cost options for a large number of industrial flows, SRS methods are more suited for highly separated flows provided the additional computational requirement is available.

11.15 Heat Transfer Models in CFD

The equation for heat transfer in turbulent flows is given by Eq. (11.9). For turbulent flows, it is formulated as

$$\frac{\partial}{\partial t}(\rho E) + \frac{\partial}{\partial x_i}[u_i(\rho E + p)] = \frac{\partial}{\partial x_j}\left(\lambda_{eff}\frac{\partial T}{\partial x_j} + u_i\left(\tau_{ij}\right)_{eff}\right) + S_h \tag{11.23}$$

where

$$\left(\tau_{ij}\right)_{eff} = \mu_{eff}\left(\frac{\partial u_j}{\partial x_i} + \frac{\partial u_i}{\partial x_j}\right) - \frac{2}{3}\mu_{eff}\frac{\partial u_k}{\partial x_k}\delta_{ij} \tag{11.24}$$

11.16 Convection Heat Transfer in Turbulent Flows

Heat transfer in turbulent flows is modelled by the Reynolds analogy which relates turbulent momentum and heat transfers [45].

$$\frac{heatflux}{convected\ heat\ flux} = \frac{momentum\ flux}{convected\ momentum\ flux} \tag{11.25}$$

$$\frac{q_w}{\rho_\infty c_\infty c_p(T_\infty - T_w)} = \frac{\tau_w}{\rho_\infty c_\infty^2} \tag{11.26}$$

The formulation of λ_{eff} in the RNG model allows good prediction across Reynolds number regimes.

11.17 Conjugate Heat Transfer (CHT)—Coupled Conduction and Convection

Estimating the temperatures in a solid component adjacent to a fluid flow region involves solving for conduction in the solid region and convection in the fluid region. The problems are coupled as the conservation of energy has to be done across the fluid–solid boundary. The conventional way of solving this problem was by solving the flow problem to obtain heat transfer coefficients at the solid interface and then solving the conduction problem separately with the heat transfer coefficient boundary conditions. However, the resulting temperatures on the solid would affect the fluid temperatures back, changing the heat flux at the boundary. Therefore, such problems are solved in a coupled manner and this is known as conjugate heat transfer.

CHT is more accurate in the prediction of temperatures than conventional methods of solving conduction and convection as a coupled problem as per Silieti et al. [46]. Jayakumar et al. [47] found that thermal analysis of heat exchangers using constant heat flux or constant temperature can produce inaccurate results. Using CHT on the other hand can provide accurate results. The fluid–solid interface coupled with CHT gives a robust solution.

Applications

CHT has a lot of applications, the primary being prediction of temperature on the solid components of the engine. It is also used in the analysis of automotive radiators to calculate temperature and effectiveness accurately.

Coupling in CFD Solvers

There are two approaches for finding the solution of coupled analysis: Desrayaud et al. [48] have used a method where the same equations are in both the solid and fluid domains. A K-parameter is introduced in the momentum equation. For solid region cells, large values of K are chosen and for the fluid region, they are set to values very close to zero. This method ensures smooth transition across the fluid–solid interface. It is known as **strong coupling** and it is a very robust method. The downsides are increased computational time and it can only be solved in the first order.

Another method involves both the solid and fluid equations to be solved separately, and the fluid–solid interface being solved by a coupled boundary condition. The solid and flow parameters are solved separately, and the temperature and heat flux is calculated. This consumes lots of time and computational power and involves sequential iterations for solid and fluid flows. It has the freedom of using the existing solver with minimal modifications. These are called **weak coupling** methods. As stated by Alonso et al. [49], the time scale in this coupling is different for both fluid and solid regions and it results in an increased computational time.

The conventional serial stage (CSS) procedure is a partitioning solver approach that is most popular in recent times, Fig. 11.11. In this approach, there are four main steps, which are:

- The temperature value that is obtained from a thermal solution at the nth time step is carried to the fluid solver.
- Then at time step n + 1 the heat flux at the fluid–solid interface is calculated.
- The heat flux from the n + 1 level is then used as a boundary condition for the solid region for the time level, n for the conduction equations.
- Finally, the conduction solver is then calculated to n + 1 time steps and the solution is obtained till convergence.

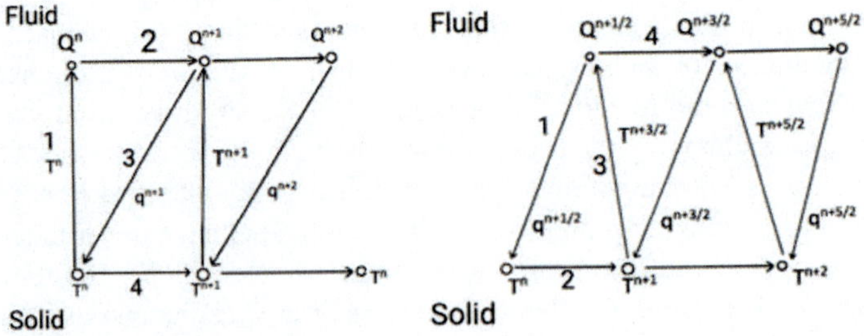

Fig. 11.11 CHT Approaches: Conventional Serial Staggered (CSS) scheme (L); Improved Serial Staggered (ISS) Scheme (R). Recreated [50]

This conventional method requires small value time scales than a strongly coupled scheme and this may cause instability and more computational time.

A **modified or improved** version (ISS) of this CSS method is proposed by Crowell [50], Fig. 11.11. In this improved version, the time step is increased by $n + ½$ and not $n + 1$. The steps are:

- The heat flux at the boundary for the $Q_{n+½}$ time step is transferred to the thermal solver as heat supplied $q_{n+½}$.
- The thermal solver then calculates temperature at the T_{n+1} time step and then it sets it as a boundary condition for the fluid region as $T_{n+3/2}$.
- The heat flux is then calculated again at the time step of $Q_{n+3/2}$. This process keeps on repeating with $n + ½$ increments, till convergence is achieved and we have a final solution.

11.18 Boiling Models

Boiling as a phenomenon may occur in the engine coolant resulting in completely different heat transfer characteristics compared to that observed in its absence. One of the important sources of Turbulent Kinetic Energy is buoyancy. Two models exist for the source term Simple Gradient Diffusion Hypothesis (SGDH) and

$$S_k = -\frac{\mu_t}{\sigma_t}\frac{1}{\rho^2}\frac{\partial\rho}{\partial x_j}\left(\frac{\partial\tau_{ij}}{\partial x_j} + \rho_\infty g_j\right) \tag{11.27}$$

Generalized Gradient Diffusion Hypothesis (GGDH) [51]

$$S_k = -\frac{3}{2}\frac{\mu_t}{\sigma_t\rho^2 k}\left(\overline{u'_j u'_k}\frac{\partial\rho}{\partial x_k}\right)\left(\frac{\partial\tau_{ij}}{\partial x_j} + \rho_\infty g_j\right) \tag{11.28}$$

Thus the Turbulence generation due to buoyancy depends on Temperature and variation of density.

Some areas where efficiency can be improved in the cooling system are heat exchangers, condensers, coolant passages on top of the cylinder heads, and coolant flow and its properties. By using computational flow dynamics with accurate models and tools to analyse the case we can save a lot of resources and time to get close enough to accurate results. There are factors such as surface tension between the coolant and the pipe, freezing and boiling point, anti-corrosive properties, density, viscosity considered for coolant flow. The boiling taking place in the cooling chambers should be controlled as well. The radiator, fans, and water pumps influence the heat transfer the most and are engineered as per the requirement.

11.18.1 Significant Properties of Coolant

Surface Tension

Surface tension is the ability of the liquid surface to behave like a flexible film. It can also be defined as the amount of work needed to stretch a liquid surface. It is measured in dyne/cm. Most engine coolants use a mixture of water and ethylene glycol in ratios of 3:1 or 1:1 respectively which has a surface tension of 56 dyne/cm. Plain water is rarely used since it has a very high surface tension of 72 dyne/cm. This effect can be ignored in most of the simulations as stated by Jafarabadi et al. [52].

Freezing and Boiling Point

Boiling occurs inside the engine cooling chamber in the nucleate region and a coolant with low surface tension ensures that the bubbles are allowed to move away or detach from the surface quickly. This allows the cooler liquid to come in and carry more heat. CFD solvers include surface tension as an input. It can be applied between different fluid–fluid or fluid–solid interfaces. Almost all the commercial CFD software allows us to manually set the freezing and boiling point of the materials as per the need. They all have different names for terms such as solidus temperature, liquidus temperature, etc.

Density

The different temperatures inside the engine lead to the boiling phenomenon. This causes a phase change. The increase in the temperature of the coolant causes a decrease in its density. Operating density needs to be given to the case along with the operating pressure as a boundary condition. When variable density is needed to be modelled, we need to introduce various temperature points in the software to specify the density at those points. For all the cases, gravity needs to be enabled.

Flow Rate

The flow of coolant must be uniform and therefore the holes and vents must be designed with care as they have a large effect on the heat transfer coefficient (HTC) in the engine. The lower flow rate may cause the overheating and inefficient performance of an engine as seen in Liu et al. [53].

11.18.2 Boiling Model for Coolant Flow

Drag Force

At the point when an air bubble ascends in a liquid, it is quickened because of its lightness i.e., buoyancy. Be that as it may, because of the interaction between the outside of the air bubble and the encompassing liquid, it is additionally decelerated due to friction or drag between the contact surfaces. It plays an important role in boiling simulation and is modelled in every case. The Ishii and Zuber model is used commonly to study wall boiling [54–56].

Lift Force

Lift forces push the particles in the perpendicular direction to the flow of the liquid and are formed due to the different velocity gradients. It is a crucial part of the quenching process in boiling. The mode as proposed by Tomiyama et al. [57] is commonly used for this effect. It has further been studied by Krepper et al. [54], and Tomiyama [58].

Wall Lubrication Force

When there is an interaction of liquid and gas flows, boiling forces the air bubbles to rise close to a wall and it feels some drainage which generates a difference in pressure fields. This pushes the bubble further than the wall. This force is referred to as wall lubrication force. The Antal et al. [59] model is widely used for calculating the wall lubrication forces where. the lubrication force is inversely proportional to, the distance from the wall. Tomiyama [58] proposed a model where it's inversely quadratic proportional to y. When more than two-phase or multiphase flows are included, the implementation of the wall lubrication force is necessary as it gives a good void fraction near the wall but its use in wall boiling where high-pressure conditions are present is questionable [54, 60]. Nevertheless, for low-pressure boiling, the force due to lubrication is still considered advantageous [61].

Turbulence Modeling

The turbulence parameters that are widely used to simulate boiling are the k-ω SST (Shear Stress Turbulence) model [54, 56] or the k-ε Realizable model [62]. Enhanced wall treatment can be turned on for smaller Y Plus values. The turbulence intensity can be varied.

Bubble Parameters

The bubble diameter size is given a default value by the solver based on the subcooling temperatures. If the bubble size is smaller than the mesh size, then the boundary between both phases is not predicted and bubble size is estimated [63].

Nucleation Site Density

The rate of boiling directly depends on the nucleation site density and it is one of the fundamentals. An accurate prediction of this has still not been made over the years. For most simulations, Lemmert and Chawla's model can be used for calculating the nucleation site density widely [56]

Heat Transfer Coefficient (HTC)

A high HTC ensures that all the heat is transferred to the coolant. Ranz-Marshall [64] model gives accurate HTC in most cases and Nusselt number values [54, 56], for the simulation.

Multiphase Flow

After the heat applied locally is equal to the latent heat of vaporization then the phase change occurs in the liquid and small bubbles start to form in the subcooled nucleate boiling region. There is a single-phase approach that uses only one phase for the simulation and a multiphase approach where the liquid and vapor phases are solved separately by two different governing equations. The drawback of the single-phase approach is that it causes a temperature rise on an increase in energy rather than a change in the phases. The multiphase approach provides more accuracy in HTC and results but needs more computational power and is time-consuming as per Tao Bo [63].

Homogenous and Non-homogenous

The energy equation needs to be turned on for the simulation. The mixing of the bubbles and liquid can be homogenous or non-homogenous. The homogenous method assumes that proper mixing is done and assigns the same velocity field to the mixture. The non-homogeneous model accounts for different velocity fields for different phases. The void fraction is modelled by the governing equations and interfacial dynamics need not be modelled to study the cavitation. There are areas near bends or walls where a thin film of vapor may be formed and be in the departure from nucleate-boiling or film boiling region.

Boiling Model Application

The boiling model increases calculation time by 10% [63]. The boiling model can be applied to the water cooling jacket and structural parts of an engine as stated by Fontanesi et al. [65]. The analysis without modelling boiling is insufficient as shown by Tao Bo [63].

11.19 Radiation Model

Radiation is an important mode of heat transfer in engine thermal performance and cooling. In underhood analyses, the exterior of the engine, particularly the turbocharger, being at a high temperature radiates heat to nearby components necessitating its modelling. There are two different approaches to model radiation in CFD analysis. While most CFD solvers have an integrated radiation solver, a commonly used approach to solve radiation is to couple separate conduction and radiation solvers with the CFD solver where heat transfer coefficients are transferred from the CFD solver to the thermal solver and temperatures fed back from the thermal solver. This is similar to weak coupling employed in the CHT problem where separate conduction and convection solvers are employed.

The equation defining radiation in CFD software is the Radiative Transfer Equation (RTE). It is a conservation equation for radiation energy where incident radiation is absorbed, emitted, or scattered, Fig. 11.12.

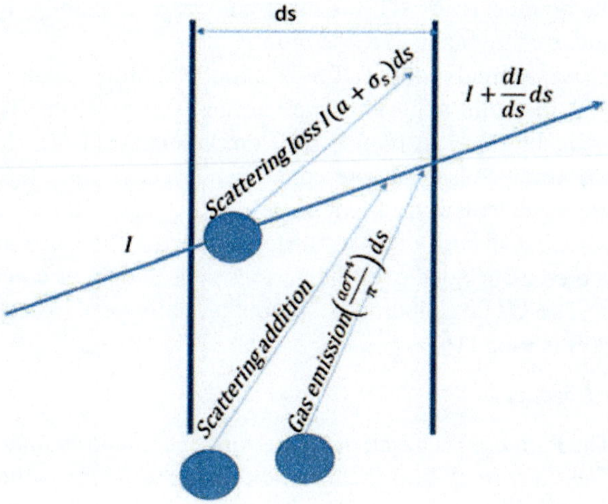

Fig. 11.12 Radiation transfer

$$\frac{dI(\vec{r},\vec{s})}{ds} + (a+\sigma_s)I(\vec{r},\vec{s}) = an\frac{\sigma T^4}{\pi} + \frac{\sigma_s}{4\pi}\int_0^{4\pi} I(\vec{r},\vec{s}')\phi(\vec{s}.\vec{s}')d\Omega' \tag{11.29}$$

$q_r = radiationheatflux$

$I = incidentradiation$

$T = temperature$

$\sigma = StephenBoltzmann'sconstant$

$n = refractiveindexofmedium$

$a = absorptioncoefficient$

$\vec{r} = positionvector$

$\vec{s} = directionvector$

$\vec{s}' = vectoralongscattereddirection$

$\phi = phasefunctionthatdecidesthescatteringinadirection.$

The radiation models commonly used in CFD analyses are the Discrete Transfer Radiation Model (DTRM), P1 Model, Rosseland, S2S, Discrete Ordinate (DO), and Monte Carlo (MC) models.

Discrete Transfer Radiation Model (DTRM)

This is a ray-tracing model, which neglects the scattering. It works by approximating the radiation from a surface over a finite solid angle using a single ray. Starting from the source of the ray where the radiation intensity is obtained based on the nature of the boundary; the RTE is integrated at points along the length of the ray to obtain radiation intensity at those points.

An optimal number of rays is modelled to emanate from each surface distributed over the solid angle that the surface is exposed to. To reduce the number of rays emanating and to reduce the computational effort, the radiating surfaces and the absorbing cells are clustered to form larger composite surfaces or volumes. The ray-tracing calculations are done before the start of the CFD calculations and the length of the ray in each of these clusters is stored. The calculated heat flux due to Radiation is then distributed over the various cells and faces constituting the cluster.

The DTRM method is computationally very expensive when a large number of rays is used [16].

P1 Model

The P1 model is based on expressing the radiation intensity as a series of harmonics. Such a formulation incorporated into the RTE results in the heat flux term being expressed as

$$-\nabla.q_r = aI - 4an^2\sigma T^4 \tag{11.30}$$

$q_r = Radiationheatflux$

$G = IncidentRadiation$

$T = Temperature$

$\sigma = Stephen\ Boltzmann's\ constant$

$n = RefractiveIndexofMedium$

$G = incidentRadiation$

$a = absorptioncoefficient.$

This heat flux is incorporated into the energy equation to calculate heat sources.

Scattering and spectral radiation can also be incorporated into the P1 model. The P1 model is computationally less intensive than the DTRM model. The model does not predict directional radiation and is accurate for regions with high optical density. It could be inaccurate when the optical thickness is small [16].

Rosseland Radiation Model

The Rosseland model is derived from the P1 model, hence, is suitable for optically thick media. It is ideally recommended for media with an optical density greater than 3.0.

In this model, the incident radiation *G* is assumed to be the intensity of blackbody radiation at the temperature of the gas instead of solving a transport equation for *G* as in the P1 model. It is faster than the P1 model [16].

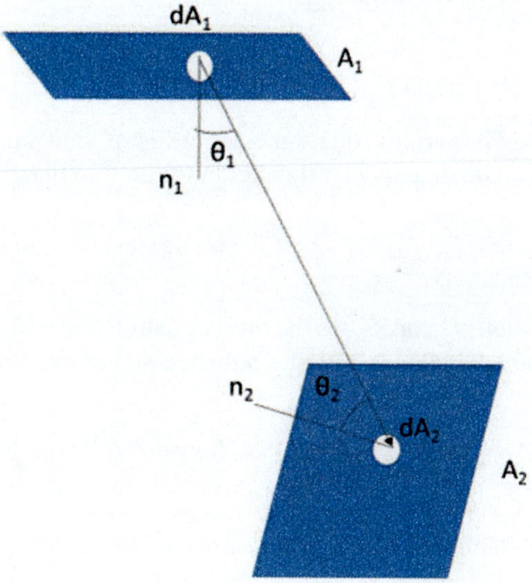

Fig. 11.13 View factor between surfaces

Surface to Surface Radiation Model

The Surface to Surface radiation model uses the analytical method of solving radiation between surfaces also known as the radiosity method. This model neglects scattering and absorption by the medium. Heat transfer through radiation happens between surfaces and heat transfer between the surfaces and the fluid happens through convection.

The radiation heat flux between any two surfaces depends on their temperatures and their relative distance and orientation. The distance and orientation decide the proportion of radiation leaving a surface that is absorbed by the other surface. In radiation analysis, this is captured by the term view factor.

View factor of radiation (Fig. 11.13) from A1 to A2 is given by

$$F_{1\to 2} = \frac{1}{A_1}\int\limits_{A_1}\int\limits_{A_2} \frac{\cos\theta_1 \cos\theta_2 dA_1 dA_2}{\pi s^2} \tag{11.31}$$

Two rules relating to view factors are
Summation Rule

$$\sum_{j=1}^{n} F_{i\to j} = 1 \tag{11.32}$$

Reciprocity rule

$$A_1 F_{1\to 2} = A_2 F_{2\to 1} \tag{11.33}$$

Superposition rule

$$F_{1\to 2,3} = F_{1\to 2} + F_{1\to 3} \tag{11.34}$$

These rules follow the definition of view factor.
The radiant heat transfer between two black body surfaces is given by

$$Q_{1\to 2} = A_1 F_{1\to 2}\sigma(J_1 - J_2) \tag{11.35}$$

where J_1 and J_2 are the total radiation emanating from surfaces 1 and 2 respectively which includes both the reflected and emitted radiation.

$$J_i = \epsilon_1 \sigma T_i^{\,4} + \rho_i \sum_{j=1}^{n} F_{j\to i} J_j \tag{11.36}$$

This gives a set of linear equations of the form

$$KJ = E$$

where E is the emitted energy dependent on the temperature. Solving this set of equations provides surface temperatures based on radiation.

Discrete Ordinate Model

The discrete ordinate model is similar to the DTRM as it assumes radiations over sectors emanating from a surface. However, instead of using a ray-tracing methodology, it formulates a transport equation for each direction or sector and solves it like all the other transport equations solved in the problem. The RTE is used in the form shown in Eq. (11.27) and the DO model works well for any optical density and thicknesses. This is one of the main advantages of the DO model. All other models have some limitations; while the DTRM model ignores scattering, the S2S model neglects both absorption and scattering. The P1 or Rosseland model on the other hand requires the optical thickness to be high. The computational requirements depend on the level of discretization. Discretization over very small angles can lead to high computational costs.

Monte Carlo (MC) Radiation Model

This is a stochastic model that tracks photons emitted from radiating surfaces in different directions and these photons may hit a surface or are absorbed or scattered. The events for the photon are probabilistic and radiative flux divergence is the statistical average of the outcomes of several energy packets. This makes the MC method computationally expensive and the time to achieve results with a low standard deviation high as it could involve a large number of packets. However, the MC method is amenable to parallelization increasing its adoption in the current days.

11.20 Component Specific CFD Models

Specific models are required to simulate the functioning of the fan and the heat exchangers in an underhood analysis. The fan being a rotating component requires specific models to simulate its effect of driving air through the heat exchangers. A heat exchanger model would not be required if all tubes and fins in the heat exchanger are resolved in the mesh. However, such a method would be practically infeasible due to the size of elements required to capture the passage through the fins. A heat exchanger model helps avoid such fine resolution of the mesh by modelling the effect using empirical data of the heat exchanger.

11.20.1 Fan Model

The rotation of the fan results in a low-pressure region behind the heat exchangers resulting in a flow through them. This effect can be simulated by having a mesh with the boundary for the fan rotating. Such a moving mesh approach would involve a transient analysis of several minutes duration till the flow reaches the steady-state. The moving mesh approach is usually implemented by creating a separate domain around the fan and the entire volume mesh in this domain rotates along with the fan boundaries. The surface elements of the moving domain with the adjoining domain slide over each other at the interface. So this method is also called a Sliding Mesh method. Sliding Mesh approaches are computationally intensive as at the typical speed of fans, time steps of the order of 10^{-6} s would be required to capture the motion and at the same time, typical underhood analysis meshes are of the order of 10 s of millions of elements. Meshes may even exceed 100 million elements. Simpler models of fans allow steady-state modelling of the fan reducing the computational cost.

Fan Zone/Momentum Source Approach

The simplest fan model is a 2D or 3D fan zone approach where empirical data in terms of the fan performance curve is used. A 2D fan region may be set up as an interface at the location of the fan while a 3D fan zone is modelled as a cylindrical or toroidal (excluding the hub region of the fan) volume enclosing the fan. Based on the fan curve, a momentum source is applied in the cells in the zone corresponding to the flow rate. One drawback of the method is that the momentum addition is in the axial direction and the rotational effect of the fan on the flow is not predicted. This drawback can however be reduced by adding a swirl at the exit.

Moving Reference Frame (MRF) Approach

The most commonly used method to model a fan is the Moving Reference Frame (MRF) model, Fig. 11.14.

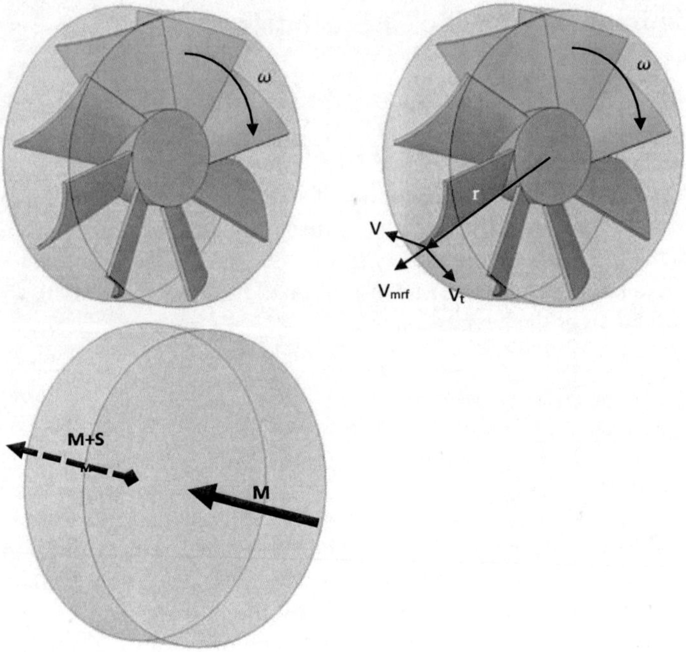

Fig. 11.14 Top Left—Moving Mesh, Top Right—MRF, Bottom—Fan Zone

The MRF model involves creating a separate domain that encloses the fan and solves the conservation equations in the rotating frame of reference attached to the fan. Solving equations in the rotating frame of reference involves transforming the velocities by the equation [66],

$$\vec{v}_{mrf} = \vec{v} - \omega \times \vec{r} \tag{11.37}$$

Across the interface, momentum conservation is solved based on absolute velocities. The actual CAD model of the fan is used in the MRF approach and the momentum change in the flow is computed using conservation equations. No empirical data such as the fan performance curve is required as in the case of the fan zone approach. Being a steady-state problem, the computational effort is significantly lesser than the sliding mesh approach.

11.20.2 Heat Exchanger

The phenomenon by which heat transfer occurs from the coolant to ambient air includes convection from the coolant to the tube inner surface, conduction across the fins, and convection from the fin surface to the ambient air. Simulating these phenomena would involve simulating the flow of the coolant, simulating the flow of air through the fin passageways, and conduction using conjugate heat transfer. However as explained earlier, the size of the gap between the fins and in the interior of the tubes is extremely small compared to the size of the vehicle which is being analysed. So resolving these geometries in the mesh would result in a huge number of elements making it computationally intensive. So the effect of the heat exchanger is modelled using empirical data.

Modeling the heat exchanger involves capturing these effects—the resistance offered to flow due to the dense fin structure and heat transfer to the air. The former is modelled using a porous medium formulation whereas effectiveness equations are used for the latter.

Porous Medium

When a porous medium model is used, the region occupied by the core of (region occupied by the tube and fins) the radiator is modelled as a separate domain with a momentum loss source given by

$$S_m = C_v \mu V + 0.5 C_i \rho V^2 \tag{11.38}$$

where,

C_v *viscous coefficient*
C_i *inertial coefficient.*

The values of C_v and C_i are obtained by fitting experimentally determined velocity v/s pressure drop curves.

Typically, the inertial and viscous coefficients are anisotropic with the values from the curve fitting used in the direction of flow whereas in the perpendicular directions these values are kept to a value several orders higher to avoid flow in those directions.

Heat Exchanger Model [16]

The heat exchanger model eliminates the need to simulate the flow of the coolant and the conduction through the tube and fin material using experimentally determined effectiveness at different flow rates of coolant (auxiliary fluid) and ambient air (primary fluid).

Heat balance is given by the equation:

$$C_{p,primary}\left(T_{p,out} - T_{p,in}\right) = C_{p,aux}\left(T_{a,in} - T_{a,out}\right) = Q_{actual} \tag{11.39}$$

By the definition of the effectiveness of a heat exchanger,

$$C_{p,min}\varepsilon(T_{a,in} - T_{p,in}) = Q_{actual} \tag{11.40}$$

The effectiveness depends on the auxiliary and primary flow rates because the heat transfer coefficient which decides the effectiveness depends on those rates. Therefore, to find the effectiveness of the auxiliary and primary heat exchanger at any flow rate, the heat exchanger is tested against a few combinations of flow rates with a reference value of $T_{a,in} and T_{p,in}$ and the heat rejection for these flow rates is tabulated. From this, effectiveness at various operating conditions in terms of primary and auxiliary flow rates is determined.

For the analysis, the heat exchanger is divided into smaller regions called macros, Fig. 11.15. The number of macros in a heat exchanger simulation is fixed such that every variation of an appropriate magnitude can be captured by all the macros.

For any heat exchanger, it can be shown that:

$$\varepsilon = f\left(NTU, \frac{C_{min}}{C_{max}}\right)$$

where NTU is the number of transfer units given by the formula

$$NTU = \frac{UA}{C_{min}}$$

Here, U is the overall heat transfer coefficient and A is the area of heat transfer. The overall heat transfer coefficient is property, intensive to the heat exchanger, i.e. it does not change with the size of the heat exchanger. Therefore, the effectiveness of a macro can be calculated by scaling the number of transfer units (NTU) of the entire heat exchanger. This NTU, which is based on the full heat exchanger and uniform conditions, is scaled for each macro using the ratio of their volumes and minimum

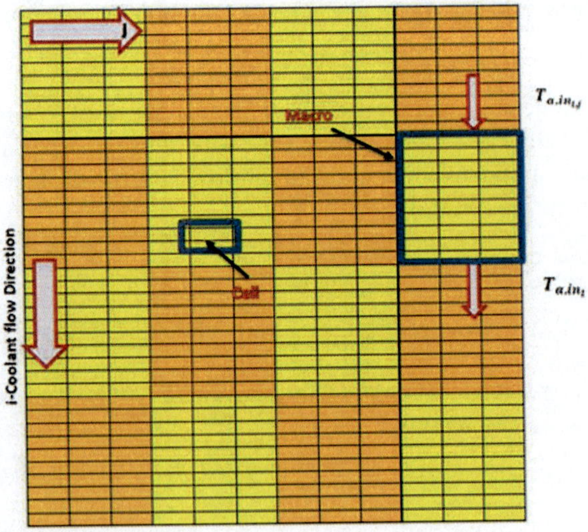

Fig. 11.15 Exchanger macro model

heat capacities. The equation that is used in scaling the NTU is as follows

$$NTU_{macro} = \frac{NTU C_{min} A_{macro}}{A C_{min,macro}}$$

The relation between NTU and effectiveness depends on the type of heat exchanger. Since we are concerned only with a crossflow heat exchanger (unmixed) the only relation pertinent to us is:

$$\varepsilon = 1 - \exp\left[-\frac{1}{C_r} NTU^{0.22}\left(1 - e^{-C_r NTU^{0.78}}\right)\right] \quad (11.41)$$

Using the scaled NTU and $C_{r,macro}$ we can determine the effectiveness of the macro.

11.20.2.1 Constant Top Tank Temperature

In the constant top tank temperature problem, we first solve for the heat rejection of the first row macros and then compute the inlet temperature to the next row macros using this heat rejection value.

$$Q_{i,j} = C_{p,min_{i,j}} \varepsilon_{i,j} (T_{a,in_{i,j}} - T_{p,in}) \quad (11.42)$$

$$C_{a_{i,j}} (T_{a,in_{i,j}} - T_{a,in_{i+1,j}}) = Q_{i,j} \quad (11.43)$$

Thereafter we use the calculated value of inlet temperature to the next macro and compute the heat rejection of the next macro, and so on. This is continued till the heat rejection values of all the macros are obtained.

11.20.2.2 Constant Heat Rejection

In the constant heat rejection problem, the equations remain the same as the constant top tank temperature problem. In addition to these equations, we can also write,

$$\sum_{\forall i,j} Q_{i,j} = Q_{total} \quad (11.44)$$

One can solve all these sets of linear equations simultaneously and find all the unknowns.

11.21 Mesh Generation

The flow of the fluid and the heat transfer is governed by the various partial differential equations. To solve the various flow domains, they need to be broken down or split into smaller finite volumes that may consist of one or more types of geometric shapes. The governing equations are then solved using a discretization process where each finite volume is solved individually. These volumes are **cells** or **elements** and the collection of these small cells is a **mesh** or a **grid**. The creation of mesh is called meshing or pre-processing. The first step in meshing is the creation of geometry. The geometry describes the region of interest in the analysis. A digital representation of this region is created by modelling the exact shapes of its boundaries using a CAD tool. The creation of geometry is also a very important task in grid/mesh generation. All the problems are first started by selecting geometry and it describes the shape that is to be analysed in the problem. It consists of volumes, edges, vertices, or faces. The grid has a very big impact on the **rate of convergence**, the **accuracy** of the solution, and the **computational time** needed. Proper care is also needed for the interface regions to ensure continuity of the solution and get a complete picture.

11.21.1 Geometry Clean-Up

It is, necessary for CFD for the model to consist of a fully enclosed volume defining the domain of interest. There may be gaps present within a part or between two parts in a CAD model used for production purposes. The main approach must be to remove these gaps in the modelling phase if they are not of interest. If the gaps are to be included in the study or simulation, then it has to be captured using very small elements in it. Using very small elements is next to impossible due to today's computational limitations. There are certain CFD methods known as Mesh-free methods that do not require us to eliminate gaps in the features that do not affect flow. We can take the example of an automobile Underhood analysis- here the region under the hood and around the vehicle is separated from the region from regions far away from it. The production model of the vehicle would have gaps between the body panels that are not needed to be studied and hence they are removed. This saves a lot of computational time and power usage. The bolts, nuts, screws, or rivets are removed if small or defeatured and simplified into simple shapes such as cylinders, beams, or cuboids. This is done because **defeaturing** them does not make much of a difference to the analysis and it saves time. This process of making a watertight model or defeaturing geometry is known as **clean-up** and is done in every study or analysis.

One way CAD tools allow for checking if a properly enclosed region without unwanted gaps has been achieved is by checking edge connectivity. There may be free edges, double edges, or multiple edges in geometry. To check for this, the geometry checking tool is used to do a topology to find these edges. A single closed volume must only consist of a double edge. In case of a baffle meeting a wall or at the interface of two or more volumes there may be multiple edges present. If not, then proper connectivity is not present and the geometry needs to be modified.

11.21.2 Structured Mesh

The **structured grid,** Fig. 11.16 has blocks of smaller elements and it is in the shape of a quadrilateral or hexahedral where one face cannot have more than one neighbouring cell face attached to it. The 3d Cartesian geometry is mapped across the region of study. Each face has the location data of the neighbouring cell and it need not be stored separately for each cell in the database because there is a proper order to it. The grid lines pass through all domains in this type of mesh. It can contain one block or be of multiple blocks of a rectangular grid. The faces, edges, and vertices of these blocks are mapped to appropriated surfaces, curves, and points in the geometry. It can be also stored as an unstructured grid as mostly done nowadays. The structure may be a **Cartesian** in which the grid lines are parallel to the coordinate axes. Each element has to be exactly connected to one face. The other type is **curvilinear** where the grid lines are fitted to go along the boundaries.

Structured grids have a high degree of control. The user can control the node locations and sizes. This helps to produce the mesh as per our requirements. They are aligned along the direction of flow. The alignment is implicit because the grid lines follow the contours of the geometry. This allows a better convergence in CFD solvers and this leads to an accurate result. The mesh data allows the use of algorithms that can help in better convergence. When there are grids normal to walls or wake like in the case of structured mesh, the turbulence and boundary conditions work more appropriately. The nodes, elements, and the connectivity table require a large amount of memory to link them in the unstructured mesh. On the other hand, in the case of structured mesh, no connectivity table is required because the mesh is in a specified pattern by the user.

Fig. 11.16 Structured mesh

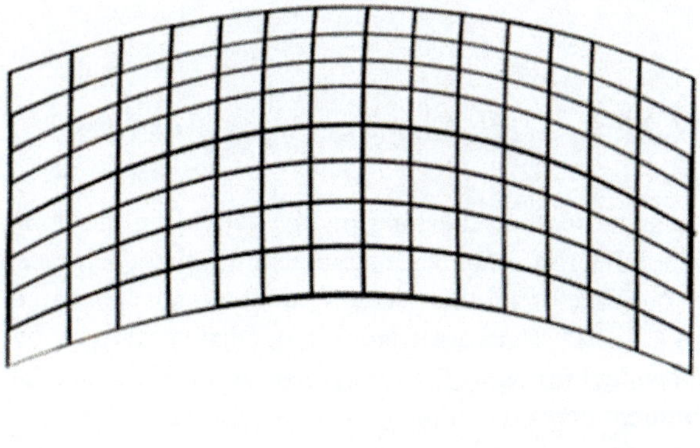

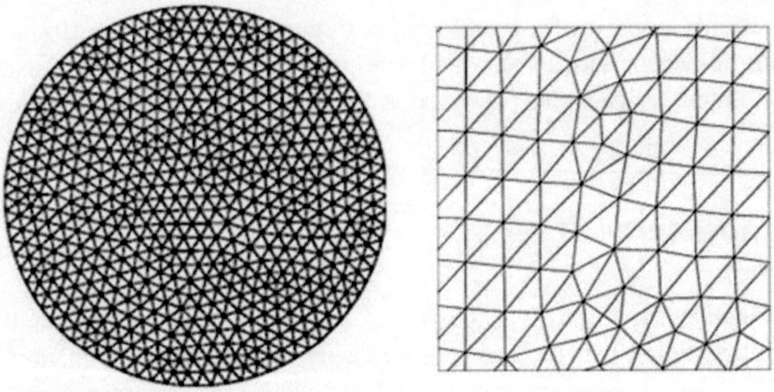

Fig. 11.17 Unstructured mesh

11.21.3 Unstructured Mesh

The **unstructured mesh**, Fig. 11.17 has elements that can be any n-polyhedron. It needs to store the neighbouring cell location data to its data for every cell. It can have an irregular or regular arrangement.

They store a list of connectivity information that has the data of a set of vertices that make up an element. Tetrahedral and triangles are used along with quadrilateral and hexahedral. Various algorithms are used for the generation of unstructured mesh.

These grids require very little manual labour since the process of generation is highly automated and only minimal specifications are required. This automation helps to reduce the time involved in meshing. Due to the various algorithms that are used, this method is suitable for very complex geometries that cannot be meshed with structured mesh. Modern-day CFD solver converts and stores even structured meshes as unstructured because the memory and storage capacity has increased a lot over the years. Industries mostly prefer these types of meshes because of their applicability to problems of complexity typically encountered in the industry. Refinements can also be done to the mesh as per our requirements. This is needed in specific areas of interest such as the boundary regions or walls, the region of the wake, or wherever the effect is to be studied in detail.

11.21.4 Mesh Element Types and Uses

As stated earlier, meshing involves dividing the domain into finite volumes of any n-polyhedron shape with the constraint that each face of every volume is connected to one face of another element or a boundary element. The minimum size of n must be 4. In 2D **elements**, triangles, quadrilaterals, and polygons are used widely for all meshing purposes. They may be used on a 2D face, plane, or baffle surface. The 2D element depends on the kind of 3D element. 3D Cells can be Tetrahedrons (n = 4),

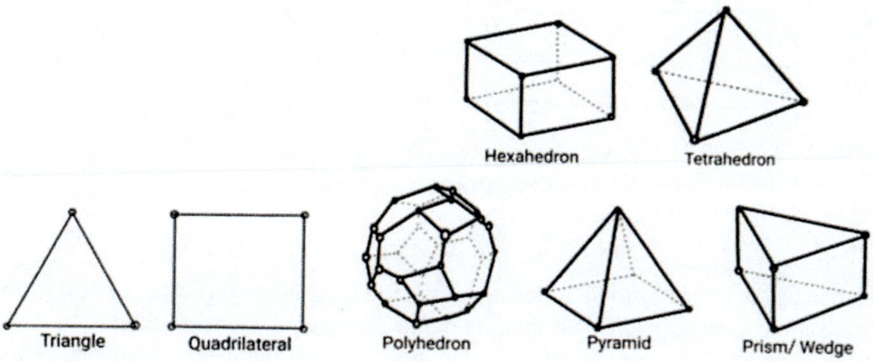

Fig. 11.18 Types of elements

Pyramids, Triangular prisms (n = 5), Hexahedrons (n = 6), or any n-Polyhedron, Fig. 11.18.

The Quad/hex grid can give a better quality solution with fewer cells than an equivalent tri/tet work for fluid work. They show lesser mathematical diffusion when it is lined up with the flow stream. But the effort needed for a Quad/hex work is more as compared to tri/tet work. This is because tri/tet has fewer amounts of vertices (5) and it allows a higher degree of control compared to quad/hex (8). For complex geometries, the unstructured mesh is easier to create and one can spare the time and effort for the mesh generation. **Hybrid meshes combine different types of 3D elements in the same mesh**. They mostly couple tri/tet components with different ones in selected areas and are more effective than tri/tet meshes alone. One more type of meshes called Prism/Wedges are also present. They are used to obtain an accurate solution at boundary layers, near walls, or regions with wake.

11.21.5 Advancing Front Method [67]

As discussed earlier, meshing is a very crucial task for the analysis of any geometry or part. The structured mesh and unstructured mesh are created with the help of computer algorithms. Various methods or techniques take place behind unstructured mesh generation. One such method is the advancing front method in which the boundaries around the whole geometry are selected and the first step in mesh generation begins from here, Fig. 11.19. The entire surface is triangulated to create a surface mesh where each element has 3 points or vertices on them and one more point in a different plane is the need to form a tetrahedral element. Now, an additional new point is created inside the domain and for each triangle. This results in 4 points or vertices so the edges are joined together to form a tetrahedral element. This is done for all the triangles in the boundary and now a new layer is created above a boundary and this is called a front. This front creates a border between the meshed and unmeshed regions.

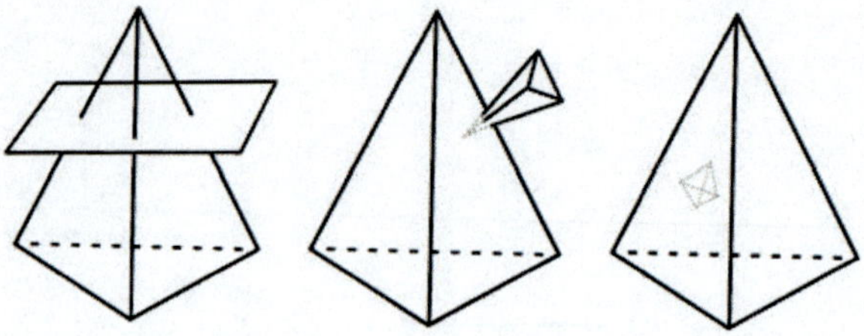

Fig. 11.19 Advancing front method: checks to be done as front progresses

Further, a new point has to be added again and the front keeps on advancing till there are no more elements left, and thus its name **advancing front method**. The new points are added at each layer for the creation of more tetrahedral elements. This decreases the size of the front and as parts of the front keep merging, the size of the front keeps reducing till no active front is left. The new point created needs to be in the correct location and this is dependent on various factors such as the point shall be within the domain, quality of the resulting elements must be acceptable, it should adhere to sizing set, neighbouring parts like the boundary, or the spacing in a mesh. Near the regions where two boundaries meet, there may be a point already present for the creation of a tetrahedral element, and in this case, a new point is not needed to be created. It is always preferred to avoid the creation of a new point and look for points already present in the region. The surrounding region needs to be analysed thoroughly. The final criteria that need to be met are that the created edges connected to the new front should not intersect any edge/facet of the 2D or 3D front. Also, the facets with the new point in them should not intersect with any edge of the 3D front. Thus, it is a very complex task to adjust the final location of this new point to be added. Lastly, the new element should not have any other points in it. There are various ways to study the boundary regions in detail with this method. One of which is capturing critical boundaries by creating fronts from them with very small step sizes. This way we have a very high control near the boundaries and this is one of the main advantages of the advancing front method, although there are downsides to this method. For example, good orthogonality cannot be guaranteed while meshing complex parts. The solution to this is to do changes/ modifications to the original advancing front algorithm in the order to increase orthogonality.

11.21.6 Delaunay Triangulation Method [68]

Another method for the unstructured grid generation used widely is the Delaunay triangulation. It is a very robust method for tetrahedral mesh creation. To understand more about this method, we need to know about topography and topology. The arrangement of an area and its features are called topography. Now the same applied to a mesh is termed **mesh topography**. **The topology** on the other hand is the properties that stay unaffected even after their shape or size is changed such as when the mesh is squeezed and stretched. For creating a mesh using Delaunay's method, there are mainly two tasks. The first task includes addressing the mesh topography which is done by placement of mesh points and the other task is to create mesh topology for a set of points by Delaunay triangulation. The order/sequence in which the tasks are to be done is a matter of preference. Here as well internal new points are needed to be added for the creation of new elements. The mesh points are added smartly based on the location and size of the elements in the boundary mesh. The topology is updated after the insertion of every new point and is done by various Delaunay algorithms. The tetrahedral elements are generated very quickly in this method compared to the advancing front because the topology is globally defined [69–71].

11.21.7 Cartesian and Octree Methods [72]

The octree method uses an algorithm that fills the interior of any geometry with structured grids to the boundaries, Fig. 11.20. The surface of the mesh is a quadrilateral mesh in the boundaries. Minimum space is maintained between the boundary and the last mesh layer. The hexahedral for the final layer is generated by connecting the nodes of the last layer to the quadrilateral on the surface. This method in which the mesh is connected to the boundary of the geometry is known as an isomorphism. The edges of the geometry shall match the edges of the mesh for ensuring good quality. There are three steps involved in the creation of a mesh by the octree method. Taking the example of generating an internal mesh, The first step is to ensure that the size of the initial octant or mesh is enough to capture the whole geometry. Then, the main

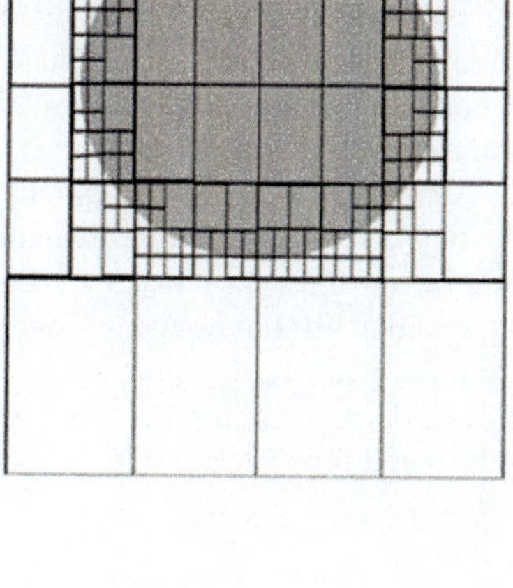

Fig. 11.20 Octree meshing

domain is divided into smaller subdomains that contain cells in evenly sized smaller hexahedral and tetrahedral elements. The final step is to apply isomorphism and then connect the mesh to the boundaries. An octree data structure helps store the mesh data upon subdivision.

One of the limitations to this method as the isomorphism results in the element quality in the interior of the mesh being superior to the boundary surface. There is a property termed an order of irregularity which indicates the size of cells with their neighbouring cells. An order of one implies that cell edges cannot be more than twice or less than half the length of the neighbour cell.

11.21.8 Mesh Quality

The accuracy and the stability of a solution are highly dependent on the mesh quality. There are certain criteria to define the quality such as the number of nodes and how they are distributed, the smoothness of the mesh, the skewness, the aspect ratio, orthogonality. Other criteria include whether refinements are done in certain areas and boundary layer inflation is present or not. The above criteria are evaluated based on the type of elements—whether they are tetrahedral, hexahedral, etc. It also depends on how they are laid out, stretched, or squeezed to fit the boundaries. The mesh must be such that it captures the whole domain and this is done by breaking the larger mesh blocks down into smaller elements. Smaller elements help to capture the minute details around a boundary or any region that is of interest to us. This distance between the points of two cells in a mesh is known as **mesh resolution** and it directly affects the mesh quality. Increasing the mesh resolution gives a more accurate solution but is also very demanding of computational power. On the other hand, having a mesh resolution too low may result in an inaccurate solution and even cause solver instabilities. Therefore, a mesh study must be done to find the perfect spot between accuracy and time.

Let us understand some of the important factors that affect the mesh quality directly. The first is mesh **skewness**, Fig. 11.21. It may be defined as the difference between an optimal cell size (an equilateral/equiangular) and the original cell size. The skewness of a mesh is always between the ranges 0 and 1. Where 0 indicates the mesh consists of all optimal cells and 1 indicates that all cells have very large variations to the optimal cell and is unacceptable. The high skewness values cause a decrease in accuracy and destabilize the solution. Values above 0.95 may also cause some convergence issues and may need one to change solver settings and controls.

Another measure of cell quality is its **aspect ratio**, Fig. 11.22. It is the ratio of the maximum to the minimum value between the centroid of a cell and the centroid of a face. An aspect ratio of 1 indicates that the cell is ideal and is not stretched or squeezed at all. For boundary layer regions 10:1 is acceptable and for regions away from the wall. In the bulk or away from walls the aspect ratio should not exceed 5:1. High aspect ratios affect the convergence of the energy equations that are being solved and thus a maximum of 35:1 is the limit for any mesh.

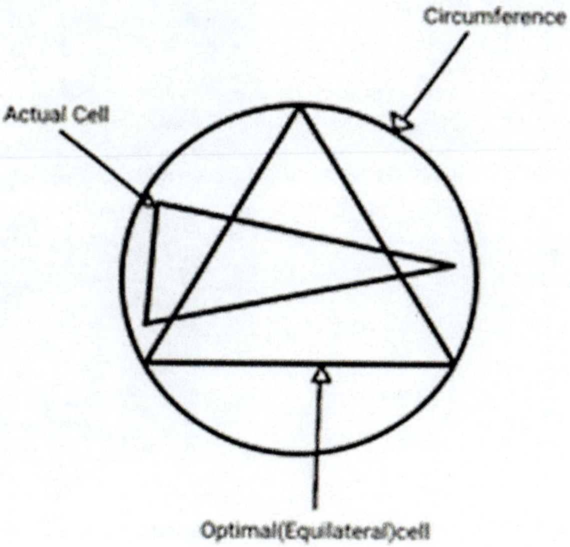

Fig. 11.21 The skewness of a cell

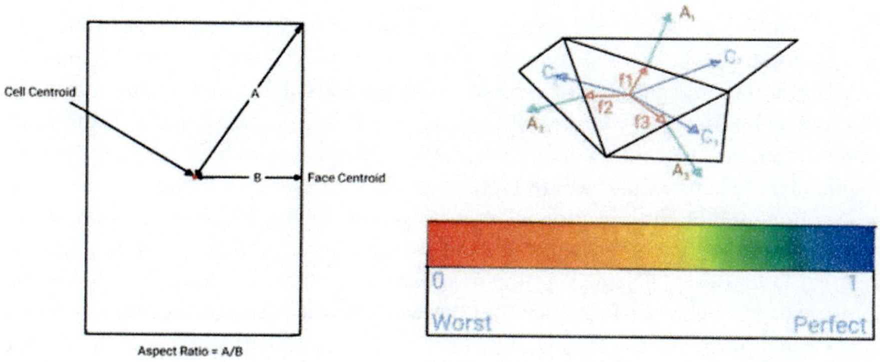

Fig. 11.22 Aspect ratio and orthogonality

Another factor that affects the mesh quality directly and is very important in a CFD simulation is mesh **orthogonality.** There is an optimal angle that each cell has to its adjacent cell. Orthogonality tells us how similar the angles in the original cell are to the optimal angle. Here 0 indicates the angles are completely dissimilar to optimal angles and 1 indicates that the angles are perfectly similar to it.

In addition to the above factors that affect the mesh quality, some further steps are done to improve the accuracy of the solution. For example, in the case of a CFD study in an aerofoil, the area around the boundaries and in front and behind the geometry is of importance. Here adjusting the cell size of the whole mesh is not needed and consumes too much power and time. **Mesh refinement** is a method that is employed here to get around the situation. In this method, the area of interest is

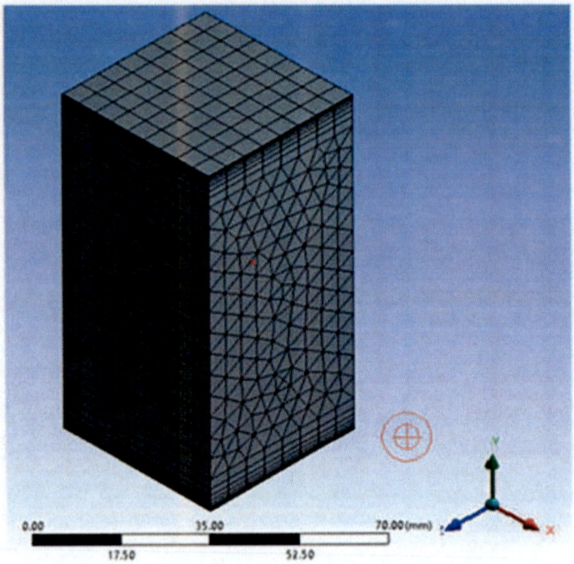

Fig. 11.23 Prism layers

selected and the cell sizes are decreased in these areas. This reduces the time and computational power needed by a drastic amount. It is a widely popular method in CFD simulations and can be applied to simulations that have not implemented this previously. **Boundary layer refinement or inflation** is also a method similar to mesh refinement. It works by having elements that are thin in a direction normal to the boundary where variations are larger and with a large area along the boundary surface along which the variation is lesser as shown in Fig. 11.23. The width of the layers, number of layers, growth rate, and the first layer thickness can be adjusted in the CFD software. There is a dimensionless wall distance called Y + which is the distance between the first node and the wall. It is a critical factor that helps us decide which turbulence model is to be used for a simulation. The Y + values must be within the range for the required turbulence model. Standard or scalable Wall functions are used for Y Plus of 30–50 $< Y + < 300$ whereas a Y + ~ 1 is required for using Enhanced wall treatment function.

11.22 Parallel Computing

As seen in the previous sections, solving CFD problems involves calculating derivatives of the conservation variables at each finite volume or grid point and solving multiple equations based on the level of physics that is modelled. Since CFD involves large volumes in space requiring millions of elements to capture the necessary flow physics in that region, industrial CFD problems are computationally intensive.

They require computational infrastructure that allows carrying out several hundred million operations over and over again till the problem converges within a reasonable time. It also requires enough memory to store the value of all the variables at each of these elements or nodes. Growth in computing power has therefore been an essential enabler of the applicability of CFD to industry-level problems. Higher processor speeds achieved by incorporating more transistors in a chip have contributed in a good measure to the power of the current day computers used for scientific computation. The number of transistors in a chip has, to date, been following Moore's law which states that the number of transistors in processor chips would double once every two years as can be seen in Fig. 11.24.

However, a bigger role in adding to meeting the compute requirements has been that of parallelization. High-performance computing which involves parallel computing is essential for practical CFD problems. Parallel computing involves aggregating multiple processors and distributing the computing workload across these processors. The aggregation of these processors can be done by having multiple cores on a chip and aggregating multiple processors over a bus in which case it is called **Symmetric multiprocessing** (SMP) as they consist of the same type of processor and all these processors have equal access to all shared resources such as memory and input/output devices. A larger aggregation is possible by using the Massively Parallel Processing (MPP) architecture which involves integrating several single or multicore processors, usually with their dedicated memory, over a network and such a system is called **distributed computing** or **cluster**. In such systems, it can be done over a network in which case it is called Computing power for calculations primarily comes from processing power and memory. In such systems, the communication speed between the processors is of utmost importance. While MPP architecture may also be established on Ethernet-based communication systems, better technologies such as Infiniband or torus interconnect are used for improved communication speeds.

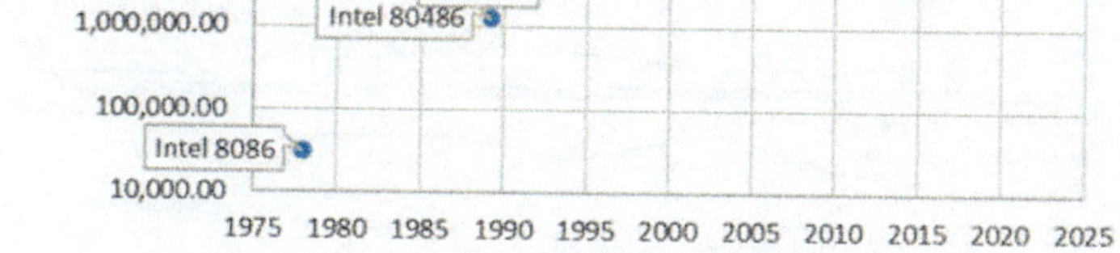

Fig. 11.24 Growth in the number of transistors on processor chips over time [73]—recreated with modifications

11.22.1 Speed and Scalability

The purpose of parallelizing solvers is to improve the speed of computation to obtain quick results. While the addition of compute nodes enhances processing power it also introduces an additional cost of information exchange between the different compute nodes. An ideal implementation of parallelization would be one where the time taken for a calculation with n processors is 1/nth time taken with one processor. An important consideration in parallelizing a code is its scalability with an increase in the number of processors. Performance close to linear scalability has been seen with CFD software with 55,000 cores [74].

11.22.2 Partitioning

For carrying out parallel computation on a code, the computational mesh has to be partitioned between the different compute nodes/processors i.e. portions of the mesh have to be assigned to each of the compute nodes in the High-performance computing (HPC) system. The strategy employed in partitioning greatly determines the speed of the code and it should ensure matching of computation time between the different nodes and reduces the communication time between different nodes. The development of best practices for partitioning is based on graph theory. Graphs are mathematical concepts that help to analyse the relationship between entities where the entities occupy vertices and the relationship represented by edges. To model the information exchange between cells on different partitions, each cell is modelled as a vertex, and the vertices of neighbouring cells that exchange data are connected by an edge. Such a graph for a mesh is shown in Fig. 11.25. Partitioning the mesh involves

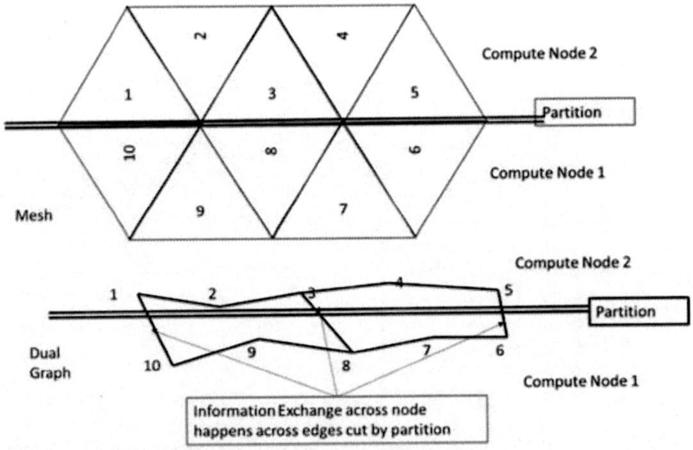

Fig. 11.25 Mesh and corresponding Dual Graph

cutting the graph along edges such that all vertices of the graph lie on one side of the cut or the other. The information exchange graph may also be created with nodes as the vertices in which case the original mesh is also the graph. However, a dual-graph with edges connecting the cell helps when calculations are done on the cell center as it keeps a cell in a particular partition. Communication across partitions happens along each of the edges cut. Thus a partition scheme that introduces a large number of cut edges increases the amount of information communicated across partitions.

To identify a good partitioning strategy, it is important to understand the parameters on which the calculation time depends. The overall time taken by any compute node is the sum of time taken for computation within the node and the time taken for communication between that node and other nodes. The communication time between two compute nodes *i* and *j* are given by [75].

$$t_{ij}^{comm} = t_s + \beta d_{ij} + \gamma m_{ij} \tag{11.45}$$

where

t_{ij}^{comm}	time for communication between the ith and the jth compute node
t_s	cost of startup latency time
β	time for communication between nearest neighbors of the topology
d_{ij}	numberof hops from node i to node j
γ	time required to communicate 1 byte
m_{ij}	numberof bytes communicated between the ith and the jth node.

The total time for computation on a single node is given by

$$t_i^{tot} = \alpha K^a + t_s|N_i| + \sum_{j\epsilon N_i} \beta d_{ij} + \sum_{j\epsilon N_i} \gamma m_{ij} \tag{11.46}$$

where

K	*No of elements in node i*
N_i	*Set of compute Nodes that exchange information with Node i*
α	*time taken for computation on one cell.*

The first term in Eq. (11.46) represents the time for computation inside a partition. For most CFD solvers the time taken for computation over multiple cells is directly proportional to the time taken for one cell and so *a* can be considered as 1.

|Ni| is the number of compute nodes in Ni. Thus reducing the number of neighbouring partitions would therefore reduce this time. The most effective method of doing this is to make parallel slits as seen in partitioning method 1 in Fig. 11.26 by which the maximum number of neighbours for any compute node is two. On the other hand, the same nine partitions done as a 3×3 array have one of the nodes having four neighbouring partitions which would increase the start-up time by four. The unpartitioned domain has n elements along each edge. The maximum number of cut edges for any partition on the strip method is 2n while that in the 3×3 method is n. In

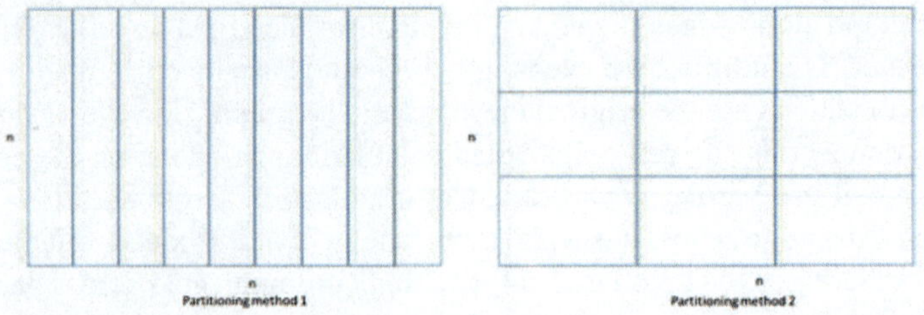

Fig. 11.26 Partitioning strip wise vs partitioning as an array

most modern machines the start-up time is usually significantly smaller than the time taken for information exchange across partition and partitioning algorithms would primarily focus on reducing the latter. The partitioning method should however take care that the number of neighbours is not increased excessively. The time taken for information exchange across partitions is captured in the last term of Eq. (11.46). Reducing this time depends on making partitions in a manner that reduced the number of cut edges in the graph.

The third term in Eq. (11.46) depends on the physical architecture of the distributed computing system. Once the mesh has been partitioned, it has to be assigned to the different processors such that the physical path of communication between neighbouring partitions, which exchange information, is the least. If two neighbouring partitions are assigned to compute nodes between which data transfer happens through multiple other nodes, the time of communication would be high.

11.22.3 Partitioning Methods

Several graph partitioning methods are available in typical commercial CFD software. The Recursive Coordinate Bisection (RCB) method involves splitting the graph repeatedly along the coordinates such that there is an equal number of nodes on each partition. While the algorithm is simple, it does not ensure minimum cut edges [76]. Other partitioning methods include the Reverse Cuthill McKee method and the Gibbs-Poole-Stockmeyer method. Recursive Spectral partitioning and Multi-level k-way are some of the commonly used partitioning methods. Kernighan-Lin is a partition refinement technique that moves one vertex across the partition and records the gain or loss in terms of the number of cut edges. It moves edges with maximum gain.

11.22.4 Parallel Processing

There are two methods for parallel processing used in HPC systems—Parallel Virtual Machine (PVM) and Message Passing Interface (MPI). software PVM helps a collection of heterogeneous systems work as a single parallel processor. MPI is a standard to be used in the development of libraries for parallel processing.

11.22.5 Parallel Grid Generation/post-Processing

Parallel processing is not only used in solving but is also used in meshing and post-processing. Full vehicle analyses require meshes of size starting from a few tens of millions of elements to even 100 s of millions of elements depending on the physics to be captured. With the increasing use of automatic methods such as wrapping, meshing time has become primarily processor-driven rather than human drive. Even when the surface mesh is done by a manual process, volume meshing is essentially automated and is a very computationally intensive process as seen in different unstructured grid generation methods. Using multiple processors, therefore, makes the meshing process significantly faster.

Post-processing involves several computations such as calculating derived variables from the raw results such as surface integrals, volume integrals, sums, gradients, etc. Visualization tools such as streamlines require an integral of velocity, to find the direction of movement of the streamline for several points till the required length of streamline. Several such calculation or visualization activities become significantly faster when parallel processed. Commercial CFD tools are increasingly parallelizing all pre and processing activities.

11.23 Common CFD Analyses for Automotive Cooling System

Some of the common analyses done for automotive cooling systems are the Underhood analysis and the coolant flow analysis.

11.23.1 Underhood Analysis

The underhood analysis involves studying the flow under the hood due to the rotation of the fan. The heat exchangers are modelled as porous media. The heat exchanger model is used to identify the temperature at which the coolant can generate the heat produced by the engine to the cooling air drawn through the heat exchanger.

The entire vehicle geometry is considered to simulate the resistance caused by the vehicle to the airflow through the cooling system. A common problem in automotive cooling systems is the recirculation of hot air back into the heat exchanger. The underhood simulation helps identify the zones of recirculation and provides design solutions to reduce or eliminate them. The temperature of air that would enter the engine is also determined by using the heat exchanger model on the Charge Air Cooler. The underhood analysis also helps predict effects such as heating of the floor and discomfort caused to passengers. The flow through the exhaust system may also be included in the analysis to see if it heats any of the critical components. Incorporating the intake system helps predict if flow entering the air intake system is heated. Hot air entry into the air intake system would lead to lower density and volumetric efficiency. Figure 11.27 shows the temperature distribution on the inlet of the radiator and CAC for a vehicle. It has been compared with Test results. A higher temperature is seen on the top portion in both eth CFD and test results though it is more pronounced in CFD. The bottom and left portions show lower temperatures. The flow path of the exhaust gas is plotted in the lowest image of Fig. 11.27.

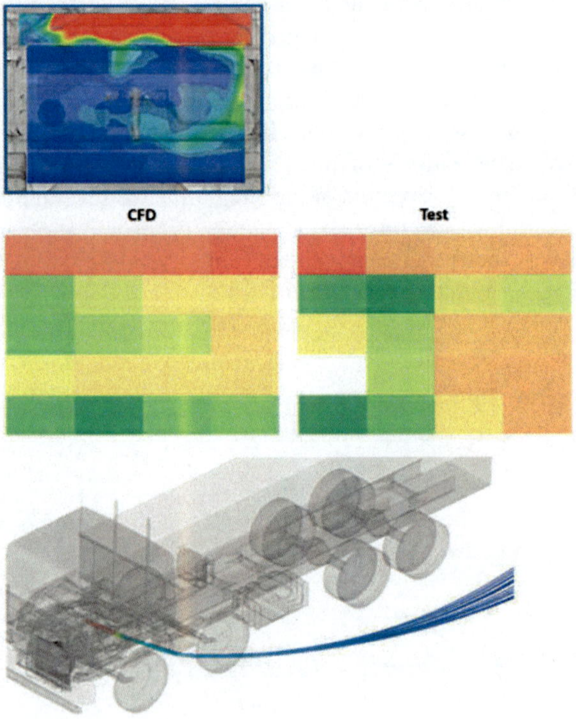

Fig. 11.27 Results of an underhood analysis

11.23.2 Coolant Flow Analysis

Coolant Flow analysis involves carrying out a CHT analysis on the entire cooling circuit including the engine or a part of it. Typical components on which it is done are the engine block and the head, the heat exchanger tubes, EGR cooler, etc. The coolant might undergo boiling due to which boiling models could also be used.

11.24 Optimization

Computational tools perform high fidelity calculations that can help simulate and predict the performance of an engineering component being designed under its service conditions. They provide a methodology for virtual testing verification/testing of the components. CFD takes significantly less time and costs for testing than physical testing. At the same time, it provides a high level of repeatability and complete control over the conditions. The lower time and cost involved in CFD allow multiple iterations to be carried out to not just find a design that meets the requirements but also arrive at designs that perform better or reduce the weight or cost of components. While such iterations can be carried out manually by analysing the results of a simulation, hypothesizing the change in design that would improve the results, implementing the design change in the CAD or the mesh, and repeating the analysis. One approach to carrying this out is to identify a set of critical design parameters, change them one by one, see the effect of the change to find out the "direction" of change in parameters that gives favourable results, and modify the value of the parameters in those directions. Numerical optimization methods automate thus allowing a larger number of parameters to be considered for the study. Several optimization algorithms help identify the "direction" and move the design in those directions so that the best value of the desired performance is achieved.

11.24.1 Formulation of the Optimization Problem

An optimization problem is essentially a maximization or minimization problem i.e. it involves finding a set of design parameters that take the desired performance metric to its highest value or keep an undesirable performance parameter at its lowest value. In an underhood problem, a desirable metric would be the Limiting Ambient Temperature while the Inlet Manifold Temperature Difference (IMTD) is a value that would have to be minimized. Similarly, in an engine Conjugate Heat Transfer Analysis, it is usually desirable to minimize the temperatures on the solid.

The mathematical function that expresses the dependence of the performance metric on the design parameters is called an objective function. Optimization involves finding a set of design parameters that result in the maximum or minimum value of the objective function. Usually, a maximization problem can be reformulated as a minimization problem by forming a new objective function which is the inverse of the original objective function. The design parameters which are the independent variables are also called design variables. The design variables are usually constrained to assume a limited set of values due to practical considerations. The equations that describe the allowable values each design variable can take are called constraint equations. The optimization problem may thus be formulated as

$$Min\left(F_{obj}(\boldsymbol{x})\right);\ x = \{x_1, x_2, x_3 \ldots\ldots x_n\} \tag{11.47}$$

Subject to

$g_i(\boldsymbol{x}) \geq 0 g_i(\boldsymbol{x}) \geq 0$ (Inequality constraints)
or $h_i(x) = 0$ (equality constraints)

In a design problem, the design variables are parameters that define the shape, properties of the materials, or characteristics of specific components. The parameters that define the shape are usually key dimensions in the geometry. Since the shape of the engineering system is defined by the boundary elements in a CFD analysis, the location of nodes on the boundary elements could also be taken as design variables of the optimization problem. In case optimization is desired to be carried out by identifying the right material, the material properties may be used as design variables. Characteristics of components such as porous media, fan curves, the effectiveness of heat exchangers, etc. can also be used as design variables if a modification in these components is considered for optimization.

11.24.2 Optimization Algorithms

Optimization algorithms used with CFD usually follow one of the following types.

- Gradient-based algorithms
- Probabilistic or heuristic algorithms the most common of them being a set of algorithms called evolutionary algorithms, more particularly genetic algorithm.

11.24.3 Gradient-Based Algorithms

A change in the objective function is achieved by modifying the design variables by incrementing their values (positively or negatively). The "direction" of change of the design is the ratio of these increments (+ve or –ve). The gradient of the objective function is a direction where the increments to the design variables are proportional to the partial derivatives of the objective function concerning them. It can be mathematically proved that the direction of change that gives the greatest improvement in the objective function for an infinitesimal change in the design is along the gradient of the objective function. Gradient-based algorithms are therefore also called steepest descent algorithms considering that the gradient is the direction of movement in which the objective function would have the steepest rise or fall (depending on whether the design variables are being incremented or decremented). Hence moving along the gradient is ideally the fastest way to reach the optimum. It should be noted that even with an infinitesimal movement of the design, the gradient would change. Therefore, the preferred direction of design change would change. However, gradient-based algorithms usually make finite changes to the design along the gradient and recompute the gradient at the new point. This is done recursively and the design progressively reaches towards the optimum. At a maximum or minimum, the gradient takes a value of zero. So the algorithm keeps repeating till the gradient reaches a very small value or there is no improvement in the objective function.

The gradient gives a deterministic method to find the direction of movement for optimization and thus reduces the number of trials. However, since the gradient method requires the calculation of partial derivatives with respect to each of the design variables it involves evaluation of the objective function. If the function evaluation is done using CFD simulations, a large number of design variables would involve a large number of simulations. Gradient-based methods also result in the optimizer getting stuck in a local minimum. A local minimum is a set of values of the design variables where the value of the objective function does not decrease further for infinitesimal small changes in any direction but a large deviation from the current set of values could result in lower values. To avoid this problem, gradient-based methods are started from multiple initial design points.

Adjoint Solver Method

The adjoint solver method helps calculate the partial derivatives of the objective function to a set of design variables which could include the location of the mesh point based on a single converged run of the solver. The adjoint solver thus helps a computationally cheap implementation of the gradient method for CFD solvers. Running the adjoint solver provides the sensitivity of the objective function to different locations on the boundary of the domain/points on the engineering system. The location with high sensitivities can then be modified such that the objective function is reduced.

11.24.4 Genetic Algorithm

Genetic algorithms (GA) are metaheuristics that use a probabilistic approach to exploring the design space and reaching the optimum. They are inspired by the evolution of living beings where traits are passed on from parents to offspring; desirable traits are retained over generations whereas traits that do not enhance survival are not carried over to the next generation. Genetic algorithms involve selecting an initial seed design candidate called a population. The initial population can be selected randomly. The algorithm creates subsequent generations of populations by three operations—selection, crossover, and mutation. Selection is an operation by which a candidate design is retained unchanged in the next population. Selection is done such that designs with a desirable value of objective function have a higher probability of getting carried over to the next generation. Crossover is the operation that primarily allows exploration of the design space by identifying new design points. The cross-over operation identifies two new points in the current generation and applying some rule of combination between them generates two new points for the next generation. Different GA algorithms have different methods of crossover. The mutation is the creation of a design for the next generation by introducing a random change in any of the designs of the current generation. The mutation is carried out with a very low probability to address situations whereas crossover alone done on the current population does not result in significantly new designs. The mutation is done very rarely to ensure a completely random exploration of the design space. Every design of the current generation undergoes one of the three operations resulting from a selection procedure such that designs with a favourable value of objective function have a higher probability (not a certainty) of being retained as such and the designs with a less favourable value have a higher probability of being modified through crossover or mutation.

Genetic algorithms require a large population for effective exploration of the design space and are not directed along the path of steepest descent. Therefore, compared to gradient-based algorithms GAs require a large number of iterations to reach the optimum. The number of function evaluations required is also significantly larger than gradient-based algorithms particularly when the design variables are less. However, GAs do not converge to a local minimum and instead are Global minimization algorithms. This makes GA popular for several practical engineering applications which might have several local minima. Since GA does not involve gradient computation it can also work when the design space is not continuous i.e. between the minimum and maximum permissible values of a design variable there are some points with infeasible values of the objective function.

11.24.5 Surrogate Models

Since CFD analyses are time-consuming, barring very small problems, function evaluation for an optimization study is rarely done by coupling full-fledged CFD software to an optimizer. The usual practice is to create a low fidelity mathematical model of the objective function based on a limited sample of designs that are evaluated using CFD. Such mathematical models are called surrogate models. Once the mathematical model is created, it is used to evaluate the objective function for the optimizer. After the optimum is reached, the value of the objective function is verified by running a high-fidelity CFD analysis for the design. Some of the common surrogate models are Response Surfaces and Kriging. A response surface is a mathematical function—usually a polynomial function—of the design variables such that its value is close to that obtained by evaluating the objective function using a CFD solver. Kriging is an interpolation model that estimates the value of the objective functions by interpolating between points at which it has been evaluated using CFD. Using surrogate models reduces the computation effort significantly.

11.25 Design Modification for Analysis

CFD analysis of multiple designs is required in both cases—using the CFD solver as a function evaluation tool directly for the optimizer or for the creation of data points to build the surrogate model. In either case, an easy and automated approach is required for modifying the mesh used for the analysis. Two approaches may be used for mesh modification.

- CAD level modifications
- Mesh morphing.

11.25.1 CAD Level Modifications

CAD level modifications involve identifying dimensions in the geometry that would be the design variables of the optimization problem and make them parameters to create the shape. Most CAD tools allow defining certain dimensions as parameters and changing them from an external source such as a text file. CAD tools also have the option of creating or modifying geometries based on a script—a set of instructions for the CAD tool on the operations to perform to create the desired geometry. To evaluate the objective function for a modified design, the optimization code would need a computer program that (1) generates the script required by the CAD tool to create a new geometry file, (2) calls upon the CAD tool to run the script, (3) calls upon the meshing tool to mesh the new geometry, (4) calls upon the solver to run the analysis, (5) extracts relevant results from the solution and (6) feeds it to the optimizer. Even when surrogate models are used it becomes necessary to automate the analysis as the objective function needs to be evaluated at several sample points for the generation of the surrogate models. Many of the modern CFD analysis software has inbuilt process integration tools for repetitive activities as those involved in optimization. These reduce the requirement for a CAE analyst to know to program. Many commercial optimization tools for CAE purposes also have a process integrator that helps a couple of multiple tools or steps that are required for objective function evaluation.

11.25.2 Mesh Morphing

Since the CAD geometry is only a template to build a mesh, an easier way of making design modifications for analysis is by modifying the mesh directly. Modification of mesh is typically done by moving the nodes of the mesh using any of the standard transformation tools such as translation, rotation, and scaling. For a CFD analysis, though a volume mesh is used, the design of the component is captured in the boundary mesh. So morphing is done by moving the boundary nodes. It is possible to morph with the volume mesh being present, and some mesh morphing algorithms allow for adjustment of the volume mesh according to the change in the boundary. However, due to a lack of robustness, it could still lead to the penetration of cells into each other. Poor quality of the resultant mesh and very high time taken for morphing a volume mesh makes it an approach that is not favored. It is usually faster to morph the boundary mesh alone and regenerate the volume mesh from the boundary mesh. This also results in better mesh quality.

There are broadly two approaches to mesh morphing.

- Applying transformation to the nodes directly
- Applying transformation to handles associated with the mesh.

Mesh morphing allows very local changes as they are not assigned to regular geometrical shapes.

References

1. Walsh JW (2021) Engine coolant deaeration system. https://patents.google.com/patent/US3255740A/en. Accessed on 12 Mar 2021
2. https://www.howacarworks.com/basics/how-an-engine-cooling-system-works
3. Double Valve Automotive Thermostat, https://commons.wikimedia.org/wiki/File:Double_valve_automotive_thermostat.jpg. Accessed on 12 Mar 2021
4. Fan Clutch, From Wikipedia, the Free Encyclopedia, https://en.wikipedia.org/wiki/Fan_clutch
5. https://www.cashcarsbuyer.com/radiator-fan/
6. Krieger RB (1966) The computation of apparent heat release for internal combustion engines. ASME Paper 66-WA/DGP-4
7. Challen B, Baranescu R (eds) (1999) Diesel engine reference book. Butterworth-Heinemann Limited
8. Sharma CS, Anand TNC, Ravikrishna RV (2010) A methodology for analysis of diesel engine in-cylinder flow and combustion. Progr Comput Fluid Dyn Int J 10(3):157–167. https://doi.org/10.1504/PCFD.2010.033327
9. Is it Okay to Use Water Instead of Car Coolant? http://www.mechanicsatwork.com.au/blogit.php?blogid=4. Accessed on 12 Mar 2021
10. What is a Car Coolant? Universal Technical Institute. https://www.uti.edu/blog/automotive/car-coolant. Accessed on 12 Mar 2021
11. Schmidt S, McIver DM, Blackburn HM, Rudman M, Nathan GJ (2001) Spectral element-based simulation of turbulent pipe flow. In: 14th A/Asian Fluid Mech. Conference
12. Zienkiewicz OC, Taylor RL, Nithiarasu P (2014) Chapter 8—Turbulent flows. In: Zienkiewicz OC, Taylor RL, Nithiarasu P (eds) The finite element method for fluid dynamics, 7th

edn. Butterworth-Heinemann, pp 283–308. ISBN 9781856176354. https://doi.org/10.1016/B978-1-85617-635-4.00008-X, http://www.sciencedirect.com/science/article/pii/B978185617635400008X). Accessed on 1 Nov 2020
13. Canuto C, Yousuff Hussaini M, Quarteroni A, Zang TA (2007) Spectral methods: evolution to complex geometries and applications to fluid dynamics. Springer Science & Business Media (2007)
14. Allmaras SR, Johnson FT (2012) Modifications and clarifications for the implementation of the Spalart-Allmaras turbulence model. In: Seventh international conference on computational fluid dynamics (ICCFD7), pp 1–11
15. Launder BE, Spalding DB (1974) The numerical computation of turbulent flows. Comp Meth Appl Mech Eng 3
16. Manual UDF (2009) ANSYS FLUENT 12.0. Theory Guide. Canonsburg, PA
17. Durbin PA (1995) Separated flow computations with the k-epsilon-v-squared model. AIAA J 33(4):659–664
18. Wilcox DC (1998) Turbulence modeling for CFD. DCW Industries, Canada, vol 2
19. Menter FR (1994) Two-equation eddy-viscosity turbulence models for engineering applications. AIAA J 32(8):1598–1605
20. Germano M, Piomelli U, Moin P, Cabot WH (1991) A dynamic subgrid-scale eddy viscosity model. Phys Fluids A 3(7):1760–1765
21. Zang Y, Street RL, Koseff JR (1993) A dynamic mixed subgrid-scale model and its application to turbulent recirculating flows. Phys Fluids A 5(12):3186–3196
22. Menter FR, Schütze J, Gritskevich M (2012) Global vs. zonal approaches in hybrid RANS-LES turbulence modelling. In: Progress in hybrid RANS-LES modelling. Springer, Berlin, pp 15–28
23. Spalart PR (2000) Strategies for turbulence modelling and simulations. Int J Heat Fluid Flow 21(3):252–263
24. Spalart PR, Deck S, Shur ML, Squires KD, Kh Strelets M, Travin A (2006) A new version of the detached-eddy simulation, resistant to ambiguous grid densities. Theor Comput Fluid Dyn 20(3):181–195
25. Strelets M (2001) Detached eddy simulation of massively separated flows. In: 39th Aerospace sciences meeting and exhibit, p 879
26. Travin A, Shur M, Strelets MM, Spalart PR (2002) Physical and numerical upgrades in the detached-eddy simulation of complex turbulent flows. In: Advances in LES of complex flows. Springer, Dordrecht, pp 239–254
27. Auza R, McNamara J, Kimbrell A, Rupp T, Kang P (2020) High-fidelity modeling and prediction of hood buffeting of trailing automobiles. No. 2020-01-5038. SAE Technical Paper
28. Mendonça F, Allen R, De Charentenay J, Lewis M (2002) Towards understanding LES and DES for industrial aeroacoustic predictions. In: International Workshop on 'LES for Acoustics', DLR Gottingen, Germany
29. Korzani MG, Galindo-Torres SA, Scheuermann A, Williams DJ (2017) Parametric study on smoothed particle hydrodynamics for accurate determination of drag coefficient for a circular cylinder. Water Sci Eng 10(2):143–153
30. Nie X, Chen L, Xiang T (2015) Real-time incompressible fluid simulation on the GPU. Int J Comput Games Technol 2015:1–12
31. Gourlay MJ (2012) Fluid simulation for video games (Part 1). Intel Developer Zone
32. Takeda H, Miyama SM, Sekiya M (1994) Numerical simulation of viscous flow by smoothed particle hydrodynamics. Progress Theoret Phys 92(5):939–960
33. Johnson GR, Stryk RA, Beissel SR (1996) SPH for high-velocity impact computations. Comput Methods Appl Mech Eng 139(1–4):347–373
34. Monaghan JJ (1988) An introduction to SPH. Comput Phys Commun 48(1):89–96
35. Wendland H (1995) Piecewise polynomial, positive definite and compactly supported radial functions of minimal degree. Adv Comput Math 4(1):389–396
36. Morris JP, Fox PJ, Zhu Y (1997) Modeling low Reynolds number incompressible flows using SPH. J Comput Phys 136(1):214–226

37. Hu XY, Adams NA (2015) An SPH model for incompressible turbulence. Procedia IUTAM 18:66–75
38. Yang X, Kong S-C (2017) Smoothed particle hydrodynamics method for evaporating multiphase flows. Phys Rev E 96(3):033309
39. Farrokhpanah A, Bussmann M, Mostaghimi J (2017) New smoothed particle hydrodynamics (SPH) formulation for modeling heat conduction with solidification and melting. Numer Heat Transf Part B: Fundam 71(4):299–312
40. Amdahl DJ, Amdahl DL (1993) Modeling aerodynamic problems using smoothed particle hydrodynamics (SPH). SAE Trans 1743–1750
41. Wang G, Gardner K, DeHoff E, del Pin F, Caldichoury I, Yreux E. Applications of ICFD/SPH solvers by LS-DYNAŽ to solve water splashing impact to automobile body
42. Teixeira CM (1998) Incorporating turbulence models into the lattice-Boltzmann method. Int J Mod Phys C 9(08):1159–1175
43. Kotapati R, Keating A, Kandasamy A, Duncan B, Shock R, Chen H (2009) The lattice-Boltzmann-VLES method for automotive fluid dynamics simulation, a review. No. 2009-26-0057. SAE Technical Paper
44. Wang, Z., J. Han, A. Alajbegovic, T. Kuthada, D. STOLL, C. MAKENS, and T. SCHWENN. "Long-term transient cabin heating simulation with multiple fluid node approach." In Proc. Veh. Thermal Manag. Sys. Conf. Proc., pp. 319–333. 2013.
45. The Reynolds Analogy. https://web.mit.edu/16.unified/www/FALL/thermodynamics/notes/node122.html. Accessed on 12 Mar 2021
46. Silieti M, Kassab AJ, Divo E (2005) Film cooling effectiveness from a single scaled-up fan-shaped hole: a CFD simulation of adiabatic and conjugate heat transfer models. In: Turbo Expo: Power for Land, Sea, and Air, vol 47268, pp 431–441
47. Jayakumar JS, Mahajani SM, Mandal JC, Vijayan PK, Bhoi R (2008) Experimental and CFD estimation of heat transfer in helically coiled heat exchangers. Chem Eng Res Design 86(3):221–232
48. Desrayaud G, Fichera A, Lauriat G (2007) Natural convection air-cooling of a substrate-mounted protruding heat source in a stack of parallel boards. Int J Heat Fluid Flow 28(3):469–482
49. Alonso J, Hahn S, Ham F, Herrmann M, Iaccarino G, Kalitzin G, LeGresley P et al (2006) CHIMPS: a high-performance scalable module for multi-physics simulations. In: 42nd AIAA/ASME/SAE/ASEE joint propulsion conference & exhibit, p 5274
50. Miller B, Crowell A, McNamara J (2011) Computational modeling for conjugate heat transfer of shock-surface interactions on compliant skin panels. In: 52nd AIAA/ASME/ASCE/AHS/ASC structures, structural dynamics and materials conference 19th AIAA/ASME/AHS adaptive structures conference 13t, p 2017
51. Končar B, Kljenak I, Mavko B (2004) Modelling of local two-phase flow parameters in upward subcooled flow boiling at low pressure. Int J Heat Mass Transf 47(6–7):1499–1513
52. Jafarabadi M, Chamani H, Malakizadi A, Jazayeri SA (2008) A fast coupled CFD-Thermal analysis of a heavy-duty diesel engine water cooling system. ASME Int Mechan Eng Congr Exposit 48715:663–670
53. Liu CB, Zhang C (2011) CFD analysis of engine water jacket and its optimization. Adv Mater Res 308:2308–2312
54. Krepper E, Končar B, Egorov Y (2007) CFD modelling of subcooled boiling—concept, validation, and application to fuel assembly design. Nucl Eng Des 237(7):716–731
55. Ishii M, Zuber N (1979) Drag coefficient and relative velocity in bubbly, droplet or particulate flows. AIChE J 25(5):843–855
56. Wang C, Cong T, Qiu S, Tian W, Wu Y, Su G (2014) Numerical prediction of subcooled wall boiling in the secondary side of SG tubes coupled with primary coolant. Ann Nucl Energy 63:633–645
57. Tomiyama A (1995) Effects of Eotvos number and dimensionless liquid volumetric flux on lateral motion of a bubble in a laminar duct flow. In: 2nd International Conference an Multiphase Flow, vol 3

58. Tomiyama A (1998) Struggle with computational bubble dynamics. Multiph Sci Technol 10(4):369–405
59. Antal SP, Lahey RT Jr, Flaherty JE (1991) Analysis of phase distribution in fully developed laminar bubbly two-phase flow. Int J Multiph Flow 17(5):635–652
60. Lucas D, Shi JM, Krepper E, Prasser HM (2004) Models for the forces acting on bubbles in comparison with experimental data for vertical pipe flow. In: 3rd international symposium on two-phase flow modelling and experimentation, Pisa, Italy, pp 22–24
61. Bartel MD, Ishii M, Masukawa T, Mi Y, Situ R (2001) Interfacial area measurements in subcooled flow boiling. Nucl Eng Des 210(1–3):135–155
62. Lee J-I, Cho N-H (1995) Numerical analysis of gasoline engine coolant flow. No. 950274. SAE Technical Paper
63. Bo T (2004) CFD homogeneous mixing flow modelling to simulate subcooled nucleate boiling flow. No. 2004-01-1512. SAE Technical Paper
64. Ranz WE (1952) Evaporation from drops, Parts I & II. Chem Eng Prog 48:141–146
65. Fontanesi S, Giacopini M (2013) Multiphase CFD–CHT optimization of the cooling jacket and FEM analysis of the engine head of a V6 diesel engine. Appl Therm Eng 52(2):293–303
66. Franzke R, Sebben S, Bark T, Willeson E, Broniewicz A (2019) Evaluation of the multiple reference frame approach for the modelling of an axial cooling fan. Energies 12(15):2934
67. Advancing Front Methods, https://www.iue.tuwien.ac.at/phd/fleischmann/node39.html. Accessed on 12 Mar 2021
68. Delaunay Methods. https://www.iue.tuwien.ac.at/phd/fleischmann/node40.html. Accessed on 12 Mar 2021
69. Joe B (1989) Three-dimensional triangulations from local transformations. SIAM J Sci Stat Comput 10(4):718–741
70. Joe B (1991) Delaunay versus max-min solid angle triangulations for three-dimensional mesh generation. Int J Numer Meth Eng 31(5):987–997
71. Marcum DL, Weatherill NP (1995) Unstructured grid generation using iterative point insertion and local reconnection. AIAA J 33(9):1619–1625
72. Schneiders R, Schindler R, Weiler F (1999) Octree-based generation of hexahedral element meshes. Lehrstuhl fur Angewandte Mathematik, insb. Informatik RWTH Aachen Ahornstr. 55, 52056 Aachen, F.R. Germany. email: robert@feanor.informatik.rwth-aachen.de
73. Moore's Law Transistor Count 1971–2018 https://commons.wikimedia.org/wiki/File:Moore%27s_Law_Transistor_Count_1971-2018.png, Max Roser, CC BY-SA 4.0 https://creativecommons.org/licenses/by-sa/4.0. via Wikimedia Commons. Accessed on 1 Nov 2020
74. CD-adapco and HLRS push the boundaries of HPC performance with STAR-CCM+, New York, and London. May 13, 2015, http://mdx2.plm.automation.siemens.com/pr/cd-adapco-and-hlrs-push-boundaries-hpc-performance-star-ccm. Accessed on 1 Nov 2020
75. Venkatakrishnan V (1995) Implicit schemes and parallel computing in unstructured grid CFD. Institute for Computer Applications in Science and Engineering Hampton VA
76. Karypis G, Kumar V (1995) Analysis of multilevel graph partitioning. In: Supercomputing '95: Proceedings of the 1995 ACM/IEEE conference on supercomputing, San Diego, CA, USA, pp 29–29. https://doi.org/10.1109/SUPERC.1995.242800

Chapter 12
Estimation of Skin Temperature on Surfaces of Exhaust Line

Sudharsan Annur Balasubramanian and P. A. Lakshminarayanan

Nomenclature

A, F	Flow area (cross-sectional)
A_s	Heat transfer surface area
A_{eff}	Effective area for flow
C_f	Friction coefficient
$C_{f,\mathrm{rough}}$	Friction coefficient of rough pipe
C_p	Pressure loss coefficient
C_{p}	Specific heat at constant pressure
C_V	Specific heat at constant volume
a, c	Speed of sound
C_p	Fluid-specific heat capacity
D	Equivalent diameter
D, d	Pipe diameter
K	Ratio of specific heats
dx	Length of the mass element in the flow direction (discretization length)
dp	Pressure differential acting across dx
e	Total internal energy (internal energy plus kinetic energy) per unit mass
f	Friction constant for the frictional loss
H	Total enthalpy
h	Heat transfer coefficient
$h_{g,\,\mathrm{rough}}$	Heat transfer coefficient of rough pipe

S. Annur Balasubramanian (✉)
CFD Engineering Analyst, MEDA Engineering and Technical Services, Windsor, Canada

P. A. Lakshminarayanan
Indian Institute of Technology Kanpur, Kalyanpur, Kanpur 208016, India

P. A. Lakshminarayanan and A. K. Agarwal (eds.), *Handbook of Thermal Management of Engines*, Energy, Environment, and Sustainability,
https://doi.org/10.1007/978-981-16-8570-5_12

$h_{external}$	Heat transfer coefficient for external convection
h_{user}	The "External Convection Coefficient" as input by the user
k	Thermal conductivity
l	Length of a cell
$\dot{m}$	Boundary mass flux into volume
m	Mass of the volume
Nu	Nusselt number
P, p	Pressure
Pr	Prandtl number
q	Heat flux, gain in energy due to heat transfer across pipe walls per unit mass of gas
Re	Reynolds number
T	Temperature, gas temperature
t	Time
Δt	Timestep
T_w	Temperature of wall
u	Velocity at the boundary
Δx	Minimum discretization element length
y	Property to be computed at any time

Greek Symbols

ε	Emissivity
μ	Dynamic viscosity
ρ	Density
σ	Stefan-Boltzmann constant

Subscripts

atm	Atmospheric conduction
end	Boundary
j	Cell number
m	Boundary cell number
n	Index for time

12.1 Introduction

The exhaust system transports the hot gases away from the combustion chamber of the engine to the atmosphere. Figure 12.1 shows a vehicle underbody with the exhaust system consisting of many parts depending on the size of the engine:

- Exhaust manifold: from multiple cylinders into one pipe.
- Catalytic converter: to neutralise harmful gases like nitric oxides, carbon monoxide, unburnt hydrocarbons
- Muffler: to reduce the noise level in the exhaust by cancelling or damping pressure waves
- Exhaust pipes connect the components and the tailpipe providing a sealed pathway for the exhaust gases. Flexible connections are used to take up the thermal expansion of the system.
- Tailpipe: Final pipe that opens to the atmosphere.

The automotive exhaust system is important for the control of emission, noise, vibration and harshness (NVH), and fuel economy. The tight emission norms saw the emergence of technologies such as exhaust-gas-recirculation (EGR), sequential turbochargers (twin-turbo), selective catalytic reduction (SCR), and silencers. Hardware components from these technologies are exposed to intense heat affecting the skin temperature and thereby the performance and life of other components in the surrounding zone. At the same time, catalysts must be warmed up quickly and should stay above light-off temperature but within dangerous limits, to avoid aging as well as catastrophic failures. Therefore, the study of the thermal management of the system is important in the design and optimization phases.

Estimating the skin temperature of the system helps in the estimation of the maximum operating temperatures, routing, design, and development of heat shields to thermally protect the underbody components. So far, the time and cost-intensive physical testing allow only the verification of temperatures [2]. In addition, several tests are needed to completely understand the system.

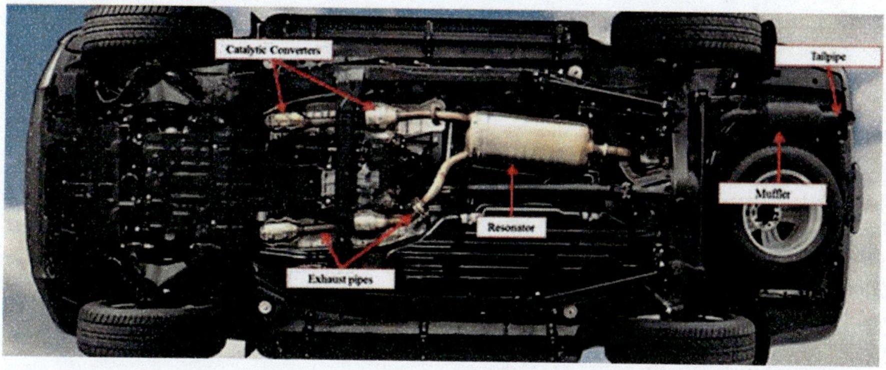

Fig. 12.1 The underbody of a vehicle showing some components of the exhaust system [1]

Computational Fluid Dynamics (CFD) techniques are very effective during the design phases in vehicle development and physical tests. Some key applications of CFD in automotive development include engine cooling, optimizing power-train and exhaust system components, aerodynamics, and thermal management. This topic is dealt with in detail in Chaps. 11 and 13.

Most of the manifold and the catalyst at high skin temperatures are less exposed to the ambient. Components such as the silencer with a higher surface area are exposed to the flow underbody and the skin temperatures are somewhat lower. Thus, both the internal and the external flows determine the skin temperatures. The complex internal flow could be simplified to a one-dimensional (1-D) model, though three-dimensional at the bends and flow restriction devices. The external flow, however, cannot be simplified. The external heat transfer coefficients must be obtained from a 3-D CFD model and coupled with the 1-D model iteratively to obtain the skin temperatures. The iterative calculation is tedious and takes days even a modern computer is used when 1-D flow calculation in the engine is not decoupled from CFD. The following decoupling procedure developed in [25] is explained during the course of the chapter to achieve:

- Simulation using 1-D models of compressible exhaust flow for mass flow rate, velocity, pressure, and temperature of exhaust gas
- Perform 1-D simulation for steady-state Wide Open Throttle (WOT) cases.
- Extend the model to transient heat transfer studies by simulating a complex drive cycle.
- Discuss coupling of 1-D and 3-D models for efficient estimation of skin temperature.
- Expedite computational solution and reduce the CPU time
- Contribution of radiation to external heat transfer.
- Offer an effective computational methodology for similar problems.

12.2 Heat Transfer in Pipe Flow

Empirical Heat Transfer Calculation using Convective Augmentation Factor

Experimental studies were performed by Malchow et al. [3] in the straight section of an exhaust port of a four-stroke spark-ignition engine. Time-averaged heat transfer based on the effects of location in the exhaust and engine variable was correlated as Nusselt-Reynolds number relationships. They showed that the usual steady-state heat transfer relationships for developed flows underpredict the heat transfer rates inside the engine ports, predominantly due to high-frequency periodic flow and geometrical effects. The constants equation is found to be different:

$$Nu = 0.0483\, Re^{0.8}\, Pr^{1/3} \tag{12.1}$$

where Re and Pr are Reynolds and Prandtl numbers respectively. Wendland [4] used both single and double-walled pipes to arrive at a correlation. The Nusselt numbers were found to be higher than that for the fully developed flow:

$$Nu = 0.027Re^{0.8}Pr^{1/3} \tag{12.2}$$

This enhancement of the measured Nusselt number from the ideal is termed the Convective Augmentation Factor (CAF). The CAF is found relatively constant across the range of engine loads and speeds: 2.3, 3.0, and 1.6 for the manifold, takedown, and tailpipe. Of the total external heat transfer, the share of thermal radiation from the tailpipe and the manifold is 12–18% and 20–30% respectively.

Condie and McEligot [5] examined the effects of pulsating flow, heat transfer parameters on a heated takedown pipe for a commercial V-6 engine and demonstrated a higher heat transfer rate with CAF varying between 2.7 and 8 with pulsation frequency and axial position.

Alkidas et al. [6] obtained the following correlations for the Reynolds number ranging from 7000 to 32,000.

$$\text{For takedown pipe: } Nu = 0.209\ Re^{0.60} \tag{12.3}$$

$$\text{For tailpipe: } Nu = 0.453\ Re^{0.51} \tag{12.4}$$

The convection coefficient in the takedown portion of the exhaust is was significantly higher than the one in the tailpipe. Further, the CAF agreed closely with Shayler et al. [7] and was less than Wendland's [4]. The external heat transfer characteristics are difficult to estimate due to free and forced convection flows as well as radiative interactions underhood and underbody of a vehicle. Free convection heat transfer contributed to nearly half of the total heat flux, while the forced convection and radiation heat transfer contributed to the other half [6].

To solve for both exhaust gas and skin temperatures, Zhang et al. [8] developed a computer model based on the finite volume method by including forced convection inside the exhaust component and natural convection and radiation on the outside. The multi-dimensional heat transfer was simplified to a one-dimensional model for exhaust pipe with both single wall and double wall. They arrived at a similar Nusselt number correlation which was validated with tests. They indicated the minimal variation of skin temperature with respect to the pipe diameter. Liu et al. [9] developed a model for the transient thermal responses of the exhaust system. Convective heat transfer between exhaust gas and pipe wall, axial conduction within pipes, and radiation to the ambient are considered. Four FTP tests on a 2.5 L 4-cylinder engine are conducted using both single-wall and double-walled pipes with an air gap of the order of a few mm. The measured parameters are used for initializing the computer model that uses the CAF given by Wendland [4]. The results over-predicted the gas temperature by 20–50 °C in the single-walled pipe.

A computer model by Konstantinidis et al. [10] covered different exhaust piping configurations. The study of the effect of added mass such as welds and flanges on heat transfer concluded that the loss in temperature response due to the concentrated mass would be significant only if the added masses are positioned at the inlet section of the pipe since, in a hotter pipe section, the thermal energy would be accumulated and reject heat to ambient more intensely.

Grose and Austin [11] developed software that interacted with the one-dimensional and three-dimensional fluid flow commercial simulation packages in a heterogeneous environment. This retained the details of internal flow whilst conserving the overall mass flow in the network, and eliminated uncertain values in the boundaries in a 3-D model. The study of a simple network without an actual power-train or exhaust model was for a very short duration and hence has the potential for use in complex drive cycles was not analysed.

Depcik and Assanis [12] surveyed the correlations in the literature for the gas side heat transfer in the intake and exhaust systems to develop a universal correlation. The correlations differ only by constants *a* and *b*:

$$Nu = a\, Re^{b} \tag{12.5}$$

The authors studied turbulence microscales and finalised the exponential factor, *a*, thereby reduced the number of adjustable correlations to one using the least-square curve fitting method:

$$Nu = 0.07\, Re3/4 \tag{12.6}$$

The simulations by Zhang et al. [13] involved 3-D compressible turbulent fluid flow in conduits with natural and forced convection, heat conduction, and radiation targeting more towards the catalytic converter.

Out of the Nusselt-Reynolds correlations based on turbulent pipe flow, the Sieder-Tate relationship [14] with the augmentation factor, CAF calculated by Wendland [4] is the most commonly used. It is denoted as Wendland Heat transfer Correlation (WHT). The universal correlation developed by Depcik and Assanis [12] is with the CAF values proposed by Wendland [4] is designated as Universal Heat transfer Correlation (UHT), Table 12.1.

Table 12.1 Heat transfer correlations

	Correlation	Empirical correlation	CAF values used
1	Universal heat transfer (UHT)	$Nu = 0.07\, Re^{\frac{3}{4}}$	2.3, 3, and 1.6 for the manifold, takedown, and tailpipe respectively
2	Wendland heat transfer (WHT)	$Nu = 0.027\, Re^{0.8}\, Pr^{\frac{1}{3}}$	2.3, 3, and 1.6 for the manifold, takedown, and tailpipe
3	No user defined correlations	n/a	N/A

Colburn Analogy for Heat Transfer; Calculation in the Commercial Code, Without CAF

This section describes the theory used for modelling heat transfer in pipes and flow splits. The heat transfer from fluids to the inner surface of the walls is calculated using a heat transfer coefficient, at every time step from the fluid velocity, the thermo-physical properties, and the wall surface roughness. The coefficient, h_g is calculated using the Colburn analogy [15] for smooth pipes for the turbulent, laminar, and transitional flows as follows:

$$h_g = \frac{1}{2} C_f \rho U_{eff} C_p Pr^{\frac{2}{3}} \tag{12.7}$$

Here:

C_f Friction coefficient of smooth pipe
U_{eff} Effective velocity outside the boundary layer
C_p Specific heat
Pr Prandtl number
ρ Density of the flow.

A Nusselt number, 3.66 is used for Reynolds number less than 2000 and the heat transfer coefficient is given by:

$$h_g = \frac{Nu.k}{d} \tag{12.8}$$

Here:

Nu Nusselt number
k Thermal conductivity, W/m-K
d Pipe diameter, m.

The surface roughness in a pipe can strongly influence the heat transfer coefficient. The heat transfer coefficient, in this case, is initially calculated using (12.8) and then increased using Eq. (12.9). The effect of roughness has been dealt with in detail in Ref. [28].

Correction for Surface Roughness:

$$h_{g,rough} = h_g \left(\frac{C_{f,rough}}{C_f} \right)^n \tag{12.9}$$

$$n = 0.68\, Pr^{0.215} \tag{12.10}$$

Here:

$h_{g,rough}$ Heat transfer coefficient of rough pipe
$C_{f,rough}$ Friction coefficient of rough pipe.

Estimating friction coefficient from heat transfer coefficient

The higher heat transfer coefficient for pulsating flows than what is given by the textbook for a steady pipe flow is more easily understood than the increase in friction. Therefore, the enhanced heat transfer can be used to estimate the realistic friction coefficient by invoking the Colburn analogy.

12.3 Numerical Methods for General Flow Solution

The flow in an exhaust system is considered as a pipe flow considering the effects of flow splits, orifices, etc. Every pipe is divided into many cellular volumes. Navier–Stokes equations of continuity, momentum, and energy equation in one dimension are solved by considering all quantities as averages in the cells. To solve these equations numerically, initial discretization of space and time is required. These volumes are connected by means of boundaries. The scalar variables like pressure, density, temperature, internal energy, and enthalpy are assumed to be uniform in each volume.

Governing Equations

The left sides of the following equations represent the derivatives of primary solution variables. The right side represents the secondary variables.

Continuity equation

$$\frac{dm}{dt} = \sum_{boundaries} \dot{m} \tag{12.11}$$

Energy equation, (Explicit Solver)

$$\frac{d(me)}{dt} = -p\frac{dV}{dt} + \sum_{boundaries} (\dot{m}H) - hA_s(T_{fluid} - T_{wall}) \tag{12.12}$$

Enthalpy equation, (Implicit Solver)

$$\frac{d(\rho HV)}{dt} = \sum_{boundaries} (\dot{m}H) + V\frac{dp}{dt} - hA_s(T_{fluid} - T_s) \tag{12.13}$$

Momentum equation,

$$\frac{d\dot{m}}{dt} = \frac{dpA + \sum_{boundaries}(\dot{m}u) - 4C_f \frac{\rho u|u|}{2} \frac{dxA}{D} - Cp\left(\frac{1}{2}\rho u|u|\right)A}{dx} \tag{12.14}$$

Here:

$\dot{m}$	Boundary mass flux into volume, $\dot{m} = \rho A u$
m	Mass of the volume
V	Volume
p	Pressure
ρ	Density
A	Flow area (cross-sectional)
A_s	Heat transfer surface area
e	Total internal energy (internal energy plus kinetic energy) per unit mass
H	total enthalpy, $H = e + p/\rho$
h	heat transfer coefficient
T_{fluid}	Fluid temperature
T_{wall}	Wall temperature
u	Velocity of the fluid
C_f	Skin friction coefficient
C_p	Pressure loss coefficient
D	Equivalent diameter
dx	Length of the mass element in the flow direction (discretization length)
dp	Pressure differential acting across dx.

There are two integration methods, marching with the discretized time, explicit and implicit. The two methods affect the solution with the time step limited by the stability criterion. The primary solution variables in the explicit method are mass flow, density, and internal energy.

Explicit Method by Commercial Software

The right-hand side of the Eqs. (12.11–12.14) are calculated using values from the previous time step. This yields the derivative of the primary variables and allows the value at the new time to be calculated by integration of that derivative over the time step. The explicit solver uses only the values of the sub-volume in question and its neighbouring sub-volumes, if necessary. To ensure numerical stability, the time step must be restricted to satisfy the Courant condition [16, 20] mentioned later while deliberating the stability of calculations. The small-time steps required by this method make the explicit method undesirable for simulations that are relatively long (on the order of minutes in real-time) but are well suited for highly unsteady flow where a high degree of resolution is required to capture the extremes of the flow behaviour. This method produces more accurate predictions of pressure pulsation that occurs in engine air flows and fuel injection systems and is required when the prediction of pressure wave dynamics is important [15].

At each time step, the pressure and temperature are calculated in the following way:

Continuity and energy equations yield the mass and energy in the volume.

- With the volume and mass known, the density is calculated yielding density and energy
- The equations of state for each species define density and energy as a function of pressure and temperature. The solver iterates the pressure and temperature calculations until they satisfy the density and energy already calculated for this time step.

The relation between the time step and the discretization length is determined by the Courant number when the explicit solver is used for first-order integration. The discretization length is the length of a sub-volume in a pipe. The solution remains stable by choosing its time steps such that the Courant–Friedrichs–Lewy condition is met:

$$\frac{\Delta t}{\Delta x}(|u| + c) \leq 0.8M \tag{12.15}$$

Here:

Δt	Timestep
Δx	Minimum discretization element length
c	Speed of sound
M	Time step multiplier specified by the user.

12.3.1 Historical Explicit Methods

The analysis of unsteady flow in the exhaust and intake systems of an internal combustion engine is of considerable importance as they describe the scavenging and volumetric capacities of the engine. Among the available methods, the oldest and well known is the method of characteristics [16, 18]. This method can be used to solve problems graphically or on a computer. This method has been treated by Benson et al. [16] extensively for different pipe flow configurations.

Finite difference schemes [17] can also be applied to the problems of unsteady pipe flow. To fix boundary conditions either the method of characteristics [16, 18] or iterative schemes [19, 20] can be used. Here, higher-order accuracy is possible whereas the method of characteristics using a computer is only first-order accurate. In the following paragraphs, a highly stable finite difference method is described in which the boundary conditions are applied by a novel approach.

Equations of flow

Though the unsteady flow through a pipe is practically three-dimensional due to sharp discontinuities, friction, and heat transfer, the problem is assumed to be unidimensional when the effects due to the above are taken into account considering proper empirical equations.

Equation of conservation of mass:

$$\frac{\partial \rho}{\partial t}+\rho\frac{\partial u}{\partial x}+u\frac{\partial \rho}{\partial x}+\frac{\rho u}{F}\frac{\partial F}{\partial x}=0. \tag{12.16}$$

Here,

ρ density
u gas veolicty
F area of cross section of the pipe
t time.

Equation of conservation of momentum:

$$\frac{\partial u}{\partial t}+u\frac{\partial u}{\partial x}+\frac{1}{\rho}\frac{\partial P}{\partial x}+\frac{4fu^2}{2D}\bullet\frac{u}{|u|}=0. \tag{12.17}$$

Here,

D diameter of the pipe
f friction constant for the frictional loss
D diameter of the pipe.

Equation of conservation of energy:

$$q\rho F dx=\frac{\partial}{\partial t}\left[\rho F dx\left(C_v T+\frac{u^2}{2}\right)\right]+\frac{\partial}{\partial x}\left[\rho u F\left(C_p T+\frac{u^2}{2}\right)\right] \tag{12.18}$$

Here,

q gain in enrgy due to wall heat transfer per unit mass of gas
T gas temperature
C_v specific heat at constant volume
C_p specific heat at constant pressure.

If $f=0$, $q=0$ the entropy of the flow is constant and flow is homentropic. In the present method, the pipe is divided into several cells. The properties in each are assumed to be uniform. The difference equations for homentropic flows are written in their corresponding difference forms. Referring to Fig. 12.2,

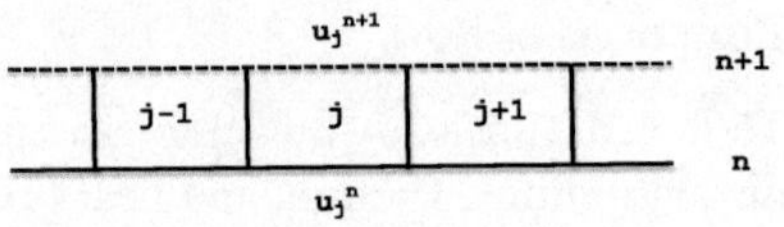

Fig. 12.2 Finite difference scheme in the time-distance plane

$$\Delta\rho_{j,n} = \left[-\rho_{j,n}\frac{u_{j+1,n} - u_{j-1,n}}{2l} - u_{j,n}\frac{\rho_{j+1,n} - \rho_{j-1,n}}{2l}\right]\Delta t \tag{12.19}$$

$$\Delta u_{j,n} = \left[-u_{j,n}\frac{u_{j+1,n} - u_{j-1,n}}{2l} - \frac{1}{\rho_{j,n}}\frac{P_{j+1,n} - P_{j-1,n}}{2l}\right]\Delta t \tag{12.20}$$

and

$$\Delta(\rho e)_{j,n} = -\frac{(hu\rho)_{j+1,n} - (hu\rho)_{j-1,n}}{2l}\Delta t \tag{12.21}$$

Here,

Variable:

h enthalpy
e energy per unit mass
P gas pressure
l length of a cell.

Subscripts:

j cell number
n index for time.

Here, $h = C_pT + u^2/2$, and $e = C_vT + u^2/2$, if divisions are uniform. If the divisions are not uniform, the gradient of any property in the neighboring cells is assumed concentrated at the centre of the cells. In all the calculations performed here, only uniform divisions are made. In principle,

$$\rho_{j,n+1} = \rho_{j,n} + \Delta\rho_{j,n} \tag{12.19a}$$

$$u_{j,n+1} = u_{j,n} + \Delta u_{j,n}, \text{ and} \tag{12.20a}$$

$$(\rho e)_{j,n+1} = (\rho e)_{j,n} + \Delta(\rho e)_{j,n}. \tag{12.21a}$$

The above method which is the simple Euler's integration yields unstable solutions. On the contrary, the third and fourth-order Runge–Kutta methods yield stable solutions. The properties matrix $\overline{y} = (\rho, u, \rho e)$ at t is denoted by $\overline{y_0}$. From Eqs. (12.19–12.21) we can write various increments in the Runge–Kutta fourth-order method, where $f(\overline{y})$ denote the right-hand sides of Eqs. (12.19–12.21).

$$\Delta\overline{y_1} = f(\overline{y_0})$$

$$\Delta\overline{y_2} = f\left(\overline{y_0} + \frac{1}{2}\Delta\overline{y_1}\right)$$

$$\Delta\overline{y_3} = f\left(\overline{y_0} + \frac{1}{2}\Delta\overline{y_2}\right)$$

$$\Delta\overline{y_4} = f\left(\overline{y_0} + \frac{1}{2}\Delta\overline{y_3}\right) \quad (12.22)$$

$$\overline{y_{t+\Delta t}} = \overline{y_0} + \frac{\Delta\overline{y_1} + 2(\Delta\overline{y_1} + \Delta\overline{y_3}) + \Delta\overline{y_4}}{6} \quad (12.23)$$

The scheme is given in Fig. 12.3. It may be observed that the increments for properties, $\Delta\overline{y_1}$, $\Delta\overline{y_2}$, $\Delta\overline{y_3} and \Delta\overline{y_4}$ after calculating y_1, y_2, y_3, and y_4.

Different boundary conditions can be impressed at the pipe ends depending on the physical configurations. In a well-posed boundary condition, some properties are explicit functions of the others. The latter properties can be assumed to be equal to the corresponding values in the boundary cells. Using this approximation many examples are solved and compared with accurate solutions [20, 23]. Two examples are given here.

12.4 Examples

In the first example, a straight pipe in which air is filled uniformly is considered. Initially, the properties are $P = 101.3$ kPa and $T = 300$ K with one side open to the atmosphere having the same properties. Suddenly, at the other end, a pressure pulse in the shape of a half sine wave is applied for a time (t_0) duration equal to that taken by sound to travel the length of the pipe once, and then the end is closed. At the boundary communicating to the atmosphere,

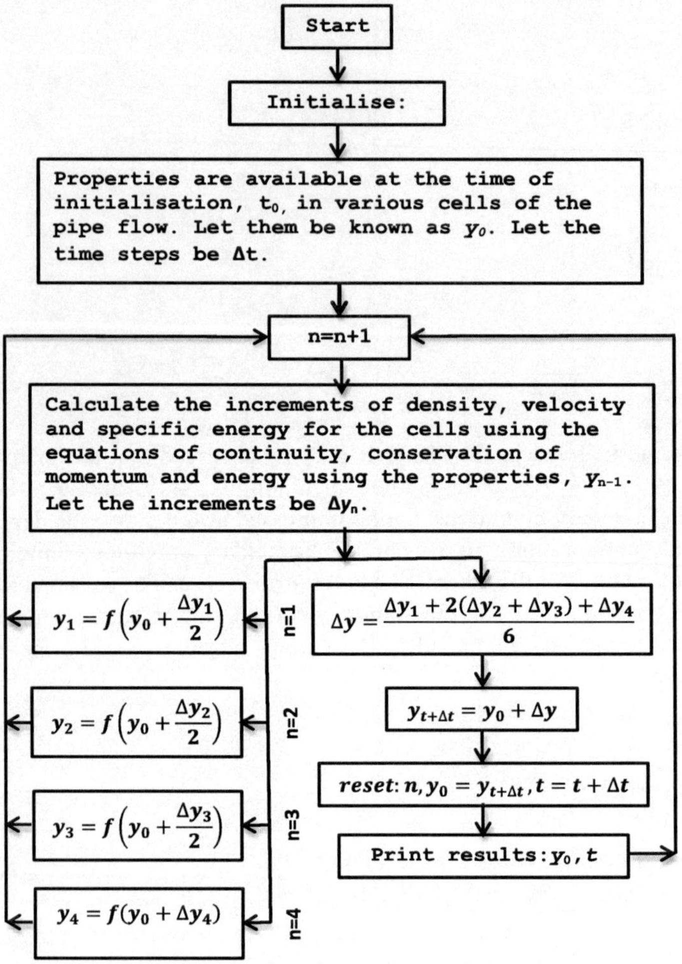

Fig. 12.3 Flow chart for calculation of unsteady pipe flow using the scheme involving Runge–Kutta 4th order method

$$if u_m > 0 or P_m > P_{atm}, then P_{end} = P_{atm}, u_{end} = u_m, \text{ and} \tag{12.24}$$

Here, subscripts:

m boundary cell number
atm atmospheric conduction
end boundary.

$$if u_m < 0 or P_m < P_{atm}, then P_{end} = P_m \quad (12.26)$$

$$u_{end} = -a_{atm}\left[\frac{2}{K-1}(1 - P_m/P_{atm})/\left(1 + \frac{K+1}{K-1}P_m/P_{atm}\right)\right]^{1/2} \quad (12.27)$$

Here,

$$T_{end} = T_{atm} - u_{end}{}^2/(2C_p), \quad (12.28)$$

$$\rho_{end} = \rho_{end}/(RT_{end}), \text{ and} \quad (12.29)$$

Here,

R gas constant.

$$(\rho h_{end}) \cong \rho_{end} h_{atm}; h_{atm} = C_p T_{atm}. \quad (12.30)$$

At the boundary cell (m), the gradient of any property y,

$$\left[\frac{\Delta y}{\Delta x}\right]_m = \frac{y_{end} - (y_{m-1} + y_m)/2}{l} \quad (12.31)$$

The equation given above can be substituted in Eqs. (12.19–12.21). On the side at which the pulse is applied, till it is closed,

$$P_{end} = 1 + 0.11 sin(t/t_0) atm \quad (12.32)$$

$$u_{end} = u_1, \text{ the velocity of the air in the boundary cell} \quad (12.33)$$

$$\rho_{end} = (P_{end}/P_{atm})^{1/K}\rho_{atm} \quad (12.34)$$

Here 1 refers to the boundary cell. After it is closed,

$$u_{end} = 0, P_{end} = P_1, \rho_{end} = \rho_1, h_{end} = h_1 = C_p T_1 + u_1^2/2 \quad (12.35)$$

The solutions are shown in Fig. 12.4 for the point at which the pressure pulse is applied and the midpoint.

In the second example, discharge of air from a cylinder with a nozzle fitted at one end and the other end closed is considered. The cylinder is initially charged to

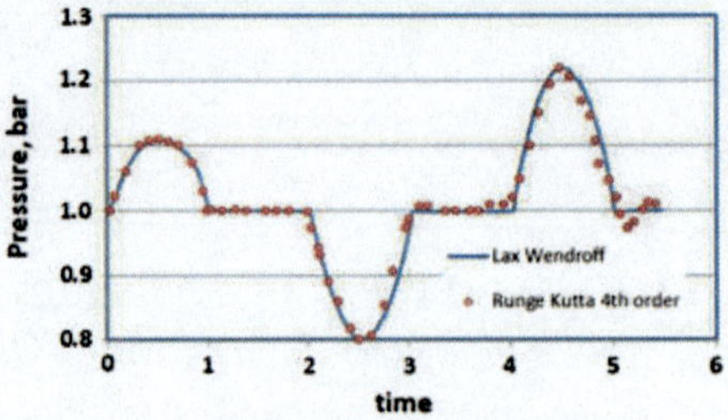

Pulsations at closed-end

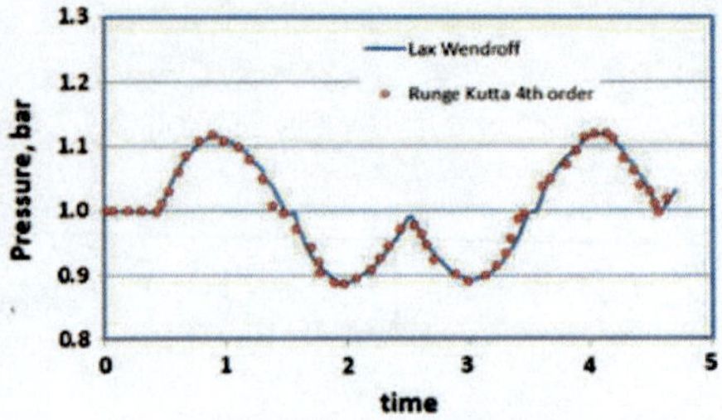

Pulsations at the mid-point

Fig. 12.4 A pipe of 0.5 m length with one end open and the other end closed is initially at 1.0 bar and 300 K. At the closed end, a half sinusoidal pressure of peak 0.1 bar gage is applied. The pressure pulsation in the pipe is studied by explicit Runge–Kutta 4th order integration and, Lax-Wendroff method

a pressure of 303.96 kPa at a temperature of 412 K with the nozzle kept initially closed. The area ratio of the nozzle, $\Phi = 0.5$.

The nozzle is suddenly opened to the atmosphere. The pressure waves travel up and down the cylinder to cause more and more rarefaction till the cylinder pressure is uniformly atmospheric. This is an unsteady pipe flow problem. The cylinder is divided into many cells and to each of them Eqs. (12.19–12.21) are applied, except at the ends. At the end, the conditions are fixed according to the boundaries. At the closed boundary end Eq. (12.35) is used. At the nozzle end,

$$P_{end} \cong P_m, \rho_{end} \cong \rho_m, h_{end} = h_m \tag{12.36}$$

$$u_{end} = f(P_{end}, P_{atm}, \phi) \tag{12.37}$$

The boundary conditions at the nozzle end are given below.

Nozzle Boundary:

In the following equations, atm refers to the atmosphere and the end refers to pipe boundary.

When, $P_{end}/P_{atm} > P_{critical} = P_{cr,}$ then the flow is choked:

$$u_{end} = a_{end}\phi\cdot(1/P_{cr})^{(K+1)/2K}. \tag{12.38}$$

When $P_{end}/P_{atm} < P_{cr}$, the flow is subsonic:

$$u_{end} = \left[\frac{a_{end}}{P_r^{(K-1)/(2K)}}\right] \left[\frac{2}{K-1}\left\{\left(\frac{a_{end}}{a_{end.atm}}\right)^2 - 1\right\} / \left\{\left(\frac{a_{end}}{a_{end.atm}}\right)^{4/(K-1)} / \phi^2 - 1\right\}\right]^{0.5}. \tag{12.39}$$

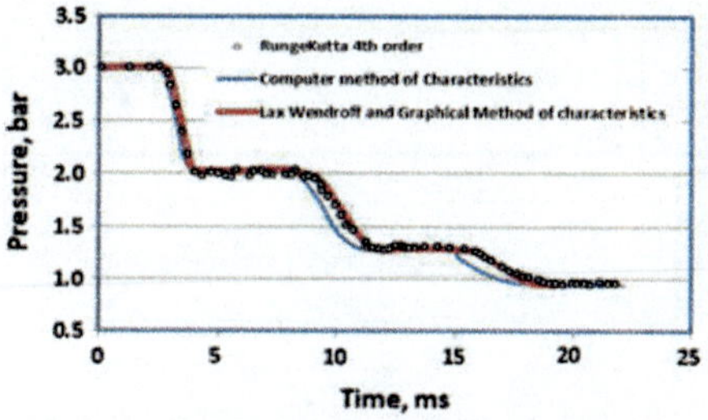

Pressure history at the closed-end

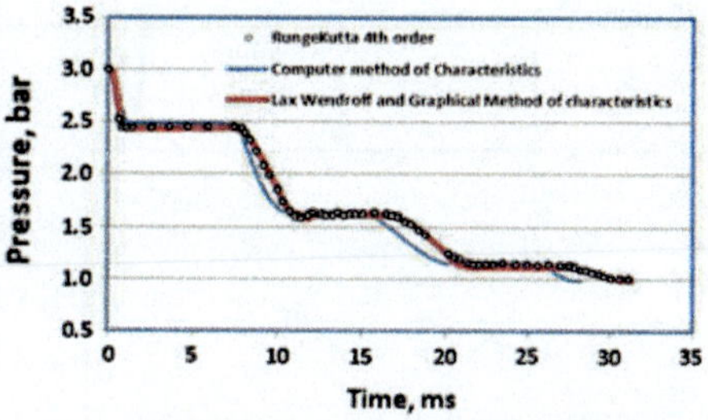

Pressure history at the nozzle end

Fig. 12.5 A cylinder with one end closed and the other end with a nozzle of area ratio 0.5 is initially at 3 bar and 412 K. Also, the nozzle which is closed initially is suddenly opened. The pressure history is studied by explicit Runge–Kutta 4th order integration, Lax-Wendroff, and graphical method of characteristics. In the backdrop, the results of the computer method of characteristics are plotted

Here,

a	sonic velocity
$a_{end.atm}$	sonic velocity of gases after isentropic expansion from pipe conditions to atmospheric pressure
P_r	The ratio of pressures before and after the nozzle and
P_{cr}	is given by $6P_{cr}^{10} - 5P_{cr}^{12} - \phi^2 = 0$ for gases with $K = 1.4$.

These are applied to find the gradients according to Eq. (12.31) and the solution is obtained as shown in Fig. 12.3. In Fig. 12.5, the pressure diagrams for the nozzle end and closed-end are shown. They compare favorably with the solutions obtained using the graphical method of characteristics and the Lax-Wendroff scheme.

Stability of the calculations

The equations of flow form a set of difference equations according to the method of solutions. The set of difference equations form a system whose stability is of interest. Stability describes the capacity of the system not to amplify any error introduced. In other words, it must damp the effect of error as the solution is being generated. Von Neumann's stability criterion [18, 20] states that the amplification factor for a system of equations for any type of error at any frequency/wavelength of the Fourier component of the solution should be less than unity. By application of z-transform, the stability criterion can be rigorously derived as follows [19–21]:

Runge–Kutta fourth-order method:$r < \sqrt{8}$
Runge–Kutta third order method:$r < \sqrt{3}$
Lax-Wendroff method:$r < 1$
Method of characteristics [16, 20] $r < 1$ (Courant condition).

Here, $r = a\frac{\Delta t}{\Delta x}$, $\Delta t = calculation\, time\, step$, $\Delta x = length\, of\, the\, cell.$
Thus, the fourth-order method is stable for a larger mesh size than the rest of the methods.

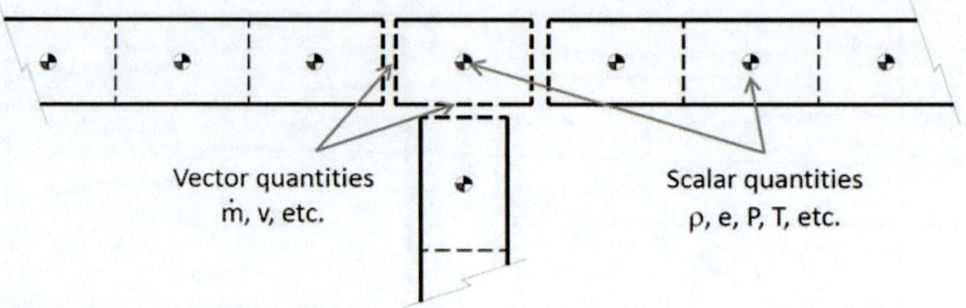

Fig. 12.6 Schematic representation of splits [15]

Splits

In the examples given so far, the exhaust pipes are straight without any splits. However, in the case of multi-cylinder engines splits and joints cannot be avoided. In such cases, each flow split is represented by a single volume. This section describes the method of flow solution used in pipes and flow splits. This type of discretization is referred to as a "staggered grid" as shown in Fig. 12.6.

The time step in flow-splits uses a similar methodology as pipes and is given as

$$\Delta t \propto \frac{V}{A_{eff} c}$$

Here, A_{eff} is the effective flow area and V is the volume.

This indicates that the time step must be small enough such that only a fraction of mass could be emptied in a given time-step if the fluid is leaving the flow-split traveling at the speed of sound.

To determine A_{eff} each port of the flow split is checked. The area for flow at each port is defined as the smaller of the areas based on the flow-split expansion diameter at that port and the diameter of the connection to the flow split at that port. The port with the largest area for flow determines the value of A_{eff}, and ultimately controls the time step requested by the flow split [15].

12.5 Application of the Finite Difference Scheme to Engines

In the practical case of engines, the gas exchange processes are not very simple. The effect of friction and heat transfer cannot be neglected in the pipes as done in the few examples. When the engine is under firing conditions heat transfer plays a very important role. A fully instrumented engine, Table 12.2, using an ingenious ultrasonic velocity cum thermometer is developed to study the exhaust and intake flows with heavy background noise [22].

Table 12.2 Engine specifications

Parameter		Parameter	
Name	W.5 Krishi	Bore	75 mm
Cylinders	Single	Stroke	64 mm
Power output	From camshaft	Compression ratio	6
Rated power	5.5 hp at 1400 cam rpm	IVC	43.5° Crank
Maximum power	5.5 hp at 1450 rpm	IVO	528.5° Crank
Speed range	500–1700 cam rpm	EVO	313.5° Crank
Cooling	Water-cooled	EVC	547.5° Crank

The heat transfer inside the pipe is considered by using the Nusselt-Reynolds number equation with a CAF = 1.2 and the friction coefficient from Moody's graph [23]. The important heat transfer outside the pipe as described later is avoided in the present calculations by considering the experimentally measured skin temperatures. To the engine specified in Table 12.2, a two-meter long exhaust pipe is fitted with a plain opening at the atmospheric end. The exhaust pressure pulsation is measured using a cooled strain gauge type transducer at a distance of 0.3 m from the engine end, under motoring conditions. The pressure is predicted well when friction is taken into account, Fig. 12.7. In the absence of friction, the calculated pressure waves are not damped sufficiently.

The same engine is studied for the velocity of the gases in the exhaust when the engine was fired, Fig. 12.8. The engine is fitted with a 2.4 m long pipe with a plain ending at the atmosphere. The engine is run at 1600 cam rpm at ¾ of wide-open

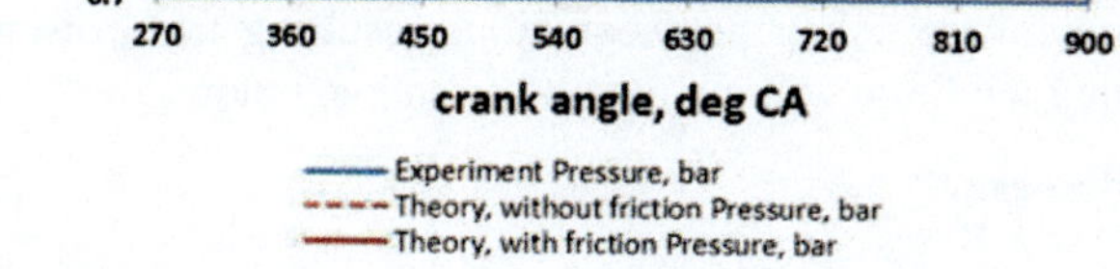

Fig. 12.7 Pressure fluctuations in the 2 m long exhaust pipe at 0.3 m from the engine (Table 12.2) under motoring conditions. The other end is plain and open to the atmosphere. Experiment and the calculation results from the 4th order Runge–Kutta method are presented. The calculations are performed when friction is **a** neglected and **b** taken into account

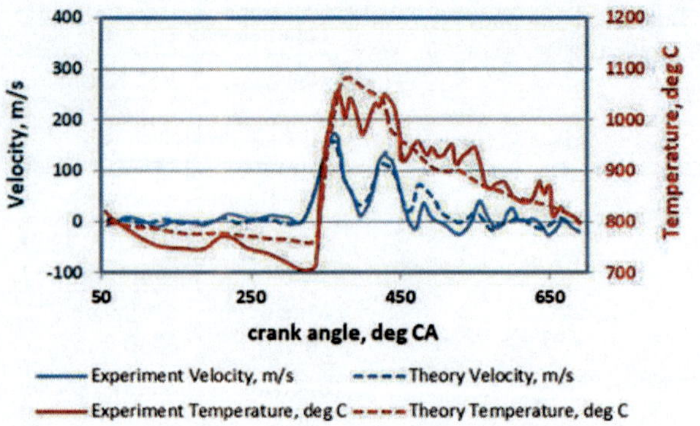

Fig. 12.8 Pressure fluctuations in the 2.4 m long exhaust pipe at 0.6 m from the engine (Table 12.2) under firing conditions. The other end is plain and open to the atmosphere. Experiment and the calculation results from the 4th order Runge–Kutta method are presented. The calculations are performed by taking into account friction and heat transfer

throttle when the airflow is 73% of the wide-open throttle (WOT) flow. The pressure and velocity are measured at a distance of 0.6 m from the engine. The calculation scheme worked satisfactorily predicting the velocity and temperature.

The scheme is successfully applied for many boundary conditions like a nozzle end, and silencer end at different speed and load conditions. The parameters namely, pressure, velocity, and temperature are satisfactorily predicted both at the inlet and exhaust [20, 23].

12.5.1 Wall Thermal Solution Considering Heat Transfer External to the Pipe

The wall temperatures are iteratively calculated using the internal heat transfer from 1-D, the external heat transfer from the 3-D simulation STAR CCM+ [24], the thermal capacitance of the walls, and the initial wall temperature entered by the user. The external heat transfer (from outer walls to the environment) in turn is calculated from the data describing forced convection, free convection, and/or radiation. See Chap. 13 for the details of the simulation technique.

When a steady-state simulation is performed, the final wall temperatures are reached in a few iterative cycles. The transient simulation for studying the warm-up characteristics of the system considering the effect of structural heat capacity, however, takes more cycles to converge.

The wall temperature solver uses equations resulting from discretized energy conservation using the finite volume method as shown below:

Conservation of Energy:

$$\int \frac{\partial(\rho C_V T)}{\partial t} dV = \int (-\nabla . q) dV \qquad (12.40)$$

Here:

C_V Specific heat
q Heat flux.

After integration we have:

$$\rho C_V \frac{\Delta T}{\Delta t} = \sum_{faces} (-qA) \tag{12.41}$$

Each of the wall layers for a pipe sub-volume is used as a control volume for the energy equation. Heat transfer across the radial faces of the volumes is calculated using resistance to conductive heat transfer as well as radiation between the surfaces viewing the pipe walls [15, 24]. The heat transfer from the gas to the pipe is calculated at every time step using the equations shown below.

Heat flux due to conduction:

$$q_{conduction} = -k\nabla T \tag{12.42}$$

Here:

T Surface Temperature.

Heat flux due to convection:

$$q_{convection} = h(T_g - T_w) \tag{12.43}$$

Here:

h heat transfer coefficient
T_g, Tw Temperature of gas and wall, respectively.

External convection coefficient:

$$h_{external} = max\left(Nu\frac{k}{D}, h_{user}\right)$$

Here:

$h_{external}$ Heat transfer coefficient for external convection.

$$Nu = 0.5 + 0.10(RayleighNumber)^{\frac{1}{3}}$$

If the geometric view factor is unity between two surfaces (say, 1 and 2) then the heat flux due to radiation is given as:

$$q_{radiation} = -\varepsilon\sigma(T_1^4 - T_2^4) \tag{12.44}$$

Here:

ε emissivity of the surface
σ Stefan-Boltzmann constant.

With many surfaces at different temperatures surrounding the pipe walls, the radiation heat transfer is solved using a network of resistances derived from the emissivities, view factors, and radiosity of surfaces. For complicated geometric arrangements like the engine exhaust systems, a computer with the software comes to our rescue.

12.6 1-D Model

Commercial software

For the problem associated with the internal flow and heat transfer, the computational analysis is done using GT-Power, a licensed product, and a tool within GT-SUITE from Gamma Technologies is used. It is a comprehensive code based on one-dimensional fluid dynamics, representing flow and heat transfer in piping and other components of a system [15] by solving the Navier–Stokes equations in 1-D. It has a built-in vehicle/engine simulation for calculating thermal loads under a driving cycle. A thorough work on the prediction of exhaust skin temperature integrating the 1-D Model with vehicle level CFD Model is accomplished by Sudharsan [25]. This work is explained in the following paragraphs with the interesting results obtained.

Steady-State Simulation

Before launching a detailed investigation into the heat transfer characteristics and skin temperature prediction, the 1-D model has to be validated with available test results. As a first stage of the work, a set of steady-state WOT simulations are carried out to validate the build of the 1-D model, and then the model is then modified for the simulation of a transient drive cycle. The results from the transient tests are then validated again with the aerothermal vehicle test data.

A complete description of a V6 engine from the intake to the tailpipe is made by feeding the software with the large information regarding engine characteristics, intake and exhaust systems, cylinder geometry, valves, and fuel injection system. Further information about heat losses as well as combustion is entered. The base model of the engine is shown in Fig. 12.9 with dual exhaust. The vehicle is equipped with a 4-wheel drive and hence the engine is mounted in the North–South configuration. The manifold and the catalyst are in the underhood of the vehicle whereas the components after the catalyst are in the underbody of the vehicle. The exhaust pipes connecting the catalyst and the resonator are called the downpipe. The pipes between the resonator and the muffler are called the intermediate pipe.

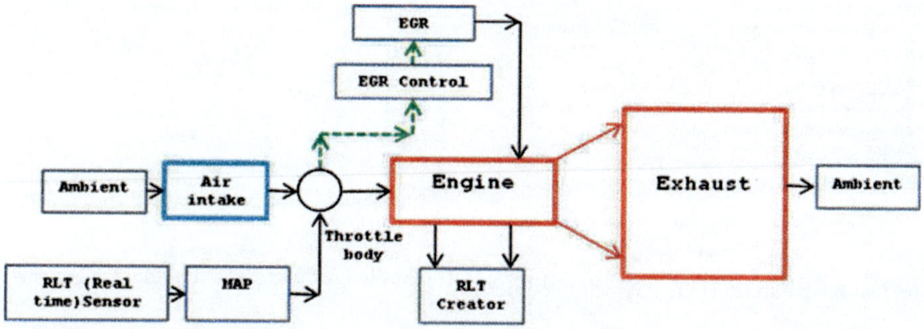

Fig. 12.9 Engine layout [25]

The region consisting of manifold, catalyst, and the exhaust downpipe is denoted as the hot-end. The region after the resonator is denoted as the cold-end. The components are divided into several sub-volumes. Each sub-volume is then discretized further to solve the Navier–Stokes equations in one dimension. Further, the flow in the exhaust is treated as a pipe flow. Some of the details required for the computation include the discretization length for the exhaust, the diameter of the exhaust pipes, heat input rate for the catalyst, and material thermal properties, etc. Thermal masses such as flanges and bell rods are not included in the current model.

For the WOT simulations, an explicit solution method is chosen. This option is used when capturing the wave dynamics in the system is important, since the complete engine is used. Further, this provides crank angle resolved results since the simulations are run for steady-state. Both the flow and the thermal solver are set for steady-state. 15 cases have been considered, and they cover different engine speeds ranging from 1200 to 6800 rpm, all at WOT. The rpm range considered in the simulation to generate is independent of the mass flow rates from the bench test. This is because no RPM data was available from the bench test. Hence, the range of the mass flow rates produced by the RPM range fall between the extremities of the mass flow rates generated in the bench test. The steady-state simulations are validated with the skin temperatures obtained from the bench test.

Three different cases are simulated: with UHT, WHT, and default heat transfer coefficient derived from the Colburn analogy. The software controls the convergence of mass flow, pressure, and temperature. When the fractional variation with respect to the previous cycle is lower than a certain value repeated for two or more iterative cycles, convergence is said to have been reached.

12.7 Results and Discussion

Figure 12.10 shows the predicted gas temperature at different important points in the exhaust line, comparing the simulations using the empirical correlations with the experiments for WOT. There is a good match with experiments over the range of mass

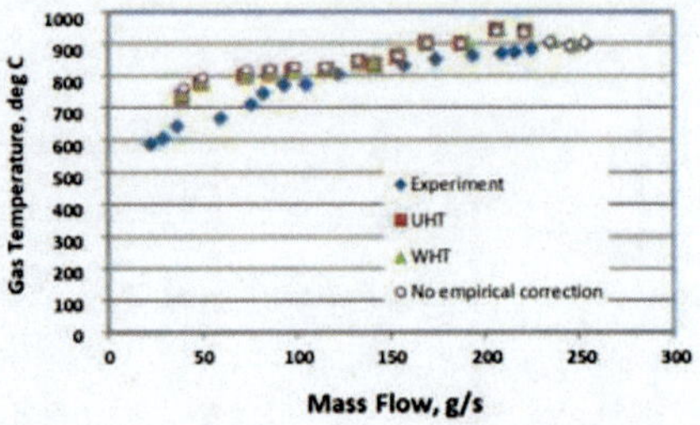

(a) Gas temperatures at the left manifold inlet

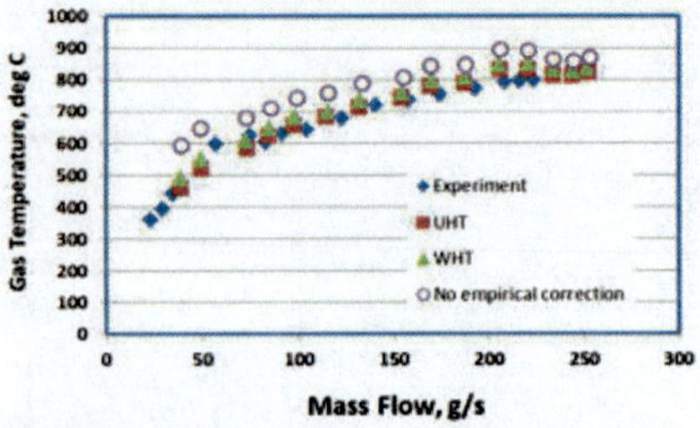

(b) Gas temperatures at left before the resonator

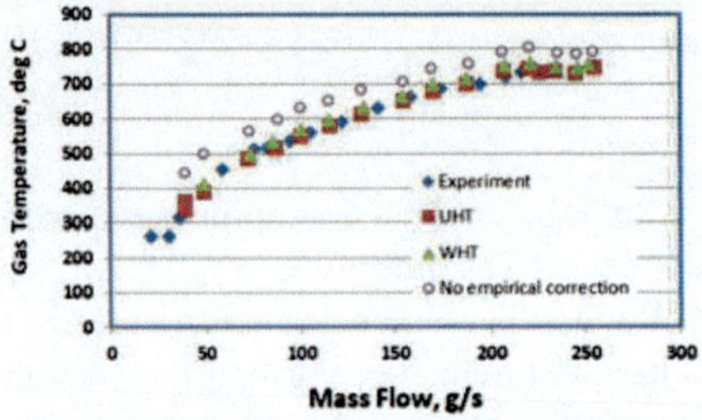

(c) Gas temperatures at left after the resonator

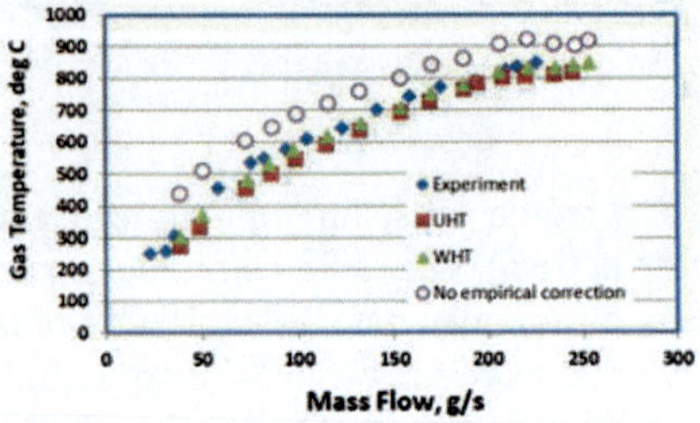

(d) Gas temperatures at left before the muffler

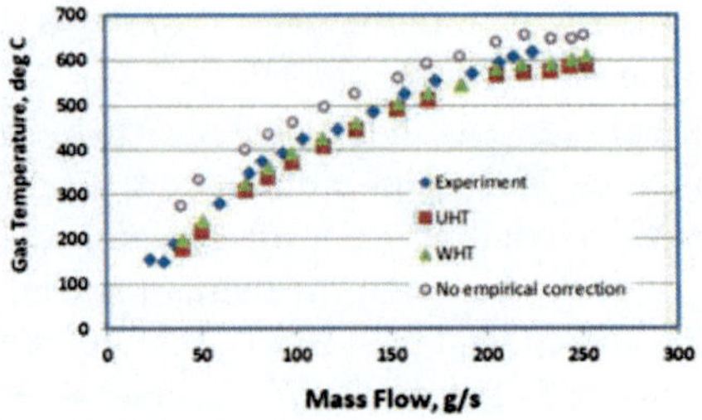

(e) Gas temperatures at left after the muffler

Fig. 12.10 Gas temperatures at various cardinal points in the exhaust line [25] at steady-state

flow rates with the correlation. Without the correlation, the use of the steady-state heat transfer coefficient (using the Colburn analogy) over-predicts the gas temperatures.

The UHT correlation provides a good match with the experiment. The WHT correlation provides a good match at the lower rpm ranges but over-predicts the gas temperatures by 10–15 °C.

Independent exercise indicated that applying CAF values to heat transfer coefficient from Colburn analogy instead of to 'the Nusselt-Reynolds number correlations over predicts the gas temperatures by 20–25 °C up to the muffler.

12.7.1 Transient Simulation

The same engine and the exhaust models used in the study of the steady-state performance are used for the transient simulation. The variable CA50 and CA10 to CA90 data, manifold absolute pressure (MAP) values, and variable valve timing data for intake and exhaust valves are input as functions of speed and load of the transient driving cycle.

Simulation Setup

The transient drive cycle used for the current study is a Davis dam test cycle similar to the SAE J2807 standard [26] with a small difference; the towing and trailer conditions are not accounted for, Fig. 12.11.

The cycle consists of an approach segment where the speed is more or less constant. The grade is from 15 and 38 min where the speed and the load vary significantly. During the cooldown, the vehicle is standstill. To simulate the entire transient drive cycle, information about the actual vehicle from the drive cell is used as boundary conditions: e.g., engine speed, fuel–air ratio, MAP values, intake and exhaust cam centreline data, variable valve-timing data, spark timing, and coolant temperature. The simulation accounts for the heat transfer from the engine to the coolant as well as the engine oil.

Due to the lack of drive cell data for all the locations on the exhaust, a select few are compared and presented. For brevity, only the results of the left side of the exhaust are presented here, as they were found to be similar to the right side. The initial objective is to match the gas temperatures at the manifold so that the exhaust system could be studied in detail without having to refine the engine side of the model.

To reduce the computational time and increase efficiency, the engine and exhaust were decoupled after the initial engine exhaust coupled run. This ensured all the dynamic effects including backpressure are captured between the engine and exhaust. Before the decoupling, quantities such as velocity, pressure, temperature, and mass flow rate were exported and then used as inlet conditions for the decoupled exhaust.

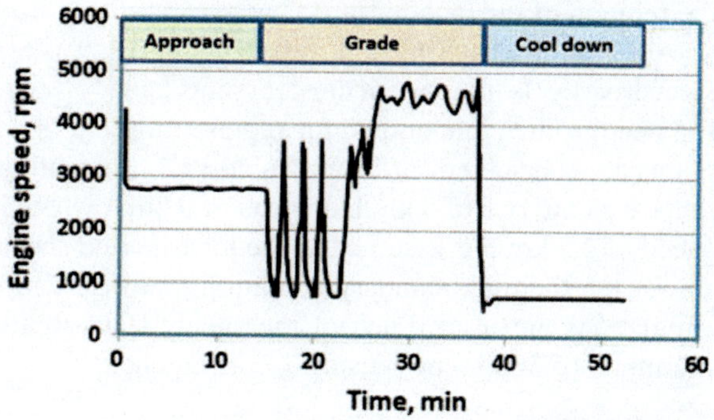

Fig. 12.11 Engine speed representing Davis dam drive cycle [25, 26]

Thus, the simulation time was reduced from 8 days to 3.5 h for the entire transient drive cycle.

12.8 Skin Temperature Estimation Coupling 1-D Model and Vehicle Level 3-D Model

3-D simulation of a full-scale vehicle level model (Chaps. 11 and 13) is investigated for the heat transfer in detail. The first iteration of the 3-D model is carried out using the preliminary skin temperatures data from the 1-D simulation as boundary conditions. The heat exchanger model for the radiator used a moving reference frame (MRF) for the front-end fan flow and heat transfer. The three modes of heat transfer are accounted to model the heat transfer: forced and free convection, and radiation.

A *K-Means Clustering* algorithm was used to identify the best operating points to represent the drive cycle accurately. These operating points are then used for steady-state simulations.

Using the results from the simulation, a new process was identified to predict the exhaust skin temperature [25]. The simulations were initially run for WHT and UHT correlations.

12.8.1 3-D Steady-State Simulation Setup

The computational domain of the model representing the actual drive cell test section is split into six regions consisting of the condenser, the radiator, the radiator fan, the transmission oil-cooler (TOC), the power-steering oil-cooler, and the airstream. The airstream region and the radiator fan are modelled as a fluid region meshed with hexahedral cells and polyhedral cells respectively, while the rest as porous regions meshed with hexahedral cells. Three prism layer cells are engaged for determining the wall effect. A no-slip condition was assumed for all the other walls within the airstream such as the vehicle. A Moving Reference Frame (MRF) is used to consider the rotation of the radiator fan.

A K-means clustering algorithm, Fig. 12.12, was used to partition the entire transient drive cycle into a set of cluster points (steady-state wind speed points) to be used for steady-state simulations, for representing the transient cycle [27]. Nine cluster points are considered sufficient, Table 12.3. Six additional speed points are selected to have an interval of 8 km/h. Instead of 0 km/h which is difficult to simulate, a wind speed of 2.5 km/h is assumed as the lowest wind speed.

For the thermal boundary condition, the skin temperatures from the initial 1-D simulations are used. Each of the steady-state simulations is run iteratively to a minimum of 5000 times ensuring convergence.

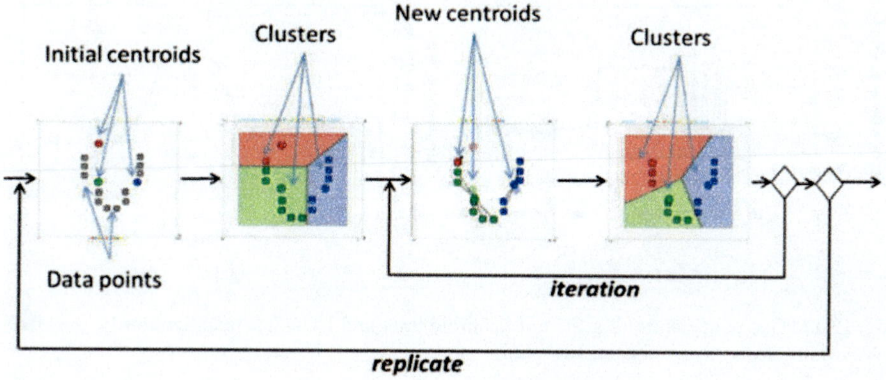

Fig. 12.12 K-Cluster methodology for steady-state speed points [25, 27]

Table 12.3 Wind speed points for 3-D simulation based on K-Mean clustering algorithm

Cluster#	1	2	3	4	5	6	7	8	9
Selected wind speeds (km/h)	74.4	40.1	92.2	76.1	88.5	78.9	57.5	8	84
Additional speeds (km/h)	48.3, 60.7, 70			31				2.4, 15.4, 27.4	

Results: Skin and Gas Temperatures Using Combined 1-D and 3-D Models

Only a few typical results are presented here. Figure 12.13 shows the simulated HTC values for the exhaust downpipe and intermediate pipe. A preliminary investigation into the distribution of HTC over the exhaust system shows that the HTC values for the upstream of the resonator are lower than those seen downstream. Further, the

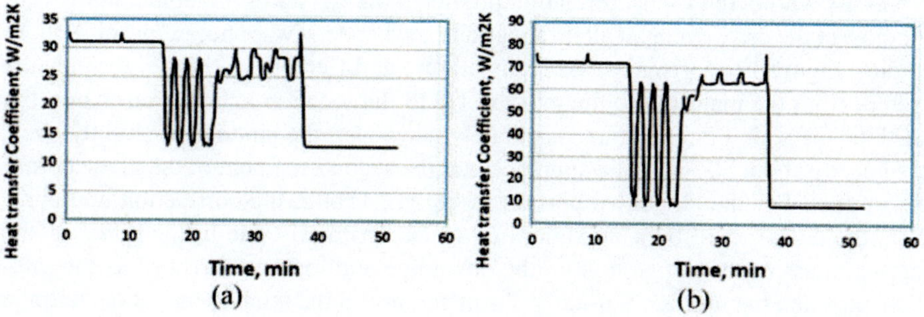

Fig. 12.13 HTC values from steady-state simulations representing transient conditions for **a** downpipe and **b** intermediate pipe [25]

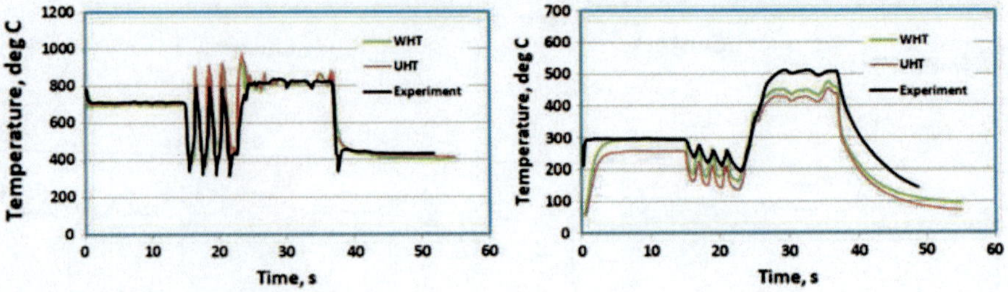

(a) (b)

Fig. 12.14 Gas temperature in **a** the left manifold inlet and **b** the left tailpipe; mostly, the UHT gave results closer to experimental data [25]

presence of a skid plate[1] underbody, starting from the air dam up to the resonator contributes to lower convection rates seen in the downpipe portion. The HTC values obtained from the steady-state simulations are then used in the 1-D simulation as a boundary condition for external heat transfer.

Figure 12.14 shows the gas temperatures at the inlet to the manifold and tailpipe respectively. The temperatures in the inlet section are predicted reasonably well by both WHT and UHT. The WHT predicts the gas temperature in the tailpipe correctly only during the approach segment. In all other cases, UHT has shown results closer to the experiment.

At the start of the grading segment, the lower peaks for both UHT and WHT agree well with the test. Still, the higher peaks are overpredicted. In the second part of the grading segment, both correlations give an accurate prediction of the skin temperatures.

The second part of the core segment is predicted well because the engine speeds are similar to the WOT conditions. In all the locations along in the exhaust system, the case without any empirical correlation performs poorly in predicting the skin temperatures and these simulation results are not shown.

On the whole, the UHT correlation predicted the gas and skin temperatures well for most of the drive cycles at all the locations and hence, it was chosen for subsequent studies. Figure 12.15 gives the skin temperature at different locations of the exhaust system from the manifold to the muffler. Up to the location after the resonator, the UHT predicts skin temperatures accurately. However, the simulation slightly overpredicts by about 30 °C, in the approach and the second segment of the grade before the muffler. For the cooldown portion in Fig. 12.15b, d, the correlation underpredicts the temperature by a maximum of 12 °C, primarily due to the effect of the thermal mass of the flange joining the downpipe and resonator as well as intermediate pipe and the muffler. Similarly, the difference in the temperature at the tailpipe (Fig. 12.15d), is due to the effect of external radiant emissivity being different from the initial assumption.

[1] Skid plates protect the underbody and its components shown in Fig. 12.1, by absorbing shock, impact, and strikes from external objects such as potholes, stones, branches, and uneven pathways.

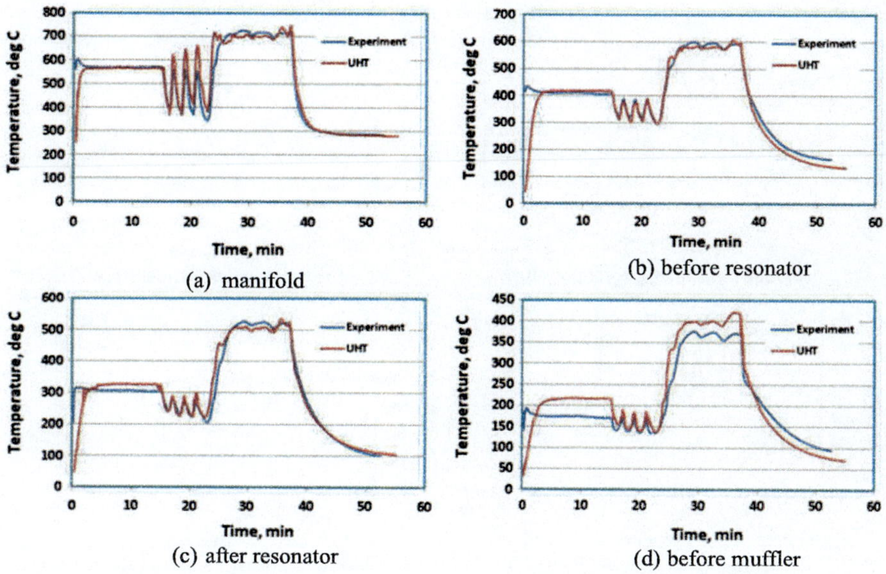

Fig. 12.15 Skin temperatures of the left side of the exhaust system, at different locations from the manifold to the muffler [25]

12.8.2 Parametric Study Involving the Effect of External Emissivity

To investigate the effect of external emissivity, a parametric study was conducted by varying the external emissivities of the components from 0.2 representing a well-polished or shiny exhaust to 0.8 representing a painted exhaust. The boundary conditions representing the emissivity values are used for the 1-D model from the above-discussed results.

Figure 12.16a, b show the gas temperatures in the manifold and tailpipe. Except for the tailpipe, $\varepsilon = 0.7 \sim 0.8$ gives good results, Fig. 16c–f, because up to the location before the resonator, the exhaust is partially covered with the skid plates. The plates delay the oxidation or wear of the exhaust while shielding it from harsh environmental conditions. However, a lower value of 0.2–0.4 predicts the tailpipe gas temperatures correctly due to the presence of a chrome tip on the tailpipe.

12.9 Summary

A novel approach and a numerical methodology utilizing both 1-D and 3-D models are presented to estimate the skin temperature of the exhaust system, after eliminating the heat rejection rates for the catalyst. Considering improper heat rejection by the

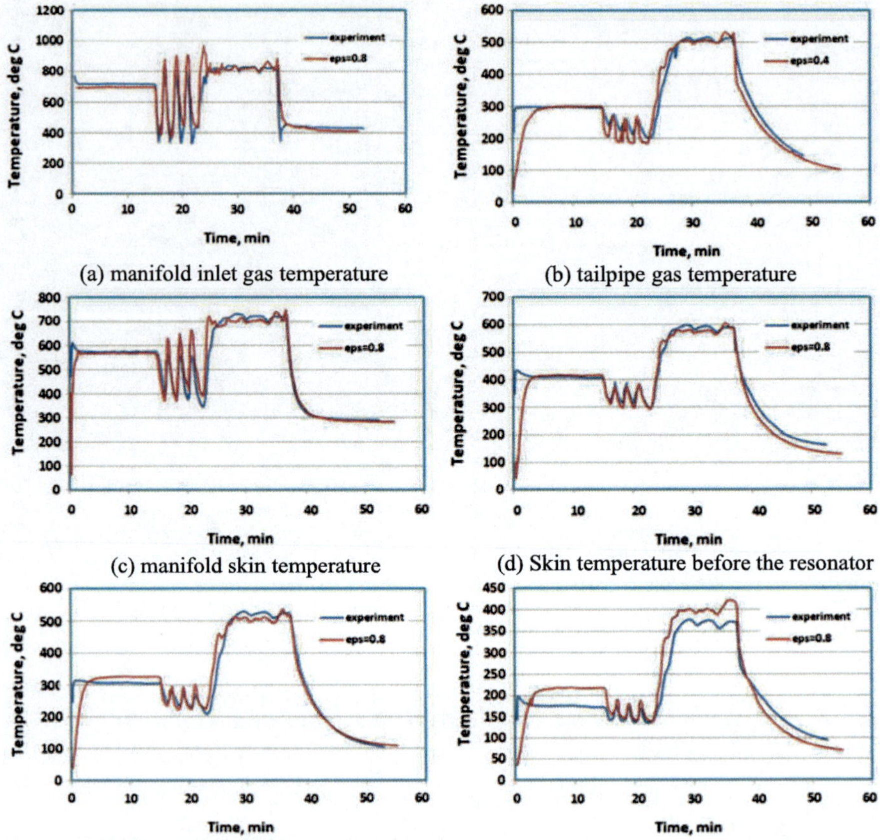

(a) manifold inlet gas temperature
(b) tailpipe gas temperature
(c) manifold skin temperature
(d) Skin temperature before the resonator
(e) Skin temperature after the resonator
(f) Skin temperature before the muffler

Fig. 12.16 Skin and gas temperatures of the exhaust system on the left side considering UHT for convection and suitable emissivity, ε; the emissivity of the exhaust system is essentially 0.8 except for the tailpipe [25]

catalyst without involving chemical reactions in the model could overpredict the skin temperature on the exterior surface of the catalyst and at the downstream of the catalyst. The vehicle level 3-D model yields the external HTC accounting for free convection and radiation. The output is used as the boundary conditions in the 1-D model. Three iterations by transferring data between the 1-D and 3-D models are found optimal for convergence. The difference between the second and third loop is found to be less than 5%. The UHT correlation gives an accurate prediction of skin temperature. Further, the contribution of external emissivity in the radiation heat transfer characteristics is demonstrated.

- The heat transfer coefficient with the CAF is superior since high-frequency flow fluctuations are taken into account.

- The friction coefficient estimated from the Colburn analogy with heat transfer coefficient considering CAF is more accurate than the steady flow friction coefficients. However, the use of these friction coefficients resulted in slightly less accurate predictions than the user-defined correlations namely, the universal (UHT) [4, 12] and Wendland's (WHT) heat transfer correlations
- In the segment after the catalyst, the temperatures could be overpredicted if improper heat rejection by the catalyst is considered as well as no chemical reaction models included.
- The use of constant heat transfer coefficients and lack of external heat transfer effects for the transient drive cycle contribute to significant underprediction of skin temperatures in the cooldown segment of the drive cycle.
- Decoupling the engine and exhaust simulation accelerates the CPU time from 8 days to 3.5 h without any loss of accuracy.
- The 3-D model provides external heat transfer coefficients required as the boundary condition in the 1-D model.
- A K-means clustering algorithm obtains a set of cluster points based on engine speed and torque to represent a transient drive cycle and to obtain the wind speeds for steady-state 3-D simulations.
- The presence of a skid plate accounts for lower HTC values of the exhaust system.
- The pulsation effects are predominant in regions such as the manifold and before the resonator and muffler. Therefore, the prediction of skin temperature using the UHT correlation is more accurate than the rest, since the exponent for Reynolds number is arrived at based on microscales of turbulence.
- The radiation heat transfer contributes 25–50% of the total external heat transfer. A parametric study shows the importance of assigning a proper emissivity for radiation heat transfer.
- Integrating 1-D and 3-D models gives a good prediction of exhaust skin temperature with efficient use of computational resources.

References

1. Toyota-land-cruiser-underbody. https://tfltruck.com/2021/03/expert-opinion-will-the-new-2022-toyota-tundra-have-a-land-cruiser-rear-suspension-design/toyota-land-cruiser-underbody. Accessed on 25 July 2021
2. Srinivasan K, Woronowycz G, Zabat M, Tripp J (2005) An efficient procedure for vehicle thermal protection development. SAE Trans:2418–2425
3. Malchow GL, Sorenson SC, Buckius RO (1979) Heat transfer in the straight section of an exhaust port of a spark-ignition engine. No. 790309. SAE Technical Paper
4. Wendland DW (1993) Automobile exhaust-system steady-state heat transfer. No. 931085. SAE Technical Paper
5. Condie KG, McEligot DM (1995) Convective heat transfer for pulsating flow in the takedown pipe of a V-6 Engine. SAE Trans:1091–1099
6. Alkidas AC, Battiston PA, Kapparos DJ (2004) Thermal studies in the exhaust system of a diesel-powered light-duty vehicle. SAE Trans:164–181

7. Shayler PJ, Hayden DJ, Ma T (1999) Exhaust system heat transfer and catalytic converter performance. SAE Trans:213–228
8. Zhang Y, Phaneuf K, Hanson R, Showalter N (1992) Computer modeling on exhaust system heat transfer. SAE Trans 101(5):199–208
9. Liu Z, Hoffmanner AL, Skowron JF, Miller MJ (1995) Exhaust transient temperature response. No. 950617. SAE Technical Paper
10. Konstantinidis PA, Koltsakis GC, Stamatelos AM (1997) Transient heat transfer modelling in automotive exhaust systems. Proc Inst Mech Eng C J Mech Eng Sci 211(1):1–15
11. Grose DJ, Austin K (2001) Coupling of one-dimensional and three-dimensional simulation models. No. 2001-01-1770. SAE Technical Paper
12. Depcik C, Assanis D (2002) A universal heat transfer correlation for intake and exhaust flows in a spark-ignition internal combustion engine. SAE Trans:734–740
13. Zhang X, Meda L, Keck M (2005) Numerical study on skin temperature and heat loss of vehicle exhaust system. No. 2005-01-1622. SAE Technical Paper
14. Sieder EN, Tate E (1936) Heat transfer and pressure drop of liquids in tubes. Ind Eng Chem 28:1429
15. GT-Power Flow Theory Manual, Gamma Technologies (2014)
16. Benson RS, Garg RD, Woollatt D (1964) A numerical solution of unsteady flow problems. Int J Mech Sci 6(1):117–144
17. Lax PD, Wendroff B (1964) Difference schemes for hyperbolic equations with a high order of accuracy. Commun Pure Appl Math 17(3):381–398
18. MacLaren JFT, Tramschek AB, Sanjines A, Pastrana OF (1975) A comparison of numerical solutions of the unsteady flow equations applied to reciprocating compressor systems. J Mech Eng Sci 17(5):271–279
19. Ledger JD (1975) A finite-difference approach for solving the gas dynamics in an engine exhaust. J Mech Eng Sci 17(3):125–132
20. Lakshminarayanan PA, Janakiraman PA, Gajendra Babu MK, Murthy BS (1979) A finite difference scheme for unsteady pipe-flows. Int J Mech Sci 21(9):557–566
21. Morton KW (1971) Stability and convergence in fluid flow problems. Proc R Soc London A Math Phys Sci 323(1553):237–253
22. Lakshminarayanan PA, Janakiraman PA, Gajendra Babu MK, Murthy BS (1979) Measurement of pulsating temperature and velocity in an internal combustion engine using an ultrasonic flowmeter. J Phys E Sci Instr 12(11):1053
23. Lakshminarayanan PA, Janakiraman PA, Gajendra Babu MK, Murthy BS (1979) Prediction of gas exchange processes in a single-cylinder internal combustion engine. No. 790359. SAE Technical Paper
24. CD-Adapco (2016) Star-CCM + user manual
25. Annur Balasubramanian S (2018) Prediction of exhaust skin temperature integrating 1-D model with vehicle level CFD model
26. SAE J2807_202002, performance requirements for determining tow-vehicle gross combination weight rating and trailer weight rating
27. k-means clustering. https://en.wikipedia.org/wiki/K-means_clustering. Accessed on 25 July 2021
28. Vedavarz A, Kumar S, Nawaz MH (2007) HVAC: handbook of heating, ventilation and air conditioning for design and implementation. In Industrial Press Inc.

Chapter 13
The Role of Computational Fluid Dynamics in Designing Thermally Safe Vehicles in a Competitive and Aggressive Development Landscape

Devadatta Mukutmoni, Young-Chang Cho, Han Li, Huhu Wang, Chin-Wei Chang, and Satheesh Kandasamy

13.1 Introduction

Vehicle development and manufacturing is a very challenging engineering process that is both time-consuming and expensive. Recently the challenges are aggravated even further due to model proliferation, reduced development time, more regulations, and increased pressure from markets.

For instance, over 15 years the number of vehicle models offered by major automotive OEMs has increased by 40% while the vehicle development time from design to production has shrunk to a maximum of 24 months [1]. Furthermore, the thermal performance of a vehicle and its subsystems is difficult to assess in the early design stages since prototypes for testing are not available until late in the design process. This makes it difficult to implement design changes early in the design cycle at a time when the design opportunities are considerably higher and as a result, the design solutions can be much more impactful and cost-effective.

Some of the vehicle design imperatives come from regulations requiring continuous improvements in fuel efficiency and range. Furthermore, since a thermal failure could lead to a catastrophic impact and safety issues, it can adversely impact the brand image and market share.

In addition to pressure from customers and competitors, it is imperative to reduce warranty costs. It is estimated that worldwide every year up to thirty million vehicles

D. Mukutmoni · Y.-C. Cho · H. Li · H. Wang · C.-W. Chang · S. Kandasamy (✉)
Dassault Systemes, Simulia, Waltham, USA
e-mail: Satheesh.KANDASAMY@3ds.com

P. A. Lakshminarayanan and A. K. Agarwal (eds.), *Handbook of Thermal Management of Engines*, Energy, Environment, and Sustainability,
https://doi.org/10.1007/978-981-16-8570-5_13

are recalled [2] and the cost of the recall is 22 billion dollars [3]. Reducing warranty costs requires that one evaluates and fixes thermal failures early in the design process. Furthermore, some of the considerations to improve thermal performance is to use cheaper and lighter materials and maximize the use of space in the engine bay.

As more electrical and electronic devices are packaged in modern vehicles, there is a greater challenge for thermal management since the maximum temperature requirements are usually lower for the electrical and electronic components.

There are limitations to the traditional testing methods as well. Thermal wind tunnels for instance may not be representative of open road driving conditions. Furthermore, test tracks and other exotic locations used for testing drive cycles may not be available, leading to expensive delays. All these factors could potentially delay vehicle launch leading to loss of market share, increased warranty costs, damage to the brand name, and reputation leading to lower sales and profitability.

13.2 Simulation Workflows

Thermal simulations offer a cost-effective alternative to vehicle testing. Also, the simulations can be applied early in the design cycle and can therefore significantly reduce warranty issues and late-stage failure. They also offer an integrated design where competing targets and trade-offs can be evaluated and balanced design decisions can be made.

In this article, we describe four different simulation workflows relevant to thermal design and testing. The first workflow referred to as cooling airflow is targeted for analyzing various cooling system layout concepts. An example of a cooling system to be simulated is shown in Fig. 13.1.

Fig. 13.1 Example of a fan, fan shroud, and heat exchangers

In the cooling airflow simulations, the various cooling airflow options are analyzed, the heat exchanger and fan performances are evaluated to provide an optimal design that also accounts for drag, noise, and energy consumption. With this workflow, we ensure that there is no overdesign; the cooling system is right-sized to meet extreme cooling demand that could be experienced by the vehicle leading to a reduction in vehicle weight.

As shown in Fig. 13.2, the various cooling airflow options can be analyzed in the installed conditions and provides a more robust and accurate design solution option.

The second thermal workflow referred to as thermal protection is used to identify at-risk components that exceed design temperatures. With this workflow, we can provide the temperature map of all vehicle components. The thermal protection workflow simulates a vehicle under steady operating conditions such as a vehicle

Fig. 13.2 Flow and thermal analysis of fan in installed conditions

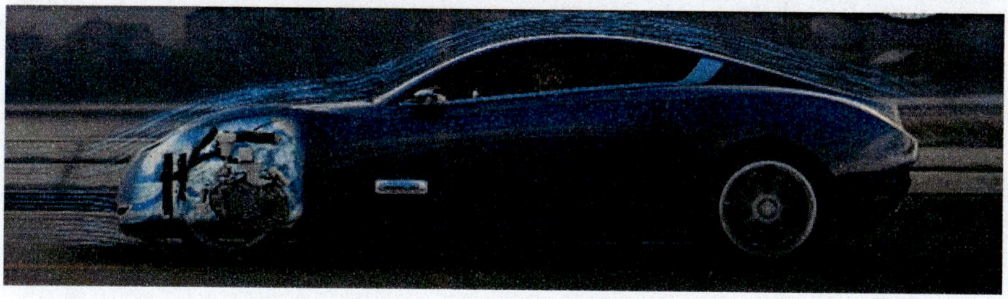

Fig. 13.3 Vehicle on road under steady operating conditions

Fig. 13.4 Temperature and velocity distribution in the underhood region

driving under open road conditions at 90 kph and with a fan at 2000 RPM. Figure 13.3 shows a vehicle on the road under steady operating conditions.

The simulation would then provide a temperature map for the entire vehicle as shown in Fig. 13.4 where the temperature and velocities in the under-hood region are shown.

The thermal protection workflow enables designs of more optimal shielding with lighter materials and the option of quickly evaluating different designs. This leads to reduced warranty costs, development costs, and vehicle costs/weight.

The third workflow referred to as thermal energy management enables one to simulate a key-off/Soak test procedure. In a Key-Off/Soak test procedure, a vehicle moving at constant speed comes to a stop followed by the shutdown of the fan and the engine. Reduced cooling, as a result, leads to increased component temperatures at the beginning of the shutdown. The thermal energy management workflow provides a transient temperature map of all vehicle components for the entire duration of the test and can therefore identify the duration and the extent of the temperature excursions.

The key-off/soak spans the natural convection regime leading to a plume-like velocity field as shown in Fig. 13.5.

With the simulation of an important vehicle thermal test procedure, one can assess the root cause of failures of parts, propose a solution, and confirm the fix with a design change. Since the simulation pertains to a more general thermal test, the thermal energy management workflow identifies thermal issues early on can further reduce thermal failure and hence warranty costs.

The fourth workflow referred to as the thermal drive cycle is a simulation of a more general and long-term vehicle thermal test such as in a test track or city traffic. The test environment and the simulation are the most thermally challenging that occur under real-world conditions and can uncover thermal issues not discovered in

Fig. 13.5 Plume like velocity field in the soak regime

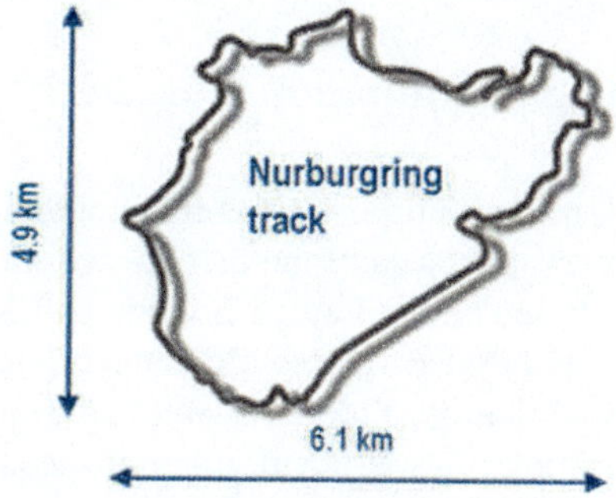

Fig. 13.6 Nurburgring test track

other test procedures. One example test track is the Nurburgring test track shown in Fig. 13.6.

Another instance would be city driving with start and stop driving schedule as shown in Fig. 13.7.

The thermal drive cycle workflow can be therefore be used to assess the thermal design of the vehicle under the most extreme conditions. Since the temperature results are available anywhere and not limited to the thermocouples, it is easier to determine the root cause of thermal failure, provide quick fixes and verify the improved design leads to improved component durability and reduced warranty costs.

Fig. 13.7 City driving for the thermal drive cycle

13.3 Simulation Numerics

A general-purpose CFD (Computational Fluid Dynamics) solver, or PowerFLOW, forms the basis of the thermal workflows to be discussed in this article. This solver is based on the Lattice Boltzmann Methodology (LBM).

For the LBM, the Boltzmann Equation is used to solve for the velocity and pressure field. The dependent variable is the distribution function $f(\mathbf{x}, \mathbf{v}, t)$, the number density of molecules at position $\mathbf{x}$ and speed $\mathbf{v}$. The Boltzmann equation can be written in the following form:

$$\frac{\partial f}{\partial t} + \mathbf{v} \cdot \nabla f = C(\mathbf{x}, \mathbf{v}, t) \tag{13.1}$$

The left-hand side represents the convective motion of particles and the right-hand side the collision term that accounts for intermolecular interactions. The flow solver uses the BGK collision operator as described in Bhatnagar et al. [4] that can be expressed as follows:

$$C(\mathbf{x}, \mathbf{v}, t) = -\frac{f - f_{eq}}{\tau} \tag{13.2}$$

where τ is the relaxation time of the fluid and f_{eq} is the Equilibrium Maxwellian Distribution Function. The Boltzmann Equation along with the BGK operator is discretized on a three-dimensional cubic lattice using a D3Q19 model as shown in Fig. 13.8.

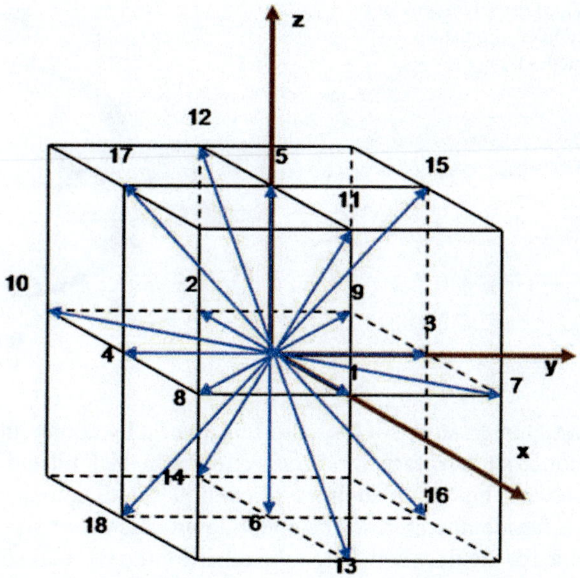

Fig. 13.8 Finite difference scheme for Boltzmann equation

It can be shown rigorously that the Lattice Boltzmann formulation thus outlined is equivalent to solving the time-dependent compressible Navier–Stokes Equation as described in Chen et al. [5] and that the formulation can be easily used to impose fluid boundary conditions on solid surfaces as detailed in Chen et al. [6]. A review of the fundamentals of the Lattice Boltzmann Method can be found in Chen and Doolen [7].

The turbulence model employed in the flow solver is conceptually similar to a large eddy approach to turbulence modeling; further details on turbulence modeling as implemented in PowerFLOW are available in Chen et al. [8]. Recent improvements in modeling the temperature scalar equation have further enhanced the accuracy of the temperature calculations in the flow solver as noted by Zhang et al. [9].

13.4 Simulation Methodology

The simulation methodology used in this study involves a combination of the flow and thermal solver. The flow solver calculates the flow and temperature fields in a volumetric flow domain and the thermal solver calculates the temperature in the solid surfaces by taking into account radiation from the surfaces, conduction within the solid surfaces, and convection at the surface.

This process is enabled by the coupling process methodology shown in Fig. 13.9.

The flow solver calculates the near-wall temperatures T_f and heat transfer coefficients H and exports them to the thermal solver. The thermal solver calculates the

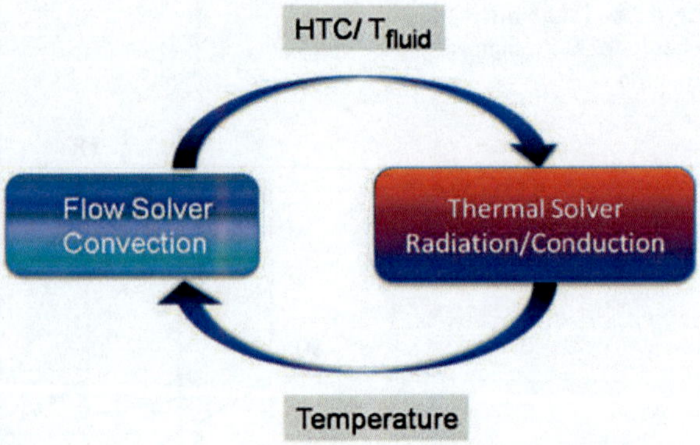

Fig. 13.9 Overview of coupled simulation methodology

wall temperatures of the solid surfaces T_w by taking into account radiation, conduction in solid surfaces, and convection from the fluid and exports them back to the flow solver. This "coupled process" continues until convergence is achieved for the solid surface temperatures. The coupling process uses two different geometrical models, one for the flow solver and the other for the thermal solver that has the same underlying geometry but different element types and sizes. The transfer of T_w, H, and T_f between the flow and thermal solver is achieved by interpolation between different geometries.

The thermal solver employs a finite-volume formulation that exactly conserves energy. The surface to surface radiation is calculated by solving a grey body radiation enclosure problem and calculating the radiation view factors by a ray-tracing algorithm. The conduction is calculated in the context of a finite volume formulation.

The discussion of the thermal workflows now follows. We begin with the cooling airflow workflow as it pertains to the analysis and optimization of the cooling module of the vehicle which includes the heat exchangers and the fans.

13.5 Cooling Airflow

Optimizing airflow under the hood is essential in ensuring that equipment and vehicles operate safely and reliably. Additionally, the optimization process can also reduce vehicle drag and improve performance and fuel efficiency.

For a complete evaluation of performance, it is not enough to test cooling in the laboratory or wind tunnel, equipment needs to be able to operate under challenging real-world conditions such as in hot, cold, and humid environments. As a result, designers often choose to be very conservative in their approach and the process sacrificing weight and performance. Powertrain cooling is not just a performance issue, it is a critical safety issue if issues are uncovered during the late stage of the vehicle development, and manufacturers will resort to using a larger radiator. This in turn will increase vehicle cost and weight.

Fig. 13.10 Simulating cooling airflow in the underhood region of a car

With simulation, engineers can visualize the flow of air and the temperature distribution under the hood as shown in Fig. 13.10 where the flow field is visualized using streamlines.

Accurate simulation of vehicles and equipment for instance can then be used to optimize fan and vent design for both cooling and fuel efficiency targets. The number of physical prototypes required for physical testing can then be significantly reduced with more virtual testing under real-world weather scenarios. This approach can drastically reduce both development time and cost while improving performance.

13.6 The Challenges of Underhood Cooling

Engines generate heat, and if their energy release is not managed properly, it can impact performance, and lead to catastrophic breakdowns and extreme events like fires. Appropriate cooling designs maintain thermal efficiencies, keep parts from overheating and prevent safety and durability issues. In Fig. 13.11, the complex flow field in the underhood compartment is shown the driving force being vehicle motion and the fans.

The air that flows through the grille openings and below the front of the vehicle as a result of vehicle motion is referred to as ram air. Air cooling is an essential part of thermal management and the flow of air around the engine and through the body needs to be considered during design. Broadly speaking, there are two different categories of under-hood cooling analysis:

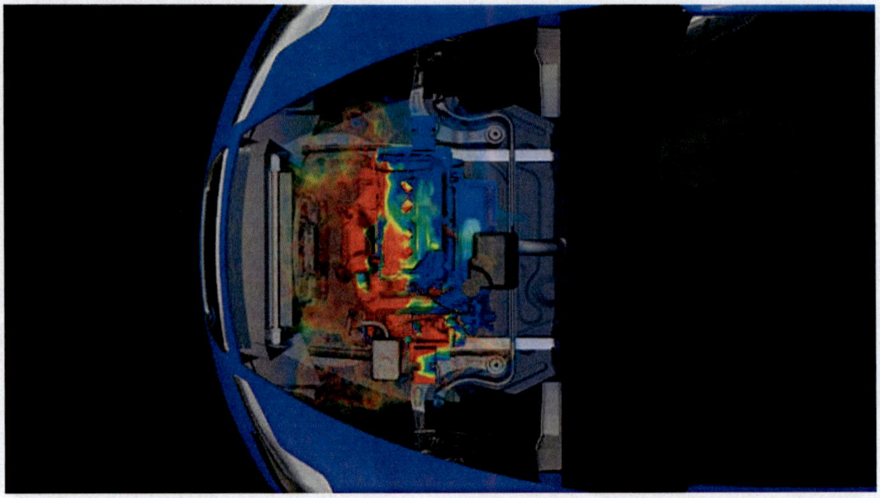

Fig. 13.11 Ram air in the underhood region and temperature distribution on the engine block

- For on-road applications, the vehicle is usually moving at sufficient speed that ram air can be used to cool the engine. This is the airflow forced through the vehicle body in motion. Using ram air reduces the need for fan cooling needed, but also increases aerodynamic drag and therefore reduces fuel efficiency. Care also needs to be taken to ensure that sufficient cooling is still available when ram air is not available, such as when driving slowly uphill in low gear.
- The other category is off-highway equipment where the equipment is either stationary as in an air compressor or moving slowly as in a tractor. In both cases, the flow contribution for ram air is limited. Therefore, the airflow needs to be driven by fans. These fans take up space inside the equipment body and add to the total weight of the equipment. The fans, therefore, need to be optimized to provide sufficient cooling under all possible operating conditions while staying within design constraints of size, weight, and noise and while using minimal power. Figure 13.12 shows cooling airflow through an excavator's cooling package.

Heavy machinery such as the excavator shown, operates under very challenging conditions such as the hot and humid environment in tropical climes or cold and dry at high altitudes. For vehicles, the open road conditions have different aerodynamics than wind tunnel testing. Therefore, wind tunnel testing cannot accurately replicate real-world conditions for either machinery or vehicles. Furthermore, the availabilities of prototypes for testing and tunnel time could seriously delay development time and vehicle launch. Accurate simulations backed up by standardized workflows can address these issues.

Important design consideration for the cooling package is the sizing of the cooling fans and the heat exchangers. The sizing is a strong function of the most extreme operating condition along with the effect of the cooling package installation. This is

Fig. 13.12 Airflow through the cooling package for off-highway equipment

usually not known beforehand and is difficult to determine by testing in what would be a trial and error process. Simulations characterize the performance of the cooling components and therefore pinpoint the extreme operating conditions and the required sizing.

With this approach, one can come up with a sizing that is optimal and in general a design of lighter fans and heat exchangers than what is observed where a larger factor of safety in design leads to unnecessarily larger cooling components. Other than the savings in total weight, smaller fans and heat exchangers also free up more space in the engine bay.

13.7 Methodology Description

As discussed earlier, the design of fans and heat exchangers are the most important design deliverables of the cooling airflow workflows. In this section, we discuss the modelling aspects of fans and heat exchangers.

13.7.1 Fan Modeling

There are mainly three modeling methods used for modelling fans. These will now be discussed in more detail.

- Sliding mesh method

- Multiple reference frame (MRF)
- In-line fan model.

13.7.2 Sliding Mesh

Sliding mesh models the true rotation of where the fan geometry is rotating relative to other parts. The domain in this case is divided into two separate regions, rotating and stationary. The two domains are solved independently with information exchanged at the interface including the velocity transformation. This coupling mechanism between the two domains happens at each time step to ensure accuracy. This method enables the user to capture transient effects created by the complex rotating geometry such as fan blades. Centrifugal and Coriolis forces, which are consistent with rotation, are applied to the fluid in the sliding mesh region. Fluid velocities are transformed across the boundary between the rotating frame and the global reference frame.

Figure 13.13 illustrates the sliding mesh region enclosed by the purple dashed line. The outer region stays stationary relative to the global reference frame in simulation.

The outer region stays stationary relative to the global reference frame in simulation. The fluid information is exchanged between the two regions at the boundary during the simulation.

The LBM methodology can handle arbitrary meshes while maintaining exact conservation or equilibrium properties as detailed in Chen [10]. Figure 13.14 shows a mesh view of the created voxels using LBM based sliding mesh approach.

Although this method is more computationally demanding compared to that of the other two methods, it offers the highest accuracy among the three methods as

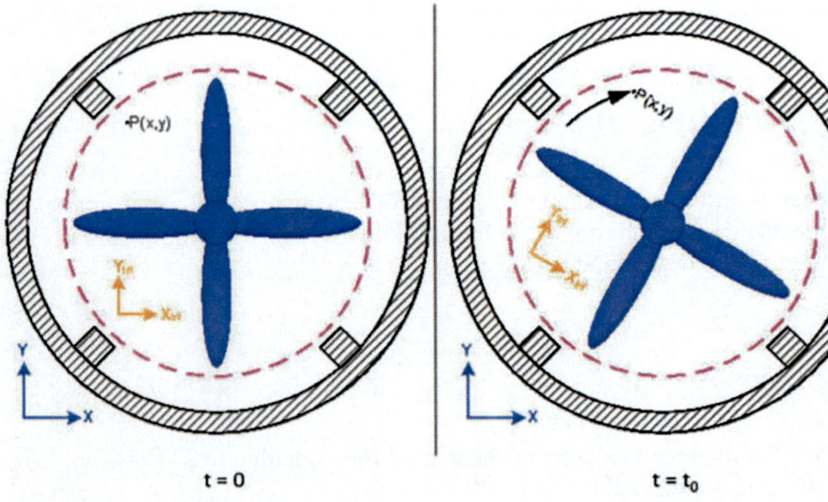

Fig. 13.13 Rotating fan inside and sliding mesh region

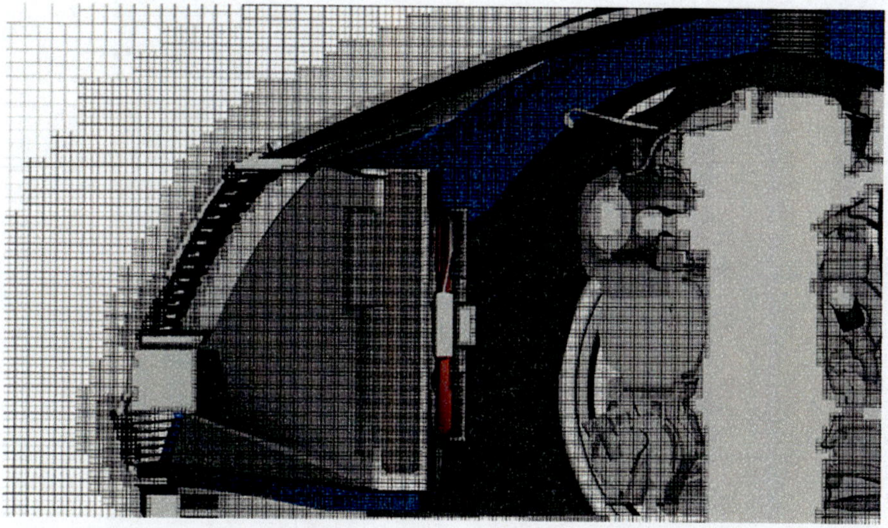

Fig. 13.14 Mesh view in the sliding mesh region for a cooling fan (shown in red)

shown in a validation study by Gulberg and Sengupta [11]. In Fig. 13.15, the flow field visualized by vortices around a cooling fan using the sliding mesh method is shown.

13.7.3 Multiple Reference Frame (MRF)

The multiple reference frame (MRF) models the rotation implicitly in the rotating frame region. In the MRF approach, all solids with a no-slip boundary condition within this local reference frame (LRF) region are treated automatically as rotating bodies, and a rotational velocity is imparted to every voxel within the region. There is no explicit rotation in the rotating frame of reference, but while post-processing the results, the rotational velocities are accounted for.

As shown in Fig. 13.16, the rotating part is placed in the MRF region. The inner MRF region is rotated at a speed of ω while the outer region stays stationary in simulation. The MRF method is cheaper to run than the sliding mesh method in general.

13.7.4 In-Line Fan Model

The in-line fan model is the simplest approach to simulate the fan. In this model, a fan region is set up in the simulation and treated as a momentum source. For a region

Fig. 13.15 Flow around a fan using sliding mesh simulation using lambda-2 iso-surfaces

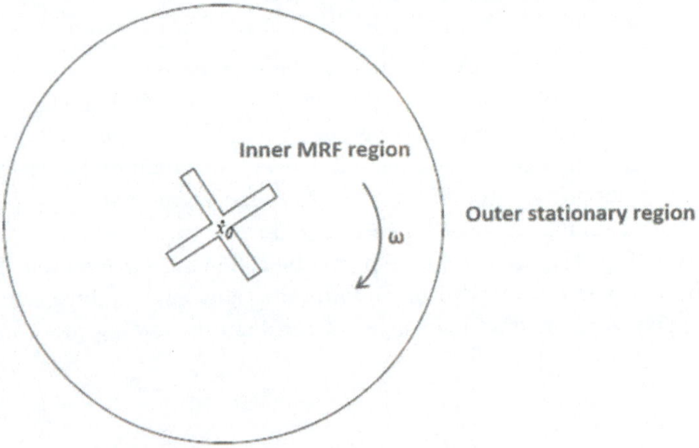

Fig. 13.16 Illustration of the MRF method

of constant cross-sectional area and planar inlet and outlet faces, the pressure rise dP through the fan can be modelled as shown in Eq. (13.3) below:

$$dP = \rho L f \tag{13.3}$$

where ρ is the fluid density, L is the axial fan length, f is the local axial force per unit mass. The force is represented by a third-order polynomial equation described in Eq. (13.4):

$$f = A_0 + A_1 u + A_2 u^2 + A_3 u^3 \tag{13.4}$$

where A_0, A_1, A_2, A_3 are axial force coefficients. Combining Eqs. (13.3) and (13.4) we have Eq. (13.5):

$$dP = \rho L\left(A_0 + A_1 u + A_2 u^2 + A_3 u^3\right) \tag{13.5}$$

where u is the velocity. The coefficients of the polynomial are calculated from test data for the fan. The variation of the pressure rise as a function of velocity is referred to as the fan curve.

The inline fan model is typically used in the early design stage when expected fan performance data is available but the geometry of the fan is not decided.

13.7.5 Heat Exchanger Modeling

Heat exchangers such as radiators and condensers used in vehicles have small fins and tubes relative to the overall size of the heat exchanger. The liquid coolant flows inside the narrow tubes while cooling air passes through heat exchangers through small gaps between each layer of fins. Such heat exchangers where the ratio of the heat exchanger surface area to volume is high are known as compact heat exchangers.

For compact heat exchangers as in automotive heat exchangers, an explicit CFD model that resolves the flow in the coolant and airside is impractical. Therefore, the heat exchanger volume is modelled as a porous medium zone where the interactions between the air and coolant sides are captured.

As shown in Fig. 13.17, the heat exchanger is represented as a rectangular box. The pressure drop created by the exchanger is calculated based on Darcy's law, which correlates pressure drop dP with the air velocity u. Two coefficients, V_R and I_R, are called viscous and inertia resistance coefficients, respectively. They are calculated based on experimentally measured pressure drop versus flow rates.

Heat transfer between the hot coolant and the cooling air is calculated using the energy balance equations. The equation for calculating the overall heat transfer coefficients is shown in Fig. 13.18. These equations are solved by an independent solver called PowerCOOL. Figure 13.18 presents the coupling mechanism between

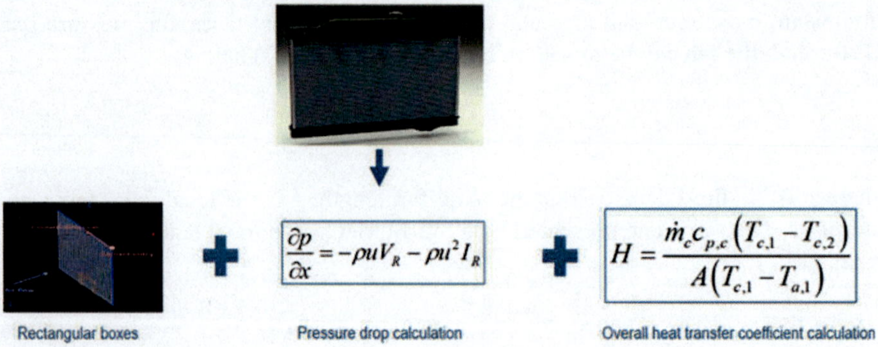

Fig. 13.17 Modeling heat exchangers with porous medium

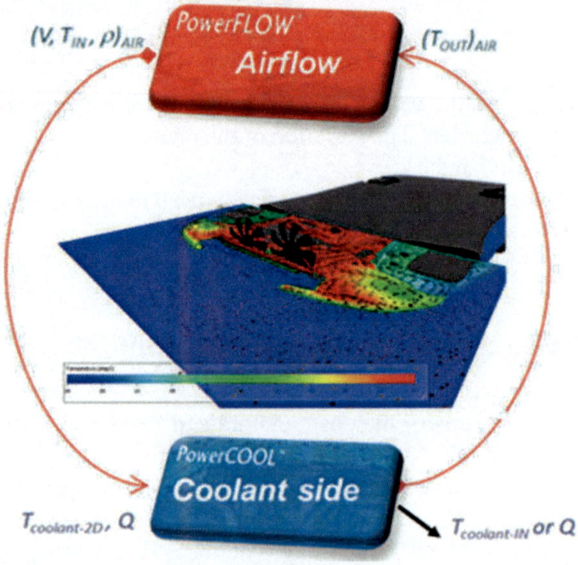

Fig. 13.18 Coupling mechanism between PowerFLOW and PowerCOOL

PowerCOOL and PowerFLOW. PowerFLOW provides the calculated air mass flow rate, density, and temperature at the inlet of the heat exchanger to PowerCOOL, and PowerCOOL computes coolant entry temperature (also called top tank temperature), coolant exit temperature (also called bottom tank temperature), and the air exit temperature. PowerFLOW uses calculated air exit temperature to update the remaining fluid domain.

13.8 Example Workflow

In this section, an example workflow that encompasses the entire simulation process including pre-processing, simulation setup, simulation run, and post-processing results will be described. The goal of this workflow development is to impart this complex, time-consuming simulation process to some degree of standardization and automation.

A more traditional and manual-based process of CFD simulation requires a great deal of expertise due to its complexity. This puts serious limits on who can work on a cooling system design, delegating tasks to the more simulation-savvy, specialists within users' organizations. For some users, those individuals or even a dedicated CAE (Computer Aid Design) division may not exist, forcing them to outsource the simulation work to other engineering services companies or to spend valuable time and money on additional training. Even worse, they may end up relying on physical testing seriously impairing the performance of the machine while increasing development costs. This can lead to delays in releasing the machinery as well, overages on product development budgets, and losing market share.

Also, the analysts would need to know fluid mechanics, heat transfer, fan modeling, and heat exchanger design. They would also have to know how to prepare mesh from the CAD model, set up the simulation parameters, run the simulation and analyze results after the simulation. This manual process is time-consuming and error-prone even for CFD experts and not suitable for a vehicle design where timelines are challenging and there is very little margin for error. Therefore, the case for automation and the resultant simplification of the workflow is quite compelling.

To address these challenges, SIMULIA has developed two powerful tools, the underhood cooing (UHC) App and the Denso Engine Cooling Module (ECM) library, aiming to reduce turnaround time including man-hours for users.

13.8.1 Underhood Cooling (UHC) App

UHC App is a cloud-based solution that is used by designers, simulation analysts, and experts. The UHC App which utilizes PowerFLOW and is hosted in a Cloud environment is fully automated and requires no prior knowledge of simulation. A user needs to input the operating conditions of the machinery; the automated process then simulates how the cooling system will function under installed conditions of the machinery.

The automated process integrates the entire simulation workflow. The process prepares the mesh from the CAD model, applies the simulation input parameters, runs the simulation, and generates the simulation results. Each step incorporates best practices developed over years and thus delivers accurate simulation results with reasonable compute costs. The workflow thus described is illustrated in Fig. 13.19.

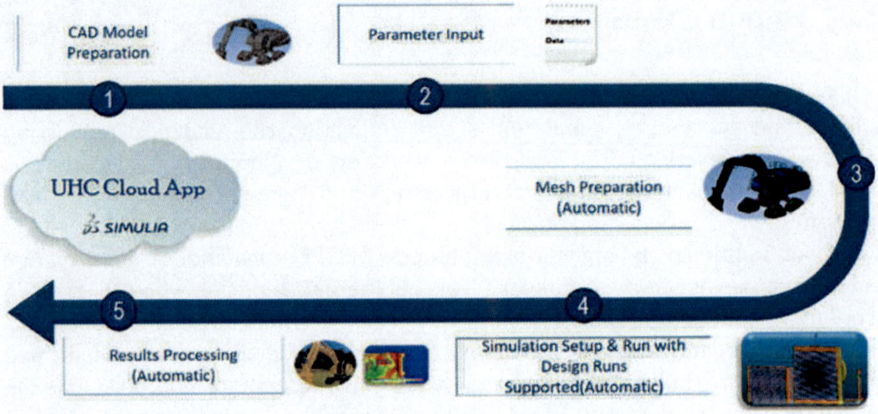

Fig. 13.19 Automated workflow with UHC App for cooling airflow simulations

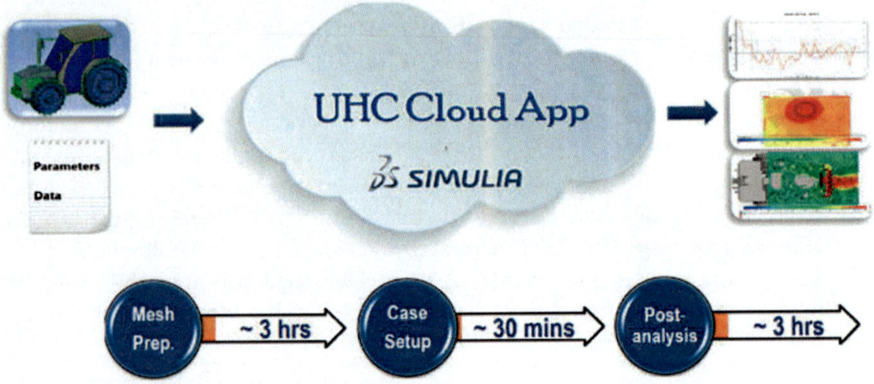

Fig. 13.20 Turnaround times for mesh preparation, case setup, and post-processing while using UHC App

It is estimated that the automation afforded by the Cloud-based App leads to total savings of 80% of man-hours. Figure 13.20 illustrates the estimated turnaround time for each simulation sub-process by using the UHC App.

13.8.2 Denso Engine Cooling Module (ECM) Digital Library

An engine cooling module as shown in Fig. 13.21 is important for a vehicle not only from the safety point of view but also from vehicle performance, durability, occupant comfort, fuel economy, and cost.

Fig. 13.21 A typical cooling module of a car

The Denso Engine Cooling Module Digital Library, a design tool for engine cooling modules, represents a paradigm shift in engine cooling design in terms of how automotive OEMs and suppliers interact with each other in the vehicle development process. Traditionally, carmakers and suppliers design the vehicle and engine cooling separately until late in the vehicle program since there is a reluctance to share proprietary data between suppliers and the vehicle manufacturers early on.

For the Denso ECM Digital library as demonstrated in Fig. 13.22, in the digital library, carmakers can select the engine cooling module from the whole Denso ECM catalogue, with a combination of different radiators, condensers, motors, and fans.

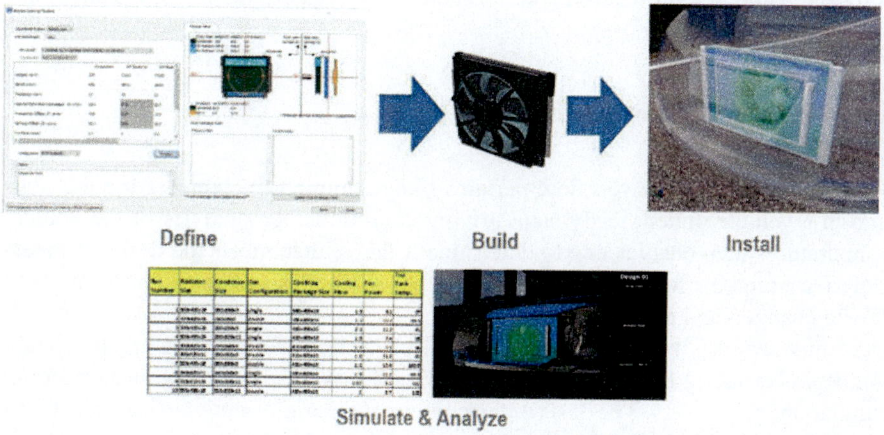

Fig. 13.22 The workflow to create a cooling module automatically with Denso ECM digital library

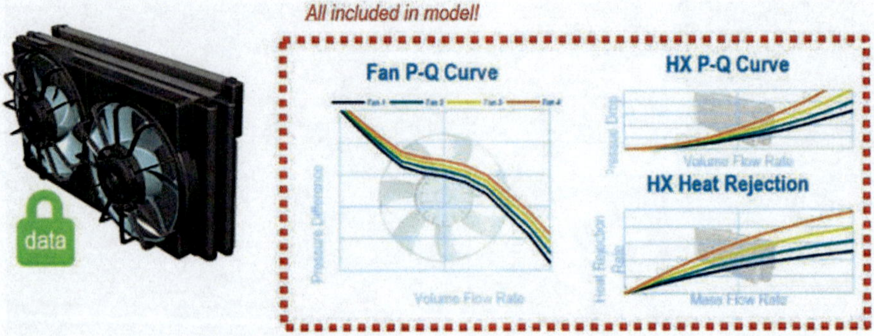

Fig. 13.23 The Denso ECM digital library encrypts all geometry data to minimize IP risk

Once selected, designers can modify the dimension, location, and relative distance between parts. Upon completion of the design, the cooling module is automatically built, and the setup is simulation-ready. Therefore, the cooling module design can be added to the vehicle very early on in the design process.

The design-to-simulate process is fully automated, with an intuitive graphical interface, which allows designers to visually confirm the design in real-time even before the model build and simulation. In addition to the automation of the design, model, and simulation, the Denso ECM Digital Library also has the manufacturing knowledge embedded in it that guarantees that each ECM created in the digital library is manufacturable.

The geometry and data encryption features are some of the building blocks of Denso ECM Digital Library as shown in Fig. 13.23.

The secured encrypted workflow was developed so that the sensitive fan geometry and heat exchanger performance data are protected while the geometry and data are used for simulation. The encryption greatly reduces the risk that a proprietary dataset is reused without authorization and thus addresses a major concern of the supplier.

13.9 Thermal Protection

The thermal protection workflow pertains to determining the temperature distribution in a vehicle under steady-state driving conditions. An analysis of the vehicle temperatures then enables one to determine if the vehicle meets the design temperature requirements for the parts and if not the vehicle design such a shield material can be changed and analyzed quickly.

In this section, we examine the thermal protection workflow in greater detail. We begin by noting that for a workflow to be effective, the accuracy of the thermal simulation is necessary. The basis of the numerical scheme as outlined earlier is LBM in conjunction with the BGK approximation for the collision term. A comprehensive

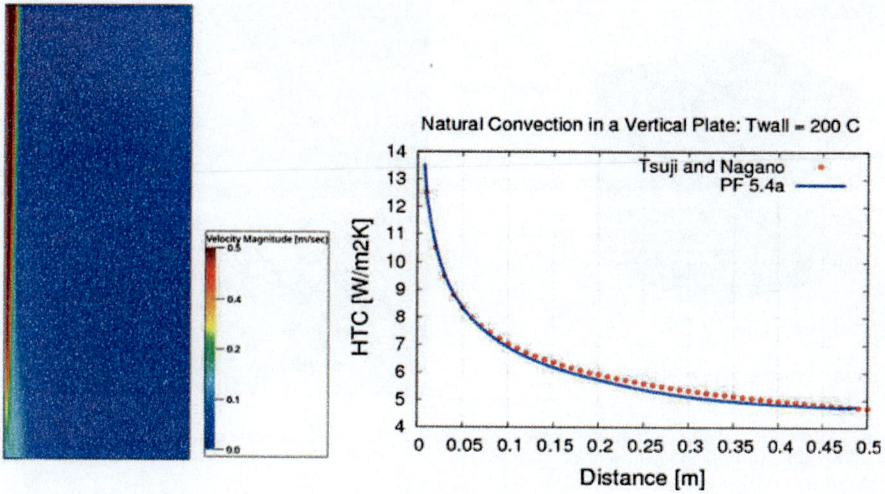

Fig. 13.24 The boundary layer flow for a heated vertical plate is shown on the left. On the right, the heat transfer coefficient is compared with experimental data

summary of the LBM methodology as applied to Fluid Dynamics is documented in Nourgaliev et al. [12].

The LBM formulation leads to a second-order finite difference like an approximation of the Navier Stokes Equation that delivers high accuracy and very low numerical dissipation enabling very high fidelity modelling of time-dependent flows. Several validation cases of different levels of geometrical complexities attesting to the high level of accuracy will be discussed.

We begin with the heated vertical plate benchmark example. In the vertical plate, a natural convection-driven boundary layer flow is formed along a vertical wall set at a constant temperature. In Fig. 13.24, we see a boundary layer flow along the vertical wall. We also compare the heat transfer coefficient along the wall with experimental data.

The experimental data which is documented in Tsuji and Nagano [13] shows that the heat transfer coefficient varies as shown in Eq. (13.6) below:

$$h_z = 0.387\frac{k}{z}\sqrt[4]{\frac{g\beta((T_{wall} - T_{ambient})z^3}{\nu a}} \tag{13.6}$$

where h_z is the local heat transfer coefficient along the vertical direction, T_{wall} is the temperature of the wall, $T_{ambient}$ is the ambient temperature, z is the vertical distance along the wall, β the coefficient of thermal expansion, g is the acceleration due to gravity, κ thermal conductivity, ν the kinematic viscosity, and α the thermal diffusivity.

As seen in Fig. 13.24, good agreement is observed between PowerFLOW simulations and experimental data.

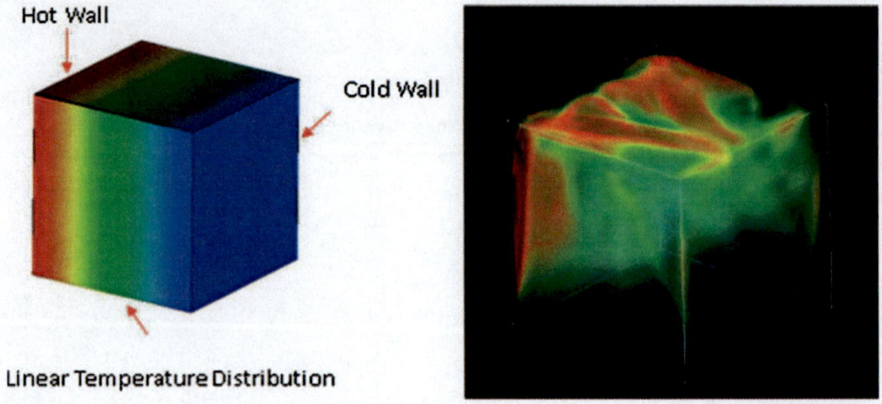

Fig. 13.25 Natural convection in a cubical cavity. Temperature boundary conditions are shown on the left, velocity field shown on the right

In Fig. 13.25, the temperature boundary conditions and the velocity field of natural convection flow in a cubical cavity are shown.

The simulations are for a Rayleigh number of 10^8. The natural convection is driven by hot and cold vertical sidewalls and linear temperature distribution is maintained for the other 4 walls. The results were compared to experimental data from experiments documented in Leong et al. [14]. For the PowerFLOW simulation, the total heat flux from the hot wall was within 5.5% of the experimental data which is well within the uncertainties of the temperature boundary conditions. The Rayleigh number is a non-dimensional parameter that quantifies the strength of natural convection in fluid flow. The Rayleigh number is defined as follows:

$$Ra = \frac{g\beta(T_w - T_{ambient})L^3}{\nu\alpha} \tag{13.7}$$

where L is the length scale of the flow domain. In PowerFLOW the buoyancy forces are handled using the equation of state for ideal gas and therefore is more general than the Boussinesq Approximation which is valid for relatively small temperature variations in a fluid.

For more complex geometries and realistic flows, physics is more complex and the simulations need to accommodate effects of turbulence, radiation exchange between surfaces, and conduction in solid parts. In the next section, the modelling aspects and more complex validation cases are discussed.

13.10 Turbulence Modelling

For most realistic flow simulations, the mesh resolution is insufficient to fully resolve the flow structures especially near the wall and a full mesh resolution of a flow field for real-world applications such as flow past an automobile would greatly exceed current compute resources. Since the length scales of the smallest flow structures are far smaller than the smallest grid size possible given current computing capabilities, turbulence models that postulates and calculates a turbulent viscosity are used to accurately model the physics of the unresolved flow field. Different turbulent models used have different prescriptions for calculating the turbulent viscosities. A brief survey of turbulence modelling can be found in Argyropoulos and Markatos [15].

As an example of resolution requirements for resolving the flow field note that the smallest length scale of fluid flow is the Kolmogorov length scale (η) calculated in the equation below:

$$\eta = \sqrt[4]{\left(\frac{\nu^3}{\varepsilon}\right)} \tag{13.8}$$

where ν is the kinematic viscosity (m^2/s), ε the rate of dissipation of turbulent kinetic energy (m^2/s^2). For a vehicle moving at 100 km/h, the rate of dissipation is around 100 and the kinematic viscosity of air is 1.5×10^{-5} leading to a Kolmogorov length scale of 0.07 mm. Given that 10 computational elements are required to resolve the smallest flow structure, we see that a prohibitively large CFD model would be required if no turbulence models are used.

A brief review of the turbulence modelling now follows. The interested reader can look into a far more detailed description in [15].

The simplest and earliest turbulence models are the zero-equation or algebraic model that such as Prandtl's mixing length model Prandtl [16] and shows reasonable results for non-separated flows with no recirculation. For algebraic models, algebraic expressions are used to calculate the turbulence quantities.

The one-equation models represent the next level of complexity in turbulence modelling. In these models, one transport equation for the computation of turbulence quantities such as turbulence kinetic energy while the length scale is prescribed algebraically. The one equation model such as the one by Spalart and Allmaras [17] works well for relatively simple geometries and flow fields such as in airfoils but not for more complex geometries and flows.

The two-equation models where two transport equations for turbulence properties are the models of choice for CFD applications because of their applicability to arbitrarily complex flow fields and geometries of flow domains. Of these, the κ-ε models are by far the most widely used. The κ-ε model such as the one described in Launder and Sharma [18] solves transport equations for two turbulence quantities and the turbulent viscosity is then calculated from these two quantities.

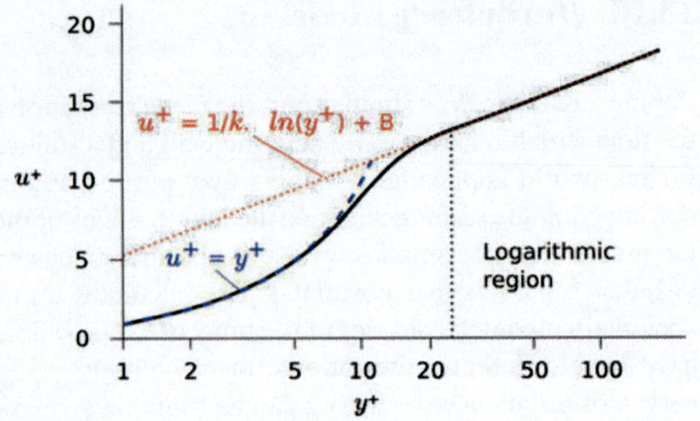

Fig. 13.26 Velocity variation parallel to the wall

Two equation models that incorporate more physics and allow for more extensive modelling of resolved scales have been developed. One such formulation is the RNG-κ-ε model where the transport equations for turbulence are modified to account for the effects of smaller scales of motion. The RNG approach proposed in Yakhot et al. [19] allows for resolving the large-scale features of the flow explicitly while modelling the small sub-grid scales of motion. This approach is sometimes referred to as VLES (very large eddy simulation).

PowerFLOW uses a customized RNG-k-epsilon two-equation model along with a wall model for the near-wall treatment suitable for LBM. The wall function is a functional variation of the velocity imposed near the wall since as noted explicit resolution of the flow near the wall is impractical due to the very small length scales involved. The functional form used near the wall is shown in Fig. 13.26.

The non-dimensional velocity parallel to the wall is u^+ and the non-dimensional distance perpendicular to the wall is y^+.

The velocity variation shown in Fig. 13.26 is as follows:

$$u^+ = y^+ \quad \text{for } y^+ < 5 \tag{13.9}$$

$$u^+ = \frac{1}{\kappa}\ln y + C \quad \text{for } y^+ \text{ between 30 and 200} \tag{13.10}$$

For standard wall function, $\kappa = 0.41$ and $C = 5.0$.

y^+ between 5 and 30 is the buffer zone.

The velocity is non-dimensionalized by the u_t, the shear velocity where

$$u_t = \sqrt{\frac{\tau_w}{\rho}} \tag{13.11}$$

$$u_+ = \frac{u}{u_t} \tag{13.12}$$

$$y^+ = \frac{yu_t}{\nu} \tag{13.13}$$

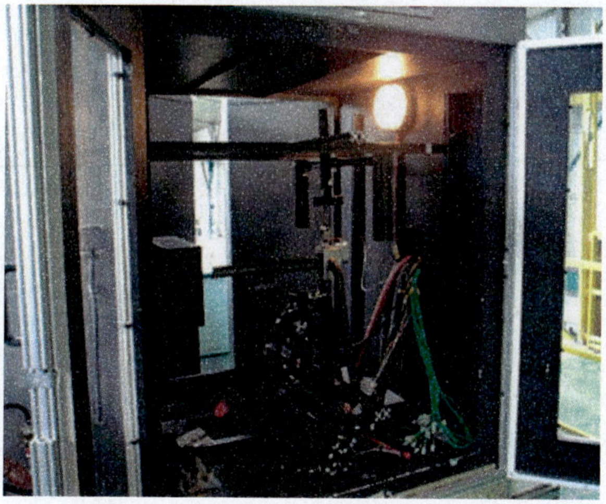

Fig. 13.27 Test Bench for tube and plate configuration

PowerFLOW uses a velocity wall model that is conceptually similar to what is shown in Fig. 13.26. Typically, the wall model is applied to impose velocity variation between the wall and the centroid of the first element next to the wall. A wall model for the temperature gradient next to the wall is also used in PowerFLOW.

The turbulence model in PowerFLOW enables one to accurately simulate flow in complex domains. An example that is documented in Mukutmoni et al. [20].

The experimental setup is shown in Fig. 13.27 is from [20].

In the experiment, a cylindrical heater heated the plate with the cooling provided by the inlet on the left. The flow field and the temperature distribution from the simulations are shown in Fig. 13.28.

The average plate temperature was compared with that of the experiments and is tabulated in Fig. 13.29.

A very important contributor to heat transfer for automotive applications is thermal radiation and one of the critical design considerations is to protect automotive components from heat sources where the mode of heat transfer is primarily by radiation. For the accurate simulation of thermal workflows, radiation modelling needs to be very accurate. In the next section, aspects of radiation modelling will be discussed.

13.11 Radiation Modelling

The energy balance equation of thermal radiation that accounts for absorption, scattering, and emission is given by the equation of radiative transfer as follows:

$$\frac{1}{c}\frac{\partial}{\partial t}I_\nu + \Omega\nabla I_\nu + (k_{\nu,s} + k_{\nu,a})\ I_\nu = j_\nu + \frac{1}{4\pi}k_{\nu,s}\int I_\nu d\Omega \tag{13.14}$$

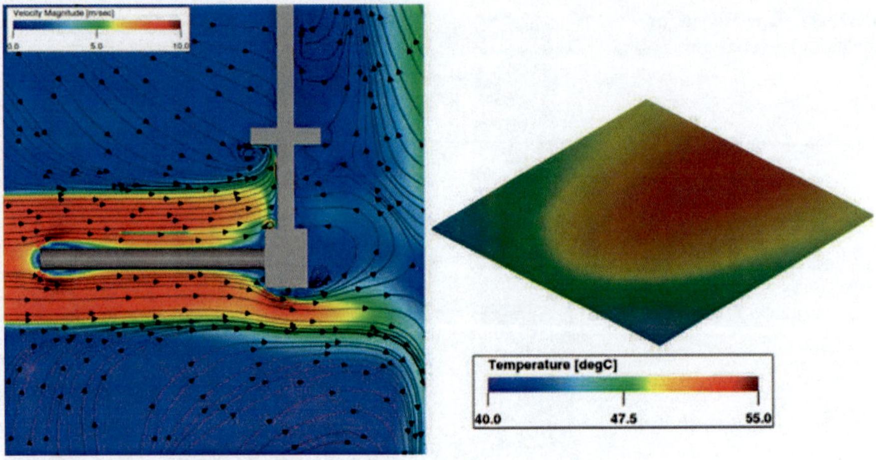

Fig. 13.28 The velocity field is shown on the left and the plate temperature distribution is shown on the right

Plate Average	Measurement	Simulation
Temperature [°c]	52.4	51.8

Fig. 13.29 Comparison between experiment and simulation for average plate temperature

The partial integrodifferential equation solves for the spectral radiation intensity I_ν with appropriate boundary conditions. The numerical solution to the radiative transfer equation is by the discrete ordinate method where the governing equation is discretized in the angular directions as detailed in the book by Chandrasekhar [21].

A brief account of the discrete ordinate method as applied to rectangular enclosures is described in Fiveland [22].

For most engineering and automotive applications, in particular, absorption, scattering, and emission by a layer of gas is not a significant factor, and the gas is considered "transparent" Under these circumstances, radiation exchange is between heated surfaces and the gray diffuse- radiation approximation is valid.

As a result, Stefan Boltzmann's law governs the radiation exchange as follows:

$$P = \varepsilon \sigma A(T^4 - T_c^4) \tag{13.15}$$

A is an effective radiating area determined by how the radiating surfaces "see" each other and is known as the view factor. The calculation of the view factors is calculated by the Monte-Carlo method and ray tracing. In this method, several particles are released from a given surface stochastically. The number of particles

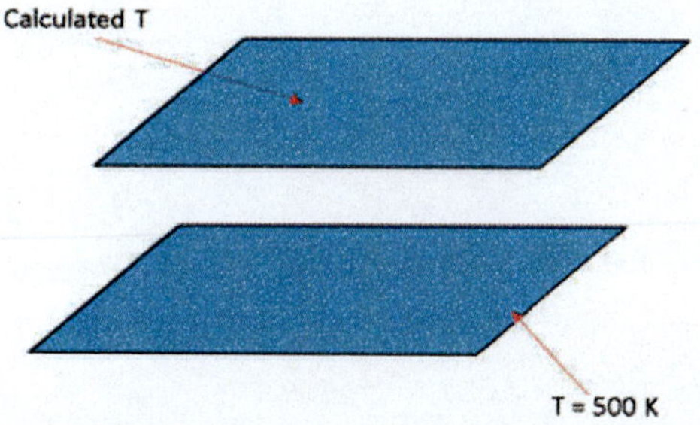

Fig. 13.30 Radiation exchange between two parallel plates

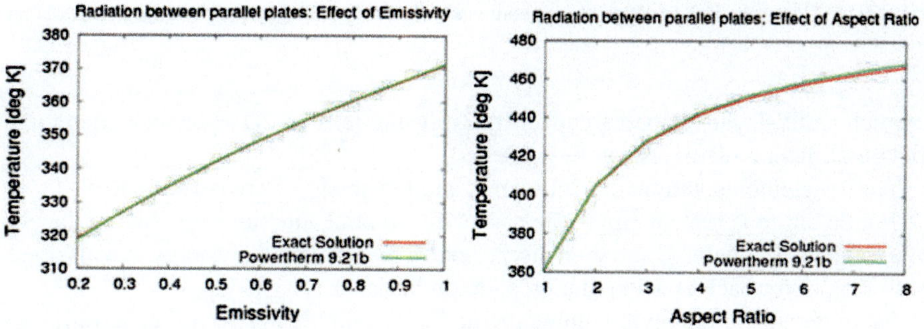

Fig. 13.31 Radiation exchange between two parallel plates

intercepted by the other surface allows one to calculate the radiation view factors that are required. The Monte Carlo Method in radiative heat transfer is reviewed in the article by Howell [23].

In Fig. 13.30, the radiation exchange between parallel plates is shown. For this simple configuration, an exact solution is available.

In Fig. 13.31, we compare the simulations with the analytical solution, and the full consistency of the results is observed.

For this academic case, the bottom plate is set to a temperature of 500 K, and the top plate temperature is evaluated as a result of radiation for different geometrical parameters and emissivities.

Another necessity of an accurate thermal workflow is to have an accurate simulation of conduction in solids. In the following, conduction modelling is discussed.

13.12 Conduction Modelling

The conduction modelling for solid parts for the thermal simulations is done using the finite volume method. The method is applied to different 2D and 3D elements. 2D elements are modelled as implicit 3D elements with assigned thicknesses.

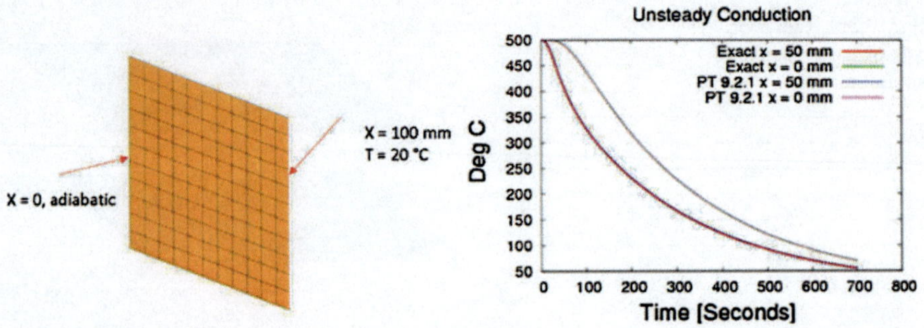

Fig. 13.32 Unsteady 1-D conduction in a slab is shown on the left. On the right are comparisons with the exact solution

Implicit multi-layer elements capability is available. For 2D elements, triangular and quadrilateral elements can be modelled.

For 3D elements, tetrahedrons, prisms, and hexahedrons are permitted.

For the finite volume method, the partial differential equations for conduction are integrated over a volume and then discretized. This ensures that energy is conserved both locally for each element and on a global basis.

More details on the finite volume method are available in Eymard et al. [24].

In Fig. 13.32, an exact solution of 1-D unsteady conduction in a slab is compared with simulation.

In Fig. 13.18, the conduction solution for a slab of thickness 100 mm is set to an adiabatic boundary condition at one end and a constant temperature of 20 °C at the other end. The initial condition for the slab is set to 500 °C.

The exact solution for the 1-D unsteady conduction problem is

$$\theta = \sum_{n=0}^{\infty}(-1)^n\left(erfc\left(\frac{2n+1-x}{2\sqrt{t}}\right)+erfc\left(\frac{2n+1+x}{2\sqrt{t}}\right)\right) \tag{13.16}$$

The exact solution is documented in several sources for instance in Carslaw and Jaeger [25].

Accurate modelling of turbulence, unsteady flow, radiation, and solid conduction enables PowerFLOW to accurately simulate flows that incorporate these complex physics. Physical problems that require a solution for both the fluid and solid domain are sometimes known as conjugate heat transfer. An example of conjugate heat transfer is documented in the paper by Wang et al. [26].

In Fig. 13.33, the experimental setup representing a simplified engine bay from [26] and the flow visualization from the PowerFLOW simulation are shown.

The fluid flow is buoyancy-driven and incorporates solid conduction in an enclosure, simplified engine, and exhaust. In Fig. 13.34, we observe that there is good agreement in the velocity field between experiments and simulation.

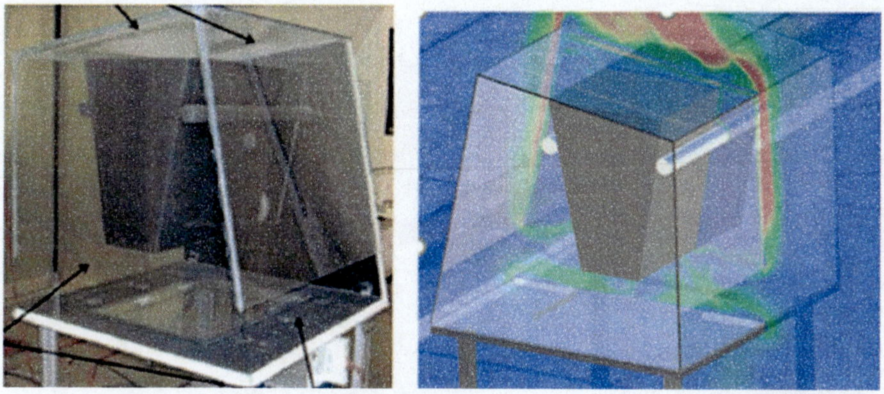

Fig. 13.33 Experimental setup of simplified engine bay shown on the left and velocity field on the right

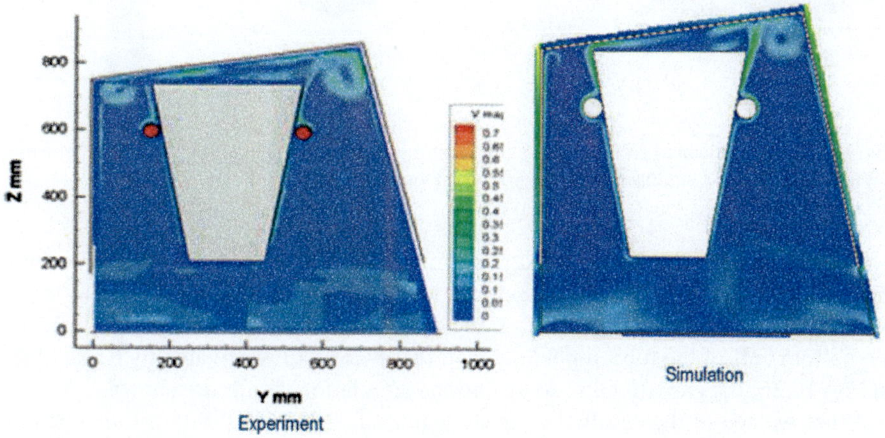

Fig. 13.34 Comparison of velocity fields. Experimental results are shown on the left and PowerFLOW simulations are shown on the right

In Fig. 13.35, the surface temperatures between the experiment and the simulation are shown. Some differences observed are the result of differences in resolution. The temperature sampling for the experiments is coarser than that of the simulations.

The comparisons shown in Figs. 13.33, 13.34 and 13.35 are documented in [26] and more details are available in that article.

As we can see in the validation cases thus far, PowerFLOW can accurately model the physics including effects of turbulence, natural convection, the surface to surface radiation, and thermal conduction in solid domains. We now touch upon another validation case that shows that the LBM numerics does model the unsteady physics very accurately as shown in the paper by Mukutmoni et al. [27].

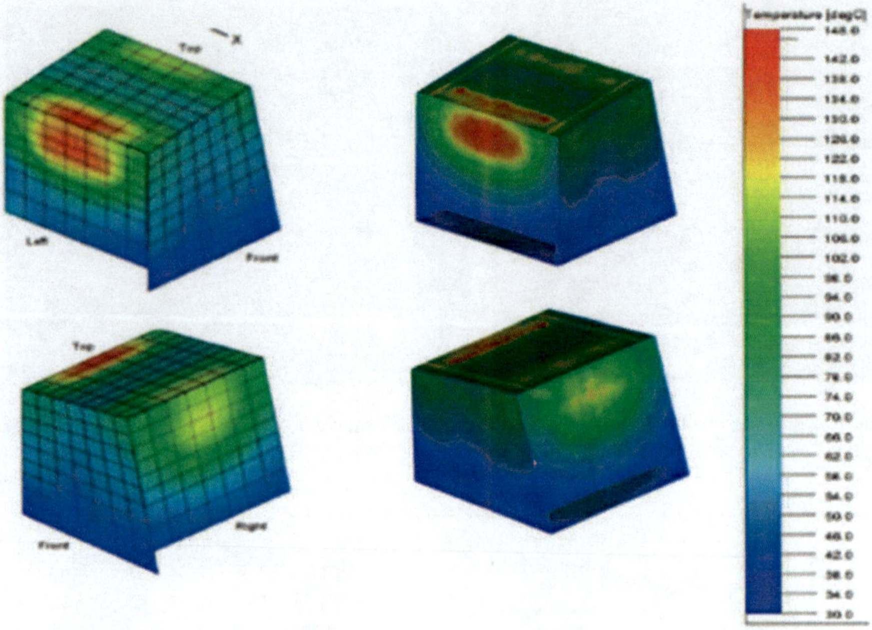

Fig. 13.35 Comparison of surface temperature distribution. Experimental results are shown on the left and PowerFLOW simulations are shown on the right

The paper [27] is an experimental and numerical study of the Rijke tube. In a Rijke tube, a vertical, hollow cylindrical tube opens at both ends, a strong tonal sound is generated by natural convection as a result of a heated wire gauze placed in the bottom half of the tube, a phenomenon that was first investigated by Rijke [28]. In Fig. 13.36, the experimental setup and the simulation results are shown.

What we see in the results is that the Rijke tube resonates with a fundamental frequency corresponding to its length. In Fig. 13.37, the measured spectrum is compared with that of the simulation.

As seen in Fig. 13.37, there is a good match in the amplitudes and frequencies of the spectrum between the experiment and the simulation thus confirming that PowerFLOW can model the unsteady flow with very high fidelity.

The accuracy of the CFD simulations enables one to propose and test a thermal protection workflow. The workflow which shows the simulation process from CAD to post-processing is shown in Fig. 13.38. The workflow has been executed, tested, and validated on mostly automotive and industrial equipment applications in general, and can be customized for any thermal test procedure.

The workflow shown in Fig. 13.38 can be applied to any vehicle under realistic conditions where for instance the entire geometry associated with the vehicle is included in the simulation model. In Mukutmoni et al. [29], simulation results of a full vehicle simulation are compared with test data from a wind tunnel test.

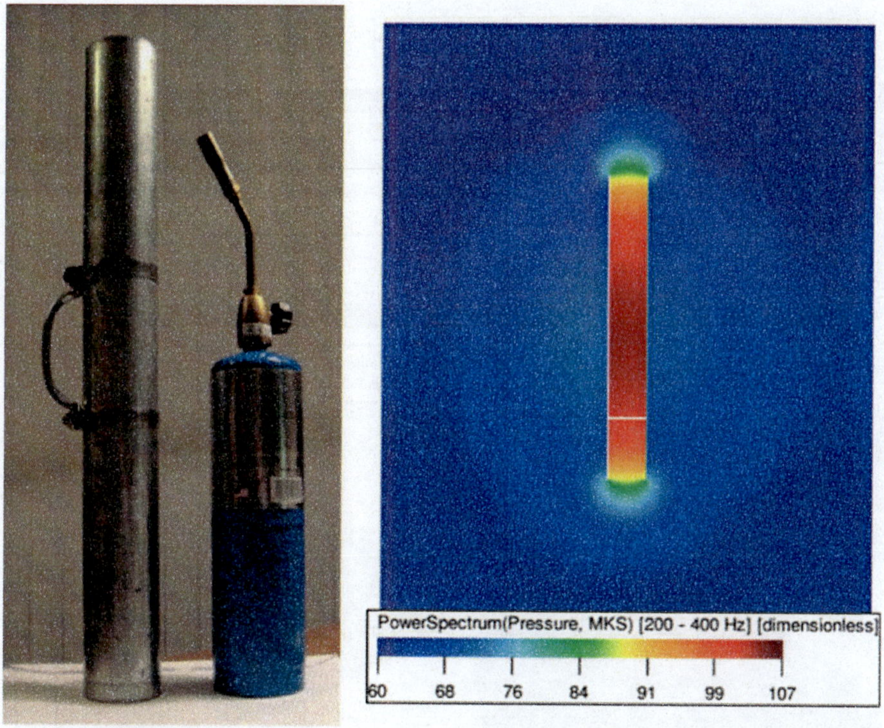

Fig. 13.36 The Rijke tube is shown on the left and the amplitudes of the pressure fluctuations are shown on the right

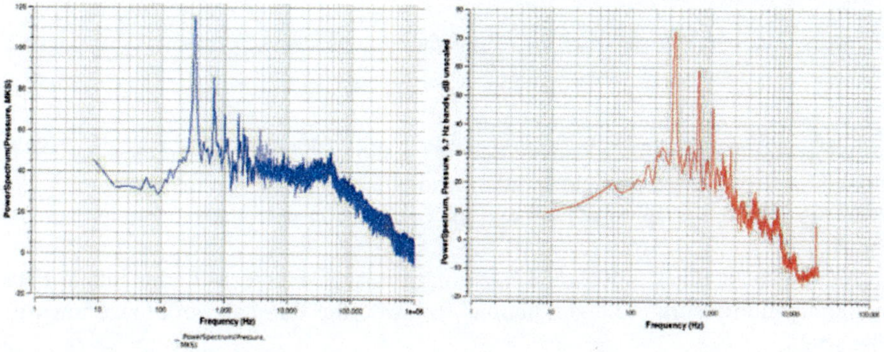

Fig. 13.37 The measured frequency spectrum is shown on the left and the frequency spectrum form simulation is shown on the right

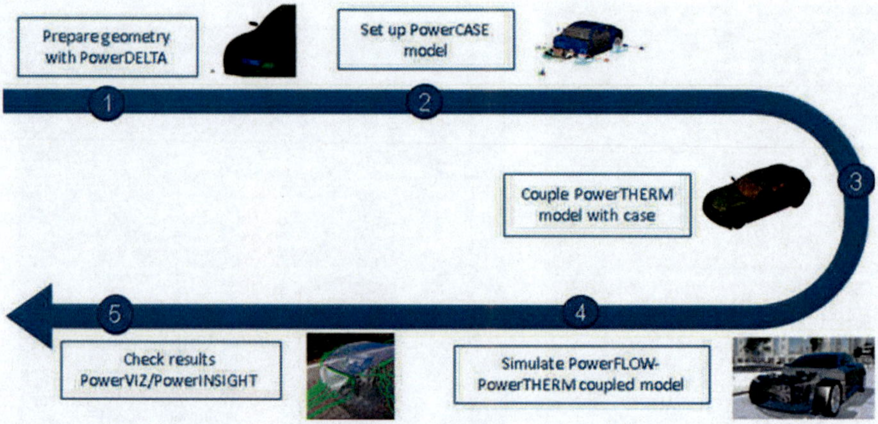

Fig. 13.38 Thermal protection workflow

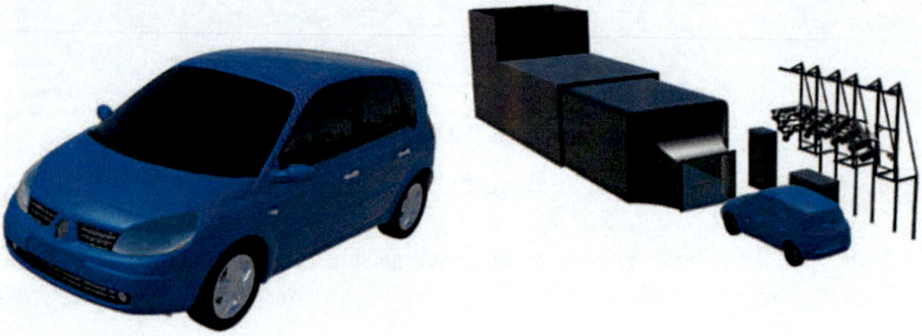

Fig. 13.39 The vehicle is shown on the left and the corresponding CFD model is shown on the right

In Fig. 13.39, the vehicle Renault Scenic2 is shown on the left, and the simulation model that includes the vehicle and the wind tunnel is shown on the right.

In Fig. 13.40, the velocity field and the comparison with thermocouple data are shown.

As shown in Fig. 13.40, good agreement is seen between thermocouple data and simulation results. The simulations followed the thermal workflow shown in Fig. 13.38.

This concludes the discussion on the thermal protection workflow. In the next section, the thermal workflow referred to as thermal energy management and thermal drive cycle will be described. These thermal workflows simulate test procedures that are time-varying and are therefore come under thermal transient workflows.

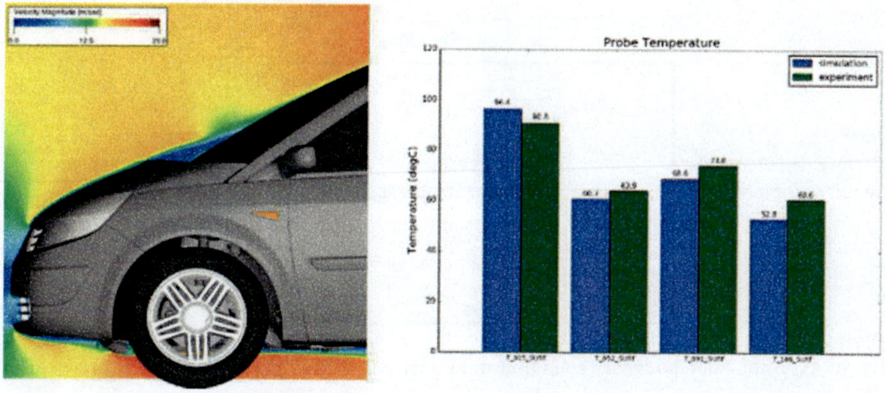

Fig. 13.40 The velocity field is shown on the left in comparison with thermocouple data shown on the right

13.13 Thermal Transient

Simulations of the transient thermal phenomenon in vehicles are important since time-varying operating conditions represent a thermally critical testing regime and must therefore be evaluated for the thermal design of vehicles. In particular, for under-hood and underbody components, temperature rise around the engine and exhaust system in the transient regime can lead to component temperatures that could well exceed design targets for material durability and reliability. Therefore, accurate predictions of their temperatures are imperative in evaluating vehicle design and providing solutions to thermal issues.

One of the main challenges of transient thermal simulations is that the total simulation time is relatively large compared to that of other simulation types, so any proposed methodology must provide a reasonable computing time and without compromising the accuracy of the results. For example, although an aerodynamics simulation is of the order of seconds of real-time, thermal transient simulations are of the order of minutes.

Typical scenarios for transient simulations are key-off (also known as a soak), idle and thermal drive cycle.

13.14 Key-Off/Soak

In a key-off/soak scenario, a vehicle under steady operating conditions of constant vehicle speed and fan rotations transitions to engine shutdown, a vehicle coming to a stop and fans shut down. In the absence of any forced convection from fans and vehicle motion, maximum velocity magnitude is typically below 3 m/s.

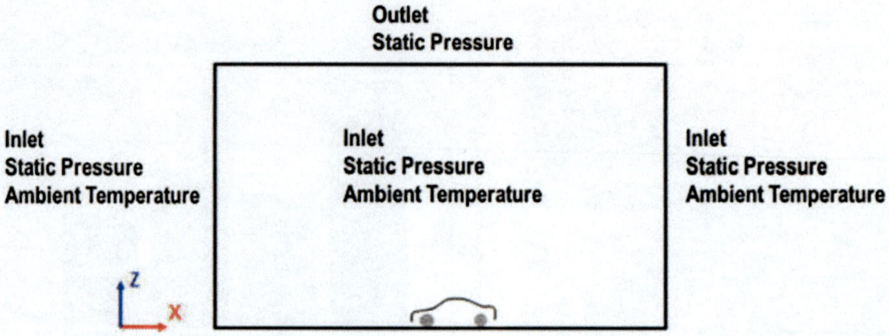

Fig. 13.41 Simulation boundary conditions for a soak

For the digital wind tunnel, the boundary conditions are shown in Fig. 13.41. The surrounding faces of the simulation domain are set as pressure inlets with zero velocity, the top face is set as an outlet. The duration of the soaking process typically lasts 30 min.

13.15 Significance of Key-Off/Soak

The key-off/soak is a very thermally significant vehicle testing scenario. When the engine is shut down, the main powertrain parts such as the engine, exhaust line, and turbocharger at the beginning are at elevated temperatures while cooling is significantly reduced since the vehicle and the fans are stationary and not contributing to the cooling by forced convection.

Under these circumstances, natural convection is the primary cooling mechanism for the engine bay and underbody. However since natural convection cooling is significantly lower than that of forced convection as a result, components such as ECU and rubber bushings could potentially see significant increases and spikes in component temperatures after the transition. For larger times since the engine is shut down, the heat source temperatures are dropping and the component temperatures eventually reduce as well.

Monitoring the history of the temperature of the components for a soak is therefore essential so that not only the extent of the temperature spikes but also the duration are known. All of which have ramifications to thermal durability of components that in term strongly correlates with warranty costs.

In Fig. 13.42, we see an example of temperature rise and drop during the soak period because initially there is a decrease in convective cooling and for later periods the effect of lower temperatures in the powertrain leads to lower temperatures.

In Fig. 13.43, we observe the plume effect of a hot excavator in the engine bay during the soaking process.

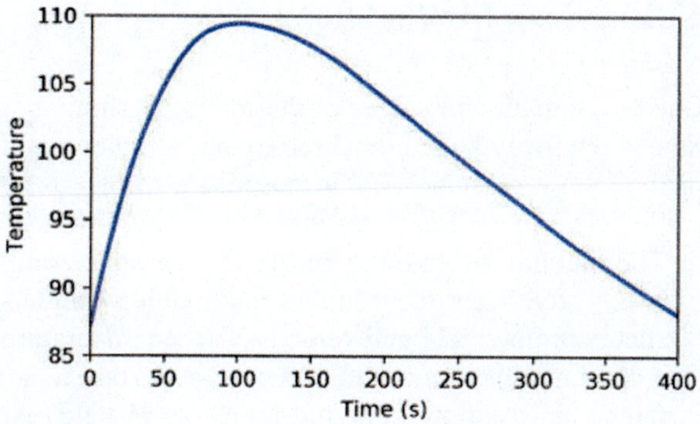

Fig. 13.42 Temperature variation typical in a soak regime

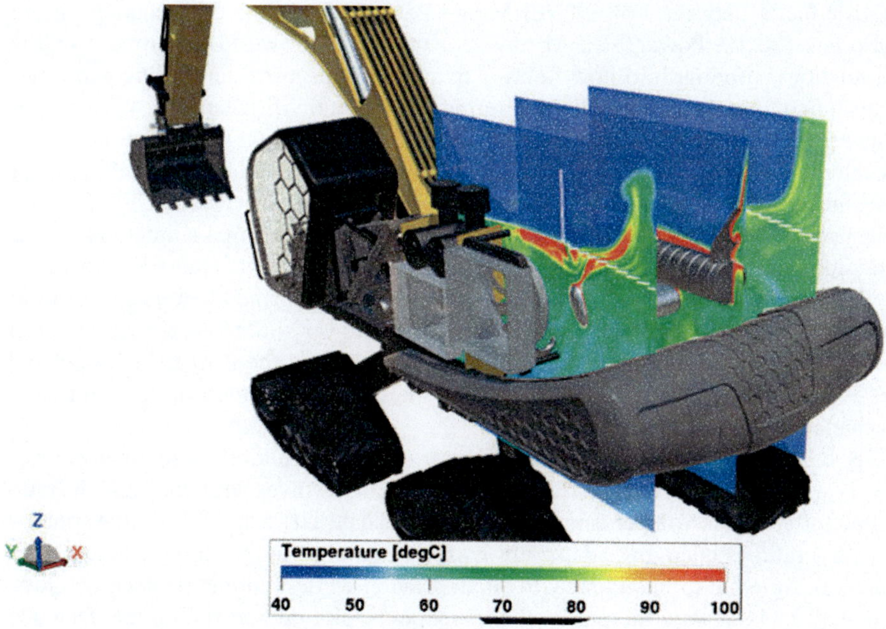

Fig. 13.43 Plume effect on an excavator machine in the soak regime

The plume flow structure which is buoyancy-driven is typically observed in a soak regime. Fluid rises to the top from heated surfaces as a result and temperature spikes observed are a result of the plume effect as well as thermal radiation.

13.16 Soak Methodology

One of the main challenges of thermal soak simulations is that the total simulation time is relatively large compared to that of other simulation types, so any proposed methodology must provide a reasonable turnaround time and compute resources while delivering accurate results.

The thermal workflow employs for the soak simulation that accomplishes this twin objective uses the coupled methodology alluded to in an earlier section to predict both flow field and vehicle surface temperatures. In this section, we emphasize the transient simulation aspects that pertain to soak. Due to the response time difference in the flow field and temperature field, especially surface temperature, PowerFLOW introduced the "PowerTHERM/PowerFLOW Time Ratio", which is a scale factor between PowerTHERM and PowerFLOW time. For example, if the ratio is set to 10, PowerTHERM time is 10 times the PowerFLOW time. In CFD simulations, solving fluid flow fields is much more resource-demanding and time-consuming compared to solving for surface temperature distributions. With proper PowerTHERM/PowerFLOW Time Ratio value setup, the CFD simulation can be accelerated. For example, PowerTHERM/PowerFLOW Time Ratio equal to 10 for 30 min period in PowerTHERM solution, will only need PowerFLOW CFD duration for 3 min. Using this ratio feature enables you to run a thermal simulation with a very long duration in a much shorter PowerFLOW time—and fewer CPU hours.

Validation studies with SIMULA and OEM have confirmed that using this ratio properly leads to much shorter simulation times with very little loss of accuracy. The fundamental reason why the ratio strategy works is that the flow field (calculated by PowerFLOW) varies more slowly than the temperature field in the solid parts (calculated by PowerTHERM).

Because the flow field keeps changing during the soak period the optimal value of the PowerTHERM/PowerFLOW Time Ratio usually evolves with time, and the ratio values often are case dependent, PowerFLOW also utilizes a smart coupling strategy that the ratio is changing adaptively with a monitored temperature gradient. The method is proven to maximize CPU hours saving while minimizing accuracy loss. Figure 13.44 shows a thermal probe temperature plot demonstrating the accuracy of enabling smart coupling strategy where the PowerTHERM/PowerFLOW Time Ratio is adaptive throughout the simulation. Figure 13.44 shows that while accuracy is maintained, simulation costs are significantly reduced as a result of adaptive coupling.

13.17 Key-Off/Soak Workflow Example

An example workflow of the soak simulation procedure is shown in Fig. 13.45. This workflow typically relied on thermal protection workflow.

Step 1, the soak scenario starts after the vehicle's normal operation. It is important to ensure the soak simulation starts from an accurate "hot" temperature. Therefore,

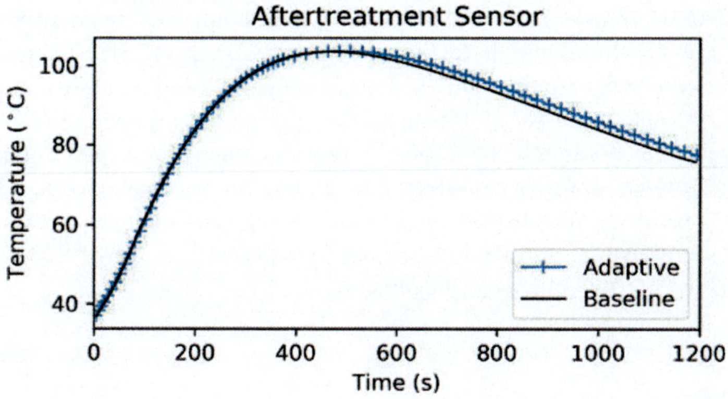

Fig. 13.44 Comparison of baseline results with reduced simulation time run referred to as adaptive

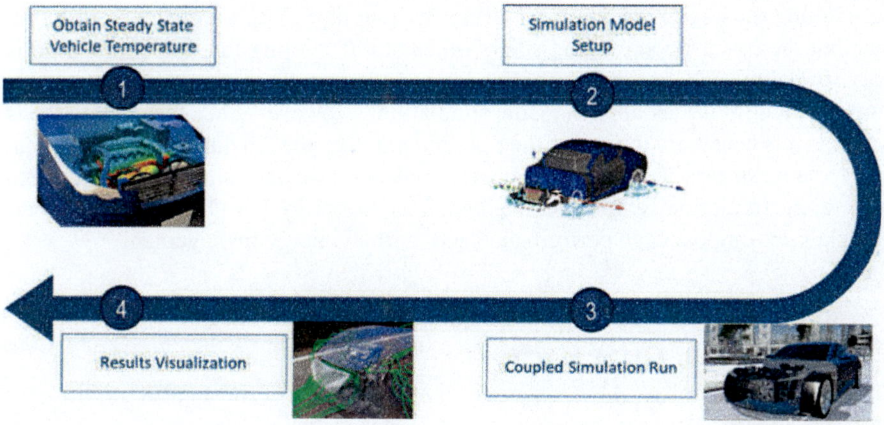

Fig. 13.45 Key-off/soak workflow

a steady-state vehicle running solution is necessary before the soak simulation. The steady-state solution will serve as the initial condition for temperature distribution on the vehicle surface and the surrounding air. This steady-state solution can be extracted from the thermal protection workflow described in the previous section.

Step 2, since the steady-state solution is readily available for a soak, it makes sense to use the existing vehicle simulation model setup and modified to a proper soak simulation scenario. Because of the low-velocity flow in soak simulation, a coarser voxel (mesh cell) size is suggested to save simulation cost. The validated study from the SIMULIA PowerFLOW solver showed that the fine voxel region around cooling fans and grills (those are typically two finest mesh levels) can be safely removed without compromise the solution accuracy. Also, vehicle velocity and cooling fan RPM should be 0. Apart from the CFD simulation setup, the corresponding thermal solver (i.e. PowerTHERM) should be configured accordingly. To observe accurate transient thermal behaviour, it is crucial to make sure all parts have the correct material and thermal mass. In some cases, OEMs have test data or existing time-dependent temperature history of some high-temperature components, i.e. engine, exhaust, turbo-scroll, for such cases time versus temperature curve could be implemented as boundary conditions for thermal management assessment.

Once the model setup is complete, the simulation is ready to run. The post-process and results visualization for soak workflow typically includes an examination of critical components' temperature history and plume-like flow visualization to evaluate proper thermal energy management.

13.18 Idle

Vehicle idle is a scenario when the car is stopped while the engine is still active and the cooling package is operating. During the idle period, the engine keeps output heat but the under-hood cooling solely relies on the cooling fan. Idle simulation is very similar to soak, except the cooling fan is operating. Therefore, most simulation strategy used for a soak applies to idle simulations. However, since the cooling fan is rotating, it is necessary to resolve the flow generated by the fan during the simulation. Thus, the mesh resolution needs to be refined around the fans similar to the thermal protection simulation setup. Idle is a significant scenario that OEMs evaluate their vehicle's cooling package performance and thermal energy management.

13.19 Thermal Drive Cycle

Thermal drive cycle simulation refers to a long-term vehicle operating virtual test to imitate real driving conditions such as test track or city driving. This type of simulation is intended to uncover thermal issues that only emerge in the complex driving scenario and evaluate part thermal behaviour over long periods. This is also helpful for thermal fatigue analysis. However, thermal drive cycle simulation is particularly challenging since the vehicle boundary condition is highly dynamic and simulation time can be very long.

A complete car CFD simulation for 30 min or one hour is often unrealistic for OEMs as the computational cost is too high, especially in the early design stage. Therefore drive cycle simulation combines 3D CFD tool with 1D system tool to drive down the simulation cost and maintains a reasonable accuracy. CFD simulations typically performed at a representative operating condition such as characteristic driving speed served as discrete point supplied to 1D system tool, which is cheap enough to perform the long-term driving simulation. Jansen et al. [30] demonstrated a method combining PowerFLOW with a 1-D system model to perform cooling pack drive cycle simulation. Yang et al. [31] presented a 3D-1D coupling method to predict charge-air-cooler performance during a race drive cycle.

13.20 Thermal Drive Cycle Workflow Example

In this example of a workflow, an exhaust drive cycle is presented to simulate a demo car in city driving. Because the hottest part in a car is from exhaust systems, accurate prediction of exhaust surface temperature is important for thermal protection and fuel economy improvement. For the efficient delivery of thermal robustness in vehicle development, exhaust surface temperature distribution needs to be well understood.

This example uses the PowerTHERM exhaust stream tool to model exhaust pipe temperature distribution. The exhaust stream tool is a 1D model to predict vehicle exhaust gas flow from the engine to the tailpipe. It takes inputs including engine speed, exhaust mass flow rate, and gas temperature. Gas temperature is calculated based on energy transfer between the exhaust stream and neighboring parts. HTCs are calculated based on correlations of Nusselt number, Reynolds number, and Prandtl number. The external temperatures are predicted by PowerFLOW where the entire vehicle is simulated.

The exhaust system shown in Fig. 13.46 is used to illustrate the drive cycle example. The exhaust is modelled from turbo to tailpipe and the internal flow is configured by PowerTHERM to predict the temperature Fig. 13.47 shows the vehicle with the exhaust system highlighted.

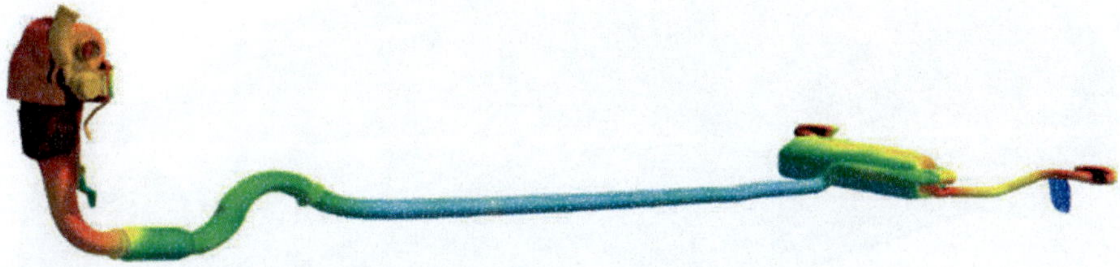

Fig. 13.46 Exhaust system example

Fig. 13.47 Exhaust system and vehicle example

In Fig. 13.48, a typical temperature distribution in the engine bay is shown.

As mentioned earlier, in a thermal drive cycle test scenario, the vehicle is tested over a long period and the driving conditions are strong functions of time. As an example, consider the variations shown in Fig. 13.49 which is representative of city driving. The variations in vehicle speed, fan speed, exhaust mass flow, and temperatures are representative of the start and stop driving schedule that occurs in city driving.

In Fig. 13.50, the engine RPM as a function of time is shown.

In Figs. 13.49 and 13.50, the city driving schedule shown is for 55 min. A simulation of the 55 min would be too computationally expensive. The simulation approach is to run the simulations for a selected set of operating points and then run the entire drive cycle in PowerTHERM by interpolation from these operating points.

In Fig. 13.51, a discrete plot of the vehicle speed vs. cooling fan RPM is shown also referred to as the operation map. The numbers marked on the operation map are selected sample points for the coupled CFD simulations.

The selected sample points are tabulated in Fig. 13.52.

Coupled CFD simulations are performed for each sample point tabulated in Fig. 13.52. Following that, each CFD simulation result will be imported into

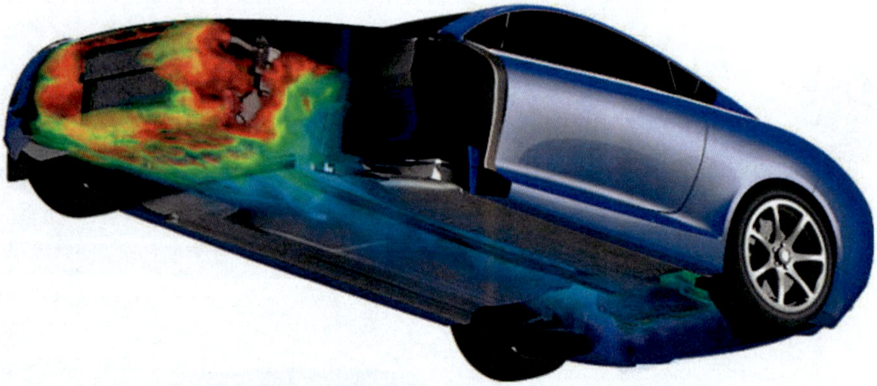

Fig. 13.48 Temperature distribution in the engine bay

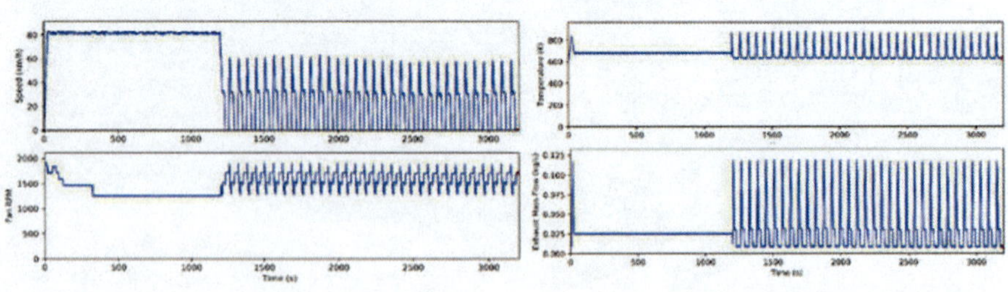

Fig. 13.49 Vehicle operating conditions for drive cycle as a function of time. Clockwise from the top left, vehicle speed, exhaust temperature, exhaust mass flow rate, and fan speed

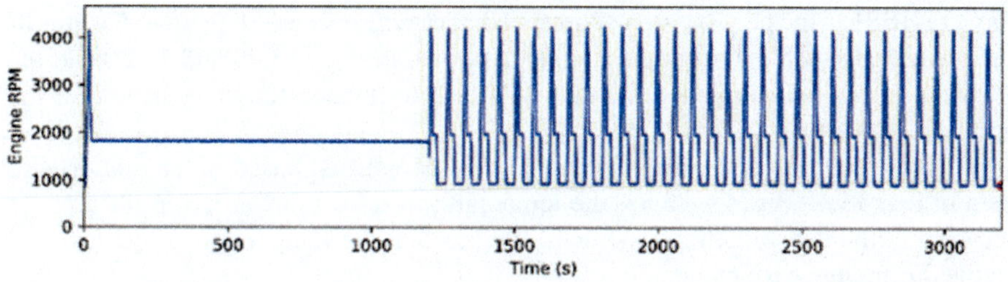

Fig. 13.50 Engine RPM as a function of time

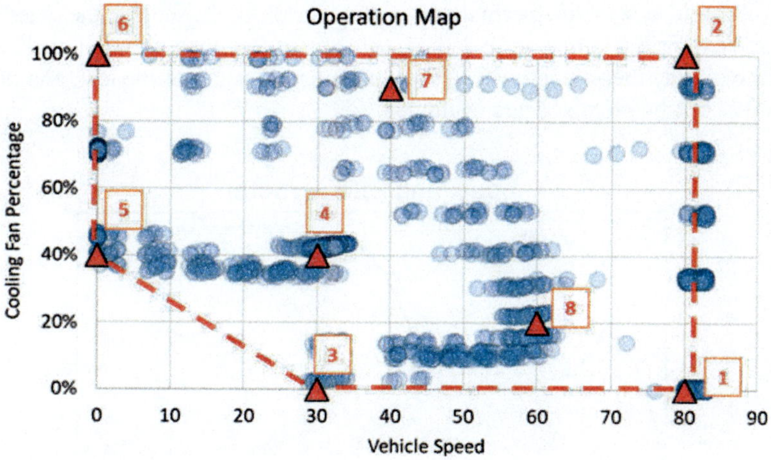

Fig. 13.51 City driving operation map and sample points selected

Vehicle Speed (km/hr)	Fan Speed (RPM)
80	1250
80	2000
30	1250
30	1550
0.01	1550
0.01	2000
60	1400
40	1925

Fig. 13.52 Selected sample points

PowerTHERM. In the next step, PowerTHERM will execute the entire 55 min of the drive cycle with the driving schedule as shown in Figs. 13.49 and 13.50 and the CFD data from the coupled simulation will be interpolated for every instant of the transient simulation.

The heat source for the drive cycle is from the exhaust based on engine driving data in Fig. 13.49. In Fig. 13.53, the temperature versus time variation for a probe location in the car undertray is shown. We see that the temperature stays low and stable during the drive cycle.

In Fig. 13.54, the exhaust shield temperature plot shows the part temperature is sensitive to the driving condition as the temperature fluctuated between 200 and 250 °C. Therefore, with the thermal drive cycle workflow, we see that the simulation can provide transient temperature data for all vehicle components and it can then be established if any thermal issues exist for the entire duration of the test, and design alternatives can be proposed and evaluated.

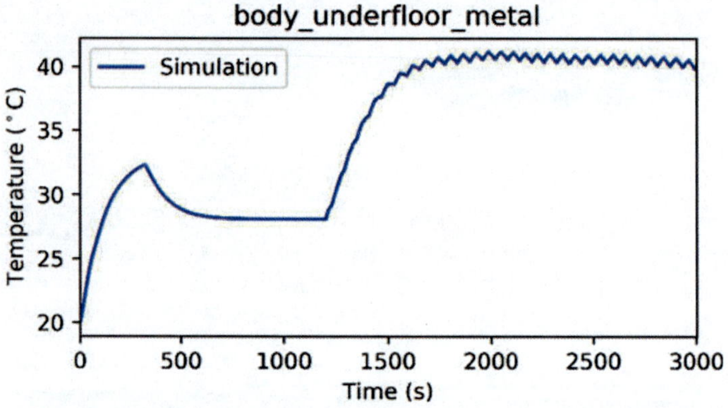

Fig. 13.53 Temperature variation on car undertray location

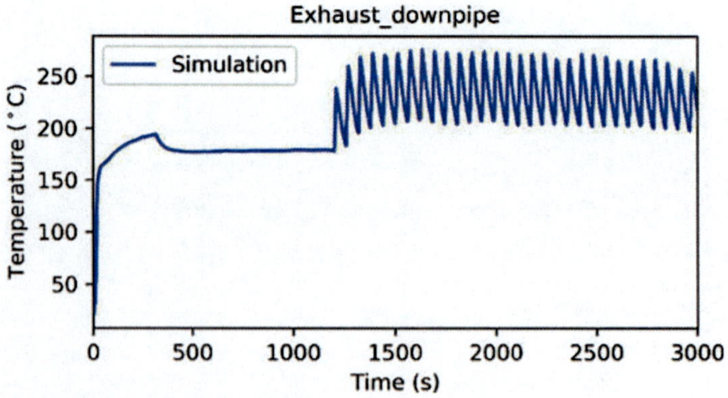

Fig. 13.54 Temperature variation on car exhaust shield

Note that one of the critical aspects of the thermal workflows that make these attractive to users aside from the accuracy of the solvers that have been discussed earlier is the level of automation of the processes and that the workflows lend themselves to quick design changes and optimization. In the next section, more details on automation and optimization will be discussed.

13.21 Automation of Simulation Process

For the numerical simulation of engine thermal management problems, it is typical to process a large set of simulation data so that the problem is numerically well defined and key results are easily accessible to the users. A typical workflow includes the processing of CAD geometry, such as sorting parts and cleaning any defects, organizing non-geometric data, preparing the simulation model including the discretized geometry model (or mesh model) as solver input, simulating the model, and post-processing the simulation result. The process can be more complicated if the workflow consists of multiple serial, parallel, or concurrent simulations involving different input models.

While computational time is the main resource required for simulation-based solutions, it is not usual that they can fully dispense with human resources. The human interventions in the execution of simulation workflow not only cost human resources but also introduce additional challenges, such as vulnerability to human errors, training burden for adopting new technologies, and scalability. This is becoming more important as more non-expert users adopt numerical simulation as part of the product design process. To minimize or remove human involvement in simulation workflows, it is critical to efficiently manage all the data and their operations. The simulation data can be categorized into a few groups.

13.21.1 Model Geometry

The simulation model usually consists of many sub-systems and a large number of components. In general, it is important to retain the geometric details as much as possible since the performance of thermal systems can be sensitive to heat transfer between parts in proximity and confined spaces. In addition to the partial or full vehicle model, sometimes the modeling of testing facilities, such as thermal wind tunnels, needs to be considered. The geometric data can also include those for abstract geometry where simulation boundary conditions, measurements, or solver-specific references are defined.

Considering the number of components (typically more than a few hundred) and levels of size (down to a few millimeter gaps) involving a typical vehicle model, the model representation needs to be structured systematically. This includes assembly structure, part/face segmentation, and tagging. As mentioned below, the abstract model data is critical for creating a streamlined simulation process that is less dependent on human operations.

The model geometry often includes components or design features more than what is needed for a specific simulation workflow. The model geometry data can incorporate the removal of these unnecessary details, namely model defeaturing, and enhance the entailing step of creating a discretized model.

13.21.2 Material Properties

Typically a large parameter set is required for the material properties of a thermal system, which are critical for simulating the relevant physics accurately. They include mass density, thermal conductivity, thermal contact resistance, specific heat, and surface emissivity/absorptivity. The complexity increases as the properties are dependent on the solution (typically temperature) and often parts consist of multiple layers of different materials.

13.21.3 Sub-System Properties

For a sub-system whose performance involves very different physical scales compared to the rest of the system, such as heat exchangers in a driving car, it is common to adopt a reduced-order model instead of simulating the original model geometry with the highest solver fidelity. This can significantly improve simulation efficiency that is the required CPU time and data storage for given solution accuracy. Then, the relevant properties, such as pressure drop due to resistance and heat rejection rates, become an important part of model data.

13.21.4 Boundary Conditions

Most thermal-fluid simulations solving discretized differential equations require boundary conditions data to be numerically well defined. Typically, the boundary conditions are flow velocity, pressure, temperature, or heat flux at simulation model boundaries. Often they are based on the design requirements (or conditions), data available from physical tests, or output of another simulation for a specific operating point or a group of them. From the workflow viewpoint, it is critical to managing the mapping between the boundary conditions data and corresponding model parts. Sometimes it may not be consistent with that of the physical model. For the simulation of cooling airflow for a driving car, for example, it is typical to take into account the rotation of flow-critical parts only, such as fans and wheels, in the simulation model instead of considering every moving/rotating component.

13.21.5 Discretized Model

The discretized geometry information of the model, also called mesh data, which is typically a set of coordinates representing the surfaces and volumes in the whole simulation domain, is another critical data set for a simulation solver. Since usually, the solution accuracy and simulation cost depends on the mesh data, the generation of appropriate mesh involves many factors to be considered, such as the element type (number of vertices and faces, interpolation order) and quality (skewness, warpage), sizing and spatial distribution and deviation from original geometry. One of the major challenges is how to make the process less dependent on the specifics of the input model, target outcome, and user expertise while achieving the desired accuracy and computational efficiency.

13.21.6 Solver Parameters

Depending on the physics equations and solver types, some parameters are not directly from input model data but explicitly needed for the solver. A popular set includes those related to flow regime (common for artificially accelerating computation), turbulence closure (typically empirical for unresolved flow scales), iterative solver parameters (number of iterations, relaxation), solution convergence criteria, and solution update frequency of coupled solvers. Since for any simulation solver, there is no universal parameter set that is optimum in accuracy and cost under all conditions, it is typical to have a default set while broad guidelines for different conditions are available to the users.

13.21.7 Simulation Result

For typical thermal-fluid problems, the solution is in the form of fluid velocity, pressure, temperature, and structural stress/strain as well as their integral or derivative quantities either in time or frequency domain. As the data size grows quickly with the mesh model size (number of grid points) times number of time steps (1 for a steady-state solution) times the number of variables and so does the data processing time, the selection of results to be processed is also important for workflow efficiency.

The decision can extend even further beyond the areas of interest. For example, even though the cooling performance of a subsystem can be represented by a single number, such as a temperature difference, integrated heat rate, or time coefficient, the result key to improving the performance may be available in a specific, localized flow structure. This means sometimes the greatest value of simulation results is accessible only with a meticulous data review. This is challenging especially for inexperienced users or those without simulation expertise.

13.21.8 Extension to Design Process

Considering design improvement as a major goal of simulation, it is natural to regard the model design data, such as design variables/space, constraints, objectives and solutions, and a broad range of operation conditions, as part of the simulation data. Here, a simulation model can be represented by ranges of all design possibilities and their target environments rather than a data set for a single design point. As simulation-based design processes become more common and the simulation workflows get more tightly connected to product design loops, the management of the extended simulation data with flexibility and consistency becomes more critical.

For the simulation of production-level vehicle models, the processing and application of the aforementioned data are not trivial and can require significant human efforts if not properly aided by simulation tools. The reduction of human interventions in simulation workflows can involve many aspects. One main challenge is to replace repetitive operations requiring data mapping with unaided programs of efficiency and flexibility. The robustness of the simulation solver to the aforementioned input data is important for a simulation workflow. Another aspect is a better user interface that can substantially improve the learning curve for adoption, number of human operations, process generalization and customization, and chances of error.

A few notable components for automating a simulation process include the following:

- Metadata extraction: The labels of model data entities, especially those of geometry data, can vary from one model to another. Also, it is typical that the taxonomy of input data does not coincide with the categories needed for the assignment of simulation data for the solver. In other words, a typical simulation model setup process requires the mapping between raw input data (CAD geometry, material properties, porous media parameters, vehicle speeds, …) and a high-level data model (rotating parts, rotation speed, parts to be modeled as conductive or insulated, components whose solutions need to be integrated, …). To keep the users from the task of data assignment, which may involve the user processing hundreds or thousands of model entities, additional data structures defining the correspondence can be introduced. A rule-based tagging system is an example of simplifying the model setup process. For this, as a prerequisite, the input model data needs to comply with a certain convention that enables effortless tagging at some point in the process. Once this is done, the subsequent operations for model setup and processing simulation data can be handled based on the metadata instead of the raw input data.
- Simulation model representation/breakdown: The simulation model breakdown applicable to all required simulation data needs to be flexible enough to support a specific workflow for any input model in the same kind without modifications or preferably multiple workflows for multiple kinds. For example, a universal simulation model structure can significantly simplify the requirements needed for a variety of simulation workflows ranging from thermal protection to vehicle aerodynamics. For a car model, an example of simulation model breakdown is

shown in Fig. 13.55. Compared to the functional or component-level model split in Fig. 13.55a, the model structure in Fig. 13.55b provides the data tailored for the solver to identify the vehicle body section (dark grey) whose surfaces are modeled as stationary solid for the ground while the relative position can be easily modified based on the ground clearance data, the suspension section (blue) which is also solid and can be positioned independently as well as along with the vehicle body depending on the loading condition, the wheel section (green) whose rotation speed changes according to the vehicle speed and effective radius, and heat exchanger sections (pink, orange, and yellow) that are modeled as porous media with variable heat rates. On the other hand, the model split for the assignment of material properties is very different as shown in Fig. 13.55c.

- Workflow exemplification as best practice: To be applicable as an industrial process, a simulation workflow needs to be generalized so that different models in the same category can be simulated with predictable resource requirements and accuracy. This involves an efficient way of preparing a simulation model, simulating the critical physical phenomena, and utilizing the most out of simulation results.
- Templatization: A pre-populated data structure based on simulation model representations and complying with solver input/output requirements can be used to execute a workflow provided the mapping from the raw input data works successfully. Such a so-called simulation template can allow somewhat loose rules for the raw input data, which is important when the input model structure cannot be strictly enforced. Since data mappings within most of the simulation workflow are based on strictly controlled simulation model representations while a loose but customized mapping for the raw data is required in the beginning, such an approach can support a wide range of input data structures. Also, as long as the simulation model representations support, a new workflow can be templatized independently.

13.22 Optimization for Engine Thermal Management

A primary goal of engine thermal management is to properly manipulate the system's input (inlet flow, surroundings) and output (exhaust flow, excessive heat generation) so that operational efficiency, as well as safety and reliability of any affected parts, can be ensured. Excessive heat generation, for example, not only means wasted energy by itself but also includes the indirect loss caused by additional weight and the cost often needed for cooling and thermal protection. Furthermore, such overdesign usually accompanies adverse impacts on other aspects of vehicle performance, such as aerodynamic drag penalty and increased noise. The design challenges involving the number of parts in engine and sub-system design, the interactions between them, and trade-offs with other performance attributes can be effectively addressed by solving an optimization problem. The optimization of engine thermal systems can involve both design optimization, where its parts or sub-system designs are intended to be

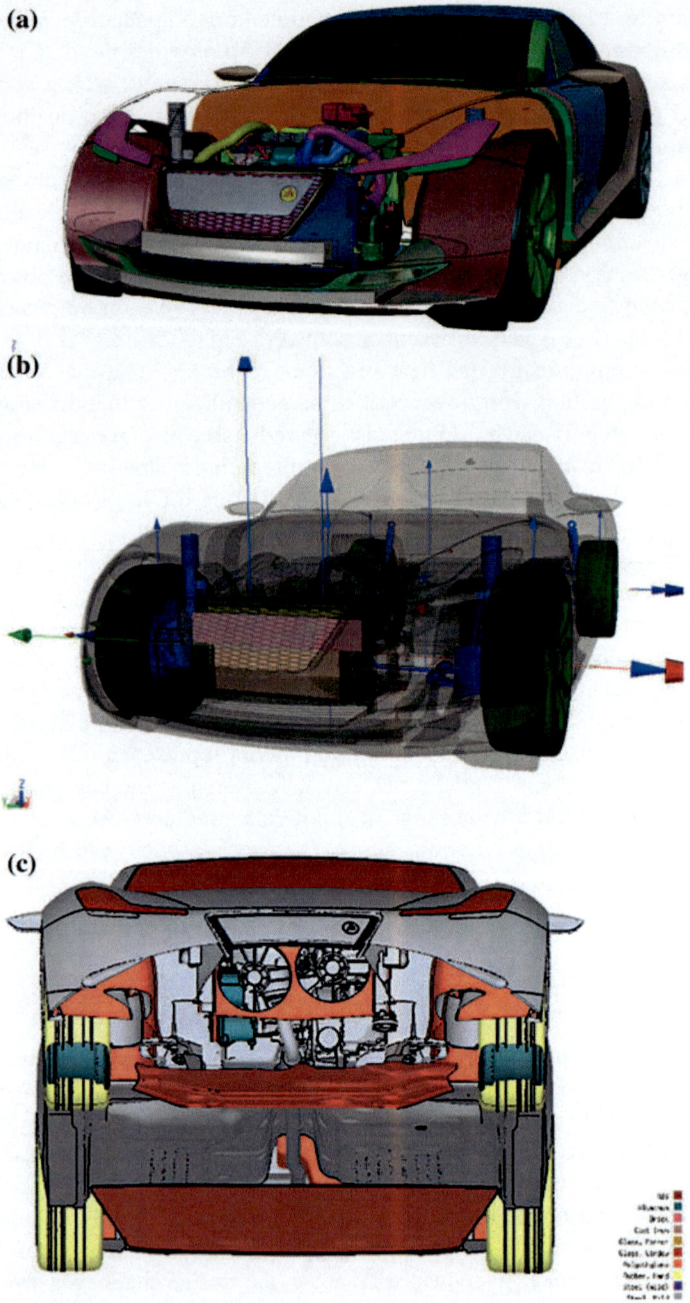

Fig. 13.55 Model breakdown comparison: **a** Input geometry model: different colors represent part segmentations in the finest level, **b** setup model for fluid solver: breakdown of the same model into the body (grey), suspension (blue), wheels (green), and heat exchangers (pink, orange, and yellow) and **c** material properties setup for thermal solver: showing only a small set while in reality there are many more different materials

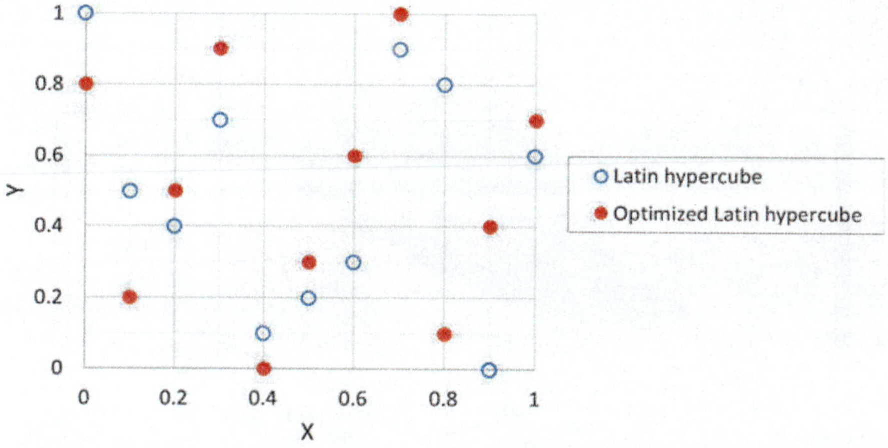

Fig. 13.56 Comparison between random Latin hypercube and optimal Latin hypercube samples

optimized under a set of representative conditions, and optimal control, where the system's operation parameters can be optimized over a given time cycle. Two distinctive approaches to general design optimization problems are mentionable depending on how the gradients in the design space are calculated.

For a small number of design variables and often with a large number of constraints, the design features can be parameterized within a design space that is relatively small in terms of the degree of design freedom, as called parametric design optimization. A popular approach is to execute a design of experiments where the parameter sets to be tested are carefully selected within the design space. This sampling process is intended to capture unbiased trends in how the performance variables change in response to the design variables. As an example, the optimal Latin hypercube sampling is very popular because it achieves spatial uniformity as well as uniformity in each dimension, as shown in Fig. 13.56, which proves to be efficient in achieving optimal solutions with limited resources as described in Jin et al. [32].

It is typical to introduce functional relations, frequently called surrogate models that approximate the true response of the system in explicit forms. In Fig. 13.57, an example of surrogate model-based prediction is shown in a 2-dimensional design space for a kriging model, one of the most widely adopted models described in Cressie [33].

The predicted mean follows the true function reasonably well where the sampled data are available while the deviation becomes significant where the samples are sparse. Once the surrogate models are available with acceptable accuracy within the entire design space or a limited region of interest (predicted optimum, for example), various approaches can be adopted to assess the performance and sensitivity (variability concerning design variables) for desired designs, locate the optimum designs, and perform trade-offs between competing objectives.

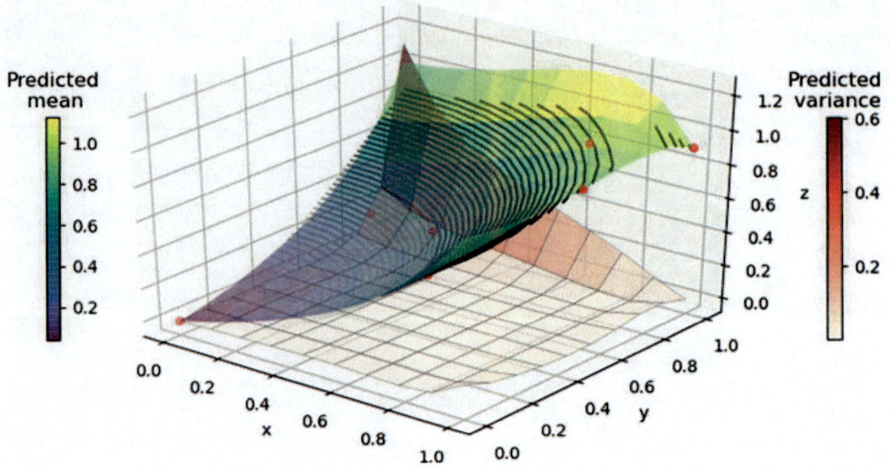

Fig. 13.57 Kriging approximation and predicted variance: black contour lines represent the true function, red circles are sampled data, and colored surfaces are kriging-predicted mean and variance

The approach of parametric design optimization can reach the global optimum more reliably based on global parametric relations compared to the other approach mentioned below. In general, this approach is also more flexible in dealing with additional constraints that may not have been considered in the initial formulation. However, as the number of design variables grows, such global design space information becomes considerably more difficult to obtain. In particular, combined with the high simulation cost generally required for engine and sub-system thermal simulations, the number of manageable design variables is typically limited to a few and rarely exceeds a few dozen.

Another approach is referred to as non-parametric optimization, where typically a much higher degree of design freedom is targeted. The approach is especially suitable for a design problem where critical design features cannot be easily defined and/or more generalized design changes are intended, such as topology optimization problems.

As an example, topology optimization tried to achieve the optimal distribution of materials that satisfies a target objective, such as maximizing the performance of a cooling device with volume or weight constraints. Based on the calculation of sensitivities that are closely linked to the primary solver simulation, it is typical to reach the design solution without a significant overhead in simulation cost. However, the approach has been largely limited to the low-Reynolds number, internal flow applications due to the severely increased complexity in flow physics and sensitivity calculations for a general external flow field at higher Reynolds numbers. Despite recent progress expanding the application boundaries as described in Dilgen et al. [34] and Kreissl et al. [35], non-parametric optimization, in general, is limited to steady-state internal flow problems.

Fig. 13.58 Engine exhaust manifold baseline (left) and optimized design for minimum pressure drop (right)

Due to its applicability to very high dimensional problems, topology optimization is adopted well for the optimization of the air intake system, engine intake/exhaust manifold, and various cooling ducts/pipes as shown in an example in Fig. 13.58.

13.23 Example 1: Cooling Module Layout Design for Cooling Efficiency and Aerodynamic Performance

An engine cooling module typically consists of one or more cooling fans and a series of heat exchangers where airflow rejects heat from high-temperature fluid circulation. The layout of the module, such as the number, sizing, and locations of fans as well as heat exchangers, can affect the cooling efficiency significantly since how the cooling airflow interacts with the components under different fan and vehicle operation conditions is subject to the design changes. Also, as mentioned earlier, a vehicle's aerodynamic efficiency can change depending on the energy loss in high-speed airflow due to cooling.

An example study shown in Fig. 13.59 shows how both the radiator cooling performance and vehicle aerodynamic drag change as the cooling module layout parameters change in a design iteration. Although the result does not bear the optimal solution, such design changes as potential design samples can be regarded as part of a typical parametric design optimization study. Starting from a baseline design (design 1), the subsequent design trails show a clear trend between the two competing objectives, i.e., maximizing cooling airflow and minimizing aerodynamic drag.

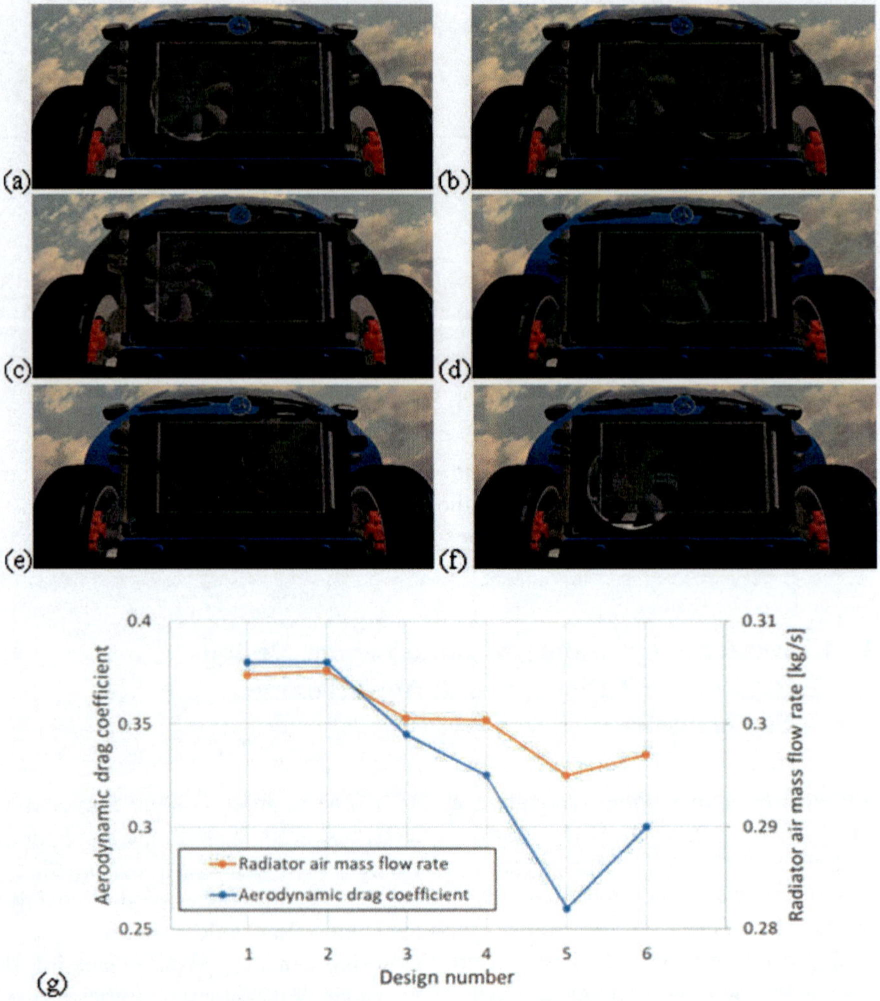

Fig. 13.59 Design iterations for fan module layout: **a–f** design 1 to 6, **g** results of aerodynamic drag and radiator cooling airflow

13.24 Example 2: Car Body Design Optimization for Cooling Efficiency and Aerodynamic Performance

It is well known that engine cooling performance competes with external vehicle aerodynamics through how efficiently the cooling airflow is generated especially in the form of cooling drag. An example of car body design is shown in Fig. 13.60 where the front-end design changes are focused. As the design features (features for front-end alone in this study) change both the aerodynamic drag and the amount of

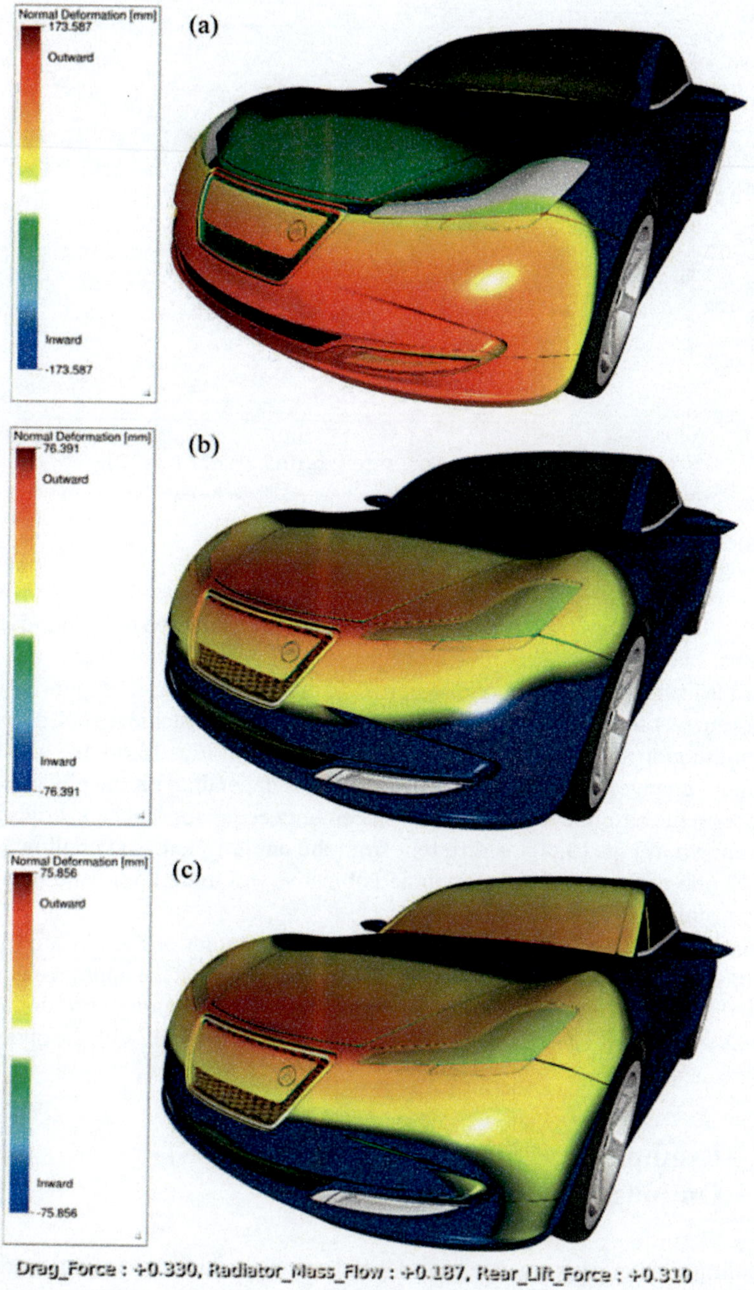

Fig. 13.60 Design surface displacement maps: **a** optimized for vehicle aerodynamic drag, **b** optimized for radiator mass flow, and **c** optimized for both, **d** objective space and Pare-to front, where the high-fidelity simulation data in red (red cross on the baseline), surrogate model predictions in grey, Pareto front designs in blue, design direction optimized for aerodynamic drag in cyan, design direction optimized for radiator mass flow in magenta, and design direction optimized for both in yellow

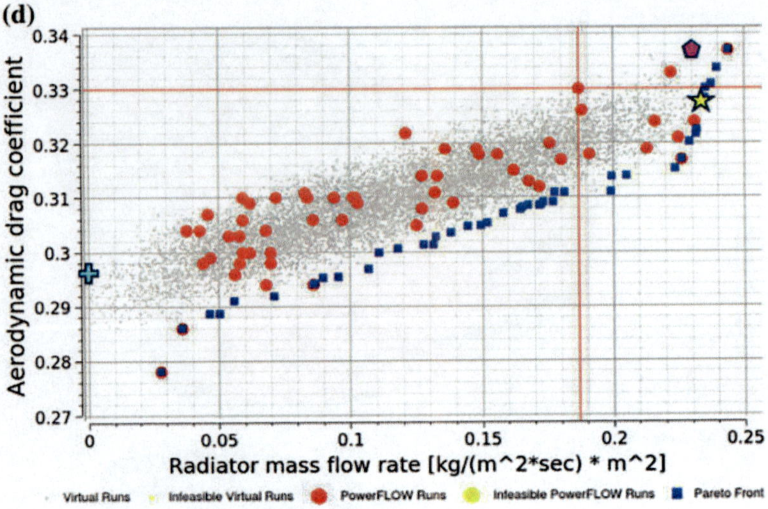

Fig. 13.60 (continued)

cooling airflow through the grille, the design object is considered to be aerodynamic drag alone, radiator cooling airflow alone, and the combination of them.

Once the trained surrogate models approximate the performance responses well, it is possible to reconstruct the desired design directions that can achieve such objectives based on the globally linearized sensitivity. As shown in Fig. 13.60, the sensitivity of geometry change along such a direction changes depending on the objective, i.e., starting from the baseline, the surface displacement needed for reduced aerodynamic drag as shown in Fig. 13.60a is different from the one for increased cooling airflow in Fig. 13.60b or the one for the combined objective (minimum drag and maximum cooling airflow) in Fig. 13.60c.

Also, Fig. 13.60d shows where the proposed design solutions are located among the designs with competing objectives. It is also noticeable that the approach based on linearized sensitivity can reliably achieve the improvements along with the desired directions while the suggested designs do not necessarily coincide with the optimums.

13.25 Example 3: Active Grille Shutters (AGS) Optimization for Fuel Economy

The cooling system performance also depends on the vehicle operating conditions, such as vehicle speed and road load, and driver input. Since they vary substantially in time during a given cycle, a desirable real-world performance is not guaranteed by just relying on the optimal designs based on a single or a few conditions.

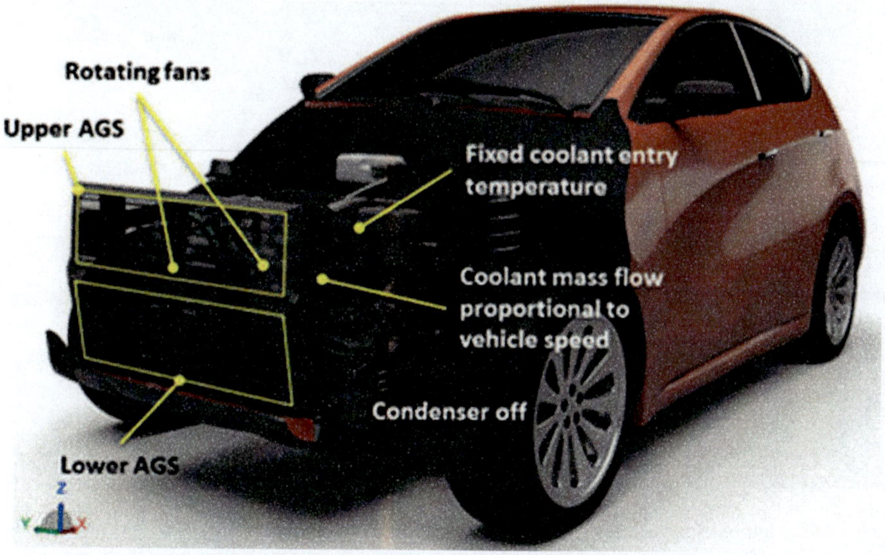

Fig. 13.61 Underhood parts and heat exchanger cooling units of the model vehicle

Active grille shutters (AGS) have been adopted widely to account for the efficiency loss that cannot be addressed by a static optimization problem. An optimization study investigation with AGS has been detailed in Cho et al. [36]. The vehicle that was the subject of the investigation is shown in Fig. 13.61.

The pressure and temperature distribution for different configurations of the AGS are shown in Fig. 13.62 taken from [36].

Depending on the need for heat rejection and aerodynamic efficiency, the amount of cooling airflow can be adjusted on-demand. However, combined with other design options, it is not clear how much gains are achievable with an AGS system for a given design. In the example shown in Fig. 13.63, we see that achievable performance gain, as well as the needed control input to AGS, can be predicted by simulating the engine cooling system along with the vehicle fuel efficiency model.

13.26 Post-processing of Simulation Results

Typical solution variables for engine thermal management include flow temperature, velocity, and pressure as well as temperature and stress/strain for solid parts. Also, the heat flux across fluid–solid or solid–solid boundaries is used frequently especially as a clue to resultant temperature fields as well as input to other simulations. Since in general both spatial and temporal variations of the variables are critical to understanding the results, it is not unusual to result in a very large data set that requires an efficient post-processing strategy. For both discrete scalar quantities, such as part

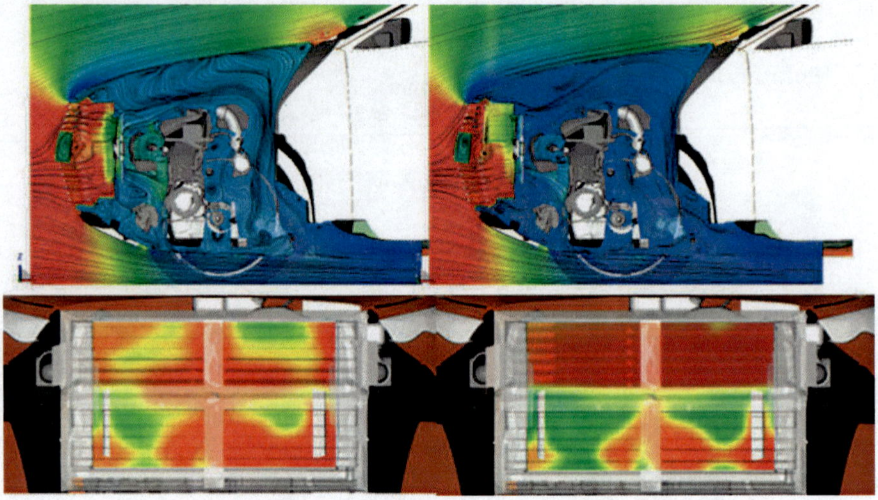

Fig. 13.62 Comparison between fully open AGS (left) and partially open AGS (right) pressure on the fan axis plane is shown at the top and air temperature at radiator exit plane is shown at the bottom

minimum/maximum temperature and average convective heat flux, and their spatial distributions, it is important to be able to efficiently manage the list of model entities (component, sub-system, and whole vehicle level) and relevant solution variables. In particular, for spatial distributions, the parts to be shown or hidden, and visualization methods, such as iso-surfaces and fluid streamlines, need to be carefully selected to effectively demonstrate the performance drivers.

An example, the solution image as shown in Fig. 13.64 visualizes the surface temperature field on engine, exhaust, and shielding parts along with underbody fluid streamlines. Such a result image can provide an overview of the impact of underbody flow on part cooling during a high-speed condition.

In Fig. 13.65, the heat flux results including the main heat sources are shown as part of vehicle thermal protection. The isolated distributions can provide some insight into how much heat energy is released and affects the surrounding parts in the form of either convective or radiative cooling. Here, for example, although convective cooling is effective on the exhaust system, the adjacent fuel tank has a relatively high impact due to radiation.

The discussion thus far has been on how thermal CFD simulation workflows using PowerFLOW can be applied to various aspects of the thermal design of vehicles and equipment. We also talked about how other tools such as automation, optimization, and post-processing have added to the effectiveness of the workflows. In the last section, we discuss aspects of another important category of tools known as a system and 1-D tools that are relevant to vehicle design at an extremely early stage where the layouts of various vehicle components are still in play.

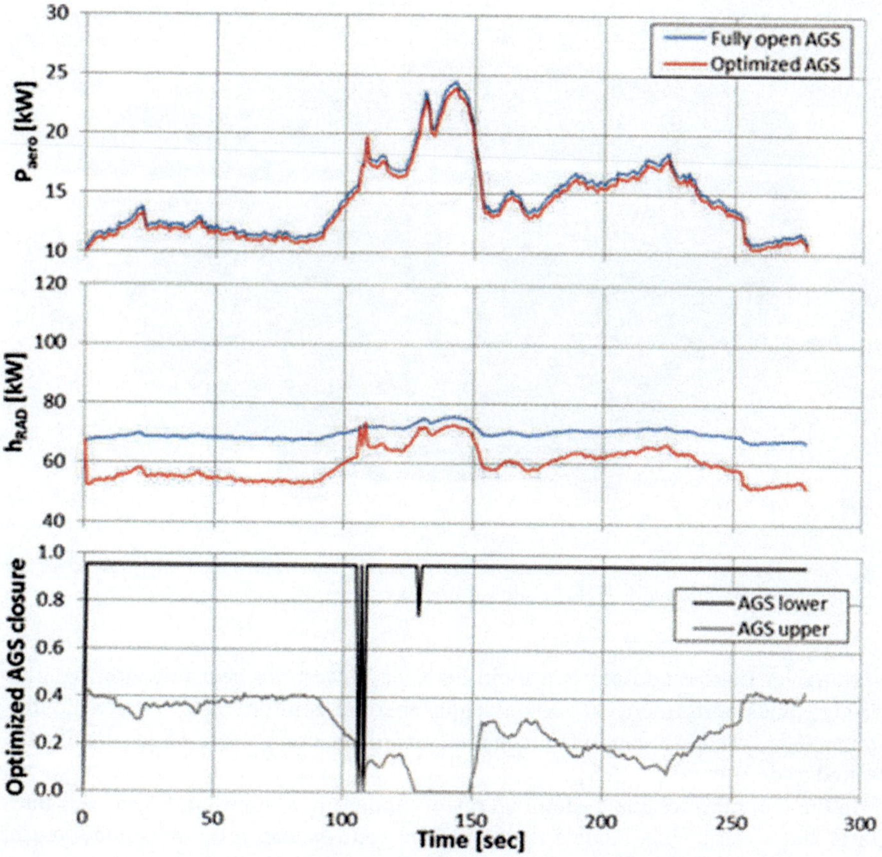

Fig. 13.63 Optimized AGS operation for minimal power consumption with radiator heat rejection rate constrained on partial SFTP-US06 [33]

13.27 System Modeling

System modeling and simulation is a commodity tool to evaluate the performance of a system with lots of interaction between different equipment and parts. In the context of engine cooling, the integration between high-temperature cooling loop, internal combustion engine, powertrain cooling, and lubrication system, and HVAC loop is a typical example. Such integration and calibration can be done before detailed design in the system modeling tool. System modeling allows us to understand the preliminary requirements of individual and overall systems. It speeds up the development, avoids late failure, and saves on development costs.

With system simulation, engine cooling system architectures can be quickly assessed based on behavior model simulation. Informed decisions are made earlier in the development process, long before a detailed design or physical prototype

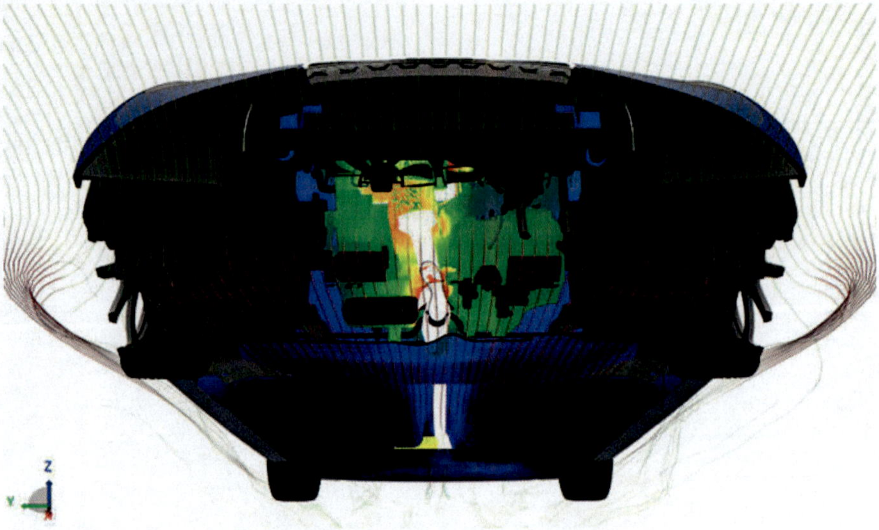

Fig. 13.64 The surface temperature on engine and exhaust systems and underbody streamlines are colored by velocity magnitude (vehicle's lower front is clipped)

is available. Engine cooling is a topic between powertrain heat rejection, vehicle aerodynamics performances, and passenger thermal comfort. The systems simulation model is the best way to balance the performance of a vehicle and identify the desired trade-offs.

In the complex vehicle system, an open standard in system model enables integrates and co-simulates models from different sources, and promotes model reuse. It reduces development costs and uncertainty.

13.27.1 System Simulation with the Modelica Open Standard

Modelica is a multi-physics modelling standard, founded on the accumulated experience of several groups in industry and academia. It is a modern and expressive language for describing dynamic systems. Particular attention has been placed on promoting the true reuse of modeling knowledge regardless of the application domain. The Modelica language and its standard library (MSL) are managed by the Modelica Association® through several design meetings. Modelica is vendor-neutral and supported by multiple tools, being developed by both users and vendors. Users are no longer locked into proprietary solutions. The open development process allows early feedback and yields well-defined semantics of the language.

Modelica is an equation-based modeling language. The dynamic behavior of systems is described by differential and algebraic equations. This allows tools to generate highly efficient simulation code, suitable for real-time hardware-in-the-loop

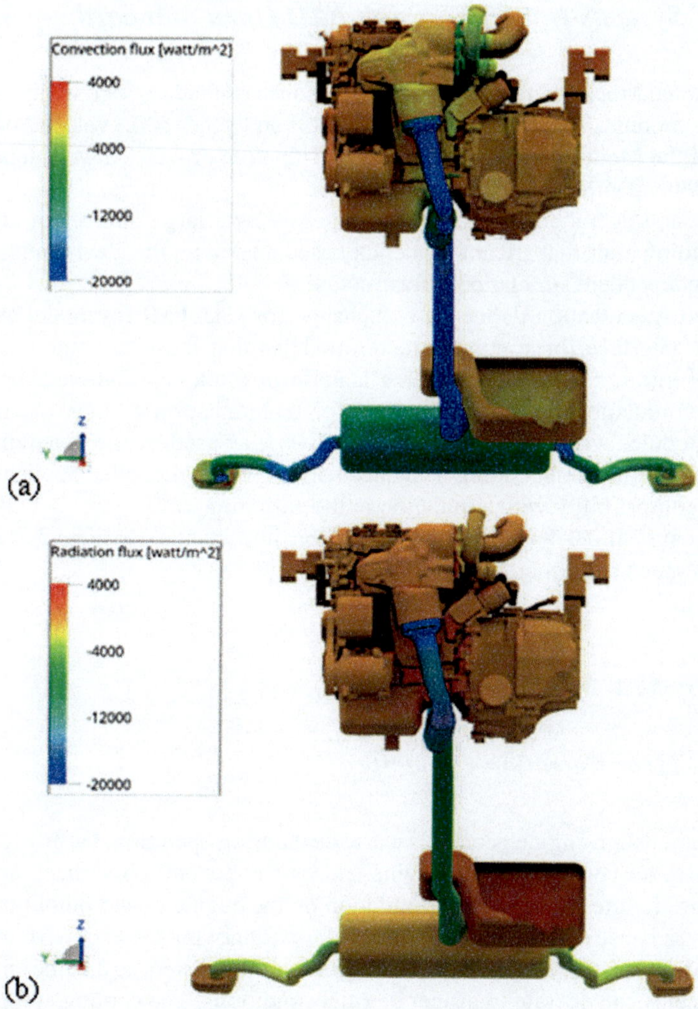

Fig. 13.65 Heat flux result example for engine and exhaust systems (the rest of the vehicle is hidden): **a** convection heat flux and **b** radiation heat flux

applications. At the lowest level of abstraction, relationships between variables are described by mathematical equations. At higher levels, components are composed graphically in Dymola thru drag-and-drop, and the externally visible variables (called connectors in Modelica) are connected by drawing a line between two connectors.

13.27.2 Systems Integration with FMI Open Standard

The Functional Mockup Interface (FMI) is an industry-standard for combining simulation code modules (FMUs) from multiple tools and vendors. Developed under the auspices of the Modelica Association, the specification provides a well-defined and vendor-independent exchange format for code (binary or source) as well as associated data and documentation. FMI is supported by a large number of authoring tools, including tools that are not Modelica-based, making it the ideal foundation for a vendor-independent simulation infrastructure.

The FMI specification defines two exchange formats. FMI for model exchange defines the interface for simulation code modules that must be combined with a common, central, solver. This ensures a uniform numeric solution and centralized strict simulation error control. FMI for co-simulation defines the interface for code modules with embedded numeric solvers, as used by the generating tool. This approach allows embedding dedicated solvers for the modelled application and facilitates compatibility with simulation in the authoring tool.

The recent FMI 2.0 specification standardizes important extensions that improve simulation speed and accuracy.

13.28 System Model Objective

13.28.1 Heat Rejection Capacity

Internal combustion engine generates waste heat during operation. Failing to remove the excessive heat will cause the engine and its component to overheat, and result in premature failure of the engine, limitation on the engine power output to protect parts, or even cause an immediate engine failure. Removing the excessive heat from the engine is the primary objective of the engine cooling system so that the engine and its components can operate in proper thermal conditions. The cooling system needs to reject enough amount of waste heat under normal and extreme vehicle operating conditions.

13.28.2 Coolant Temperature

A water-cooled engine requires an engine coolant as the cooling medium. It absorbs heat from the engine block, engine head, and other high-temperature parts, and transfers the heat to the high-temperature heat exchanger (radiator). Then, through the radiator, the heat is dissipated into the atmosphere. If the coolant temperature is too high, it can and reaches its boiling point and starts vaporization. The vaporized coolant will create bubbles or air pockets inside the coolant circuit. They will cause inefficient coolant circulation and localized hot spots in the circuit, and result in localized or general overheating in the engine system and eventually cause engine failure. It is important to verify the engine coolant temperature under different duty cycles and weather conditions, such that the maximum temperature does not exceed the design limit. The highest coolant temperature is when it just exits the engine block, or when it just entered the high-temperature heat exchanger. The coolant temperature is usually measured from those locations since they are the highest temperature point in the system.

13.28.3 Engine Efficiency

Vehicle efficiency and emission regulations have continued to become more difficult to achieve across the world. Regulations such as WLTP, EPA GHG, and VECTO set efficiency and emission requirements for vehicles operating under certification driving schedules. The emission and efficiency of internal combustion engine vehicles are primarily determined by the engine system. The engine temperature plays a significant role in engine emission and efficiency. The cooling system needs to predict and manage the engine temperature during the complete driving schedule to ensure compliance with efficiency and emission standards.

13.28.4 Cooling Power Consumption

The cooling system consumes power to reject heat from the engine. Components such as water pumps and oil pumps need the power to operate, to drive coolant or engine oil to overcome the resistance and circulate in the system. Cooling power consumption is part of the overall vehicle power consumption, and it will affect vehicle efficiency. Thus, it is important to design an efficient cooling system that not only meets the cooling requirement but also minimizes the cooling power consumption.

13.28.5 Packaging

The cooling systems include multiple parts that need to fit in tight spaces. For example, the front-end cooling package includes multiple heat exchangers, pipes, valves, containers, fans, ducts. These components need to fit into the space inside the engine compartment, with a certain location, dimension, and spacing requirements. During the design process, the packaging requirements need to be verified to ensure the cooling system components can properly fit into place.

13.28.6 Aerodynamic Efficiency

Cooling air is introduced to the cooling package to reject the heat from the cooling system to the atmosphere. More cooling airflow allows more heat rejection but reduces the vehicle aerodynamic efficiency, and negatively affects the vehicle efficiency and emissions. There is a tradeoff between the vehicle's aerodynamic efficiency and engine cooling capacity. Engineers need to ensure introducing sufficient airflow to the cooling package, but not more than what is needed, to minimize the impact on the vehicle's aerodynamic efficiency.

13.28.7 Cabin Heater Performance

In an internal combustion engine vehicle, the cabin heater relies on the heat from the engine coolant to heat the cabin air. A three-way valve is typically used to direct part of the engine coolant to the cabin heater core and dismiss some heat to the cabin air. However, too much heat directed to the cabin heater core will reduce the coolant temperature and may result in lower-than-desired engine temperature. An insufficient engine temperature will reduce engine efficiency and increase emissions. This situation commonly happens during vehicle start in cold weather, while both engine and the occupant needs to be warmed up as soon as possible. Thus, there is a tradeoff between heating the cabin versus warming up the engine, to ensure both the satisfaction on the thermal comfort and compliance to the emission and efficiency regulation.

13.28.8 Weight and Cost

The engine cooling system contributes to the vehicle weight and cost, which impacts the vehicle driving dynamics and ride comfort. It is one of the design targets to make sure the cooling system is sized properly and meets the system weight and cost requirement.

13.29 Methodology

13.29.1 General Modeling Principle

The modeling principle is based on lumped parameter approaches, i.e. a description of the cooling circuit using flow resistances (e.g. orifices) and capacitive elements (e.g. volumes). The cooling circuit behavior in terms of flow rate, pressure, temperature, representative volumes, and restrictions has to be implemented in the model. For instance, the capacitive effect can be calculated in a general chamber component. The temperature in the volume is calculated by integrating the following expression:

$$\begin{aligned} & m \cdot \text{Medium.specificHeatCapacity Cp(state)} \frac{\mathrm{d}T_{\text{balance}}}{\mathrm{d}t} \\ & = \text{port}_a m_{\text{flow}} \text{actualStream}(\text{port}_a h_{\text{outflow}}) \\ & \quad + \text{port}_b m_{\text{flow}} \text{actualStream}(\text{port}_b h_{\text{outflow}}) \end{aligned} \tag{13.17}$$

The pressure inside the volume is calculated by integrating the following expression:

$$\begin{aligned} & \frac{\text{port}_a \cdot m_{\text{flow}} + \text{port}_b \cdot m_{\text{flow}}}{\max(1, \rho)} = \frac{\mathrm{d}p_{\text{balance}}}{\mathrm{d}t} C \\ & \frac{\mathrm{d}m}{\mathrm{d}t} = \text{port}_a \cdot m_{\text{flow}} + \text{port}_b \cdot m_{\text{flow}} \\ & m = p \cdot \text{volume} \end{aligned} \tag{13.18}$$

The thermal–hydraulic restriction elements from *Claytex.The fluid* library represents the resistance in thermal–hydraulic circuits. For instance, the pressure drop can be calculated in a general restriction component. The mass flow rate [kg/s] and the enthalpy flow rate [W] are the outputs of this model. The mass flow rate is calculated as follows:

$$\begin{aligned} & \text{Re} = \frac{\text{diameter} \cdot |m_{\text{flow}}|}{\text{crossArea} \cdot \mu} \\ & \lambda_2 = \begin{cases} 64\ \text{Re} & \text{if Re} \le \text{Re1} \\ \left(\begin{cases} 0.25 \left(\dfrac{\text{Re}}{\text{Math log10}\left(\frac{\Delta}{3.7} + \frac{5.74}{\text{Re}^{0.9}} \right)} \right) & \text{if Re} \ge \text{Re2} \\ \text{interPolateInRegion2(Re, Re1, Re2, } \Delta) & \text{else} \end{cases} \right) & \text{else} \end{cases} \\ & \mathrm{d}p = \frac{\text{length} \cdot \mu \cdot \mu}{2p \cdot \text{diameter} \cdot \text{diameter} \cdot \text{diameter}} \left(\begin{cases} \lambda_2 & \text{if } m_{\text{flow}} \ge 0 \\ -\lambda_2 & \text{else} \end{cases} \right) \end{aligned} \tag{13.19}$$

The heat exchange between a metal mass and fluid is calculated with internal flow thermal convection. The basic constitutive equation for convection is:

$$Q_{flow} = h_f \times \left(T_{solid} - T_{fluid} \right) \tag{13.20}$$

13.29.2 Important Components

In this session, we will describe the detailed modeling methodology for important components in engine cooling systems.

13.29.3 Heat Exchangers

Typical heat exchangers in the engine cooling system include Low-temperature heat exchangers (LT-HX), high-temperature heat exchangers (HT-HX), charged air cooler (CAC), heater core, and condenser. They are devices to transfer heat from the cooling medium to the atmosphere. HT-HX, LT-HX, and heater core are liquid-to-air heat exchangers; CAC is air-to-air heat exchanger; condenser involves in the condensation of high-pressure refrigerant vapor and rejects condensation heat to the atmosphere.

They can be modeled by a semi-empirical approach or a purely empirical approach. In the semi-empirical approach, certain geometry parameters are specified, such as the number of pipes, fines, passes, and the geometry of each. An empirical or theoretical-based correlation model will be applied and estimate the heat rejection performance and the pressure drops. In the empirical approach, both the heat rejection performance and the pressure drops are provided directly from the experimental measurements. They are usually provided as functions of operating conditions, such as airflow rate, cooling medium flow rate, and the temperature difference between the two. The performance measured from the experiment will be used during the simulation.

Figure 13.66 shows the *Engine-cylinder block.Cooling.Components.HX* model from ClaytexFluid® library in Dymola®. In the generic heat exchanger model, the heat exchange is calculated from an equivalent Nusselt number Nu defined in Eq. 13.21:

$$Nu = \frac{h_f k_f}{L}, \tag{13.21}$$

where h_f is the heat transfer coefficient k_f is the thermal conductivity, and L is the characteristic length. Nusselt number, Nu, is estimated through Dittus-Boelter correlation, showing in Eq. 13.22:

$$Nu = ARe^B Pr^C, \tag{13.22}$$

where Re is Reynolds number and Pr is Prandtl number. A, B, and C are the tuner parameter obtained from testing, simulation, and correlation.

The friction pressure loss dp of the fluid is calculated by quadratic interpolation around nominal conditions in Eq. 13.23:

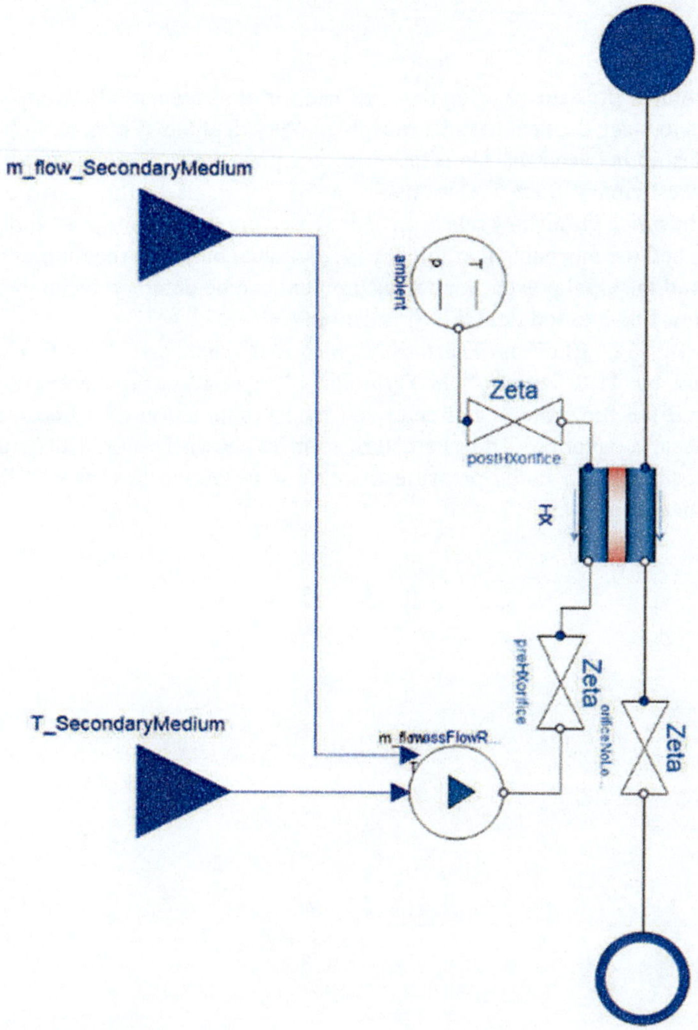

Fig. 13.66 Engines.CylinderBlocks.Cooling.Components.HX model

$$dp = dp_{nominal}\left(\frac{\rho_{nominal}}{\rho}\right)\left(\frac{\dot{m}}{\dot{m}_{nominal}}\right)^2, \tag{13.23}$$

where ρ and $\dot{m}$ are the fluid density and mass flow rate, and $dp_{nominal}$, $\rho_{nominal}$, and $\dot{m}_{nominal}$ are the fluid nominal pressure loss, nominal density, and nominal mass flow rate respectively.

13.29.4 Fan

The fan creates pressure rise in the air. The air is driven to flow through heat exchangers to bring the heat to the atmosphere. A typical fan system model captures the relationship between the fan rotation speed, pressure difference, flow rate, and mechanical efficiency. Such a relationship can be characterized through an empirical approach, in which the airflow rate is measured as a function of fan speed and pressure difference, and the mechanical efficiency is calculated based on the momentum gain of the air and the axial power. Such measurement can be obtained by an experiment text bench or full detailed fan CFD simulation.

Shown in Fig. 13.67 is *ThermalSystems.GasComponents.Fans.Fan2ndOrder* model from by TLK-Thermo® in Dymola®. The relationship between pressure increase and the flow rate is characterized by a combination of a linear equation and a quadratic equation. The characterization is shown in Fig. 13.68, in which $dp_{nominal}$ and $V_{nominal}$ is the pressure increase and volume flow rate at the given nominal fan speed.

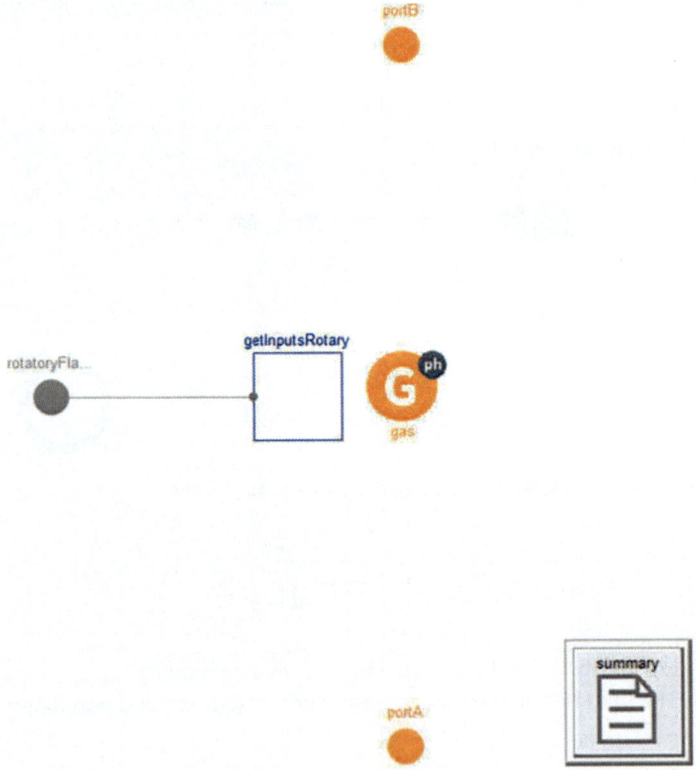

Fig. 13.67 ThermalSystems.GasComponents.Fans.Fan2ndOrder model

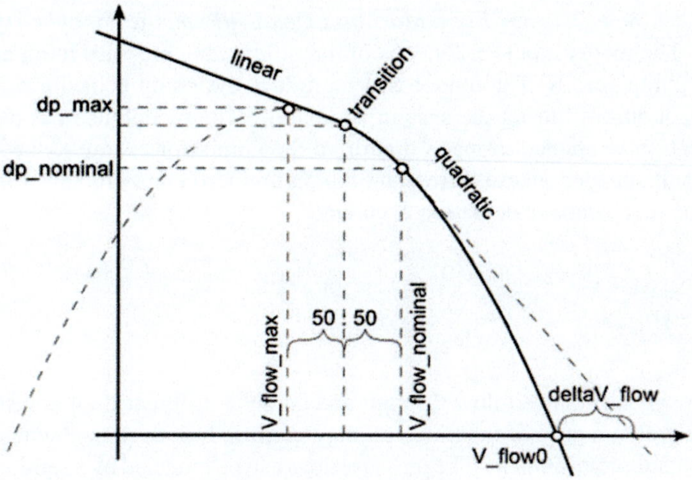

Fig. 13.68 Combination of linear and quadratic characterization of fan model

13.29.5 Pump

A pump is a power device that drives the liquid flow. Similar to the fan, an empirical modeling approach is usually taken. The volume flow rate is measured as a function of pump speed and pressure difference. The performance then is used, directly or with correlation, in the system model simulation.

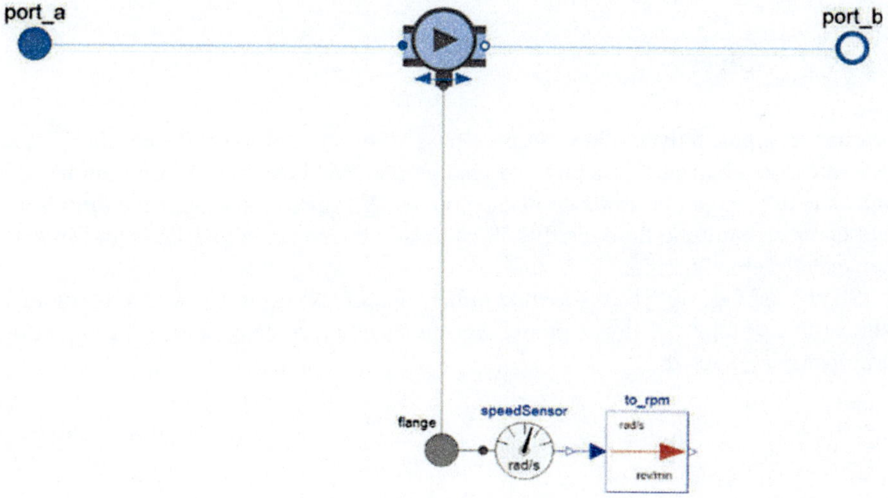

Fig. 13.69 Claytex.Fluid.Machines.DetailedPump model

Figure 13.69 is *Claytex.Fluid.Machines.DetailedPump* from by Claytex® in Dymola®. The model inputs are values of mass flow rate, pressure ratio, efficiency at several pump speeds. The model will reproduce the pump behavior at different operating conditions during the system simulation. The model also has a mechanical connection to connect to parts that drive the pump, for example, the motor or the driveshaft. In such a case, providing the pump inertia and efficiency input will improve the mechanical calculation accuracy.

13.29.6 Pipe

Pipes allow air or liquid to flow through and connect different fluid devices. Pipes may cause pressure drop of the fluid and may perform heat exchange with the inner fluid and outside environment. The pressure drop can be modeled by a semi-empirical approach, in which we specify the pipe geometry and pressure drop model, and then can correlate the pressure drop by modifying the modeling parameters. The heat exchange behavior can also be modeled similarly. We first specify the geometry and the material and then correlate the model with experimental or CFD results to improve the accuracy.

Figure 13.70 is *ClaytexFluid.Pipes.Pipe* model from ClaytexFluid® library in Dymola®. The pressure drop. The heat transfer of the pipe can be modeled as an ideal heat transfer model such that the wall temperature is equal to the internal fluid temperature. The heat transfer can also be modeled by the Dittus-Boelter correlation in Eq. 13.22.

13.29.7 Valves

Similar to pipes, valves allow fluid to pass through. However, it can change its flow-pressure characteristics and can change the flow behavior. A common use of valves is to change the flow rate of the coolant. Valves can also split the fluid flow and connects multiple fluid circuits. It can also change the connections and switch between different circuits.

Shown in Fig. 13.71 is *ClaytexFluid.Valves.LinearValve* from ClaytexFluid® library in Dymola®. A linear model can be specified to characterize the pressure drop behavior, such that:

$$dp = dp_{nominal} \frac{\dot{m}}{\dot{m}_{nominal}}, \tag{13.24}$$

where dp is the pressure drop, $\dot{m}$ is the mass flow rate, and $dp_{nominal}$ $\dot{m}_{nominal}$ are the nominal pressure drop and nominal flow rate.

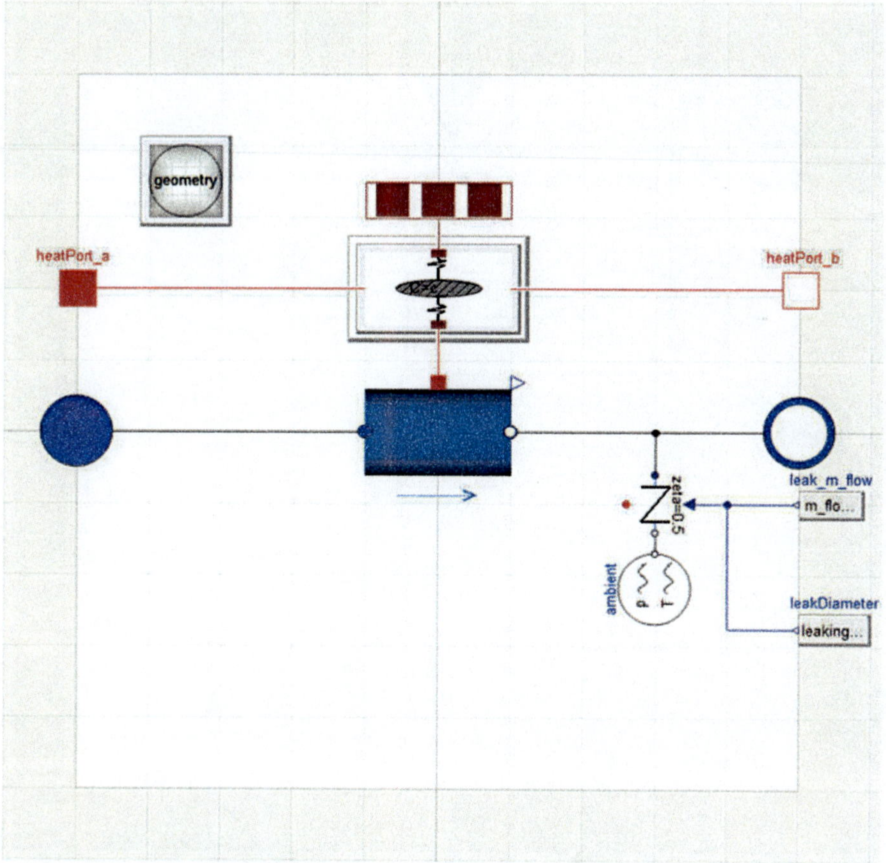

Fig. 13.70 ClaytexFluid.Pipes.Pipe model

13.30 Design Example

A typical design process includes system architecture trade-off, component selection, and sizing, as well as system integration and calibration. An example model is shown to demonstrate the design process.

13.30.1 Example Models

In this example, we first exercise a model of a vehicle equipped with a 110 kW diesel mapped based engine model using an optimal control law to follow the trajectory of a "world harmonized light-duty vehicles test procedure" (WLTP) speed cycle. The system model of the vehicle architecture is shown in Fig. 13.72.

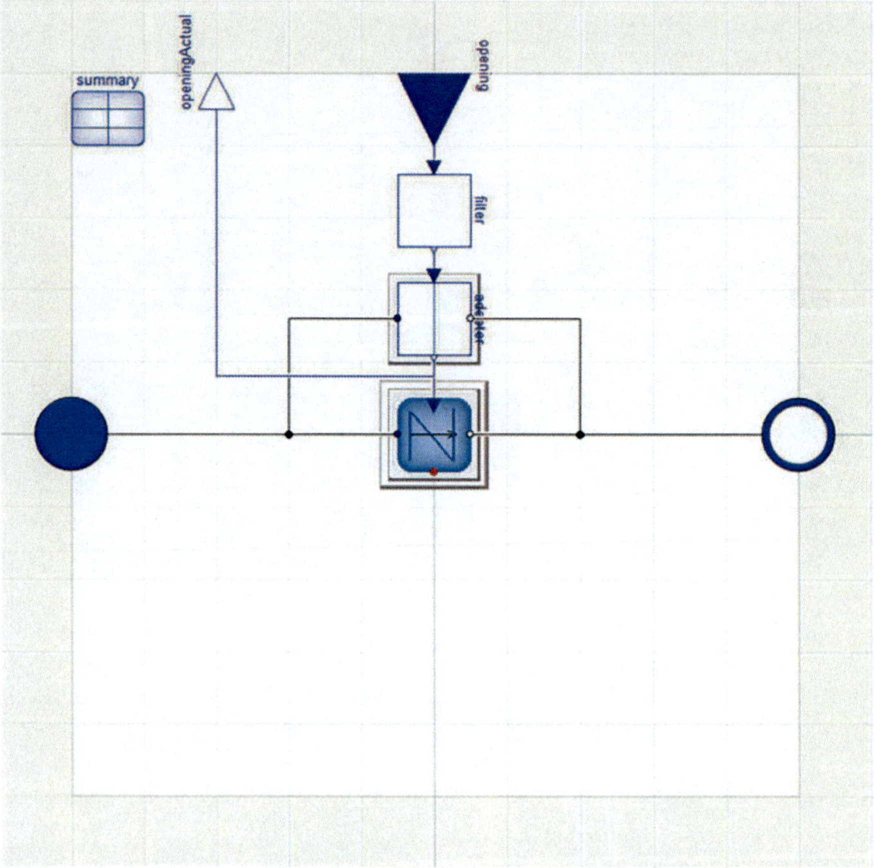

Fig. 13.71 ClaytexFluid.Valves.LinearValve model

The map-based engine model (shown in Fig. 13.73) can provide the necessary engine power (in terms of torque and speed) based on a lookup table that computes the output engine torque in dependency of needed engine speed and load. The fuel consumption of the vehicle is deduced from a Brake Specific Fuel Consumption map.

The power at the Flywheel is taken as the total generated power from the engine.

Using the map-based engine model, the WLTP speed cycle is simulated to generate a realistic lookup table of the engine speed used to drive the water pump within the engine cooling system as well as extracting a realistic heat release from the engine. The results of the engine model simulation with a focus on engine speed (signal named "*speed. w*") used to drive pump and generated heat release (signal named "*heat loss*") are illustrated in Fig. 13.74.

In a second model, an example design of a cooling system is implemented which is illustrated in Fig. 13.75. The model consists of all the components needed to model the main parts of the cooling system like oil cooling loop and water hot cooling

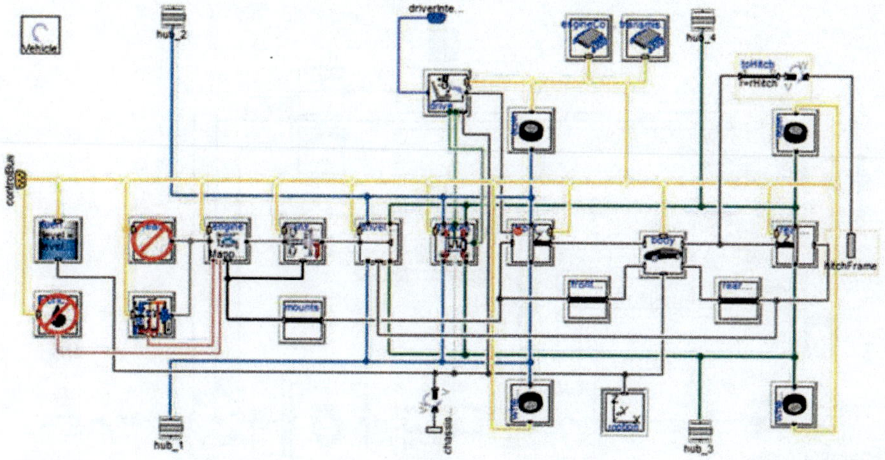

Fig. 13.72 Vehicle architecture

loop, including pumps, oil cooler heat exchanger, Air–Water heat exchanger, and a thermostat to by-pass the Air–Water heat exchanger at low coolant temperatures to decrease engine warm-up phase and increase engine efficiency. See Fig. 13.76 highlighting the main parts of the cooling system.

The fluids used in the cooling system model are:

- 50–50 glycol mix for the water loop
- Engine oil lubricant (0W30)
- Dry air as a first assumption

The boundary conditions of this model are:

- Dry Air mass flow rate computed from vehicle front area and Air density based on the vehicle speed from the WLTP vehicle cycle
- Water pump speed computed by the vehicle model at the first phase (see Fig. 13.9)
- Oil pump speed
  - Engine heat release at the water jacket and oil cooling side

From a thermal point of view, the engine model within the cooling system design experiment is composed of a heat release connected to thermal inertia, reflecting the thermal mass of the engine, and a specific convection model to release heat into the oil or the water. The total engine power needed to drive the WLTP cycle is split into 3 main consumers which are 1/3rd the wheel power, 1/3rd the exhaust, and 1/3rd the heat releases to oil and water. There is a second split of the heat release into oil and water at the ratio of 1/10th for oil and 9/10th for water. The main heat exchanger and the oil cooler are parameterized with optimal exchanges.

During running two consecutive WLTP cycles, the behavior of the previously described cooling system model is examined with particular focus on the behavior

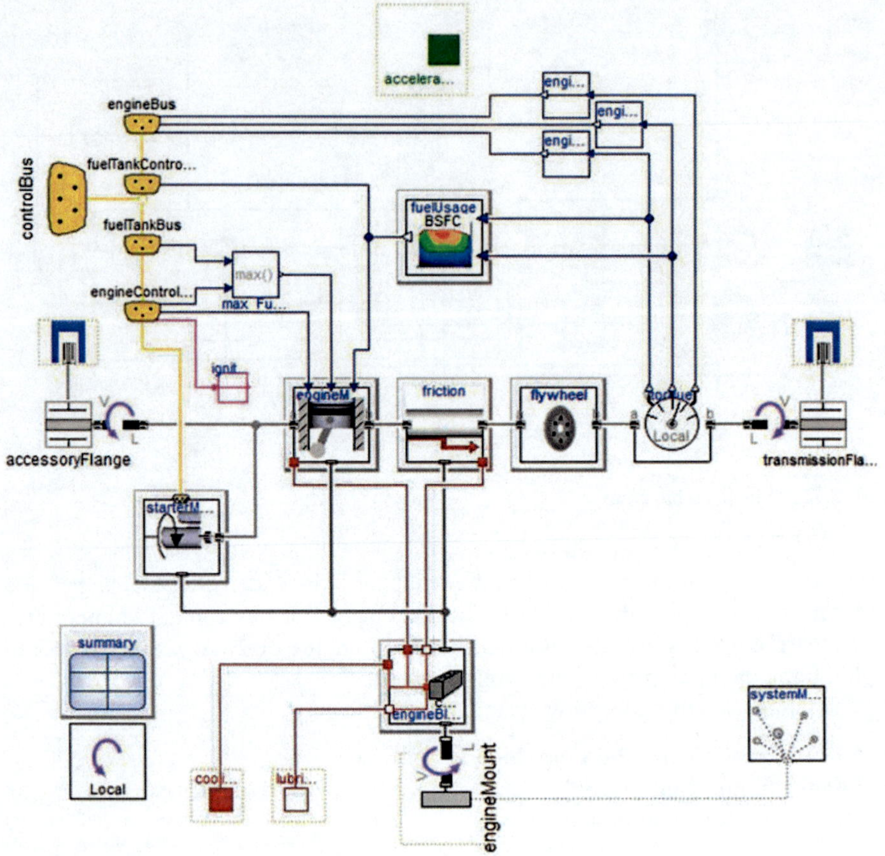

Fig. 13.73 Map-based engine model

of the heat exchanger by-pass control. When the heat exchanger is bypassed the thermostat output is directly connected to the water pump. The implemented cooling system model enables visualization of temperature gradients appearing during simulation using a color scale. An increase of the coolant temperature is visualized by changing its color from blue to red in the respective color shades. Figure 13.77 shows a snapshot of the WLTP simulation at a point in time when the main heat exchanger is bypassed.

We can observe also part of the WTLP consecutive scenario where the main heat exchanger cools down the overall cooling loop and is not bypassed (Fig. 13.78).

A more detailed look at the evolution of the coolant temperature during the two consecutive WLTP cycles is given in Fig. 13.79:

We see the influence of thermostat sizing and control as well as fan control over the two consecutive WTLP cycles on the coolant temperature:

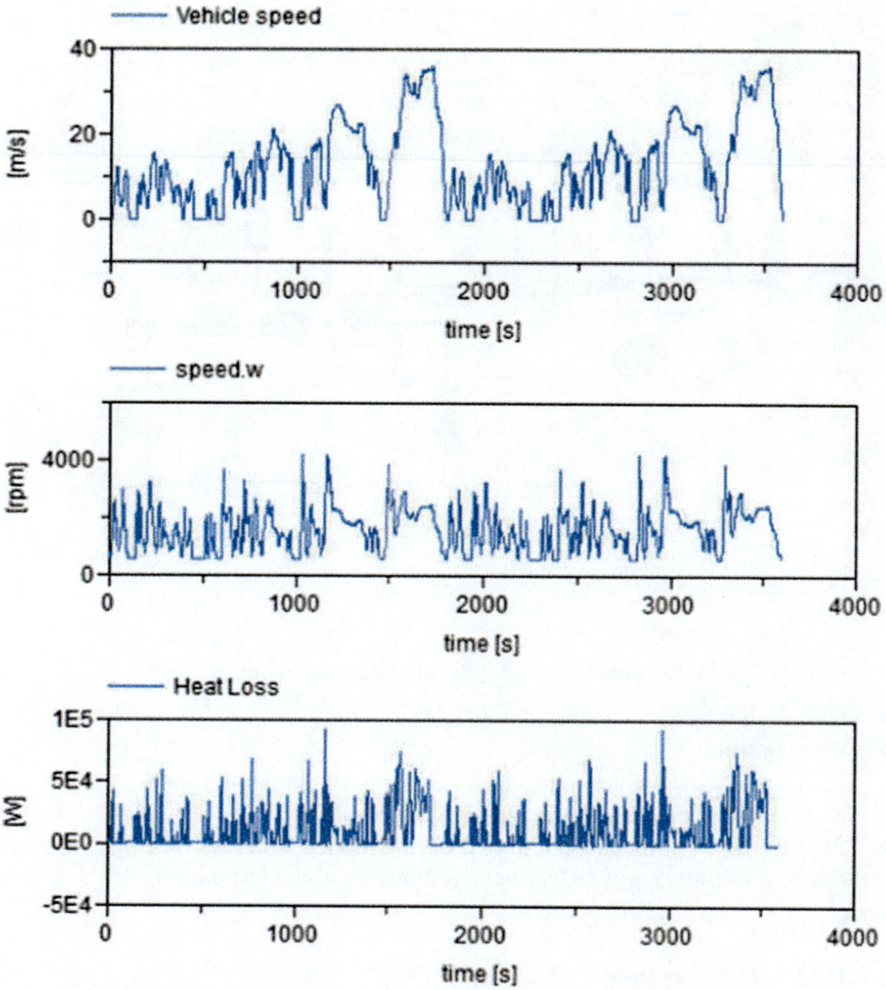

Fig. 13.74 WLTP vehicle speed, water pump speed, and engine heat release

- At the start-up, the thermostat is by-passing the main heat exchanger to reduce the warm-up time (200 s on the 1st WTLP cycle)
- The fan allows reaching optimal heat exchange when additional airflow on the heat exchanger is needed (especially at low speed)
- The heat exchanger is playing his role to cool down the coolant at high speed though it seems to be a little undersized because we slightly reach a temperature over 100 °C at the end of the WTLP cycles

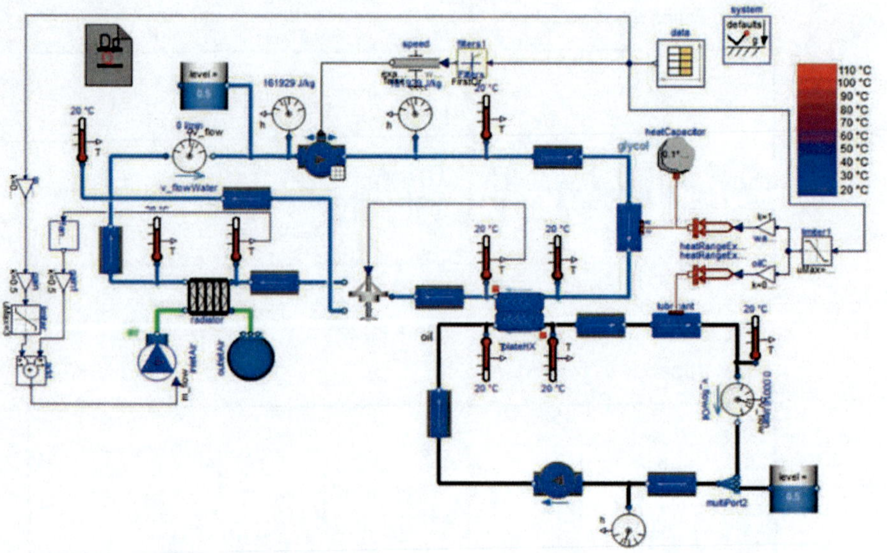

Fig. 13.75 Engine cooling system

With this predictive model, the next design step would be:

- to adjust the sizing of the thermostat
- to fan control
- The main heat exchanger sizing
- Exercise the model on a more intensive duty cycle
- Plug this model into a mapped-based engine model as well as an HVAC system to deduce overall energy balance and pollutant emission of the vehicle

13.30.2 Architecture Design and Trade-Off

In this step, multiple cooling system architectures are modeled with preliminary systems specifications. We then verify the system performance against design requirements. After that, we filtered out architectures that meet the design requirements and then study the trade-off between them. After this step, one or multiple system architectures will remain in the design process and can move on to the next step.

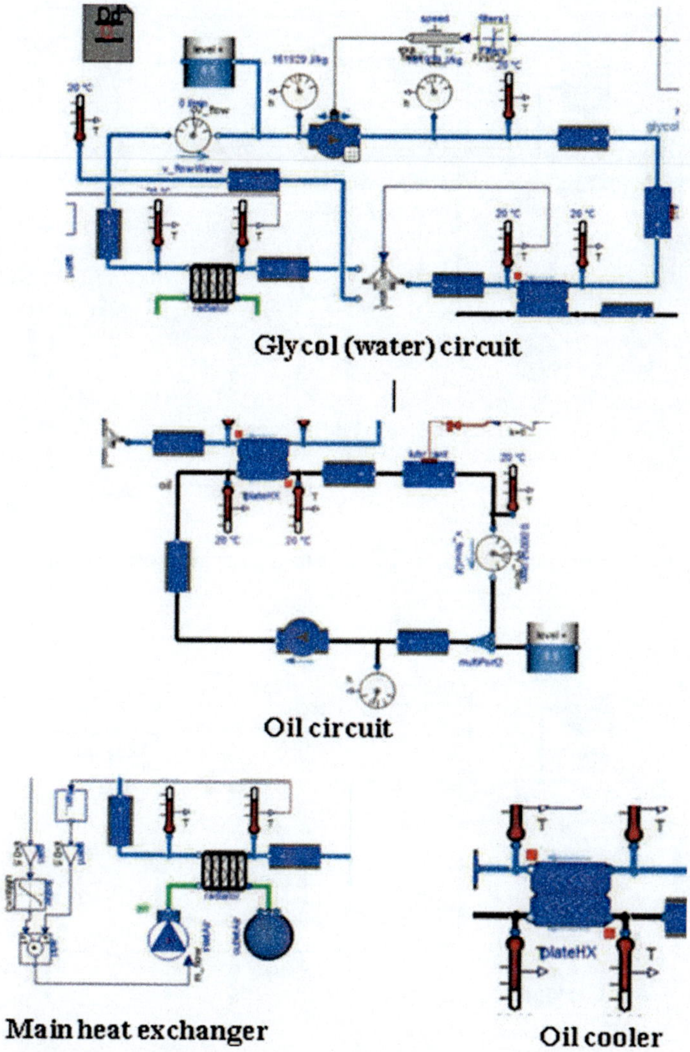

Fig. 13.76 Individual components

13.30.3 Component Design and Trade-Off

In this step, the goal is to determine the detailed component design and the trade-off between different design targets. We will conduct part selection and sizing, heat exchanger stacking, and system integration and calibration.

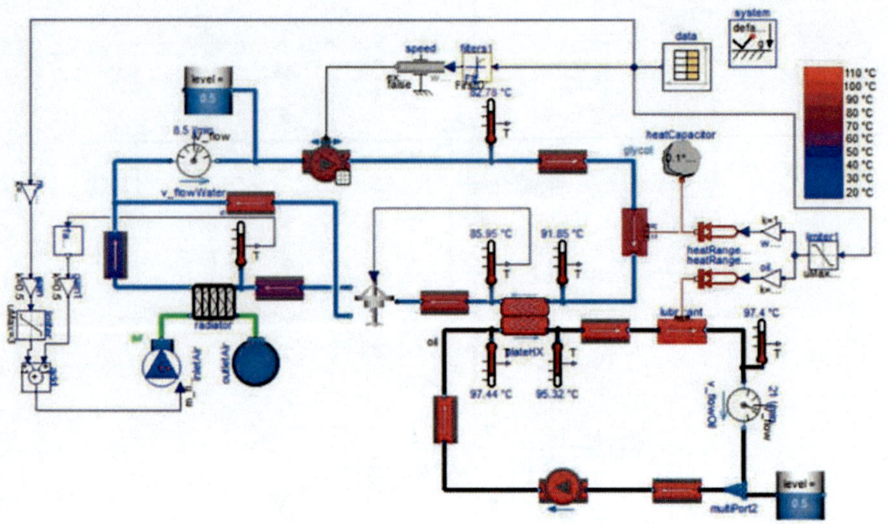

Fig. 13.77 Results of an example scenario where the thermostat bypasses the main heat exchanger

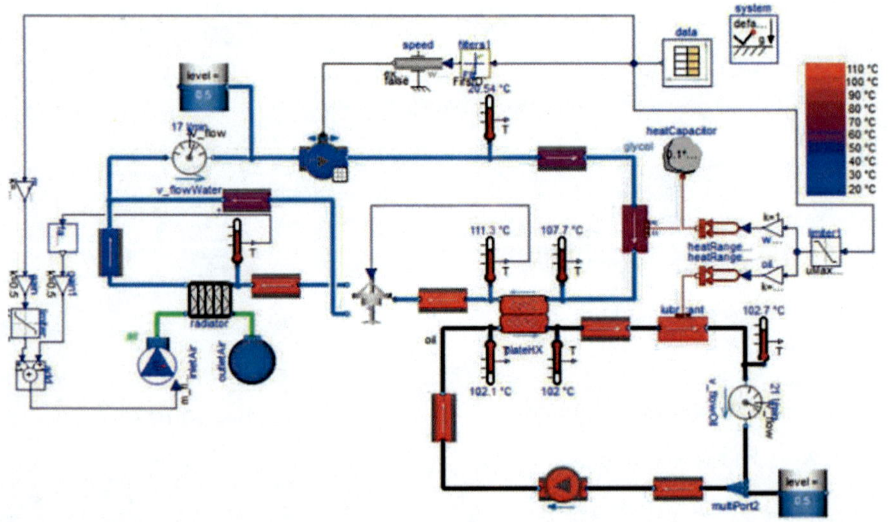

Fig. 13.78 Results of an example scenario where the thermostat is not bypassing the main heat exchanger

13.30.4 Part Selection and Sizing

The cooling package consists of multiple parts to perform its function properly. Each part has a different design specification and performance. With cooling system

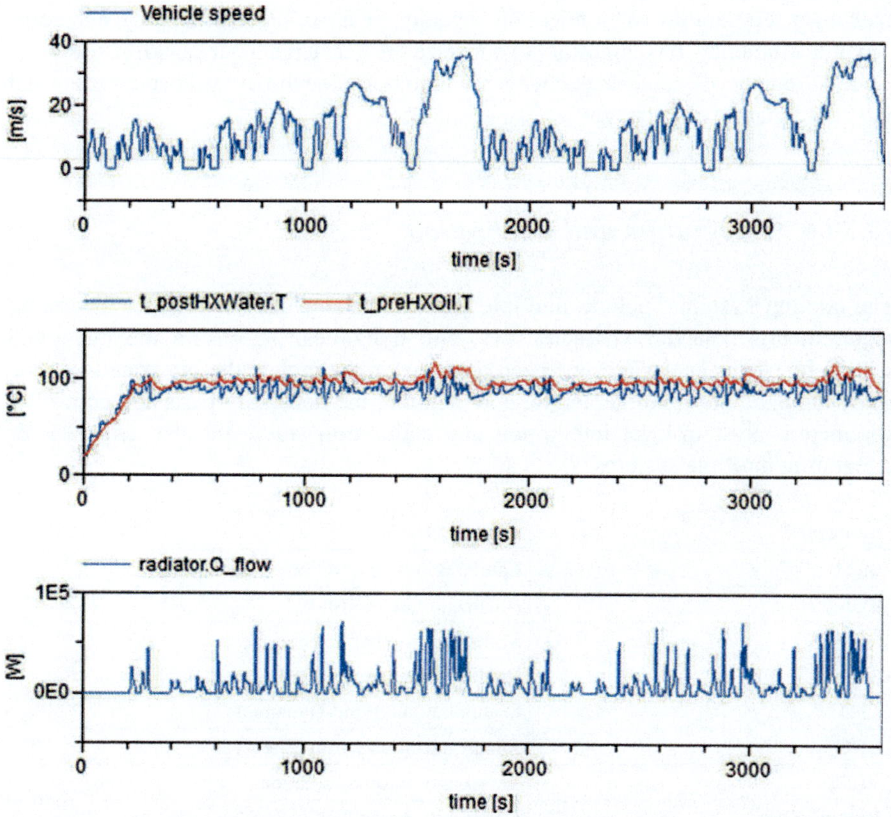

Fig. 13.79 Evolution of WLTP vehicle speed, coolant temperature, and heat exchanger heat flow

simulation, we can estimate the system performance based on different part selection and sizing. We will also verify the design requirement and the trade-off between different design objectives based on the combination of different part selection and sizing.

13.30.5 Heat Exchanger Stacking

The front-end cooling package is an important part of the cooling system. The package includes one or multiple fans. It also includes one or multiple heat exchangers, for example, the high-temperature heat exchanger (radiator), low-temperature heat exchangers, condenser, charged air cooler, and engine oil cooler. The sorting order, locations, and gaps between the components will result in the cooling airflow passing the heat exchangers in different temperatures, pressure, and flow rate. They will significantly influence the individual heat exchanger's heat exchanger temperature, heat rejection capacity, heat exchange efficiency, and aerodynamic efficiency. It is important to optimize the heat exchanger stacking, to satisfy the requirement of each heat exchanger, and to optimize the overall package thermal and aerodynamic efficiency.

13.30.6 Integration and Calibration

The cooling systems include multiple parts and controllers, with their operating requirements, behavior characteristics, and dynamics. To ensure the integrated system functions as desired, we need to model and simulate the integrated system, study the behavior and interaction, and calibrate the necessary parts and controller parameters. System-level integration and calibration study are also required for integrating multiple systems.

Acronym	Expansion
LBM	Lattice Boltzmann Method
BGK	Bhatnagar Gross Krook
f	Molecular number density
Ra	Rayleigh number
CFD	Computational Fluid Dynamics
RNG	Renormalization Group Theory
VLES	Very Large Eddy Simulation
PowerFLOW	General Purpose CFD Solver

Acknowledgements and Warranty The authors of the chapter thank Dassault Systèmes Simulia Corporation (DS), 1301 Atwood Avenue Suite 101W, Johnston, RI 02919, USA for granting to the Editor a nonexclusive, worldwide, copyright license to reproduce, publish and distribute, prepare derivative works, and display in the DS contribution and to sublicense these rights to the Publisher, reserving all rights not granted to the Editor, as per the agreement between the Editor of the book and the company DS dated 12-Feb-2021 and 13-Feb-2021. The corporation provides its contribution under this agreement "as-is" and neither makes nor grants any other warranties, express or implied. Also, the agreement excludes all implied warranties including specifically any implied warranty of merchantability or of fitness for the particular purpose or any obligation of results concerning its contribution whether transmitted in writing or orally.

References

1. Morley C. https://www.jabil.com/blog/automotive-industry-trends-point-to-shorter-product-development-cycles.html, Blogpost
2. Taylor T. https://www.motorbiscuit.com/already-over-13-million-cars-have-been-recalled-in-first-half-of-2020/, Blogpost
3. Jibrell A. https://www.autonews.com/article/20180130/RETAIL05/180139974/auto-recall-bill-grew-26-to-22-billion-in-2016-study-says, Automotive News, 2018
4. Bhatnagar PL, Gross EP, Krook M (1954) A model for collision processes in gases. I. Small amplitude processes in charged and neutral one-component system. Phys Rev 94:511–525
5. Chen H, Chen S, Matthaeus WH (1992) Recovery of the Navier-Stokes equations using a lattice-gas Boltzmann method. Phys Rev A 45(8):R5339
6. Chen H, Teixeira C, Molvig K (1998) Realization of fluid boundary conditions via discrete Boltzmann dynamics. Int J Mod Phys C 9(08):1281–1292
7. Chen S, Doolen G (1997) Lattice Boltzmann method for fluid flows. Ann Rev Fluid Mech 30:329–364
8. Chen H, Kandasamy S, Orszag S, Shock R, Succi S, Yakhot V (2003) Extended Boltzmann kinetic equation for turbulent flows. Science 301(5633):633–636
9. Zhang R, Fan H, Chen H (2011) A lattice Boltzmann approach for solving scalar transport equations. Philos Trans R Soc A Math Phys Eng Sci 369(1944):2264–2273
10. Chen H (1998) Volumetric formulation of the lattice Boltzmann method for fluid dynamics: Basic concept. Phys Rev E 58:3955–3963
11. Gullberg P, Sengupta R (2011) Axial fan performance predictions in CFD, comparison of MRF and sliding mesh with experiments. SAE Technical Paper 2011-01-0652
12. Nourgaliev RR, Dinh T-N, Theofanous TG, Joseph D (2003) The lattice Boltzmann equation method: theoretical interpretation, numerics, and implications. Int J Multiphase Flow 29(1):117–169
13. } Tsuji T, Nagano Y (1988) Characteristics of a turbulent natural convection boundary layer along with a vertical flat plate. Int J Heat Mass Transf 31(8):1723–1734
14. Leong WH, Hollands KGT, Brunger AP (1999) Experimental Nusselt numbers for a cubical-cavity benchmark problem in natural convection. Int J Heat Mass Transf 42(11):1979–1989
15. Argyropoulos CD, Markatos NC (2015) Recent advances on the numerical modelling of turbulent flows. Appl Math Model 39(2):693–732
16. Prandtl L (1925) Report on the investigation of developed turbulence. ZAMM 5:136–139
17. Spalart P, Allmaras S (1992) A one-equation turbulence model for aerodynamic flows. In: 30th aerospace sciences meeting and exhibit, p 439
18. Launder BE, Sharma BI (1974) Application of the energy dissipation model of turbulence to the calculation of flow near a spinning disc. Lett Heat Mass Transfer 1(2):131–137
19. Yakhot VS, Orszag SA, Thangam S, Gatski TB, Speziale CG (1992) Development of turbulence models for shear flows by a double expansion technique. Phys Fluids A 4(7):1510–1520
20. Mukutmoni D, Han J, Alejbegovic A, Colibert L, Helene M (2010) Numerical simulation of transient thermal convection of heated plate (No. 2010-01-0550). SAE Technical Paper
21. Chandrasekhar S (1960) Radiative transfer. Dover Publication
22. Fiveland WA (1984) Discrete-ordinates solutions of the radiative transport equation for rectangular enclosures, pp 699–706
23. Howell JR (1998) The Monte Carlo method in radiative heat transfer, pp 547–560
24. Eymard R, Gallouët T, Herbin R (2000) Finite volume methods. Handb Numer Anal 7:713–1018
25. Jaeger JC, Carslaw HS (1959) Conduction of heat in solids. Clarendon Press
26. Wang Z, Han J, Mukutmoni D (2014) Numerical simulation of unsteady natural convection in a simplified engine bay enclosure under soak conditions (No. 2014-01-0651). SAE Technical Paper
27. Mukutmoni D, Powell R, Mutnuri LR (2014) Investigation of thermo-acoustic excitations in a Rijke tube geometry. SAE Int J Passenger Cars-Mech Syst 7(2014-01-1981):175–1182

28. Rijke PL (1859) On the vibration of the air in a tube open at both ends. Phil Mag 17:419–422
29. Mukutmoni D, Alajbegovic A, Han J (2011) Numerical simulation of transient thermal convection of a full vehicle (No. 2011-01-0645). SAE Technical Paper
30. Jansen W, Amodeo J, Tate E, Yang Z (2014) Drive cycle simulation of a tiered cooling pack using non-uniform boundary conditions. SAE Technical Paper 2014-01-0654
31. Yang Z, Tate E, Day E (2018) Prediction of charge air cooler performance in a racing drive cycle by 1D-3D coupling. SAE Technical Paper 2018-01-0781
32. Jin R, Chen W, Sudjianto A (2005) An efficient algorithm for constructing optimal design of computer experiments. J Stat Plan Inference 134:268–287
33. Cressie NAC (1993) Statistics for spatial data, Revised Edition. John Wiley & Sons, New York
34. Dilgen SB, Dilgen CB, Fuhrman DR et al (2018) Density-based topology optimization of turbulent flow heat transfer systems. Struct Multi-disc Optim 57:1905–1918
35. Kreissl S, Pingen G, Maute K (2011) Topology optimization for unsteady flow. Int J Numer Meth Eng 87:1229–1253
36. Cho Y, Chang C, Shestopalov A, Tate E (2017) Optimization of active grille shutters operation for improved fuel economy. SAE Int J Passeng Cars Mech Syst 10(2):2017

Part VI
Fuels, Oils and Equipment for Managing at Ultra-Low Temperatures

Chapter 14
Low-Temperature Operation: Fuels and Lubricants for Cold Temperature Regions

A. S. Ramadhas and Hongming Xu

14.1 Introduction

Fuel economy improvement and emission reduction are the two main concerns for the automotive industries and oil companies that drive the advancement in engine technologies and fuel quality. The vehicular emission regulations become stricter and currently at Euro 6 or equivalent emission norms and a further reduction in emissions norm i.e. Euro 7 norms is proposed for the year 2025 onwards. The engine technologies such as gasoline direct injection, common rail direct injection, and after treatment devices namely, diesel oxidation catalyst, selective catalyst reduction, and diesel particulate filter are under development. The vehicles are well-calibrated to meet the legislative emission test cycle requirement, i.e. 25 and − 7 °C depending on the climatic regions. However, vehicles may not perform well at very cold temperatures as in normal temperature conditions. Vehicle manufacturers calibrate the vehicles to work better in the very cold ambient temperatures and also using the coolant & and oil heaters to support in starting of the engine. Though the hardware modification and calibration are upgraded to overcome the bad effects of cold temperature issues the properties of fuel and lubricants play a significant role in the starting and driving performance of the vehicles. This chapter covers the cold start phenomena, factors affecting cold startability, fuel, and engine oils requirement for operating vehicles in cold weather conditions.

A. S. Ramadhas (✉)
R&D Centre, Indian Oil Corporation, Faridabad, India

H. Xu
University of Birmingham, Birmingham, England

P. A. Lakshminarayanan and A. K. Agarwal (eds.), *Handbook of Thermal Management of Engines*, Energy, Environment, and Sustainability,
https://doi.org/10.1007/978-981-16-8570-5_14

14.2 Cold Start Behaviour of Internal Combustion Engines

Cold start can be defined as starting of a vehicle after overnight soaking i.e. after driving a vehicle the vehicle is parked in a garage or open space. The engine oil is settled at the bottom and the temperatures of coolant, oil, and body equalize with the ambient temperature conditions overnight or 6–8 h. Factors that influence the startability performance of the engine are ambient temperature conditions, fuel, engine oil, and battery (Fig. 14.1). Cold start emission is higher at low ambient temperature conditions and also this depends upon the hours of parking. The viscosity of the lubricant is higher at low temperatures compared to normal temperatures, and that increases the friction among the engine components and consequently decreases the efficiency of the pump. Therefore, the net power required to pump oil increases. Increased viscosity of fuel at cold temperatures affects the fuel spray pattern and thereby its combustion. Battery performance deteriorates at lower temperatures and causes issues in the cranking power. The external surfaces of the combustion chamber are exposed to ambient temperatures that increase heat loss that also causes poor burning of fuel.

The combustion behavior of an engine depends on combustion chamber design, fuel properties, fuel injection system and its calibration, and also upon the engine

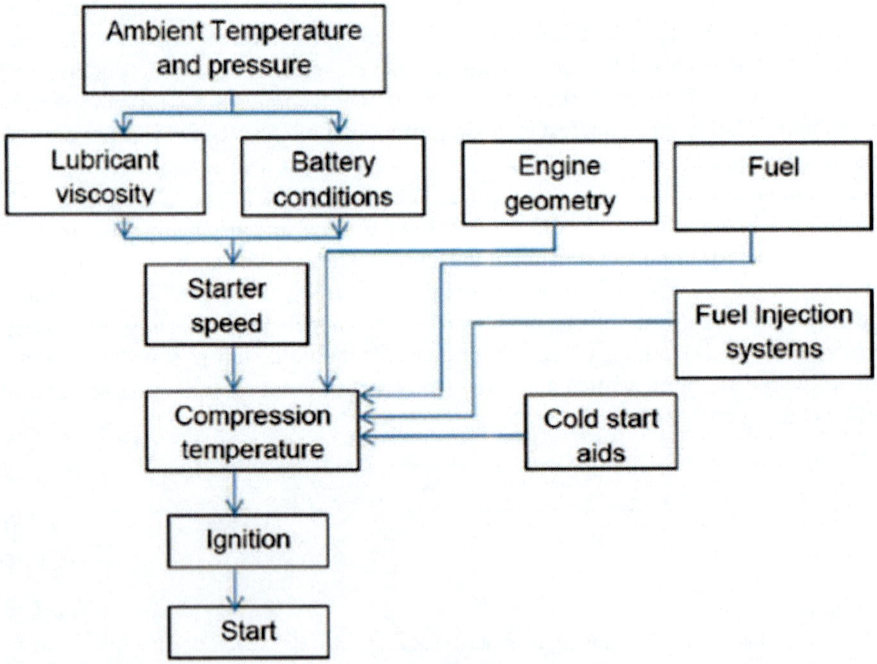

Fig. 14.1 Factors affecting the cold start: compression ratio, fuel quantity, and quality (viscosity, cetane number), coolant temperature, injection timing [1]

operating conditions. In a gasoline engine, the fuel–air mixture is ignited using a spark plug whereas in diesel engines the combustion of the fuel–air mixture is initiated by the temperature increased by compression. The fuel properties primarily depend upon its chemical composition and influence the combustion characteristics of the fuel. Misfiring of fuel due to unsuitable conditions of temperature and pressure in the cylinder at the cold start also causes combustion instability and rough idling of the engine. It is desirable to have a short time to achieve the first firing cycle, avoiding misfiring cycles as much as possible, and reach the stable idle RPM at a shorter period or a minimum number of cycles. Cold start aids such as coolant heaters, oil heaters, and block heaters are commonly used to raise the temperature of the engine at very cold ambient temperatures. See Fig. 14.A1 in Appendix for a typical coolant circuit with a block heater. The engine speed reaches a peak value in the acceleration period and then maintains its idle speed and in-cylinder pressure is also higher for the first few combustion cycles. The effect of ambient temperature on engine RPM is shown in Fig. 14.2. The engine RPM reaches a peak quickly at normal temperatures whereas at subzero temperatures it takes more cycles to initiate the combustion and the cycle-to-cycle variation in engine RPM is also higher [2].

The European Union implemented transient emissions testing at − 7 °C for gasoline passenger cars and the cold temperatures testing may be implemented for diesel passenger cars in the future. The impact of ambient temperature on the cold start performance of diesel engines is analyzed under controlled test cell conditions. This work is a part of the European Commission-funded project investigating the Diesel Engine Cold Start and Transient operation (DECOST). A 6 cylinder CRDI diesel engine (199 kW at 4000 RPM and 600 Nm at 2000 RPM, 3 L capacity) is used to investigate the startability behavior of the engine at different ambient temperatures (− 20, − 7, and + 20 °C) in the cold engine transient dynamometer test cell.

The test engine is soaked at desired test temperature for eight hours in the cold cell and ensured that the air, coolant, oil, and fuel temperatures are maintained before

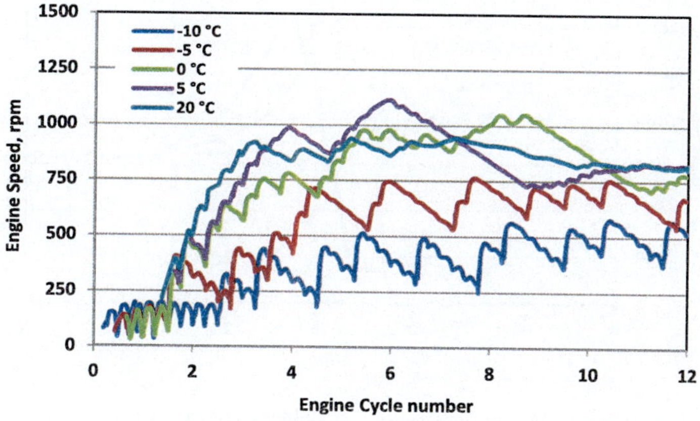

Fig. 14.2 Engine speed during the cold start [2]

the start of the test. After the engine is keyed on and started the engine is run at idle speed for three minutes. The exhaust emissions are measured upstream of the after-treatment devices from key-on. The fuel, oil, coolant, and intake air temperatures are maintained constant throughout the test to eliminate the effect of their variation on cranking speed. The engine is cranked without fuel to remove any residual gases in the engine after completing each test.

The cranking period is the time taken for the start of the combustion of the air–fuel mixture from the start of cranking. The acceleration period in which the engine speed rapidly increases after the fuel starts burning and then the speed is dropped down to maintain its idle speed. The engine RPM during cold start at different ambient temperatures is shown in Fig. 14.3. The ambient temperatures influence the start time and idle speed stability. The engine takes longer periods to get stabilized at idle speed at cold temperatures compared to normal ambient temperatures after some observable initial hesitation. It is also interesting to observe the deceleration after the peak is achieved is dependent on the viscosity; the higher the viscosity the lesser is the deceleration. The coefficient of variation (CoV) of the idle RPM is calculated after 30 s from a cold start and it is observed that CoV is higher at lower temperatures than at normal ambient temperatures (Fig. 14.4).

The engine is calibrated in such a way that the EGR valve operates only after the cold start or the engine is slightly warmed up to avoid condensation of exhaust particles in the EGR valves (Fig. 14.5). This closure of the EGR valve at the cold start does not decrease NOx emissions and it is severe at cold ambient temperatures (Fig. 14.6). Low ambient temperatures increase the frictional mean effective pressure and pumping loads. Hence, a more quantity of fuel is required to be injected during the cranking and acceleration period and this injection quantity decreases while maintaining the idle speed. The engine RPM attains a very high peak during cold start and then maintains a stable idle RPM after many cycles. This causes a longer period

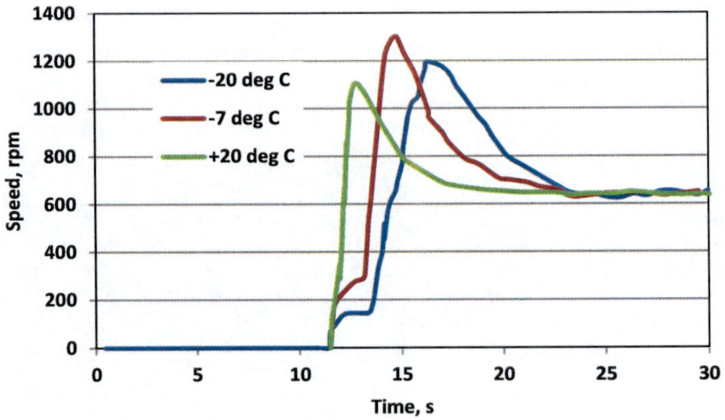

Fig. 14.3 Engine speed at the cold start at different ambient temperatures

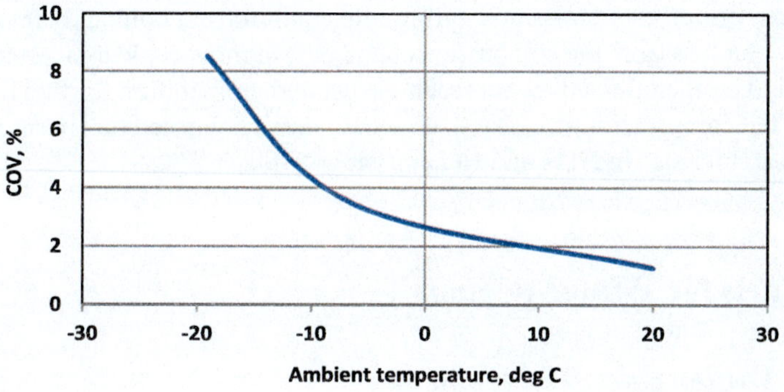

Fig. 14.4 CoV during idle

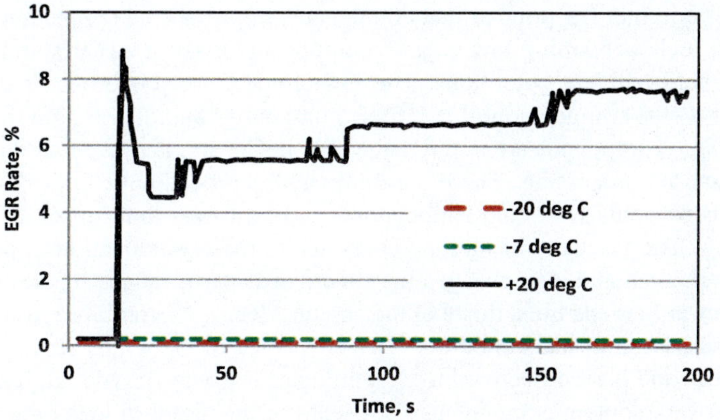

Fig. 14.5 EGR during cold start

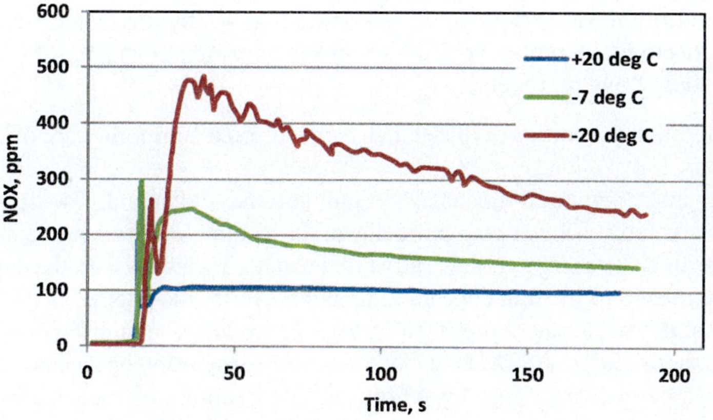

Fig. 14.6 NOx emissions during cold start

and a larger variation in emissions during idle conditions at cold ambient temperatures. The heat loss from the combustion chamber is higher at cold temperatures. The compression temperatures may not reach the desired temperature for fuel burning at the end of the compression stroke and the accumulated unburned fuel burns suddenly and cause an increase in NOx and HC emissions [3].

14.3 Fuels for Winter Regions

14.3.1 Winter Fuel Development

The source of crude oil influences fuel composition and it varies from refinery to refinery. The lighter fractions of fuel components improve cold temperature performance i.e. fuels containing low volatile components make it easier starting of the engine at cold ambient conditions. The low-temperature properties of diesel are dependent on the composition of paraffinic compounds and distribution in the fuel. The paraffinic compounds are in the range of C_{10}–C_{22} and that depends on the type of crude and its processing. These paraffinic hydrocarbons have limited solubility in the fuels particularly at cold temperatures and form wax that causes disturbance in fuel flow. The viscosity of fuel increases due to the crystallization of paraffin at cold temperatures and also changes the rheological performance of the fuel. This leads to power loss and breakdown of the engine. Hence, it is required to control the wax crystallization in real conditions. Diesel specific to the winter region is called winter diesel and is manufactured from refining of less waxy crude oil; it is a blend with lower wax content components (kerosene) in the diesel to lower the endpoint components (to avoid plugging of the filters), and doped with fuel additives. This can be achieved through the following:

1. By controlling refinery process: Iso-dewaxing—Catalytic isomerization of n-paraffins to iso-paraffins, which have lower freezing points.
2. By adding flow improvers.

The important properties of diesel that influence cold startability are discussed in this section.

Density is the mass of a substance per unit volume. In general, low-density fuels show higher volatility, lower pour point, lower flash point, lower viscosity, and higher vapor pressure. The energy content of the fuel usually increases with the density and this can be related to the fuel consumption and power of the engine.

Flashpoint is the temperature at which the fuel produces enough vapor and ignites with the ignition source. ASTM D93 method is used to measure the flashpoint of fuels. Higher flash point fuels are preferred for the safe handling of fuels and maximum limit values are set in fuel specification. The Flashpoint of diesel does not have any significance on engine combustion or auto-ignition performance characteristics.

Table 14.1 Methods to evaluate cloud point

Test method	Methodology	Cloud point range (°C)	Cooling rate (°C/min)	Sample volume (ml)
ASTM D 2500	Manual cooling and detection	+ 49 to – 60	Not specified, ~ 1	40
ASTM D 5771	Optical detection stepped cooling	+ 49 to – 60	Not specified, ~ 1	40
ASTM D5772	Linear cooling	+ 49 to – 60	1 ± 0.2	20
ASTM D5773	Constant cooling	+ 49 to – 60	1.5 ± 0.1	0.15
ASTM D7397	Miniaturized optical	+ 20 to – 60	6 ± 3	20

Cloud point is used to predict the low-temperature operability of diesel in vehicles. This gives estimated low-temperature operability of diesel that does not contain cold flow improvers. Cloud point is defined as the temperature at which the crystallization starts or cloudy/haze appears and it is visible to human eyes. These crystals can accumulate in the fuel filter and stop the fuel supply to the engine. The cold flow improver additives are added in diesel to prevent the agglomeration of crystals and that have a minimal impact on cloud point. Table 14.1 depicts the test methods are available to evaluate the cloud point of diesel.

Fuels with a higher cloud point are more likely to have problems during cold weather. The cloud point and cold filter plugging point (CFPP) results are very similar for the neat or unadditised diesel. But the cloud point for additised fuel is approximately 10 °C above the CFPP level. Moreover, rapid cooling of diesel results in the formation of smaller crystals whereas in the actual scenario the slow cooling process gives rise to larger wax crystals that restricts the flow. The cloud point is not a suitable assessor for the biodiesel blended fuel in which the crystallization forms much earlier and resulting in unexpected fuel system problems. The components in the biodiesel vary from one source to another source affecting CFPP.

Pour point is the lowest temperature at which fuel is in a liquid state or stops flow as per the pour point test method ASTM D97. It is expressed as a multiple of 3 °C. Pour point depressants reduce the freezing point and thereby improve the pumpability of fuels.

Cold Filter Plugging Point is the minimum temperature at which diesel flows through the filter that simulates the operation of diesel under cold weather conditions. The paraffin in the diesel forms wax crystals that block the fuel line and filter. This can be seen by its visual appearance—milky color and presence of reddish deposit on the bottom of the tanker surface of the fuel filter. The cold flow operability of fuel can be extended by using wax dispersants that change the diesel wax structure and also avoid agglomeration in the fuels. This cold flow improver supports the operation of diesel at lower temperatures than that of unadditised diesel. However, the cloud point is not reduced by the cold flow improver. Commonly used flow improvers are copolymers based on ethylene–vinyl acetate monomers (C18H30O6X2) [4]. When the fuel temperature decreases below the cloud point the square-like wax structures

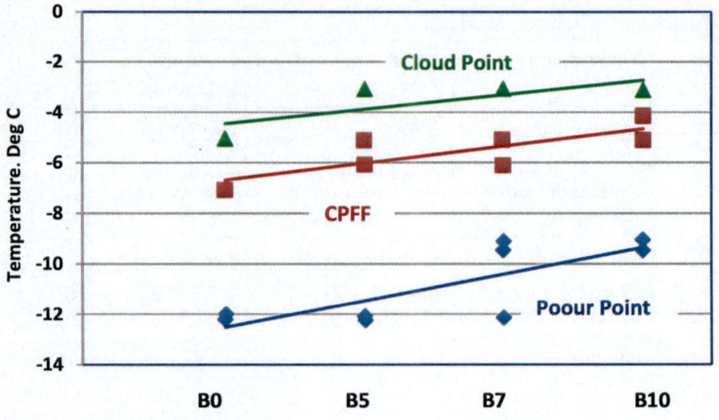

Fig. 14.7 Diesel cold flow properties [7]

get larger and in the end, plug the filters and do not allow the fuel to pass through. The additised diesel changes the wax structure of fuels to a pin-like structure and this allows the wax to pass through filters easily at cold temperatures.

If a vehicle is parked for a prolonged time in the winter season the wax crystals settle down at the bottom of the tank and obstruct the fuel flow to the engine. The dispersant additives such as amines and amides slow down the settling of additives and allow them to float in the solution [6]. Figure 14.7 shows the typical cold flow properties of biodiesel-diesel blends. The cold flow properties of diesel increase with the concentration of biodiesel in diesel [6, 7].

Low-Temperature Flow Test (LTFT) is widely used to predict cold temperatures performance of diesel fuel for use in winter regions. It is a slow cooling test used for the evaluation of fuels containing flow improver additives in a range of + 10 °C … − 30 °C and for the filterability of diesel at cold temperatures as per ASTM D4539. The LTFT test severity is better than CFPP. The test fuel is cooled at a specified rate of 1 °C/h and at every °C of cooling a vacuum of 20 kPa is applied to a filter assembly inserted in the fuel. The temperature at which the test fuel recovered in a graduated receiver vessel reaches 180 ml in 60 s is the minimum LTFT pass temperature.

Viscosity is a measure of fluid's resistance to flow. It influences atomization characterization and thereby combustion of fuel. Very low viscosity fuels affect the components of the fuel pump due to lack of lubrication and also fuel leakage in the injectors whereas too high viscosity fuel can cause damage to the fuel pump. The viscosity of diesel is usually in the range of 1.5–2.5 cSt at 40 °C.

Distillation or Volatility Characteristics of fuel are expressed in terms of temperature at which how much percentage of fuel is distilled or percentage of fuel distilled at what temperature is evaluated in a standard apparatus or distillation column following the accepted test method, such as ASTM D86. The sample is heated in the distillation apparatus and the vapors forms as the temperature increases. The vapors are condensed through the cooler and collected in a graduated cylinder. This gives information on the percentage of the initial volume of liquid collected i.e. information on initial boiling point (IBP), the temperature at which 10%, 50%, and 90% of the fuel boil off, the endpoint (EP), or final boiling point (FBP) is the temperature at which no fuel remains in the column, percentage of non-volatile residues. It is expected that the endpoint or final boiling point of the fuel should be lower for the fuel used in the winter seasons. Larger molecules that are bottom of the distillation process are high boiling points, low volatile, and do not flow easily. The boiling point range of diesel fuels is typically between 200 and 380 °C. These are streams of distillate and kerosene blends. Higher molecular weight paraffin in diesel is of high boiling point components and form wax that is accountable for low-temperature issues in the diesel. The high density high molecular weight diesel components are also favourable for better fuel economy as well. The distillation range of fuel depends on the fuel components and gives information on fuel composition, fuel properties such as viscosity, density, Cetane number, etc. The distillation characteristics of diesel influence safety, startability, warm-up, vapor locking tendency at high temperatures or high altitudes, and overall performance of vehicles. The presence of high boiling point components in fuel increases the combustion chamber deposits.

Reid Vapour pressure (RVP) is used to characterize the volatility of gasoline. The RVP of fuel is evaluated as per the ASTM D323 procedure. A sample is placed in a container which is maintained at 100 °C such that the ratio of the vapor volume to the liquid volume is 4:1 and the absolute pressure at 100 °C in the container is the RVP for the mixture. Refiners alter the composition of gasoline specific for winter and summer seasons. The primary difference between summer and winter-grade gasoline is its RVP. Higher RVP gasoline easily evaporates and in winter seasons the gasoline blend requires increased RVP. If the gasoline does not evaporate easily in cold temperatures it causes cold start and rough idle issues. If the gasoline contains more light components it vaporizes easily and causes increased emissions.

Octane number is the measure of auto-ignition or knocks resistance of spark-ignition fuel. The knocking tendency of a fuel is measured in Cooperative Fuel Research (CFR) engines as per the standard test protocols. CFR F1/F2 units are used to measure the research octane number (RON) as per ASTM D2699 and motor octane number (MON) as per the ASTM D2700 test method. RON represents mild operating conditions and MON represents severe operating conditions. The octane number of fuel is measured by comparing the knock resistance of the sample bracketed between the blends of primary reference fuels—iso-octane and n-heptane. Higher the octane numbers of fuel the better resistance to knock.

Cetane number fuel is a measure of its ignition quality and it is evaluated in CFR F4 Cetane unit as per ASTM D613. The combustion behaviour of the sample is compared or bracketed with the reference fuels at the standard test conditions. N-Cetane and alpha-heptamethyl-nonane are used as the primary reference fuels. Further secondary reference fuels are developed due to the scarcity of primary reference fuels. Currently, high Cetane T and low Cetane U fuels are used as secondary reference fuels for the evaluation of the Cetane number of diesel. In the CFR Engine, the sample is bracketed between the high and low reference fuels blends. The standard test conditions namely, the fuel injection delay 13° BTDC and injection advance

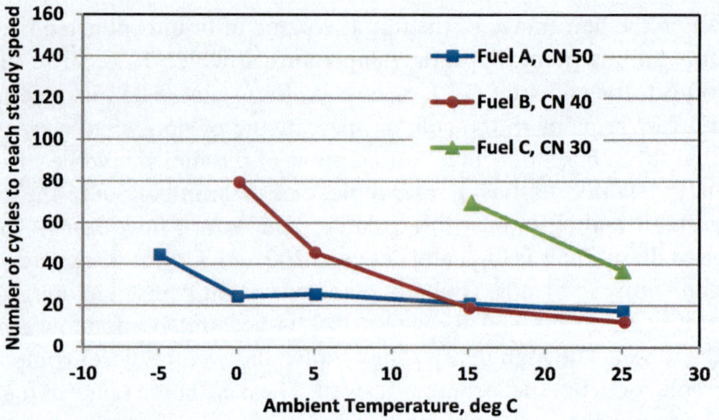

Fig. 14.8 Effect of Cetane number on startability [8]

of 13° are maintained for all the fuels, and knock intensity is adjusted by varying the compression ratio. Cetane number is linked to the ignition delay period, the time between the start of injection and the start of combustion. Fuels with shorter ignition delay are characterized as having higher ignition quality or Cetane number. The ignition delay period is also strongly dependent on the temperature and pressure of the ambient into which the fuel is injected. Hara et al. [8] evaluated the startability performance of iso-octyl nitrate (ION) on a single-cylinder direct-injection diesel engine in a cold room at ambient temperatures ranging from 25 to – 5 °C. Hara et al. [8] studied the startability performance of various Cetane number fuels on a single-cylinder direct injection diesel engine in a cold cell at an ambient temperature ranging from 25 to – 5 °C. The ION shows better performance in terms of auto-ignition. The cold startability performance of higher Cetane number diesel is better than lower Cetane number fuels [8]. The higher Cetane number fuels need a lesser number of cycles to reach the steady speed to lower Cetane number diesel, Fig. 14.8.

The desired properties of, the fuel for low and ultralow temperature applications are summarized in standards under various classes, Table 14.2. Commonly, winter-grade fuels are known as arctic fuels.

14.4 Engine Oils for Winter Regions

Lubricants are mainly used to reduce the friction between moving parts and to reduce component wear. The engine oil reduces friction between moving metal surfaces, reduces wear of engine components, dissipates heat away from the combustion chamber, sealing, anti-corrosion, and also handle contaminants such as sludge, soot deposits, etc. Automotive engine oils constitute more than 50% of the world's lubricant consumption. The engine oils used in vehicles operating at normal temperature regions do not operate efficiently in cold climate regions. Care must be taken in the

Table 14.2 The specific requirement for the Arctic or winter-grade diesel

				Limits	(Class)		
Property	Unit	(0)	(1)	(2)	(3)	(4)	(5)
CFPP	°C, max	– 20	– 26	– 32	– 38	– 44	EN 116
Cloud point	°C, max	– 10	– 16	– 22	– 28	– 34	EN 23015
Density at 15 °C	kg/m^3, min–max	800–845	800–845	800–840	800–840	800–840	EN ISO 3675
Viscosity at 40 °C	mm^2/s min–max	1.5–4.0	1.5–4.0	1.5–4.0	4.4–4.0	1.2–4.0	EN ISO 3104
Cetane number	minimum	49	49	48	47	47	EN ISO 5165
Cetane index	minimum	46	46	46	43	43	EN ISO 4264
Distillation							
% (V/V) recovered at 180 °C	% (V/V), max	10	10	10	10	10	EN ISO 3405
% (V/V) recovered at 340 °C	% (V/V), min	95	95	95	95	95	

selection of engine oils, gear oil, transmission oils, and brake oils for the vehicles operating in the low-temperature regions. Lubricants must have a sufficiently low viscosity and must also be able to reach over all surfaces requiring lubrication at cold temperatures. During engine cranking, the input torque from the starter motor has to overcome the cylinder gas pressure-induced resistance torque and friction torque from moving components such as bearings, pistons and rings, valve train and oil pump, etc. In the beginning, the starter motor output torque is greater than the static (or breakaway) torque, so the engine starts to accelerate. Both gas pressures build-up in the first cycle and friction torque under – 30 °C have a big influence on the starter motor torque requirement [9].

The engine oil should have lower viscosity to overcome the resistance offered by the engine. It is very important to formulate the application-specific engine oils considering the engine design and operating conditions so that better longevity and fuel economy can be achieved. Proper selection of base oil and additives is crucial in developing engine oils. The base oil is produced in refineries and oil additives are manufactured in the chemical industries. It involves many physicochemical properties testing, engine durability, and field evaluation testing to optimize the base oils and additives to achieve the best performance for the intended purpose.

14.4.1 Engine Oil Classification

Base oil or base stock is the main component of finished lubricants and the characteristics of base oils determine its function and reliability. A single type of base oil cannot meet the requirement of all the functions, there are trade-offs. High viscosity base oils provide adequate film strength while low viscosity oils give low-temperature fluidity and hence lower friction. So, there must be a balance in the selection of base oils.

With the advancement of refining technologies for mineral oils—distillation, solvent refining, hydrogen processing, oligomerization, esterification, and refining, and the introduction of synthetic oils in the twentieth century, the American Petroleum Institute categorized the base oils into five categories [10].

1. Mineral oils: Mineral base oils produced in refineries—Group I, II, III
2. Synthetic oil: Synthetic base oils produced in chemical or petrochemical industries—polyalphaolefins, poly-phenyl ethers, etc. Group IV and V.

Table 14.3 shows the classification of API base oils.

Paraffinic or straight-chain molecular oils are waxy and have high viscosity index. They begin to crystalize at high temperatures whereas naphthenic oils which contain ring structures have a low viscosity index and crystalize at lower temperatures. The properties of mineral oils, however, depend on the nature of the crude used. The characteristic of these mineral oils can be altered by using the VI improvers, pour point depressants, etc.

Synthetic oils are made by polymerization of various short-chain molecules and have uniform physical properties. They have a higher viscosity index, better stability, and oxidation characteristics. These oils can be tailor-made according to the specific application requirement. They have good low-temperature performance and are widely used in cold temperature conditions though it is cost more than mineral oil. The synthetic base oils have excellent capabilities for use in cold temperature conditions, – 40 °C or below and also have high viscosity index i.e. PAO—150–250 cSt; Esters—145–175 cSt; PAG—150–270 cSt; Silicone oils—190–500 cSt [11].

Table 14.3 Classification of API base oils

Type	Base oil category	Sulphur (ASTM D1552)	Saturates (ASTM D2007)	Viscosity index (ASTM D2270)
Mineral oils	Group I (solvent refined)	> 0.03	< 90	80–120
	Group II (hydro treated)	< 0.03	> 90	80–120
	Group III (hydrocracked)	< 0.03	> 90	> 120
Synthetic oils	Group IV	PAO synthetic lubricants		
	Group V	All other base oils not included in Group I, II and III or IV		

Automotive engine oils are classified as gasoline, diesel oils, gas engine oils, railroad oils, marine oils, 2-stroke oil, 4-stroke oils, tractor oils, off-highway equipment oils, etc. Lubricants are working under different operating conditions of temperature, speed, and load requires different viscosity grades to satisfy the performance characteristics.

Engine oils classified based on viscosity are:

- SAE viscosity classification for engine oils
- ISO classification for industrial oils
- NLGI (National lubricating gear institute) classification for gears etc.

Engine oils classified based on performance are:

- API Classification of gasoline engine oils—S category (Service)
- API Classification of diesel engine oils—C category (Commercial)
- API Classification of gear engine oils—GL category
- ISO 6743 classification—Industrial oils
- Two-Stroke engine oils—JASO

14.4.2 Properties of Engine Oils for Low Temperatures

The cold start phase of an engine is characterized by higher frictional losses, ineffective combustion, increase fuel consumption, and exhaust emissions. Excess emissions are released during the cold start or initial period of running the vehicle i.e., the engine combustion chamber temperature is low, this is due to poor atomization of fuel. Cold temperature conditions thicken the engine oils and the cranker requires more energy to start the vehicle. The friction between engine components is also higher during the cold start. The engine oil should flow to the various parts of the engine as quickly as possible to reduce friction and improve the fuel economy. Thermal management of the engine is essential for coolant as well as engine oils to overcome the issues associated with the cold start. Vittorini et al. reported the warming up of the engine oil via exhaust heat recovery reduced the vehicle exhaust emissions: CO_2—51.4%, HC—44.6%, and NOx—41.8% [12].

The coefficient of friction is a measure of the amount of friction existing between two surfaces. A low value of the coefficient of friction indicates that the force required for sliding to occur is less than the force required when the coefficient of friction is high [11]. From the Stribeck curve three main lubrication regimes can be distinguished, Fig. 14.9:

1. Boundary lubrication regime: where the lubricant layer between the parts in relative motion allowing direct contact between the parts and there is no load by hydrodynamic effects,
2. Hydrodynamic Lubrication Regime: one where the lubricant film layer is fully developed and the main resistance is given by the lubricant inner viscous friction.

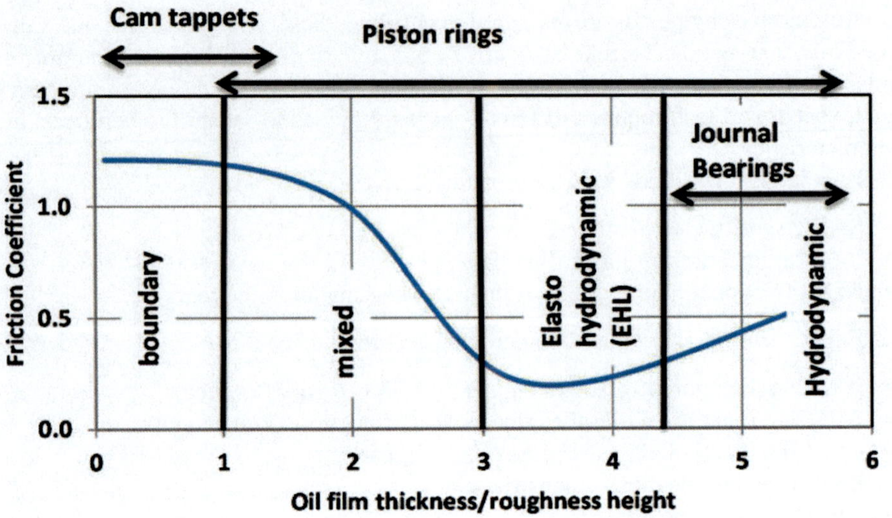

Fig. 14.9 Stribeck curve

3. A mixture of the previous two with miscellaneous characteristics of boundary and hydrodynamic regimes along the contact interface is called mixed lubrication.

Typical values for the percentages of friction power of the engine are: piston—36%, crankshaft—9%; pumping losses—27%; auxiliary—13%; valve train—16% [13]. The friction-regimes are broadly shown in Fig. 14.9.

Testing of engine oils is important to ensure how it performs in service. Measurement of physical properties such as viscosity provides the basic information on its aptness for the specific application. The performance testing of engine oils can be categorized as (1) physical properties testing—e.g. Viscosity, density, etc. (2) Chemical composition—e.g. hydrocarbon types, etc. (3) performance tests—E.g. Corrosion, anti-wear, etc. This includes rig tests, engine testbed evaluation, and field evaluation. In this chapter, we discuss only the properties of engine oils which are very relevant to cold temperature conditions.

The viscosity of engine oil is the measurement of resistance or friction offered to flow or force required to overcome the friction. It also refers to how easily oil pours at a specified temperature. It is the most important property of engine oils that influences engine starting, frictional power, power output, engine cooling, wear of engine components, oil consumption, oil leakage, and engine noise. Thin oils have lower viscosity and pour more easily at low temperatures than thicker oils that have a higher viscosity. Thin oils reduce friction in engines and help engines start quickly during cold weather whereas thick oils are better at maintaining film strength and oil pressure at high temperatures and loads. The larger the molecule structures, the thicker, or higher, or heavier the viscosity. In the case of mineral oils, the size and structure of mineral oil molecules vary whereas the synthetic oil

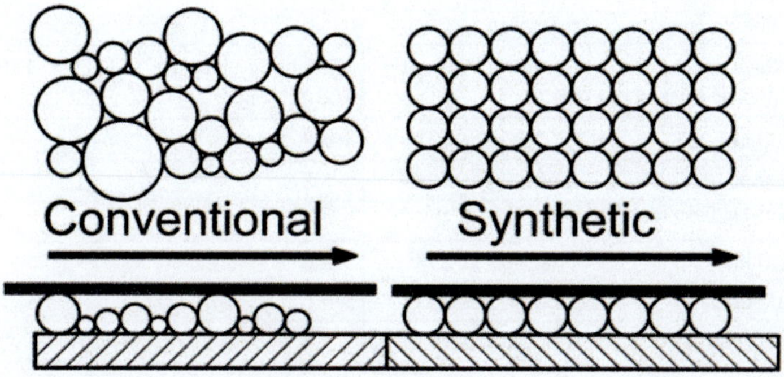

Fig. 14.10 Structure of synthetic versus mineral oils

manufacturing process results in consistently sized molecules of an identical structure [14], Fig. 14.10.

Four viscosity tests are used to predict the engine oil behavior in varying operating conditions.

Kinematic viscosity is a measure of the viscosity of lubricants by gravity where density plays a major role. It is measured by using a capillary viscometer as per ASTM D445. It is the time taken to flow a fixed volume of oil through a capillary orifice at controlled temperatures 40 and 100 °C under gravity. This is equal to dynamic viscosity divided by density at the same temperature. The unit of kinematic viscosity is Stoke or centistokes or cSt or mm^2/s. The higher the kinematic viscosity the greater is the fluid resistance to flow which results in increased fuel consumption. *Viscosity Index* (VI) is the rate of change of viscosity between two temperatures i.e. 40 and 100 °C. The lower the VI, the more the drop in viscosity as the oil warms up, and the higher the VI value, the less the drop in viscosity as the oil warms up. The engine oil's viscosity should not change with temperatures.

High-temperature high shear (HTHS) viscosity is a measure of the viscosity of engine oils when subjected to high temperature and high shear conditions. The HTHS is measured by a high-pressure capillary viscometer as per ASTM D4624. A known volume of oils is forced through a small diameter capillary by applying gas pressure at 150 °C temperature and rate of shear $10^{-6}/s$. This test method correlates the engine wear during high-temperature high shear conditions which are experienced in normal engine operation and its unit is cP.

Low-temperature Cranking viscosity (CCS): Cold-cranking simulator (CCS) following the ASTM D5293 test method assesses the apparent viscosity of engine oils under cold temperatures and this method is the basis for classifying the viscosity of J300 engine oils. This viscosity is called cranking viscosity and its unit is centipoise (cP). This defines the winter-grade engine oils that correlate with the engine speed or crank ability during low temperatures. It is a non-cylindrical rotor enclosed by a stator a motor applies a constant torque for three minutes to the rotor and the speed of rotation of the rotor depends upon the viscosity of the engine oils at that temperature. The test temperature is − 10 °C for 25 W oils and − 35 °C for 0 W

Table 14.4 SAE viscosity classification

SAE	Cold-cranking viscosity, (cP) Max	Pumping viscosity, cP	Kinematic viscosity, 100 °C, cSt	HTHS, 150 °C, cSt
0 W	6200 at – 35	< 60,000 at-40 C	> 3.8	–
5 W	6600 at – 30	< 60,000 at-35 C	> 3.8	
10 W	7000 at – 25	< 60,000 at-30 C	> 4.1	
15 W	7000 at – 20	< 60,000 at-25 C	> 5.6	
20 W	9500 at – 15	< 60,000 at-20 C	> 5.6	
25 W	13,000 at – 10	< 60,000 at-15 C	> 9.3	
20			5.6–9.3	> 2.6
30			9.3–12.5	> 2.9
40			12.5–16.3	> 2.9
50			12.5–16.3	> 3.7
60			16.3–21.9	> 3.7
70			21.9–26.1	> 3.7

oils. Lower the CCS value of the oil results in reduced engine friction and, in turn, lower fuel consumption.

Mini rotary viscometer viscosity (MRV) evaluated as per ASTM D 4684 is used for predicting the possibility of air binding in engine oils when the vehicle is standing still for a considerable period or cold start of the engine i.e. at low temperature and low pressure. The engine oils become semi-solid and fail to flow through the pump when the engine is started. This would lead to the pumping of air instead of oil and may damage the engine as well as the pump. Moreover, even if the air binding not happens the engine is highly viscous and the sufficient quantity of engine oil may not reach the bearing and rocker arm also.

SAE Viscosity grade classification for automotive oils and their different viscosities are given in Table 14.4 and the comparison of different viscosity classifications of lubricants is given in Fig. 14.11.

Six SAE winter grades—SAE 0, 5, 10, 15, 20, and 25 W are characterized by rheological properties at lower temperatures to ensure good lubrication and engine startability at cold temperatures. These engine oils are also with the minimum kinematic viscosity at 100 °C to provide sufficient lubrication when the engine running i.e. after warmup. The SAE monograde oils—SAE 20, 30, 40, 50, and 60 are characterized for high-temperature viscosity at 100 °C for minimum and maximum viscosity limits. Monograde oils do not contain viscosity improver. The multi-grade engine oils say SAE 5W30 should meet the viscosity requirement of winter-grade SAE 5 W at lower temperatures and monograde SAE 30 at higher temperatures.

Oil additives such as detergency, dispersant, and antioxidants, etc. are added in the suitable viscosity base oil to achieve the engine oils with desired properties. In the case of multi-grade oils, the low viscosity base oils are chosen to meet the requirement of winter grades and in addition to the performance additives, viscosity modifier additive

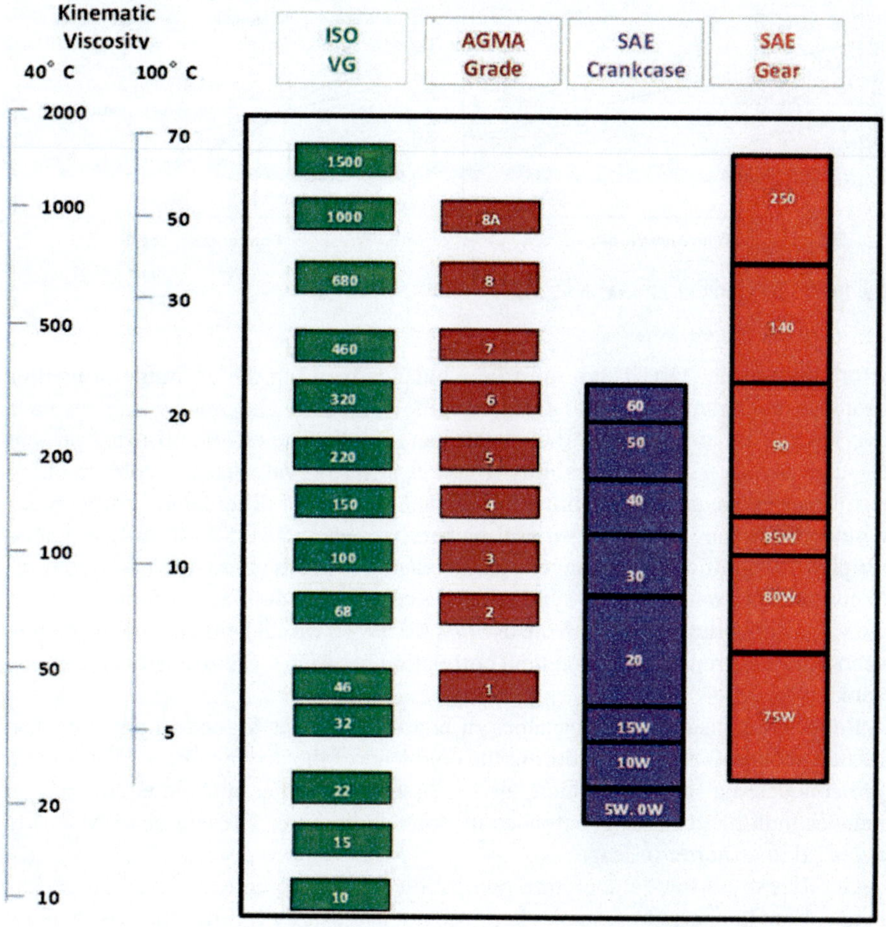

Fig. 14.11 Comparison of viscosity classification

is added to get the required viscosity index engine oils. Low viscosity oils that are less prone to thickening in low temperatures help you start your engine more quickly in winter while thick oils that are less prone to thinning at high temperatures help your engine perform better in summer. Figure 14.12 shows the temperature-viscosity behavior of multi-grade oils at different temperature conditions.

Pumpability of engine oils is crucially important at low temperatures. Circulation of engine oils through the various parts of engines reduces the friction and heat up of components. If the oil is not able to pump it cannot reach vital parts of the engine, build a protective film and protect engine parts from wear and ultimately from destruction. A range of factors affecting aged oil low-temperature pumpability is additive components and base stocks, the service cycle, and the top-ups. Some engines are fitted with oil heaters to overcome this issue and the engine oil is heated

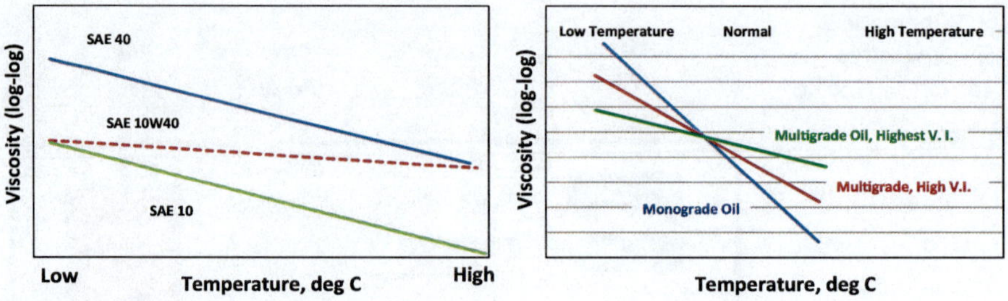

Fig. 14.12 Temperature effects on viscosity

before the engine start. There must be a balance between the desirable properties for a low-temperature start and normal engine operation. The modern engines have a very fine clearance between the contacts and if adequate engine oil is not present it causes severe issues in durability. Buck et al. [15] evaluated the cold cranking performance of synthetic and mineral-based engine oils of different viscosity grades in modern light and heavy-duty diesel-powered vehicles. SAE 5W-40 fully synthetic oils provide significantly enhanced engine cranking speeds at low-temperature start-up compared to mineral/semi-synthetic based SAE 15W-40. Faster oil pressure rise and valve train oiling times with the use of SAE 5W40 would help in reducing engine wear. They also reported an excellent correlation between CCS viscosity and engine cranking speed.

Better CCS viscosity performance of engine oils gives a good improvement in the fuel economy of vehicles during the cold start of the driving cycle. Wolak et al. [16] conducted a series of vehicle tests to assess the winter-specific engine oils. It includes multiplied starting, extended idle and short drive. The engine oil viscosity increased in a shorter mileage.

Good fresh oil low-temperature pumpability is no guarantee of aged oil performance. Low-temperature pumpability does not necessarily follow other viscometric parameters such as kinematic viscosity (KV) and cold cranking simulator (CCS) viscosity in aged oils. May et al. [17] reported that fuel dilution that caused a drop in KV and CCS viscosity resulted in increased MRV viscosity.

Faster cooling profile MRV (ASTM D3829) and slow cooling profile MRV TP-1 (ASTM D4684) tests address two mechanisms of low-temperature pumpability failure. The fast cooling profile MRV flow limited failure occurs when the viscosity of engine oils is increased and the oil could not reach different parts of the engine. Slow cooling profile MRV TP1 phenomena air binding failure occurs when the oil forms a gel-like structure/solid that prevents the oil to flow and the air is drawn into the engine oil flow circuit. Studies reported a good correlation between MRV TP1 testing and engine oil pressure build-up as a measure of pumpability [18]. Oberoi et al. [19] evaluated the pumpability performance of engine oils in heavy-duty diesel engines at low temperatures. Good fresh oil MRV TP1 viscosity did not guarantee good field performance and the aging of engine oil is critical to its performance. There is no correlation reported between the ROBO test (ASTM D7258) and the low-temperature pumpability of engine oils. Further, low emission diesel engines raised has raised concern about the high concentration of soot in the engine oils and its effect on the low-temperature pumpability of aged oils [20, 21].

Cockbill et al. [22] evaluated the low viscosity engine oils at cold temperatures (– 20, – 25, and – 35 °C) in a 2-L four-cylinder diesel engine. The dynamometer runs the engine at nominal cranking speed and then moves on to cold idle speed. The lowest viscosity engine oil reduced the energy required to turn the engine by approximately 36%. The low viscosity engine also exhibited higher gallery oil pressure and reduced bearing temperatures [22].

Engine oil is heated by using external heaters to reduce its viscosity before starting to reduce the pumping effect and circulate the engine oil to enter the oil in various parts of the engine easily. Otherwise, it increases the wear of engine components. thereby Malozemov et al. studied the effect of the engine oil temperature control system that increases the oil temperature in the main oil line by – 10 to 23 °C, in the crankcase by 5–10 °C, the average temperature of diesel by 7–8 °C before the start. This helped in starting the engine at the ambient temperatures of – 25 °C within 5–8.5 min, at – 50 °C within 21–28 min [23].

14.5 Conclusion

The ambient temperature plays a very significant influence in starting and operating the engine in winter regions. In addition to the engine design and its calibration, the fuel, engine oil, and battery influence the engine operation. The performance of these deteriorates when the ambient temperature is reduced. Careful selection of these elements is required for the efficient operation of the engine in winter regions. The winter-grade fuels have better cold temperature capabilities to flow smoothly through the fuel filter at low temperatures and give better atomization to burn the fuel. Low viscosity engine oils with flow improvers reduce the oil viscosity and thereby lower the frictional loss and pumping loss for easier cranking to start the engine. The use of the winter-specific grades of fuel and oil enables easier cranking of engines and combustion of fuel in engines for winter regions.

Acknowledgements A part of the chapter is with copyright permission obtained for using the materials in an SAE paper by the first two authors. License ID: 1107796-1, 29-Mar-202, System ID: 2014-01-2715

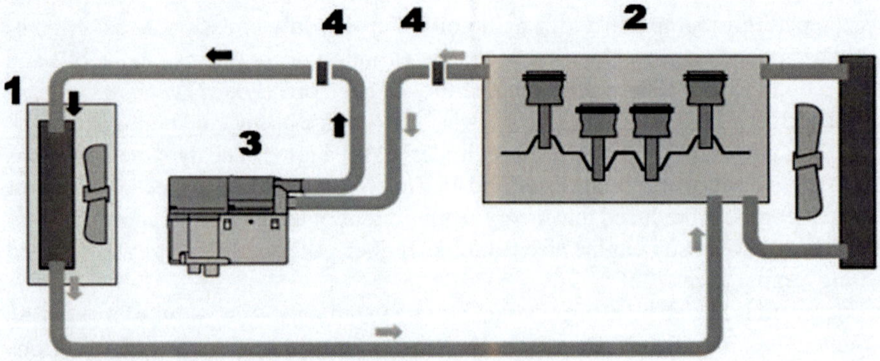

Fig. 14.A1 Block heater circuit: 1 heat exchanger, 2 IC engine, 3 block heater, 4 connection pipes

Appendix: Block Heater Circuit for Increasing Coolant Temperature at Cold Ambient

See Fig. 14.A1.

References

1. Ramadhas AS, Xu H, Liu D, Tian J (2014) Key factors affecting the cold start of diesel engines. Int J Green Energy
2. Han Z, Henein N, Nitu B, Bryzik W (2001) Diesel engine cold start combustion instability and control strategy. No. 2001-01-1237. SAE Technical Paper
3. Ramadhas AS, Xu H, Liu D, Tian J, Wyszynski M, Piaszyk J (2014) Impact of cold ambient conditions on cold start and idle emissions from diesel engines. No. 2014-01-2715. SAE Technical Paper
4. BP (2013) Changing diesel low-temperature properties by using additives. British Petroleum. www.bp.com.au
5. Chevron technical review. Diesel fuels. Chevron Products Company. www.Chevron.com
6. Zöldy M (2019) Investigation of correlation between Diesel fuel cold operability and standardized cold flow properties. Periodica Polytech Transp Eng
7. Teixeira RM, de Castro Cortás L, Viscardi SLC, Cavalcanti EHS (2013) Impact of biodiesel on filterability and cold flow properties of diesel BX. In: IASH 2013, the 13th international conference on stability, handling and use of liquid fuels, Rhodes, Greece, 2013, pp 628–644
8. Hara H, Itoh Y, Henein NA, Bryzik W (1999) Effect of cetane number with and without additive on cold startability and white smoke emissions in a diesel engine. SAE Trans:858–872
9. Zhou Q, Houldcroft J (2007) Cold engine cranking torque requirement analysis. No. 2007-01-1967. SAE Technical Paper
10. Base Oil Groups Explained, Noria Corporation (2021). https://www.machinerylubrication.com/Read/29113/base-oil-groups. Accessed on 29 Mar 2021
11. Bird J, Owen J, Chivers PJ (2014) Newnes engineering and physical science pocketbook. Newnes
12. Vittorini D, Di Battista D, Cipollone R (2018) Engine oil warm-up through heat recovery on exhaust gases—emissions reduction assessment during homologation cycles. Thermal Sci Eng Progr 5:412–421

13. Kamil M, Rahman MM, Bakar RA (2014) An integrated model for predicting engine friction losses in internal combustion engines. Int J Autom Mech Eng 9(1):1695–1708
14. Oil viscosity and oil grades, engine oil viscosity. Accessed on 28 Mar 2021. https://www.lubricants.total.com/consumers/maintenancetips/Oil-viscosity-and-oil-grades
15. Buck WH, Lohuis JR (1994) Lubricant effects on low-temperature diesel engine cold starting. SAE Trans:1–8
16. Wolak A, Zając G (2018) Cold-cranking viscosity of used synthetic oils originating from vehicles operated under similar driving conditions. Adv Mech Eng 10(11):1687814018808684
17. May CJ, Habeeb JJ (1987) Factors affecting the low-temperature pumpability of used engine oils. SAE Trans:298–306
18. Koenitzer BA, Lai PKS, May CJ (1998) A study of gel index and MRV effects on SAE 5W-30 pumpability. SAE Trans:1911–1921
19. Oberoi S, Goldmints I (2012) In-service low-temperature pumpability: field performance versus bench tests. No. 2012-01-1708. SAE Technical Paper
20. Geehan M, James A, Eiden KL, Elden KL (2000) Low-temperature oil pumpability in emission-controlled diesel engines. SAE Trans:1886–1904
21. Galbraith RM, May CJ (2003) A cold start and pumpability study of fresh and highly sooted engine oils in 1999 heavy-duty diesel emission engines. SAE Trans:2499–2507
22. Cockbill B, Bennett J (2000) The effects of crankcase oil viscosity on engine friction at low temperatures. No. 2000-01-2052. SAE Technical Paper
23. Malozemov AA, Bondar VN, Cherepanov SI (2016) Experimental research of forced diesel engine with oil temperature control system in cold ambient conditions. Proc Eng 150:1143–1148

Chapter 15
Low-Temperature Operation: Impact of Cold Temperature on Euro 6 Passenger Car Emissions

P. A. Lakshminarayanan

15.1 Introduction

High pollution during winter in urban areas [1–4] has been strongly influenced by vehicle exhaust. Emissions of THC, NMHC, CO, NOx, particle number (PN greater than 23 nm), and particle mass (PM) are controlled for Euro 6 vehicles C using the new worldwide harmonized light-duty driving test cycle (WLTC) by the Type 1 test at 23 ± 5 °C, Fig. 15.1a.

In Europe and Korea, the Type 6 test at − 7 ± 3 °C was introduced in 1998 as a measure against emissions from motor vehicles at cold ambient; however, the test is applied only to positive-ignition light-duty vehicles over the Urban Driving Cycle (UDC; first of the two phases constituting the New European Driving Cycle, NEDC, Fig. 15.1b. The second phase corresponds to the highway or extra-urban driving cycle), and CO and THC emissions must be lower than 15 g km^{-1} and 1.8 g km^{-1} respectively; the limits are 15 times those allowed during Type 1 test at a higher temperature of 23 ± 5 °C. Similar procedures exist in the USA and China for petrol and diesel vehicles.

NH_3 forms secondary inorganic aerosol in the atmosphere [5]. Nitrous oxide (N_2O) has a global warming potential[1] of 295 and it is also a very important ozone-depleting substance (ODS) [6]. NH_3 is regulated in Korea and N_2O in the USA.

[1]Global warming potential (GWP) is the heat absorbed by any greenhouse gas in the atmosphere, as a multiple of the heat that would be absorbed by the same mass of carbon dioxide (CO_2). GWP is 1 for CO_2.

P. A. Lakshminarayanan (✉)
Indian Institute of Technology Kanpur, Kalyanpur, Kanpur 208016, India

P. A. Lakshminarayanan and A. K. Agarwal (eds.), *Handbook of Thermal Management of Engines*, Energy, Environment, and Sustainability,
https://doi.org/10.1007/978-981-16-8570-5_15

Table 15.1 Euro 6 vehicles specifications

Vehicle	Engine type	After-treatment	Engine displacement (cm^3)	Engine power (kW)
D1	CI HDi	DOC + DPF + SCR	1560	73
D2	CI HDi	DOC + DPF + SCR	1997	110
D3	CI TDI	DOC + DPF + SCR	2987	140
D4	CI TDI	DOC + DPF + LNT	1422	55
D5	CI TDI	DOC + DPF + LNT	1968	110
G1	SI GDI	TWC	998	76
G2	SI GDI	TWC	999	81
G3	SI GDI	TWC + NSC	1991	155
G4	SI GDI	TWC	1242	51
G5	SI PFI	TWC	1368	57

Emissions of these gases are related to the use of catalytic converters [7–11]:

- Three-Way Catalyst (TWC),
- NOx Storage Catalyst (NSC),
- Diesel Oxidation Catalyst (DOC),
- Selective Catalytic Reduction (SCR) and
- Lean NOx Trap (LNT).

15.2 Experiments by Europe JRC

The European Commission Joint Research Centre (JRC), at Ispra, Italy conducted a study [12] on twelve passenger cars from the European market. Of these gasoline and diesel vehicles are taken here for study (Table 15.1).

The vehicles are run for a distance of 23.3 km according to the WLTC at 23 and − 7 °C ambient after soaking them inside a climatic cell for six hours. The difference in driving resistance is adjusted by decreasing the coast-down time estimated at 23 °C by 10% for the tests at − 7 °C, as indicated in the regulations.

15.3 Comparison of WLTC Emissions at − 7 and 23 °C

The rise in emissions is calculated by the ratio of emissions at − 7 and 23 °C available (see Appendix) in Ref. [12], Table 15.2. Further, extra data are available for different phases of WLTC in a companion Ref. [13].

Table 15.2 The ratio of emissions at − 7 °C and + 23 °C

	THC	CO	CO_2	PN($\times 10^{11}$)	NOx	NO_2	N_2O	NH_3
DV1	2.00	1.58	1.16	6.00	7.20	49.40	1.00	0.00
DV2	2.00	3.00	1.20	0.78	2.40	5.89	1.25	0.43
DV3	1.14	2.15	1.09	3.33	3.37	3.93	1.21	0.89
DV4	1.31	1.36	1.20	8.89	1.73	1.60	1.50	-
DV5	1.05	1.10	1.15	0.19	2.15	2.61	1.00	1.00
GV1	2.70	1.40	1.03	1.59	1.26	N.A	1.00	1.53
GV2	9.75	1.34	1.10		1.14	N.A	3.00	1.33
GV3	13.08	5.82	0.99	2.55	4.56	N.A	0.64	1.11
GV4	6.12	1.75	1.08		0.78	N.A	2.00	1.62
GV5	5.54	2.68	1.23	5.81	1.30	N.A	1.00	1.47

N.A.: NO_2 emissions from gasoline vehicles negligible

15.3.1 THC and CO Emissions

THC and CO exhaust emissions result from the incomplete combustion of fuel. Spark ignition engines are normally equipped with a three-way catalyst (TWC) which simultaneously oxidizes CO and THC to CO_2 and water and reduces NOx to N_2. Alternatively, diesel vehicles with DOC convert CO and THC to CO_2 and water.

THC and CO emissions rise higher for spark-ignition vehicles than for diesel engines. The increase of THC in phases 1 and 4 is predominant at low temperatures. THC from diesel vehicles is one order of magnitude lower than those for gasoline vehicles at − 7 and 23 °C as a consequence of the lower combustion efficiency of the spark ignition vehicles, Fig. 15.2. For this reason, diesel vehicles are excluded from the regulatory EU Type 6 test at low temperatures.

Higher CO and THC emissions at cold temperatures from spark-ignition vehicles [14] are due to incomplete combustion in the cylinder or aftertreatment systems, Figs. 15.2 and 15.3:

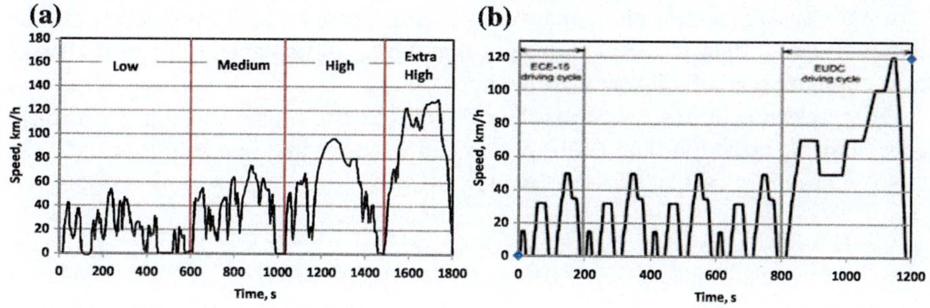

Fig. 15.1 Driving cycles **a** WLTC, **b** NEDC

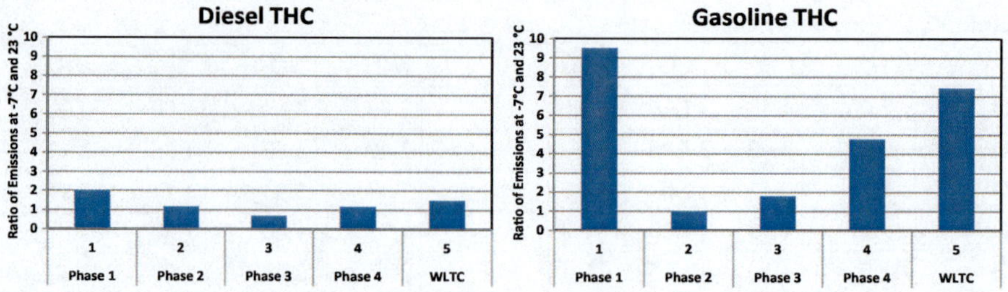

Fig. 15.2 Increase in THC emissions at cold averaged for five diesel and five gasoline vehicles during the four different phases of the WLTC as well as during the whole WLTC

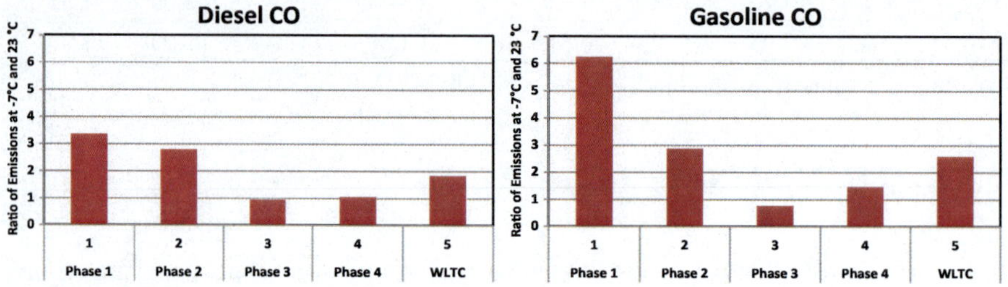

Fig. 15.3 Increase in CO emissions at cold averaged for five diesel and five gasoline vehicles during the four different phases of the WLTC as well as during the whole WLTC

- use of rich air–fuel mixtures at cold start,
- incomplete combustion near the cold walls of the cylinder during warm-up,
- lower catalytic efficiency and
- longer periods to reach light-off temperature.

This has led to monitoring emissions only during UDC and not during the extra-urban driving cycle (EUDC), where the TWC is expected to be already working at optimum temperatures, Fig. 15.1b. However, the results show that a large fraction of CO can be emitted during the high phase (phase 3) and extra-high phase (phase 4) of the WLTC, Fig. 15.3.

In most cases, the emissions measured during Phases 3 and 4 are several times higher at − 7 °C than at 23 °C indicating that the catalysts have not lighted off and the engines are not sufficiently warmed up.

THC emissions from modern gasoline vehicles lead to higher secondary carbonaceous aerosol formation than at temperate temperatures and hence higher PM [15] at − 7 °C, as discussed later in the chapter.

15.3.2 NOx and NH_3 Emissions

Spark ignition engines usually use TWC and in some cases NSC to reduce NOx. Since the introduction of the Euro 6 standards, diesel vehicles necessarily use SCR or LNT to reduce NOx. NSC and LNT adsorb NOx in the fuel-lean mode and reduce it in the fuel-rich mode to get regenerated. SCR has led to the emissions of NH_3 and N_2O, formed following different reactions depending on the catalytic system and precursors present on the catalysts.

In the TWC, NH_3 is formed via steam reforming from hydrocarbons [16] and/or via reaction of NO with H_2 (Eqs. 15.2a or 15.2b) where H_2 is produced from a water–gas shift reaction between CO and water [17] while the engine runs rich (Eq. 15.1):

$$CO + H_2O \rightarrow CO_2 + H_2 \tag{15.1}$$

$$2NO + 2CO + 3H_2 \rightarrow 2NH_3 + 2CO_2 \tag{15.2a}$$

$$2NO + 5H_2 \rightarrow 2NH_3 + 2H_2O \tag{15.2b}$$

In the SCR system, NH_3 is produced by the hydroxylation of urea and the SCR catalyst reduces NOx by reacting the NO and NO_2 with NH_3 (Eqs. 15.3, 15.4, and 15.5). The reaction with NO_2 is fast and it enables the downsizing of the SCR catalyst. To increase NO_2 in the exhaust to 50% from 5–10% available in the engine exhaust and thus to enhance the reaction rate with NH_3, NO is oxidized to NO_2 on a DOC [17]. Overdosing of urea, low temperatures in the system, or catalyst degradation may lead to NH_3 emissions [17].

$$CO(NH_2)_2 \rightarrow NH_3 + HNCO \tag{15.3}$$

$$HNCO + H_2O \rightarrow NH_3 + CO_2 \tag{15.4}$$

$$2NH_3 + NO + NO_2 \rightarrow 2N_2 + 3H_2O \tag{15.5a}$$

$$4NH_3 + 4NO + O_2 \rightarrow 4N_2 + 6H_2O \tag{15.5b}$$

$$8NH_3 + 6NO_2 \rightarrow 7N_2 + 12H_2O \tag{15.5c}$$

The increase of NOx emissions at cold is two times for spark ignition engines and more than three times for diesel engines, Fig. 15.4.

Dardiotis et al. [18] found that NOx emissions from diesel vehicles were quite low after the UDC and suggested that the vehicle test over the EUDC is not needed.

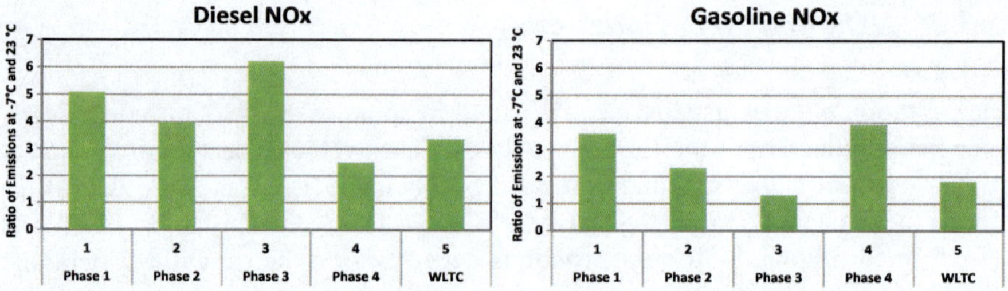

Fig. 15.4 Increase in NOx emissions at cold averaged for five diesel and five gasoline vehicles during the four different phases of the WLTC as well as during the whole WLTC

It is also stated that this applies especially to vehicles equipped with an SCR system because the device works satisfactorily only over the EUDC. However, the three SCR-equipped vehicles result in high NOx when NH_3 is absent (by software setting, to account for lower catalyst temperatures than the light-off temperature), during the entire test cycle at − 7 °C indicating that the efficiency of the SCR system at low temperatures is indeed very low. Therefore, vehicle emissions patterns at − 7 °C are very similar to those obtained at 23 °C. However, the after-treatment strategy could change with a change in the ambient temperature. SCR systems stop working or work at lower efficiency at cold temperature even in phase-4 of WLTC.

NOx emissions from spark-ignition vehicles constitute only NO and no NO_2 at all, whereas the ratio of NOx to NO_2 from diesel is about 4. The increase of NO_2 emissions in modern diesel engines in the after-treatment systems affects air quality since it leads to less O_3 being consumed in the reaction with NO and consequently higher O_3. Furthermore, NO_2 is the most harmful pollutant to human health in Europe [19].

NH_3 is emitted from modern gasoline fleets [20] and since the introduction of the SCR and LNT systems (Euro-6 vehicles), NH_3 is also existing in diesel exhaust [21, 22]. See Table 15.1 and Fig. 15.5. Moreover, NH_3 emissions from SCR-equipped diesel vehicles may increase as a consequence of a higher dosage of urea to meet NOx emission limits under the real driving emission test (RDE).

At low temperatures, NH_3 emissions from SCR-equipped diesel vehicles are in the same range as gasoline vehicles at 23 °C. SCR systems in diesel vehicles are deactivated at cold temperatures, resulting in very low NH_3 and very high NOx emissions. No NH_3 emissions are observed when LNT-equipped. However, when equipped with the latest LNT generation vehicles emit some NH_3 at − 7 °C and 23 °C.

NH_3 emissions from vehicles equipped with TWC are up to ~ six times higher than those found in the literature for Pre-Euro 6 vehicles (Euro 3, Euro 4, Euro 5, Ultra-Low Emission Vehicles (ULEV) and Low Emission Vehicles (LEV)) [8, 20]. NH_3 emissions from gasoline vehicles are up to five times higher at cold temperatures.

The high NH_3 from the spark ignition vehicles results from the emission control strategy that aims at reducing NOx at the expense of emitting NH_3, which is not

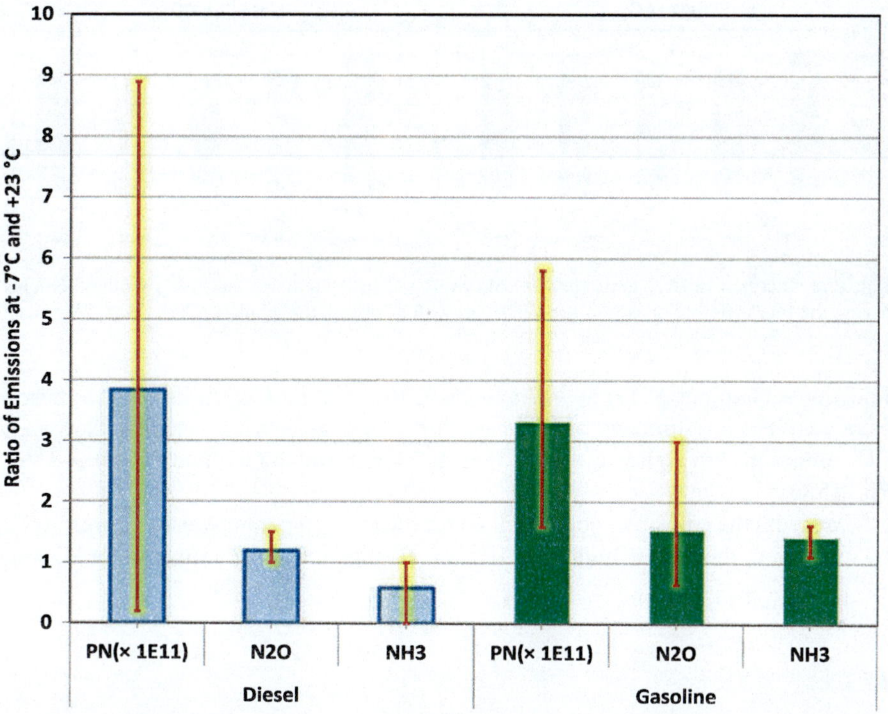

Fig. 15.5 Increase in PN, N_2O, and NH_3 emissions at cold averaged for five diesel and five gasoline vehicles during the WLTC

regulated for light-duty vehicles. NH_3 emissions from gasoline vehicles expressed in moles/km exceed NOx emissions.

NH_3 and NO_2 form ammonium nitrate (NH_4NO_3) in the atmosphere [23, 24]. Also, PM rapidly increases when NH_3 is photo-oxidized [25] and NH_3 emissions from secondary inorganic particles ($PM_{2.5}$) in NOx environments [26]. Consequently, the ratio of NO_2 and NH_3 from the new generation vehicles may strongly affect the air quality in the cold.

15.3.3 GHG Emissions

15.3.3.1 CO_2

Average CO_2 from the entire light-duty vehicle fleet of each vehicle manufacturer in the EU is mandated to be within 95 g km^{-1} by 2020 [27]. Because fuel volume and injection timing can be accurately controlled [28], gasoline direct injection (GDI) technology provides superior fuel economy and lower CO_2 emissions to Port Fuel

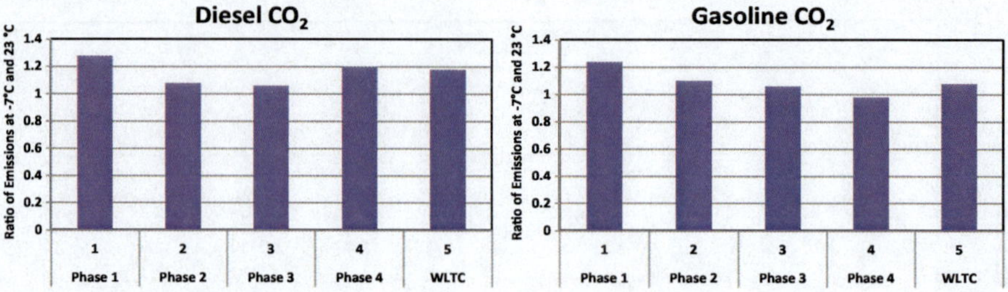

Fig. 15.6 Increase in CO_2 emissions at cold averaged for five diesel and five gasoline vehicles during the four different phases of the WLTC as well as during the whole WLTC

Injection technology (PFI) at − 7 and 23 °C and similar CO_2 emissions to diesel. Hence GDI is an attractive proposition in the new generation vehicles. However, CO_2 emissions are higher at − 7 °C than at 23 °C for both GDI and PFI by 9–16%, Fig. 15.6.

Even with the adjustment of road-load for the low-temperature tests, as prescribed by regulation, the CO_2 is higher by 10% on average when the temperature drops to − 7 °C.

15.3.3.2 N_2O

N_2O is generated in various types of aftertreatment systems over wide temperatures, depending on the catalyst type and exhaust conditions and the after-treatment operation strategies [7, 10]:

- in TWC, NSC, and LNT via a complex series of chemical mechanisms involving NO, molecular nitrogen (N_2), and atomic nitrogen.
- in DOC at low temperature as a by-product of NOx reduction by hydrocarbons depending on their type and concentration, temperature, and the DOC formulation
- in SCR:
  - by NH_3 oxidation by NO and
  - by oxidation by O_2 of NH_3 which is linked to urea dosage.

N_2O from both spark ignition and diesel vehicles increases as temperature decreases. N_2O emissions from diesel vehicles are 4 times higher than those from spark-ignition vehicles at − 7 as well as at 23 °C.

The higher N_2O emissions observed at − 7 °C compared to those at 23 °C are linked to the higher NOx concentrations present at cold temperatures.

In terms of CO_2 equivalents (N_2O has 298 times the global warming potential of CO_2 over 100 years), the N_2O emissions are 3–5 g CO_2-eq km^{-1}, which is approximately 2% of the average CO_2 emissions of the vehicles under consideration.

Figure 15.5 shows the ratio of N_2O at − 7 and 23 °C obtained over the WLTC cycle, with maximum and minimum shown as error bars.

15.3.4 *Particle Number*

The PN measurement based on the counting solid particles is indicated in the European emissions regulation: (a) diesel light-duty vehicles (Euro 5), 2011 and (b) gasoline direct injection (GDI) light-duty vehicles (Euro 6) 2014. A minimum diameter of 23 nm size is mentioned to include the primary soot particles and to avoid the volatile nucleation mode particles [29]. Refer to Fig. 15.5. A wide range of PN results from the gasoline engines (2–24 $\times 10^{11}$ # km^{-1} at 23 °C and 12–38 $\times 10^{11}$ # km^{-1} at − 7 °C). PN from the Euro 6 GDI vehicles (11 − 24 $\times 10^{11}$ # km^{-1}) are higher than Euro 6 standards (6 $\times 10^{11}$ # km^{-1}) for PN. PN is approximately two orders of magnitude lower than those from gasoline vehicles because of the good DPF technology used [30].

At cold temperatures, PN increases for both gasoline and diesel engines. PN from the diesel engines is higher at a cold temperature for the following reasons:

- semi-volatile material escaping oxidation as the catalytic converters have not the lighted off yet,
- non-volatile particle deposits blowing out during highly transient thermal and flow conditions
- local defects in the filter that reduce DPF efficiency at low temperatures.

In the case of PFI spark-ignition vehicles, higher PN at cold temperature is due to:

- THC emissions,
- incomplete combustion when a rich air–fuel mixture is supplied to compensate for the reduced fuel vaporization and elevated engine friction.
- catalytic after-treatment systems light off late.

PFI gasoline vehicles produce very low PN in standard testing or driving at 23 °C. Therefore, only diesel and GDI gasoline vehicles are required to meet a PN limit in Europe. However, PFI gasoline vehicles can result in high PN at cold.

15.4 Conclusions

Emissions from both spark ignition and compression ignition vehicles are strongly and negatively affected by low temperatures. Higher THC, CO, NOx, PN, and NH_3 are emitted at − 7 °C than at 23 °C. They affect PN, ground-level O_3, and nitrogen dioxide (NO_2) which are not mentioned in EU Type 6 test. NO_2 ratio (NO_2/NOx) in diesel exhaust and NH_3 ratio (NH_3/NOx) in gasoline of Euro 6 vehicles are involved in some fundamental chemical reactions in the atmosphere and could affect urban air quality in cold conditions. GHGs namely, CO_2 and N_2O are higher by 9–30% and 90% respectively when vehicles are tested at – 7 °C than at 23 °C, though they are neither measured nor regulated under the Type 6 test underestimating the GHG burden of the transport vehicles. Substantial emissions take place during the last two phases of the new type approval cycle (i.e., WLTC), and not only during the cold start.

Appendix: Summary of Experimental Results From Ref. [12]

See Tables 15.3 and 15.4.

Table 15.3 Average emission factors (mg km-^{1}; CO_2 g km^{-1} and PN # km^{-1}) over the WLTC at 23 °C

	THC (mg km^{-1})	CO (mg km^{-1})	CO_2 (mg km^{-1})	PN (× 10^{11}) (mg km^{-1})	NOx (mg km^{-1})	NO_2 (mg km^{-1})	N_2O (mg km^{-1})	NH_3 (mg km^{-1})
D1	4	126	138	0.05	148	5	111	24
D2	2	46	154	0.09	476	73	8	7
D3	7	41	337	0.06	238	83	14	9
D4	13	22	146	0.09	484	167	8	0
D5	19	41	173	2.4	183	28	12	2
G1	54	567	117	24	34	N.A	1	17
G2	12	154	145	-	21	N.A	1	9
G3	13	158	177	11.0	18	1	14	46
G4[a]	25	5766	142	-	9	N.A	2	34
G5	24	972	152	2.1	27	N.A	1	17 ± 0

N.A.: below the limit of detection

Table 15.4 Average emission over the WLTC at − 7 °C

	THC (mg km^{-1})	CO (mg km^{-1})	CO_2 (mg km^{-1})	PN (× 10^{11}) (mg km^{-1})	NOx (mg km^{-1})	NO_2 (mg km^{-1})	N_2O (mg km^{-1})	NH_3 (mg km^{-1})
D1	8	199	160	0.3	1066	247	11	0
D2	4	138	185	0.07	1142	430	10	3
D3	8	88	368	0.2	803	326	17	8
D4	17	30	175	0.8	839	267	12	0
D5	20	45	199	0.46	393	73	12	2
G1	146	791	120	38.2	43	N.A.	1	26
G2	117	206	160	-	24	N.A.	3	12
G3	170	920	175	28	82	N.A.	9	51
G4	153 ± 7	10,111	153	-	7	N.A.	4	55
G5	133 ± 1	2604	187	12.2	35	N.A.	1	25

N.A.: below the limit of detection

References

1. Custodio D et al (2016) A one-year record of carbonaceous components and major ions in aerosols from an urban kerbside location in Oporto, Portugal. Sci Total Environ 562:822–833
2. Wang Q et al (2017) Carbonaceous species in PM2.5 and PM10 in the urban area of Zhengzhou in China: seasonal variations and source apportionment. Atmos Res 191:1–11
3. Hofman J et al (2016) Ultrafine particles in four European urban environments: results from a new continuous long-term monitoring network. Atmos Environ 136:68–81
4. Hama SML et al (2017) Quantifying primary and secondary source contributions to ultrafine particles in the UK urban background. Atmos Environ 166:62e78
5. Kim BM, Teffera S, Zeldin MD (2000) Characterization of PM25 and PM10 in the South Coast air basin of southern California: part 1-spatial variations. J Air Waste Manage Assoc 50(12):2034–2044
6. Ravishankara AR, Daniel JS, Portmann RW (2009) Nitrous oxide (N_2O): the dominant ozone-depleting substance emitted in the 21st century. Science 326(5949):123–125
7. Guan B et al (2014) Review of the state of the art technologies of selective catalytic reduction of NOx from diesel engine exhaust. Appl Therm Eng 66(1–2):395–414
8. Suarez-Bertoa R, Zardini AA, Astorga C (2014) Ammonia exhaust emissions from spark-ignition vehicles over the new European driving cycle. Atmos Environ 97:43–53
9. Suarez-Bertoa R, Astorga C (2016) Isocyanic acid and ammonia in-vehicle emissions. Transp Res Part D Transp Environ 49:259–270
10. Wallington TJ, Wiesen P (2014) N_2O emissions from global transportation. Atmos Environ 94:258–263
11. Ko J et al (2017) Comparative investigation of NOx emission characteristics from a Euro 6-compliant diesel passenger car over the NEDC and WLTC at various ambient temperatures. Appl Energy 187:652–662
12. Suarez-Bertoa R, Astorga C (2018) Impact of cold temperature on Euro 6 passenger car emissions. Environ Pollut 234:318–329
13. Appendix A. Supplementary data related to this article can be found at https://doi.org/10.1016/j.envpol.2017.10.096
14. Heck RM, Farrauto RJ, Gulati ST (2002) Catalytic air pollution control. Wiley, New York

15. Gordon TD et al (2014) Secondary organic aerosol production from diesel vehicle exhaust: impact of aftertreatment, fuel chemistry, and driving cycle. Atmos Chem Phys 14:4643–4659
16. Whittington BI et al (1995) Vehicle exhaust catalysis: I. The relative importance of catalytic oxidation, steam reforming, and water-gas shift reactions. Catal Today 26:41–45
17. Bradow RL, Stump FD (1977) Unregulated emissions from three-way catalyst cars. SAE Technical Paper No. 770369
18. Dardiotis C et al (2013) Low-temperature cold-start gaseous emissions of late technology passenger cars. Appl Energy 111:468–478
19. EEA (2015) European Environmental Agency. Air quality in Europe. http://www.eea.europa.eu/publications/air-quality-in-Europe-2015
20. Bishop GA, Stedman DH (2015) Reactive nitrogen species emission trends in three light-/medium-duty United States fleets. Environ Sci Technol 49:11234–11240
21. Suarez-Bertoa R et al (2015) Intercomparison of real-time tailpipe ammonia measurements from vehicles tested over the new world-harmonized light-duty vehicle test cycle (WLTC). Environ Sci Pollut Res 22(10):7450–7460
22. Durbin TD et al (2004) The effect of fuel sulfur on NH_3 and other emissions from 2000–2001 model year vehicles. Atmos Environ 38(17):2699–2708
23. Aksoyoglu S et al (2016) Secondary inorganic aerosols in Europe: sources and the significant influence of biogenic VOC emissions, especially on ammonium nitrate. Atmosph Chem Phys 17(12):7757–7773
24. Petetin HS et al (2016) Assessing the ammonium nitrate formation regime in the Paris megacity and its representation in the CHIMERE model. Atmos Chem Phys 16:10419–10440
25. Liu T et al (2015) Role of ammonia in forming secondary aerosols from gasoline vehicle exhaust. Sci China Chem 58:1377–1384
26. Link MF et al (2017) Elevated production of NH_4NO_3 from the photochemical processing of vehicle exhaust: implications for air quality in the Seoul Metropolitan Region. Atmos Environ 156:95e101
27. EC, 333/2014. Regulation (EU) No 333/2014 of the European Parliament and of the Council of 11 March 2014 amending regulation (EC) No 443/2009 to define the modalities for reaching the 2020 target to reduce CO_2 emissions from new passenger cars. Off J Eur Un L 103/15.
28. Maricq MM, Szente JJ, Jahr K (2012) The Impact of ethanol fuel blends on PM emissions from a light-duty GDI vehicle. Aerosol Sci Technol 46:576–583
29. Giechaskiel B, Manfredi U, Martini G (2014) Engine exhaust solid sub-23 nm particles: I. Literature survey. SAE Int J Fuels Lubr 7:950–964
30. Braisher M, Stone R, Price P (2010) Particle number emissions from a range of European vehicles. SAE Technical Paper, 2010-01-0786